# Food-Borne Microbes

## SHAPING THE HOST ECOSYSTEM

# Food-Borne Microbes

## SHAPING THE HOST ECOSYSTEM

EDITED BY

Lee-Ann Jaykus

*Department of Food, Bioprocessing and Nutrition Sciences*
*North Carolina State University*
*Raleigh, North Carolina*

Hua H. Wang

*Department of Food Science and Technology*
*and Department of Microbiology, The Ohio State University*
*Columbus, Ohio*

Larry S. Schlesinger

*Center for Microbial Interface Biology*
*and Department of Internal Medicine, The Ohio State University*
*Columbus, Ohio*

ASM PRESS WASHINGTON, D.C.

Address editorial correspondence to ASM Press, 1752 N St. NW, Washington, DC 20036-2904, USA

Send orders to ASM Press, P.O. Box 605, Herndon, VA 20172, USA
Phone: (800) 546-2416 or (703) 661-1593
Fax: (703) 661-1501
E-mail: books@asmusa.org
Online: estore.asm.org



**Library of Congress Cataloging-in-Publication Data**

Food-borne microbes : shaping the host ecosystem / edited by Lee-Ann Jaykus, Hua H. Wang, Larry S. Schlesinger.
p. ; cm.
Includes bibliographical references and index.
ISBN 978-1-55581-405-2 (hardcover)
1. Food—Microbiology. 2. Microbial ecology. 3. Gastrointestinal system—Microbiology. I. Jaykus, Lee-Ann. II. Wang, Hua H. (Hua Helen), 1965– III. Schlesinger, Larry S. IV. American Society for Microbiology.
[DNLM: 1. Food Microbiology. 2. Ecosystem. 3. Food Contamination. 4. Host-Pathogen Interactions. QW 85 F6844 2009]

QR115.F635 2009
664.001'579—dc22

2009002100

ISBN 978-1-55581-405-2

*Cover photo: Salmonella enterica* serovar Typhi can colonize the bile-rich gallbladder, and biofilms on gallstone surfaces in this organ may help establish and maintain a carrier state. A human gallstone was incubated with serovar Typhimurium and 3% bile for 12 days, washed, fixed in glutaraldehyde, and visualized by scanning electron microscopy. The weblike, flocculent material is indicative of desiccated exopolysaccharide from the bacterially initiated extracellular matrix of this mature biofilm. Image assembled by A. M. Prouty and R. W. Crawford in the laboratory of J. S. Gunn.

# CONTENTS

# CONTRIBUTORS

**Dan I. Andersson**
Department of Medical Biochemistry and Microbiology, Uppsala University, Box 582, S-751 23 Uppsala, Sweden

**Dionysios A. Antonopoulos**
Division of Infectious Diseases, University of Michigan, Ann Arbor, MI 48109

**Jennifer M. Brulc**
Department of Animal Sciences, The Institute for Genomic Biology, University of Illinois at Urbana-Champaign, Urbana, IL 61801

**Kathleen Bush**
Department of Environmental Health Sciences, School of Public Health, University of Michigan, 109 Observatory St., 6626 SPHI, Ann Arbor, MI 48109

**Natalia Cernicchiaro**
Department of Veterinary Population Medicine, University of Minnesota, St. Paul, MN 55108

**Robert W. Crawford**
Center for Microbial Interface Biology, Department of Molecular Virology, Immunology and Medical Genetics, The Ohio State University, Columbus, OH 43210

**Francisco Diez-Gonzalez**
Department of Food Science and Nutrition, University of Minnesota, St. Paul, MN 55108

**David M. Elliott**
Division of Gastroenterology-Hepatology, Department of Internal Medicine, University of Iowa, Iowa City, IA 52240

**Timothy E. Ford**
University of New England, Pickus Hall, Room 105, Biddeford, ME 04005

**Joseph F. Frank**
Department of Food Science and Technology, University of Georgia, Athens, GA 30602

**John S. Gunn**
Center for Microbial Interface Biology, Department of Molecular Virology, Immunology and Medical Genetics, The Ohio State University, Columbus, OH 43210

**James M. Jay (deceased)**
Department of Biological Sciences, Wayne State University, Detroit, MI 48202

**Howard F. Jenkinson**
Department of Oral and Dental Science, University of Bristol, Bristol BS1 2LY, United Kingdom

**Radhey Kaushik**
Department of Biology and Microbiology and Department of Veterinary Sciences, NPB Rm. 252A, Box 2140D, South Dakota State University, Brookings, SD 57007

**Jan-Ulrich Kreft**
Centre for Systems Biology, School of Biosciences, University of Birmingham, Edgbaston, Birmingham B15 2TT, United Kingdom

**Julie Kuruc**
PepsiCo Chicago, 617 W. Main St., Barrington, IL 60010

**Karen L. Lachmayr**
Department of Organismic and Evolutionary Biology, Harvard University, 16 Divinity Ave., Room 4083, Cambridge, MA 02138

**Richard J. Lamont**
Department of Oral Biology, University of Florida, Gainesville, FL 32610

**Yung-Hua Li**
Department of Microbiology and Immunology and Department of Applied Oral Sciences, Dalhousie University, 5981 University Ave., Rm. 5215, Halifax, NS B3H 3J5, Canada

**Patrick F. McDermott**
National Antimicrobial Resistance Monitoring System, Center for Veterinary Medicine, U.S. Food and Drug Administration, 8401 Muirkirk Rd., Mod 2, Laurel, MD 20708

**Mark Morrison**
Department of Animal Sciences, The Ohio State University, Columbus, OH 43210, and CSIRO Livestock Industries, St Lucia, QLD, Australia

**Michael Otto**
National Institute of Allergy and Infectious Diseases, National Institutes of Health, Bethesda, MD 20892

**Trevor G. Phister**
Department of Food, Bioprocessing and Nutrition Sciences, North Carolina State University, Raleigh, NC 27695

**Marilyn C. Roberts**
Department of Pathobiology and Department of Environmental and Occupational Health Sciences, University of Washington, Seattle, WA 98195

**Srinand Sreevatsan**
Department of Veterinary Population Medicine and Department of Veterinary Biomedical Sciences, University of Minnesota, St. Paul, MN 55108

**Gerald W. Tannock**
Department of Microbiology and Immunology, University of Otago, Dunedin, New Zealand

**Hua H. Wang**
Department of Food Science and Technology and Department of Microbiology, The Ohio State University, Columbus, OH 43210

**Joel V. Weinstock**
Division of Gastroenterology-Hepatology, Department of Internal Medicine, Tufts Medical Center, Boston, MA 02111

**Bryan A. White**
Departments of Animal Sciences and Pathobiology, Division of Nutritional Sciences, and Institute for Genomic Biology, University of Illinois at Urbana-Champaign, Urbana, IL 61801

**David G. White**
Office of Research, Center for Veterinary Medicine, U.S. Food and Drug Administration, Laurel, MD 20708

**Chuanwu Xi**
Department of Environmental Health Sciences, School of Public Health, University of Michigan, 109 Observatory St., 6626 SPHI, Ann Arbor, MI 48109

**Anthony Yannarell**
The Institute for Genomic Biology, University of Illinois at Urbana-Champaign, Urbana, IL 61801

**Zhongtang Yu**
Department of Animal Sciences, The Ohio State University, Columbus, OH 43210

**Qijing Zhang**
Department of Veterinary Microbiology and Preventive Medicine, Iowa State University, Ames, IA 50011

**Yongli Zhang**
Department of Environmental Health Sciences, School of Public Health, University of Michigan, 109 Observatory St., 6626 SPHI, Ann Arbor, MI 48109

# PREFACE

Food microbiology has historically focused on the study of pathogens or toxin producers causing acute (and sometimes chronic) diseases, spoilage agents affecting food quality, and starter cultures essential for food fermentation or used as probiotics with potential human or animal health benefits. However, it has become increasingly apparent that the same organisms that have been so widely studied actually account for only a small proportion of the microflora associated with foods. In fact, recent evidence from molecular studies shows that microbial populations are more diverse than previously thought, with many microbes actually nonculturable. Furthermore, the interactions between these microbes are complex, multifaceted, and poorly characterized. Clearly, diverse microbial species within the environment, foods, and the host interact with one another in ways that are only beginning to be elucidated.

In recent years, with the assistance of the tools of high-throughput molecular biology and population genetics, the focus of microbiological research has been extended from single cells to complex microbial ecosystems, particularly those associated with the host. It is now recognized that many diseases result from polymicrobial interactions and that the microbial compositions of these ecosystems have direct impacts on the overall health of the host. In addition to the contribution of pathogenic organisms, the contribution of commensal organisms to ecosystem development, disease progression, and the development of resistance to antimicrobial agents is becoming increasingly recognized.

Along the same line, food microbiology research is now evolving to consider complex microbial ecosystems. The study of food-borne pathogens is no longer limited to typing or detection of specific target organisms but now encompasses an understanding of the contributions of various environmental and host factors to the persistence, virulence, and evolution of food-borne bacteria. For instance, recent evidence shows the importance of microbial interactions in biofilm formation and of the effects of the environment and the host on the stimulation of the microbial stress response and virulence gene expression. This sort of work is at the forefront of microbial ecology, and food microbiologists are part of this movement.

The composition of microbial communities in foods, be they raw, minimally, or highly processed, provides a fertile area for the study of the complexity of environmental ecosystems. It has been understood for decades that many microorganisms are not completely eliminated by using common food processing measures and that injured organisms surviving a process can recover, particularly in the absence of a competitive microflora. Such survivors may harbor stress survival determinants that also facilitate their survival in the presence of host defenses such as low gastrointestinal pH. Furthermore, various food ingredients (such as probiotics, preservatives, antibiotic residues, and homologues of host factors) may also affect, either directly or indirectly, the evolution of host ecosystems by interfering with attachment or integration of microbes to host tissues or detachment of microbes from host tissues or by stimulating host responses. Thus, while an increased understanding of oral and gastrointestinal ecosystems is essential for future disease treatment, understanding of microbial ecosystems within the context of their interactions in foods and with the host ecosystem will be important in food processing and preservation as well as disease prevention.

However, our current understanding of the contributions of food-borne microorganisms to human health and microbial interactions within various microbial ecosystems (the environment, foods, and the host) is quite limited. For example, although it is generally recognized that antibiotic treatments and probiotics can modulate the host ecosystem, there is a lack of knowledge about the overall contribution and significance of food and food-borne microfloras in host ecosystem evolution. Indeed, humans can easily consume $10^{10}$ CFU of microorganisms daily, with up to $10^{6}$ CFU harboring some sort of antibiotic resistance. The vast majority of these organisms are commensal rather than pathogenic in nature. Clearly, they have ample opportunity to shape the host ecosystem by interacting with other host microfloras by using mechanisms such as signal exchange or horizontal gene transfer or by becoming a part of the host ecosystems. The mechanisms of such interactions are, however, largely unknown.

The purpose of this book is to begin the dialogue between food, environmental, and medical microbiologists as they seek to understand microbial communities and their interactions within and between different environments. In this way, we can better understand the significance of food-borne microbes in the universe of the microbiology. The book is organized into five sections. Section I serves to introduce major microbial ecosystems associated with hosts, foods, and the natural environment and to discuss microbial interactions within the context of each ecosystem. Section II describes various manifestations of and mechanisms for microbial interaction, including biofilm formation, stress response, competitive inhibition, and the unique interactions that occur between the host and the microbial community. In section III, we use antibiotic resistance, a very complicated but hotly debated issue, as an example to illustrate the potential interactions between food, environmental, and host microbial systems and their potential impact on public health. Section IV focuses on several notable food-borne bacteria as models to demonstrate some of the concepts described in the previous chapters. Section V, the final section, introduces emerging tools which can be used in the further study and characterization of microbial ecosystems.

This book is intended to serve as a comprehensive reference for the general microbiology community, and it may be of particular interest to food and industrial microbiologists and those engaged in microbial ecology research. This includes scientists in industry, university researchers, and those affiliated with regulatory agencies. The book also provides a solid reference for microbiologists without a food microbiology background, providing them a quick primer to understand some of the emerging topics in the field, with particular emphasis on the complexity of microbial ecosystems associated with foods, the environment, and the host and the interactions between these communities. For food microbiologists without a microbial ecology background, this book provides an introduction to key concepts, mechanisms, and tools used in microbial ecology and medical microbiology research, as well as an opportunity to understand the importance of food and food-borne microfloras to public health. We hope that this book will stimulate discussion, brainstorming, and collaboration between food, veterinary, and medical microbiologists and their colleagues. To our knowledge, this is the first comprehensive attempt to emphasize the central position of food and food-borne microbes in host ecosystem development by connecting the complexity of the ecosystems from environment to the host, bridged by the food carrier. We believe that a comprehensive and integrated knowledge of microbial ecology as it relates to food microbiology, the environment, and the host is vital for developing practical approaches for establishing healthy host ecosystems.

Within days of the submission of this book for production, Dr. James Jay, professor emeritus of food microbiology and a senior contributor, passed away. We are humbled by his contributions to the field, and we recognize that his work serves as the foundation upon which many future discoveries in food microbiology and microbial ecology will be based. We dedicate this work to his memory. Jim, you will be missed.

LEE-ANN JAYKUS
HUA H. WANG
LARRY S. SCHLESINGER

# CIRCULATION AND DYNAMICS AMONG MICROBIAL ECOSYSTEMS

# I

# THE ORAL MICROBIAL ECOSYSTEM AND BEYOND

*Howard F. Jenkinson and Richard J. Lamont*

# 1

The oral cavity environment is anatomically, physiologically, and microbiologically diverse. A range of host tissues are present, including teeth, adenoids, tonsils, and mucosal epithelia, and comprise different types, including keratinized, nonkeratinized, and junctional. Oral cavity tissues are continually bathed in saliva, which contains many hundreds of proteins and glycoproteins (64). To ice the cake, in this complex ecosystem over 800 species (or taxa) of bacteria have now been identified as potential components within the oral microbiota. Although in reality it seems that microbial communities at specific sites in the oral cavity rarely contain more than 100 species, the potential for multiple host-bacterium interactions in shaping the health or disease status of individuals presents a major challenge in understanding for microbiologists and immunologists alike. It is also a challenge to microorganisms to grow and survive in this environment. Although salivary samples contain between $10^3$ and $10^6$ bacteria per milliliter, most of the organisms that enter the oral cavity, through either ingestion or reflux from the gastrointestinal tract, generally fail to become established. A combination of fierce competition from the resident oral microflora and attack by host innate defenses ensures that in noncompromised subjects, potential invading pathogens are excluded. While saliva contains an array of potent antimicrobial compounds, such as lysozyme, lactoperoxidase, lactoferrin, and antimicrobial peptides, oral bacterial communities are apparently able to survive and replicate within saliva. Since saliva is actively swallowed in the awake hours, it is apparent that in order for the microbial ecosystem to be sustained in the oral cavity, cells must adhere to the variety of surfaces present and become established as communities within biofilms.

Most of the surfaces within the oral cavity become rapidly covered by a layer (pellicle) of salivary components, mainly glycoproteins, but also with polysaccharides and proteins of host or microbial origin. Major components of saliva, such as mucins, secretory IgA, proline-rich proteins, statherin, cystatins, and α-amylase, have been well studied in their various roles as protective or antibacterial agents or as molecules supporting microbial adherence to salivary pellicle. However, new techniques in glycoprotein capture, peptide

*Howard F. Jenkinson,* Department of Oral and Dental Science, University of Bristol, Bristol BS1 2LY, United Kingdom. *Richard J. Lamont,* Department of Oral Biology, University of Florida, Gainesville, FL 32610.

*Food-Borne Microbes: Shaping the Host Ecosystem,* Edited by L.-A. Jaykus, H. H. Wang, and L. S. Schlesinger, © 2009 ASM Press, Washington, DC

separation, and mass spectrometry, applied to determine the human salivary proteome, have revealed unexpected complexities, with more than 400 components identified (49, 64). In addition to the secretory products of the salivary glands, these include many plasma-derived or cellularly derived molecules, suggesting that salivary pellicle may also provide receptors normally associated with epithelial or subepithelial tissues. The pellicles formed on enamel surfaces are different from the pellicles formed on prosthetic, e.g., acrylic, surfaces or on epithelial surfaces. Hence, differential presentation of receptors on surfaces provides for attachment of different bacterial species, and microbial communities at various sites in the oral cavity are therefore ecologically diverse (2).

## MICROBIAL DIVERSITY

Since microbial communities at different oral cavity sites can be unique in their compositions, it is prudent to be aware that generalizations about specific microbes and their activities in oral health or disease may not be always relevant. Oral microbiologists have relied on cultivating bacteria and characterizing them biochemically for much of the last century, and such studies have provided a sound basis for current ideas. More recent molecular, culture-independent techniques have vastly expanded our knowledge and revealed unrecognized complexities in genotypic diversity and intermicrobial associations. It is reassuring that early concepts about disease etiology, developed from cultivation of oral microorganisms associated with carious or periodontal lesions, have stood the test of time and are generally supported by more modern genomic and systems analyses. The tooth surfaces, coated with salivary glycoprotein pellicle, are colonized rapidly by mainly gram-positive bacteria, including cultivable species of *Streptococcus* and *Actinomyces* (12) and less easily cultivable *Granulicatella, Gemella, Abiotrophia,* and *Rothia* (2). Conversely, biofilms associated with gum margins and supragingival regions contain a greater preponderance of anaerobic microorganisms, e.g., *Fusobacterium nucleatum, Porphyromonas, Prevotella,* and *Peptostreptococcus* (46), while subgingival periodontal pockets acquire highly complex microbial populations including *Tannerella forsythia, Porphyromonas, Eikenella,* and *Treponema* (35, 46). The incidences of major bacterial genera found in surface dental plaque and subgingival plaque are shown diagrammatically in Fig. 1. It is important to note, however, that periodontal disease sites within the same mouth can carry very different communities of bacteria. This advances the idea that different combinations of bacteria can result in the same disease con-

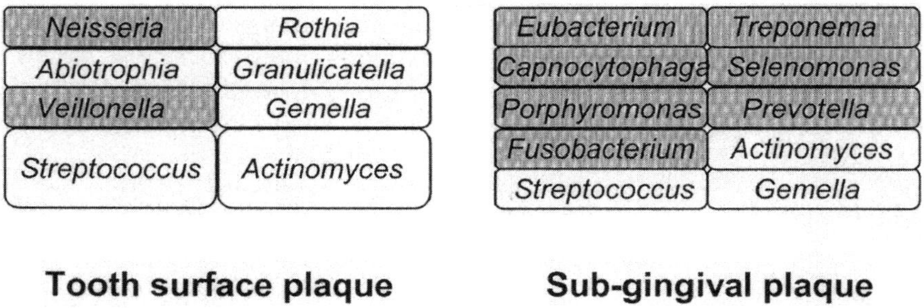

| Tooth surface plaque | |
|---|---|
| Neisseria | Rothia |
| Abiotrophia | Granulicatella |
| Veillonella | Gemella |
| Streptococcus | Actinomyces |

| Sub-gingival plaque | |
|---|---|
| Eubacterium | Treponema |
| Capnocytophaga | Selenomonas |
| Porphyromonas | Prevotella |
| Fusobacterium | Actinomyces |
| Streptococcus | Gemella |

**FIGURE 1** Diagrammatic representation of the incidence of major bacterial genera found in healthy dental plaque from the tooth surface and in subgingival plaque associated with periodontal disease. Subgingival plaque differs from tooth surface plaque in containing a greater variance of bacterial taxa and a higher proportion of gram-negative bacteria (gray shading). Data summarized and condensed from references 2 and 46.

dition and the notion that there is a spectrum of disease-associated communities. Some of the ways in which individual components of these communities interact with each other are described in the following section.

## BUILDING COMMUNITIES

Oral microbial communities exist primarily as multispecies biofilms on the surfaces present in the oral cavity, although large numbers of organisms are also present in the fluid phase of the saliva. It is useful to conceptualize the mixed oral microbial biofilm community as a programmed developmental process. This has been depicted in recent years as an interactive system wherein different bacteria have specific physical partners, supported by experimental observations in vitro and in vivo (12, 25, 26, 30, 32, 54). However, we can adjust these depictions, in light of new metabolic and metagenomic data, to consider the perspectives of temporal acquisition and disease association. The initiation and accumulation of oral biofilm communities proceed through pioneer organisms, such as streptococci and actinomycetes, attaching to pellicle and then providing a new surface for the attachment of succeeding organisms. Furthermore, an intricate network of metabolic and signaling interactions exists among the constituents of oral biofilms which helps shape the temporal and spatial relationships that characterize these complex heterotypic communities (26).

### Initial Attachment to Oral Surfaces

The adhesion of primary colonizers to saliva-coated surfaces is a multifactorial process, the precise mechanistic bases of which vary among species and even from strain to strain within species. In general, successful oral pioneer organisms present a surface decorated with multiple adhesins of differing specificities and affinities. Structurally, these adhesins are often surface proteins or are associated with fibrillar appendages (25). Such a configuration is likely to provide versatility in receptor recognition and ultimately result in more avid binding to the surface. Adsorbed salivary molecules that function as receptors for microbial adhesion also show diversity, and moreover, some attain an active configuration only after deposition on a surface. A major group of salivary receptors comprises the proline-rich proteins and statherin, proteins that evolved to maintain calcium and phosphate in a state of supersaturation and thus are tightly bound by enamel. Salivary mucins and agglutinin such as gp340 glycoprotein, which can bind to bacteria when in solution, can also adsorb to enamel and to epithelial surfaces and provide receptors for bacterial attachment (23, 36). Mechanisms of initial bacterial adhesion, which can be additive or sequential, include recognition of oligosaccharide receptors by protein adhesins in lectin-like reactions, protein-protein interactions, and ionic or hydrophobic associations between microbial surface components and the adhesion substratum. The numerically dominant colonizers of saliva-coated tooth enamel are the oral streptococci such as *Streptococcus sanguinis, S. mitis, S. oralis,* and *S. gordonii.* Widely distributed among these species is a family of major surface proteins termed antigen I/II (or antigen B) that function as adhesins through recognition of salivary gp340, fibronectin, and collagen (24). Adhesion of another common early colonizer, *Actinomyces naeslundii,* is mediated through fimbrial interactions with proline-rich proteins (55). In this manner the pioneer organisms adhere tenaciously to surfaces, even in the presence of soluble receptors in the fluid phase, and resist the shear forces that would otherwise precipitate their removal from the surface and from the mouth.

### Accumulation of Biofilm Communities

Once the pioneer colonizers have become established, interbacterial coadhesion with secondary colonizers helps drive subsequent accretion and population shifts that can be associated with disease. The mechanistic basis of interbacterial adhesion is similar to bacterium-salivary receptor adhesion and can involve lectin-like protein-carbohydrate or protein-protein interactions that are often multivalent (26, 30). One of the best-documented examples is provided by *Porphyromonas gingivalis,* a gram-negative anaerobe and periodontal pathogen (9, 11, 33). *P. gingivalis* coadheres with the antecedent colonizer *S. gordonii.* The initial interaction is mediated by the long fimbriae of *P. gingivalis,* which bind to glyceraldehyde-3-phosphate dehydrogenase (GAPDH) on the streptococcal surface. Subsequently, higher-affinity binding is effected by the short fimbriae of *P. gingivalis,* which engage the streptococcal SspA and SspB (antigen I/II family) adhesins (Fig. 2). The functional domain of the *S. gordonii* Ssp proteins is a 26-amino-acid region toward the C terminus, within which an NITVK motif is essential for recognition by the *P. gingivalis* fimbriae. Interestingly, the NITVK motif is not conserved in the antigen I/II family proteins of *Streptococcus mutans,* and consequently *P. gingivalis* does not accumulate into biofilms on substrata of *S. mutans.* Selectivity of bacterial recognition and adhesion to other spe-

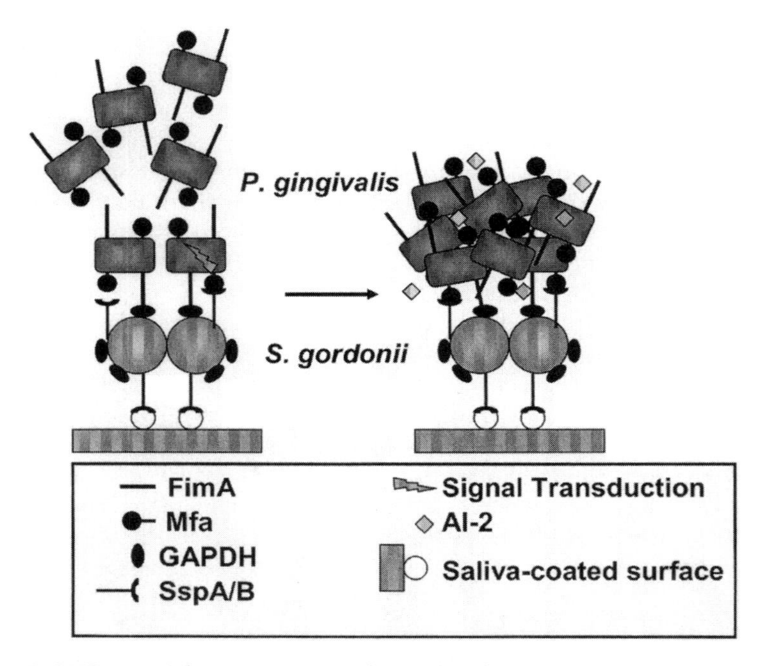


**FIGURE 2** Schematic (not to scale) model of initial events leading to *P. gingivalis-S. gordonii* heterotypic biofilm community development on the supragingival tooth surface. *S. gordonii* (spheres) is a pioneer colonizer, and cells attach to the saliva-coated tooth surface. *S. gordonii* produces multiple adhesins, many of which have cognate salivary receptors; for simplicity, only SspA/B is shown. Initial localization of *P. gingivalis* (rods) is mediated by FimA interaction with GAPDH on the streptococcal surface. Higher-affinity binding occurs through engagement of Mfa with SspA/B. This interaction initiates a signal transduction event that modulates the *P. gingivalis* transcriptome. The resulting phenotypic adaptation of *P. gingivalis,* along with the production of signaling molecules such as AI-2, is necessary for the recruitment of additional *P. gingivalis* cells from the planktonic phase and the initiation of community development. Note that this is only one developmental stage of the process that leads to a mature heterotypic biofilm.

cies is a common property; for example, *S. gordonii* coadheres with only certain subgroups of *A. naeslundii* (30). The ability to discriminate among the numerous constituents of a biofilm sets the stage for the accumulation of organisms into groups that are metabolically compatible or interdependent.

## Interbacterial Interactions in Oral Biofilms

An advantage to heterotypic biofilm accumulation is the potential for groups of organisms to utilize complex substrates with maximal efficiency (26). For example, oral streptococci and actinomycetes produce lactate as the end product of fermentation of carbohydrates. In turn, the gram-negative coccus *Veillonella,* which adheres tightly to oral streptococci, utilizes lactate as a fermentable substrate. A similar nutritional cross-feeding occurs through the close association of *P. gingivalis* with *Treponema denticola.* Here, *P. gingivalis* utilizes succinate that is produced by *T. denticola,* whereas growth of *T. denticola* is enhanced by isobutyric acids generated by *P. gingivalis* (19). In addition to optimization of metabolic pathways, success in the sessile, mixed-species environment of oral biofilms requires broadly based transcriptional and proteomic adjustment by previously planktonic organisms. Such adaptations often ensue from recognition of the surface to which an organism is attached and identification of its nearest neighbors. For example, cells of *S. gordonii* that are bound to saliva-coated hydroxyapatite beads initiate signal transduction events through a two-component system termed BrfA/B (67). This signaling system controls aspects of streptococcal adherence and adaptation required for biofilm growth. Networks of signal transduction and transcriptional regulation by one species can also facilitate colonization by another species. The interaction of *S. gordonii* with saliva results in upregulation of several genes, including those encoding the antigen I/II family adhesins that promote coaggregation interactions of *S. gordonii* with actinomycetes and with *P. gingivalis,* as discussed above (15). Sensing of *S. gordonii* by *P. gingivalis* then initiates a signal transduction cascade within *P. gingivalis* (Fig. 2) that facilitates the accumulation of organisms into the developing heterotypic biofilm (32).

A number of bacterial intercommunication systems have been identified that are based on the production and detection of diffusible external signals and that enable coordinated responses throughout a bacterial population. Quorum sensing is a system by which a population can monitor its cell density through release of specific signaling molecules called autoinducers (AIs). As the population grows, AIs accumulate in the local environment until a critical detection threshold is reached, at which point changes in gene expression are triggered. In this manner bacterial populations can synchronize expression of specific genes required for community survival under the prevailing environmental conditions. AI-2, produced through the action of the LuxS enzyme, is a major AI present in oral communities. AI-2 in fact may comprise a family of closely related molecules generated from rearrangements of 4,5-dihydroxy-2,3-pentanedione (DPD), a product of LuxS-mediated cleavage of *S*-ribosylhomocysteine. Mutants of oral organisms that are deficient in the production of LuxS often show a related biofilm phenotype. For example, both *S. mutans* and *S. gordonii* exhibit aberrant monospecies biofilm formation in the absence of LuxS (5, 41, 66). Moreover, LuxS-dependent signaling is required for the development of mixed biofilms of *P. gingivalis* and *S. gordonii* (Fig. 2) and of *S. oralis* and *A. naeslundii* (40, 53). While these studies clearly demonstrate a role for LuxS in oral biofilm formation, it is important to consider that the metabolic pathway involving LuxS is the activated methyl cycle, and discrimination of signaling events from metabolic perturbation is not always possible.

While many interbacterial signaling systems are mutually beneficial to the participating organisms, the converse situation of antagonism between the inhabitants of oral biofilms is also common. Production of bacteriocins and production of hydrogen peroxide represent some of the more lethal examples of interbacterial conflict. Bacteriocin activity of the SalA (salivaricin) type can be detected in saliva from subjects colonized by *Streptococcus salivarius* (62). Bacteria are able to sense their own or related bacteriocins in the oral environment, recognizing the presence of friends or foes in the vicinity, and respond accordingly. Contact-dependent signaling may also be involved in more subtle antagonistic interactions. Initial contact between *P. gingivalis* and *S. cristatus* results in activation of a regulatory cascade that significantly decreases the level of transcription of *fimA,* encoding the long fimbrial adhesin subunit of *P. gingivalis* (63). As a result, coadhesion between these organisms is restricted and biofilm accretion does not occur with *P. gingivalis* and *S. cristatus.*

The composition of the oral biofilm, along with its metabolic and pathogenic potential, thus derives from the collective outcome of interconnected biological processes, including adhesion to surfaces, environmental sensing and response, physiological adaptations, cell-to-cell communication, and interbacterial competition. Consequently, oral biofilms are complex and dynamic entities comprising physiologically diverse organisms that successfully coexist and adapt to the ever-changing environmental conditions in the oral cavity.

## HEALTH OR DISEASE COMMUNITIES

The microbial biofilm communities that develop in the oral cavity are clearly intrinsically driven by bacterial interactions, but they are also host driven. Factors such as age, immune status, hormone levels, salivary flow rate, smoking, and dental hygiene standards all impact on oral microbial community formation and composition. Dietary factors and antibiotic usage both have major influences. The role of fermentable sugars in the initiation and development of dental caries is perhaps the best-known example of a diet-provoked disease. Elevated sugar consumption levels, frequency and retentiveness of sucrose-containing foods and drinks, in conjunction with relaxed oral hygiene, promotes the enrichment of streptococcal bacterial populations that are tolerant to the lower pH values (pH ~5) generated through homolactic fermentation of carbohydrate. The effect of prolonged low pH levels in plaque is to cause enamel demineralization, dissolution, and tooth decay. There is also evidence that specific host factors, such as acid PRP-1 variant (60), may influence susceptibility to colonization by more cariogenic mutans group streptococci. This, and other as-yet-unidentified host molecules, may possibly go toward future explanation of why some children who have low frequency of dietary sugar intake and good oral hygiene are nevertheless more susceptible to dental caries.

As better ecological perspectives of oral microbial communities are formulated, it becomes possible to more clearly define healthy oral biofilms and thus find ways to promote their formation and retention. Although dental plaque may be regularly removed by tooth brushing, regrowth of a normal plaque is undoubtedly protective, along with salivary defenses, in thwarting attempts at colonization by the multitude of bacterial species that enter the oral cavity from the environment, and especially in foods. It is thus envisaged that oral disease conditions result from a shift in oral microbial community at a site from a health-associated profile to a disease-associated profile. In the development of dental caries, it seems possible to generalize that the more prevalent *S. sanguinis*-like organisms become outcompeted and replaced by mutans group streptococci (4), enhanced by dietary sugars and potentially influenced by extraneous factors, e.g., salivary molecules, antibiotics, and other pharmaceuticals. The ability of *S. mutans*-like bacteria to be sustained in carious lesions probably results not just from their acid tolerance but also from their ability to secrete a potentially large array of bacteriocins (47, 61) that inhibit competing organisms. In periodontal disease, a shift toward a more anaerobic gram-negative community is the stimulus for periodontal tissue destruction. Likewise, oral malodor (halitosis) subjects show a shift in composition of the tongue microflora, from an *S. salivarius*-dominated population in odor-free subjects to a more complex community in which *S. salivarius* is only a minor component (28). Loss of *S. salivarius,* which is a potent producer of bacteriocins, from the tongue community presumably removes constraints normally exerted by this organism on colonization by other gram-positive bacteria. Better understanding of these population shifts and what causes them will inform future strategies that may include attempting to recolonize disease sites with more healthy communities.

## PROTECTING THE HOST

Despite the fact that most oral disease conditions are associated with oral bacteria, the communities that develop in the mouth do largely retain harmonious relationships with the host over long periods. They play protective roles, in controlling enamel demineralization, in inhibiting colonization of pathogens, in reducing damage to epithelia or tissues by microbial and dietary products, and in mucosal homeostatic functions, e.g., epithelial regeneration and priming of immune responses.

The tooth enamel surface is subject to continuous assault, and $Ca^{2+}$ levels have to remain saturated in order to resist loss of hyroxyapatite mineral. The presence of fluoride ions in drinking water and preprepared foods impacts this process to promote the integrity of the enamel. Fluoride ions can substitute for hydroxyl ions, and the resulting calcium fluorapatite is more stable, and less soluble at low pH, than hydroxyapatite without the substitution. Fluoride ions, present in the fluid phase around enamel crystal, also promote remineralization. The pellicle and plaque associated with it are recognized as contributing to stability, and plaque fluid is supersaturated with respect to tooth mineral and other calcium phosphate phases. Plaque fluid also promotes remineralization. Compounds present in diet and saliva that influence plaque metabolism also potentially have protective effects. For example, urea metabolism and ammonia production by non-mutans group streptococci may neutralize plaque acids and lead to elevated pH levels, thus enhancing remineralization (8).

An important concept emerging in oral microbiology and immunology is that the oral microflora has a very close association with the host tissues in terms of regulation and response. The relatively high levels of bacteria present in saliva do not evoke a continuous inflammatory response in the oral cavity. Bacteria grow and survive efficiently despite the presence of high levels of secretory IgA antibodies, other salivary defense proteins, and antimicrobial peptides. The commensal species of bacteria do not induce the same levels of proinflammatory cytokines as pathogenic species do (43), yet they may prime the surface mucosae for more rapid response to pathogenic challenges. On the other hand, highly successful oral colonizers do have the ability to invade oral epithelial cells, and they can also selectively modulate cytokine responses. For example, *P. gingivalis* not only down-regulates expression of proinflammatory cytokines, e.g., interleukin-8 (IL-8), but also has the ability to degrade IL-8 (10). Oral bacteria have also evolved mechanisms that enable them to survive challenge by antimicrobial peptides that are secreted in the oral environment (27) that kill nonoral microorganisms. The oral microflora is therefore exquisitely evolved to interact with the host environment.

The role of the immune system in maintenance of a normal oral microflora is critical. A dramatic example is provided by the rampant caries that results from reduced salivary flow as a result of disease (e.g., Sjögren syndrome) or damage to salivary glands following radiation treatment for head and neck cancer. When subjects are immunocompromised, undergo surgical procedures, or are given antibiotic courses, then unusual organisms appear within the oral cavity (1). For example, with the rapidly expanding use of dental implants for cosmetic as well as disease-related reconstruction, implant failure associated with microbial infection, which can be potentially devastating, is increasingly prevalent. Bacteria associated with infections of implants that interfere with successful integration into bone include *P. gingivalis, Prevotella,* and *Staphylococcus aureus* (17). The presence of *S. aureus,* recognized as a nasal or skin organism rather than an oral bacterium, suggests that conditions for establishment are promoted by surgical procedures. Staphylococci have the ability to become internalized by osteoblasts as they grow on titanium (Fig. 3), leading to delayed maturation and inhibiting osseointegration. Staphylococci are also prevalent in periodontal conditions of the immunocompromised, e.g., human immunodeficiency virus-infected subjects (1), and *Enterococcus faecalis* is a frequent component, along with *S. aureus,* of the root canal microflora in infections of the tooth roots and pulp (3, 37). These observations collectively underpin the protective role of the normal oral microflora, in close association with the host mucosal defense system, against both biological and chemical challenges.

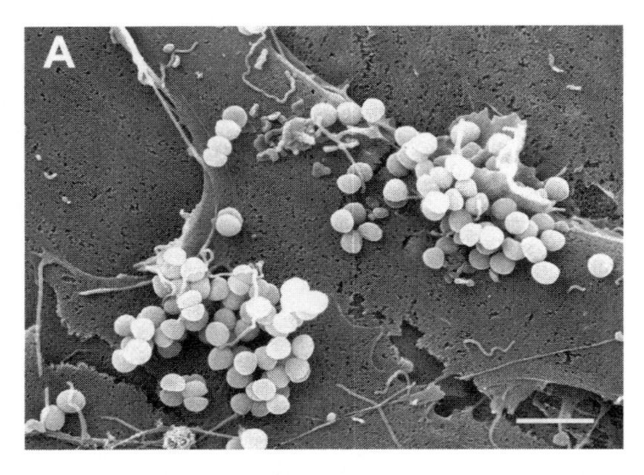


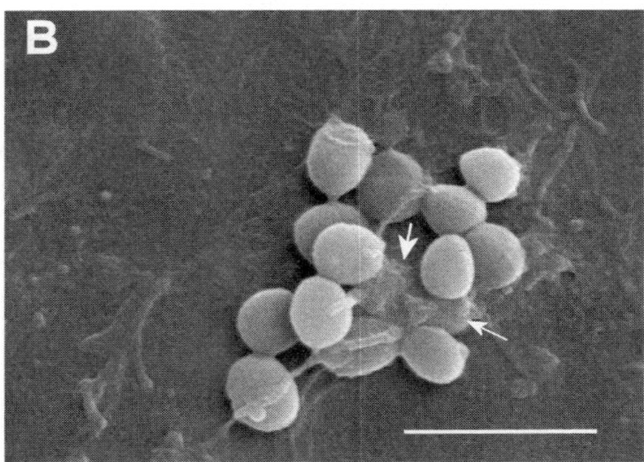


**FIGURE 3** Adhesion and internalization of *S. aureus* by cultured osteoblasts. Staphylococci adhere avidly in groups to osteoblasts growing on titanium surfaces (A). Cellular projections entrap the staphylococci, and individual bacterial cells within these groups induce membrane ruffling and become internalized (B). Arrows indicate internalized bacteria. Bars, 1 μm. Images provided by C. M. Moffatt and T. Sjöström.

## EFFECT OF DIET ON ORAL MICROBIAL COMMUNITIES

Dental caries is a disease for which diet plays a major and direct role. Diet also clearly plays a role in the development and maintenance of the gastrointestinal microflora. In both these environments the bacterial community profiles can be manipulated by major dietary changes. However, there are many more subtle effects of dietary components on the oral microflora. For example, abrasive foodstuffs are thought to be possibly beneficial in removing accessible dental plaque. Particulate foods, in contrast, can become incorporated into plaque, and those particles high in fermentable carbohydrate provide for localized feeding areas and acid production. Other common dietary components may have beneficial physiological effects. Chocolate and milk casein peptides provide increased buffering capacity within plaque, and casein phosphopeptides promote remineralization of enamel in the face of acid attack (38, 52).

There is mounting evidence for natural plant compounds in foods and drinks having antibacterial and antiplaque properties. Among these are polyphenols and catechins (in tea extracts) (21, 56) and antiadhesive factors in cranberry juice (58, 59). These, and many other cited dietary compounds that potentially influence oral microbial growth or biofilm formation, provide the basis for emerging multimillion-dollar industries marketing natural products beneficial to oral health. Dietary compounds also influence host inflammatory responses in periodontal disease. Aside from medicaments within toothpastes that are reported to have anti-inflammatory properties, there is evidence that dietary compounds may modulate periodontal health. Levels of antioxidants such as ascorbic acid are much lower in saliva or gingival crevicular fluid samples from subjects with active periodontitis, suggesting that dietary antioxidants might possibly assist in reducing periodontal inflammation (7).

## THE ORAL MICROBIOTA AND SYSTEMIC HEALTH

It is well established that systemic conditions, particularly those affecting neutrophil function, can lead to unconstrained growth of oral bacteria and exacerbation of periodontal disease. Other systemic diseases or conditions that impact the pathogenic potential of periodontal bacteria include osteoporosis, renal dysfunction, pregnancy, and diabetes. Moreover, a bidirectional relationship exists between oral organisms and systemic health, the full extent of which has yet to be established. Currently, a causal relationship between oral bacteria and infective endocarditis is well established, and evidence is emerging for links with coronary artery disease, preterm delivery of low-birth-weight infants (PDLBW), pulmonary disease, and even Alzheimer's disease (18, 57).

The diseases associated with the oral biofilms in the mouth are very tissue specific: caries that attack the mineralized tissue and periodontal diseases that attack the supporting tissues of the teeth. Organisms within the biofilm have a spectrum of potential pathogenicity; for example, *S. mutans* is the major pathogen in caries, and *P. gingivalis* contributes to periodontal disease. However, some organisms that are adapted to life in the oral cavity may be able to adapt to environmental conditions at different body sites and consequently exhibit a distinct pathogenic potential. Pathogens, and moreover commensals, from the oral cavity can thus cause disease at remote sites. Indeed, transient bacteremias with oral organisms are thought to be common. Bacteremia can arise when dental procedures such as tooth extraction, professional cleaning, or even tooth brushing create a breach in tissue integrity. Similarly, infection or ulcerative lesions in the soft tissue can provide systemic access to the oral microbiota. Most bacteremias of oral origin are rapidly cleared by the immune system of a healthy host without consequence. However, some organisms, such as *P. gingivalis,* can resist complement-mediated killing and survive inside professional phagocytic cells. Others, such as the oral streptococci, can locate in abnormal or injured anatomical sites and avoid immune surveillance. In this manner, oral microorganisms can have the opportunity to spread and contribute to serious systemic conditions.

### Infective Endocarditis

One of the best-documented examples of a systemic infection that can be caused by oral bacteria is infective endocarditis (20). In this disease, oral bacteria, frequently the streptococci, infect injured or abnormal heart valves. Conditions creating potentially susceptible heart valves include autoimmune disorders such as rheumatic fever, surgery, developmental abnormalities, or the use of drugs, for example, the formerly widely used diet drug fen-phen. These conditions promote sloughing of the valvular endothelium, and circulating platelets are then activated to help repair the damage. Bacteria bind to the platelets and exposed connective tissue and subsequently induce the platelets to bind fibrinogen, which promotes aggregation and the formation of a thrombus (clot). The mass of bacteria, fibrin, and platelets constitutes a vegetation, within which bacteria continue to grow and multiply. Without antibiotic treatment the valves can cease to function. The mortality rate for infective endocarditis is very high.

### Cardiovascular Disease

Cardiovascular diseases are a heterogeneous group of conditions, including atherosclerosis, ischemic heart disease, and stroke, that are the leading causes of death in developed countries. It is now apparent, however, that classic risk factors such as hypertension and smoking do not account for all cases of disease. There is increasing epidemiological evidence that infection with periodontal bacteria can now be considered a risk factor (57). Experimental evidence also supports such an association. Genomic DNA from periodontal pathogens can be detected in atheroma tissue. Furthermore, studies with apolipoprotein E-deficient mice, a model for atherosclerosis in humans, provide evidence for a possible role for *P. gingivalis* in the disease. Atherosclerosis is a progressive disease that results from an excessive, inflammatory-fibroproliferative response to various forms of insult to the endothelium and smooth muscle of the artery wall. Proinflammatory cytokines, such as C-reactive proteins, IL-1, and tumor necrosis factor alpha, which are involved in cardiovascular disease, are known to be stimulated by intracellular bacteria. Therefore, invasion of coronary artery cells by periodontal bacteria such as *P. gingivalis* may aggravate an inflammatory response that can contribute to atherosclerosis. Moreover, dietary intake can modulate inflammatory mediators and thus potentially contribute to the severity of both atherosclerosis and periodontal disease.

### Preterm Delivery

PDLBW occurs in about 12% of births in the United States, a rate that has not declined with improvements in overall health care. Over 70 risk factors for PDLBW are recognized, including bacterial infection. Epidemiological and intervention studies have begun to ascribe a role for periodontal disease and periodontal bacteria in PDLBW (65). Moreover, oral bacteria such as *P. gingivalis* can have a devastating effect on pregnancy in animal models. The mechanistic basis of a pathogenic role for periodontal bacteria in PDLBW is unclear, although several plausible processes have been proposed. First, while still within the confines of the periodontal tissues, periodontal bacteria and their antigens could initiate an inflammatory response with systemic consequences that could trigger preterm labor. Second, once in the bloodstream, bacterial antigens can induce a systemic immune response that could affect pregnancy. Third, direct oral-hematogenous spread of the bacteria themselves to the gestational tissues may occur. Supporting this concept are reports that the oral organisms *Fusobacterium* and *Porphyromonas* can be isolated from amniotic fluid. The ability of organisms such as *P. gingivalis* to invade cells and tissues and to disrupt cytokine networks could induce premature uterine contraction.

## CONCLUSIONS

Following the identification of *S. mutans* as a major etiologic agent in dental caries, and the association of specific species of anaerobic bacteria with periodontal disease, the oral microbial biofilm that constitutes dental plaque has been viewed in the public domain as being undesirable. It is clearer now that not all oral biofilms are necessarily bad, and fine mapping of bacterial components and metabolic interactions occurring within them has led to the view that there are health-related and disease-related biofilm communities. It is still not possible to predict simply from species or taxon analyses what constitutes an aggressive as opposed to passive biofilm community. However, better understanding of oral microbial ecology through molecular methodologies, such as the current aim to define all the genomes (the microbiome) of the human oral microflora, will undoubtedly assist in this quest. Metabolic and proteomic profiles will provide additional knowledge about the activities of different communities that are able to, nevertheless, result in similar oral disease outcomes.

Although all the combinations of oral bacteria that go toward benefiting the host are not known, there are many interesting strategies under development that may enable modifications of oral microbial biofilms to promote health. For example, it is believed that the early retention of *S. salivarius* and then *S. sanguinis* within the infant oral cavity may be beneficial to the future oral and dental health of subjects (6). Therefore, provision of selected oral bacteria shortly after birth may provide a health benefit. Although there is some evidence that colonization by bacteriocin-producing *S. salivarius* may assist in reducing the incidence of group A streptococcal pharyngitis in children (14), it is not wholly apparent that probiotics applied to the oral cavity would be useful in controlling or stabilizing the oral microflora. Application of lactobacilli to a young population resulted in a net decrease in caries incidence (42), but it is hard to judge what might be the longer-term effects of introducing acid-producing lactobacilli into dental plaque communities. The potential for controlling *S. mutans* levels by replacement therapy, introducing nonpathogenic mutans group streptococci with a selective advantage to replace pathogenic strains, is intriguing, but the efficacy when applied to human subjects remains to be established.

Biofilm formation depends on adherence and signaling processes occurring between the component microorganisms. Some promising results have been obtained with blocking adherence of *S. mutans* to salivary pellicle with peptides that mimic the major streptococcal antigen I/II surface protein adhesin (29). Variable Fc chain antibodies to this protein expressed by lactobacilli are also effective in preventing *S. mutans* colonization and caries development in rodent models of infection (31). Although not tried yet with complex oral biofilms, quorum sensing inhibitors have been shown to effectively impair normal biofilm formation by staphylococci (48, 50), so it is possible that signaling inhibitors (16) or biomimetics (39) might be found that disrupt the interactions that occur between oral bacteria in establishing biofilms (Fig. 4).

What is clear is that new approaches are required, informed by microbiology and mo-

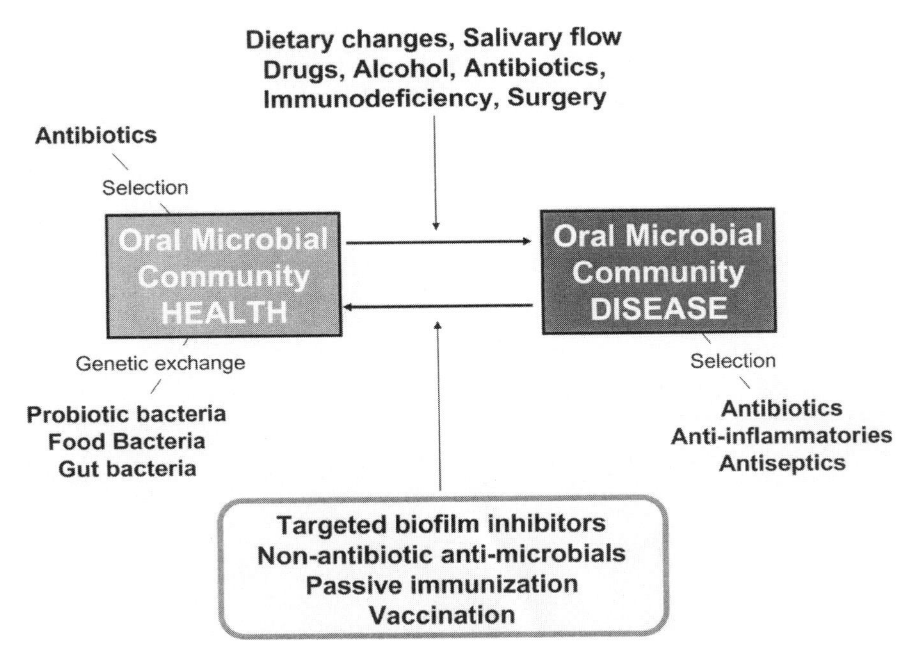


**FIGURE 4** Factors influencing the transition from health-associated to disease-associated biofilms (top) and intervention strategies potentially able to induce a reverse transition (bottom). Increased incidence of antibiotic resistance within the oral microflora is promoted by genetic exchange within biofilms, with antibiotic usage driving the development and retention of less susceptible commensals and pathogens.

lecular biology, to control the microbial communities that form on or in the human body. The continued usage of broad-spectrum antibiotics to solve infection problems will not be sustainable with the increasing incidence of antibiotic resistance. The antibiotic resistance gene pool within the oral microflora (13, 34, 44) and the ability of oral bacteria to readily exchange genetic information within the close confines of biofilms (51) reflect the potential for widespread transfer of antibiotic resistance genes (22) across commensal, pathogenic, and food-borne bacteria (45) (Fig. 4). Given the evidence for oral bacteria playing significant roles in systemic infections, it is critical that nonantibiotic measures be found to assist in controlling oral biofilms. The oral cavity and microflora provide an excellent experimental system for such studies, and it is anticipated that these might be valuable in extending similar methodologies to control biofilms at other sites and in situations where the composition and activities of the microorganisms are not as well understood as those in oral biofilms.

## ACKNOWLEDGMENTS

We thank our lab members for many contributions to the research field, and also colleagues for helpful suggestions and discussions that have informed this chapter.

Current research funding from the National Institutes of Health (NIDCR) and the Wellcome Trust, United Kingdom, is gratefully acknowledged.

## REFERENCES

1. **Aas, J. A., S. M. Barbuto, T. Alpagot, I. Olsen, F. E. Dewhirst, and B. J. Paster.** 2007. Subgingival plaque microbiota in HIV positive patients. *J. Clin. Periodontol.* **34:**189–195.
2. **Aas, J. A., B. J. Paster, L. N. Stokes, I. Olsen, and F. E. Dewhirst.** 2005. Defining the normal bacterial flora of the oral cavity. *J. Clin. Microbiol.* **43:**5721–5732.
3. **Adib, V., D. Spratt, Y. L. Ng, and K. Gulabivala.** 2004. Cultivable microbial flora associated with persistent periapical disease and coronal leakage after root canal treatment: a preliminary study. *Int. Endod. J.* **37:**542–551.
4. **Becker, M. R., B. J. Paster, E. J. Leys, M. L. Moeschgerger, S. G. Kenyon, J. L. Galvin, S. K. Boches, F. E. Dewhirst, and A. L. Griffen.** 2002. Molecular analysis of bacterial species associated with childhood caries. *J. Clin. Microbiol.* **40:**1001–1009.
5. **Blehert, D. S., R. J. Palmer, Jr., J. B. Xavier, J. S. Almeida, and P. E. Kolenbrander.** 2003. Autoinducer 2 production by *Streptococcus gordonii* DL1 and the biofilm phenotype of a *luxS* mutant are influenced by nutritional conditions. *J. Bacteriol.* **185:**4851–4860.
6. **Caufield, P. W., A. P. Dasanayake, Y. Li, Y. Pan, J. Hsu, and J. M. Hardin.** 2000. Natural history of *Streptococcus sanguinis* in the oral cavity of infants: evidence for a discrete window of infectivity. *Infect. Immun.* **68:**4018–4023.
7. **Chapple, I. L., M. R. Milward, and T. Dietrich.** 2007. The prevalence of inflammatory periodontitis is negatively associated with serum antioxidant concentrations. *J. Nutr.* **137:**657–664.
8. **Chen, Y.-Y. M., K. A. Clancy, and R. A. Burne.** 1996. *Streptococcus salivarius* urease: genetic and biochemical characterization and expression in a dental plaque streptococcus. *Infect. Immun.* **64:**585–592.
9. **Daep, C. A., D. M. James, R. J. Lamont, and D. R. Demuth.** 2006. Structural characterization of peptide-mediated inhibition of *Porphyromonas gingivalis* biofilm formation. *Infect. Immun.* **74:**5756–5762.
10. **Darveau, R. P., C. M. Belton, R. A. Reife, and R. J. Lamont.** 1998. Local chemokine paralysis, a novel pathogenic mechanism for *Porphyromonas gingivalis*. *Infect. Immun.* **66:**1660–1665.
11. **Demuth, D. R., D. C. Irvine, J. W. Costerton, G. S. Cook, and R. J. Lamont.** 2001. Discrete protein determinant directs the species-specific adherence of *Porphyromonas gingivalis* to oral streptococci. *Infect. Immun.* **69:**5736–5741.
12. **Diaz, P. I., N. I. Chalmers, A. H. Rickard, C. Kong, C. L. Milburn, R. J. Palmer, Jr., and P. E. Kolenbrander.** 2006. Molecular characterization of subject-specific oral microflora during initial colonization of enamel. *Appl. Environ. Microbiol.* **72:**2837–2848.
13. **Diaz-Torres, M. L., V. Aurelie, N. Hunt, R. McNab, D. A. Spratt, E. Allan, P. Mullany, and M. Wilson.** 2006. Determining the antibiotic resistance potential of the indigenous oral microbiota of humans using a metagenomic approach. *FEMS Microbiol. Lett.* **258:**257–262.
14. **Dierksen, K. P., C. J. Moore, M. Inglis, P. A. Wescombe, and J. R. Tagg.** 2007. The effect of ingestion of milk supplemented with salivaricin A-producing *Streptococcus salivarius* on the bacteriocin-like inhibitory activity of streptococcal populations on the tongue. *FEMS Microbiol. Ecol.* **59:**584–591.

15. **Du, L. D., and P. E. Kolenbrander.** 2000. Identification of saliva-regulated genes of *Streptococcus gordonii* DL1 by differential display using random arbitrarily primed PCR. *Infect. Immun.* **68:**4834–4837.
16. **Eckert, R., J. He, D. K. Yarborough, F. Qi, M. H. Anderson, and W. Shi.** 2006. Targeted killing of *Streptococcus mutans* by a pheromone-guided "smart" antimicrobial peptide. *Antimicrob. Agents Chemother.* **50:**3651–3657.
17. **Fürst, M. M., G. E. Salvi, N. P. Lang, and G. R. Persson.** 2007. Bacterial colonization immediately after installation on oral titanium implants. *Clin. Oral Implants Res.* **18:**501–508.
18. **Garcia, R. I., M. M. Henshaw, and E. A. Krall.** 2001. Relationship between periodontal disease and systemic health. *Periodontol. 2000* **25:** 21–36.
19. **Grenier, D.** 1992. Nutritional interactions between two suspected periodontopathogens, *Treponema denticola* and *Porphyromonas gingivalis*. *Infect. Immun.* **60:**5298–5301.
20. **Herzberg, M. C.** 2001. Coagulation and thrombosis in cardiovascular disease: plausible contributions of infectious agents. *Ann. Periodontol.* **6:**16–19.
21. **Hirasawa, M., K. Takada, and S. Otake.** 2006. Inhibition of acid production in dental plaque bacteria by green tea catechins. *Caries Res.* **40:**265–270.
22. **Jacobsen, L., A. Wilcks, K. Hammer, G. Huys, D. Gevers, and S. R. Andersen.** 2007. Horizontal transfer of tet(M) and erm(B) resistance plasmids from food strains of *Lactobacillus plantarum* to *Enterococcus faecalis* JH2-2 in the gastrointestinal tract of gnotobiotic rats. *FEMS Microbiol. Ecol.* **59:**158–166.
23. **Jakubovics, N. S., S. W. Kerrigan, A. H. Nobbs, N. Strömberg, C. J. van Dolleweerd, D. M. Cox, C. G. Kelly, and H. F. Jenkinson.** 2005. Functions of cell surface-anchored antigen I/II family and Hsa polypeptides in interactions of *Streptococcus gordonii* with host receptors. *Infect. Immun.* **73:**6629–6638.
24. **Jenkinson, H. F., and D. R. Demuth.** 1997. Structure, function and immunogenicity of streptococcal antigen I/II polypeptides. *Mol. Microbiol.* **23:**183–190.
25. **Jenkinson, H. F., and R. J. Lamont.** 1997. Streptococcal adhesion and colonization. *Crit. Rev. Oral Biol. Med.* **8:**175–200.
26. **Jenkinson, H. F., and R. J. Lamont.** 2005. Oral microbial communities in sickness and in health. *Trends Microbiol.* **13:**589–595.
27. **Joly, S., C. Maze, P. B. McCray, Jr., and J. M. Guthmiller.** 2004. Human β-defensins 2 and 3 demonstrate strain-selective activity against oral microorganisms. *J. Clin. Microbiol.* **42:**1024–1029.
28. **Kazor, C. E., P. M. Mitchell, A. M. Lee, L. N. Stokes, W. J. Loesche, F. E. Dewhirst, and B. J. Paster.** 2003. Diversity of bacterial populations on the tongue dorsa of patients with halitosis and healthy patients. *J. Clin. Microbiol.* **41:**558–563.
29. **Kelly, C. G., J. S. Younson, B. Y. Hikmat, S. M. Todryk, M. Czisch, P. I. Harris, I. R. Flindall, C. Newby, A. I. Mallet, J. K. Ma, and T. Lehner.** 1999. A synthetic peptide adhesion epitope as a novel antimicrobial agent. *Nat. Biotechnol.* **17:**42–47.
30. **Kolenbrander, P. E., R. J. Palmer, Jr., A. H. Rickard, N. S. Jakubovics, N. I. Chalmers, and P. I. Diaz.** 2006. Bacterial interactions and successions during plaque development. *Periodontol. 2000* **42:**47–79.
31. **Krüger, C., A. Hultberg, C. van Dollenweerd, H. Marcotte, and L. Hammarström.** 2005. Passive immunization by lactobacilli expressing single-chain antibodies against *Streptococcus mutans*. *Mol. Biotechnol.* **31:**221–231.
32. **Kuboniwa, M., G. D. Tribble, C. E. James, A. O. Kilic, L. Tao, M. C. Herzberg, S. Shizukuishi, and R. J. Lamont.** 2006. *Streptococcus gordonii* utilizes several distinct gene functions to recruit *Porphyromonas gingivalis* into a mixed community. *Mol. Microbiol.* **60:**121–139.
33. **Lamont, R. J., A. El-Sabaeny, Y. Park, G. S. Cook, J. W. Costerton, and D. R. Demuth.** 2002. Role of the *Streptococcus gordonii* SspB protein in the development of *Porphyromonas gingivalis* biofilms on streptococcal substrates. *Microbiology* **148:**1627–1636.
34. **Lancaster, H., R. Bedi, M. Wilson, and P. Mullany.** 2005. The maintenance in the oral cavity of children of tetracycline-resistant bacteria and the genes encoding such resistance. *J. Antimicrob. Chemother.* **56:**524–531.
35. **Ledder, R. G., P. Gilbert, S. A. Huws, L. Aarons, M. P. Ashley, P. S. Hull, and A. J. McBain.** 2007. Molecular analysis of the subgingival microbiota in health and disease. *Appl. Environ. Microbiol.* **73:**516–523.
36. **Loimaranta, V., N. S. Jakubovics, J. Hytonen, J. Finne, H. F. Jenkinson, and N. Strömberg.** 2005. Fluid- or surface-phase human salivary scavenger protein gp340 exposes different bacterial recognition properties. *Infect. Immun.* **73:**2245–2252.
37. **Love, R. M., and H. F. Jenkinson.** 2002. Invasion of dentinal tubules by oral bacteria. *Crit. Rev. Oral Biol. Med.* **13:**171–183.
38. **Malkoski, M., S. G. Dashper, N. M. O'Brien-Simpson, G. H. Talbo, M. Macris,**

**K. J. Cross, and E. C. Reynolds.** 2001. Kappacin, a novel antibacterial peptide from bovine milk. *Antimicrob. Agents Chemother.* **45:**2309–2315.

39. **McDougald, D., S. A. Rice, and S. Kjelleberg.** 2007. Bacterial quorum sensing and interference by naturally occurring biomimics. *Anal. Bioanal. Chem.* **387:**445–453.
40. **McNab, R., S. K. Ford, A. El-Sabaeny, B. Barbieri, G. S. Cook, and R. J. Lamont.** 2003. LuxS-based signaling in *Streptococcus gordonii:* autoinducer 2 controls carbohydrate metabolism and biofilm formation with *Porphyromonas gingivalis*. *J. Bacteriol.* **185:**274–284.
41. **Merritt, J., F. Qi, S. D. Goodman, M. H. Anderson, and W. Shi.** 2003. Mutation of *luxS* affects biofilm formation in *Streptococcus mutans*. *Infect. Immun.* **71:**1972–1979.
42. **Meurman, J. H.** 2005. Probiotics: do they have a role in oral medicine and dentistry? *Eur. J. Oral Sci.* **113:**188–196.
43. **Nobbs, A. H., B. H. Shearer, M. Drobni, M. A. Jepson, and H. F. Jenkinson.** 2007. Adherence and internalization of *Streptococcus gordonii* by epithelial cells involves b1 integrin recognition by SspA and SspB (antigen I/II family) polypeptides. *Cell. Microbiol.* **9:**65–83.
44. **Nyfors, S., E. Könönen, R. Syrjänen, E. Komulainen, and H. Jousimies-Somer.** 2003. Emergence of penicillin resistance among *Fusobacterium nucleatum* populations of commensal oral flora during early childhood. *J. Antimicrob. Chemother.* **51:**107–112.
45. **O'Connor, E. B., O. O'Sullivan, C. Stanton, M. Danielsen, P. J. Simpson, M. J. Callanan, R. P. Ross, and C. Hill.** 28 March 2007. pEOC01: a plasmid from *Pediococcus acidilactici* which encodes an identical streptomycin resistance (*aadE*) gene to that found in *Campylobacter jejuni*. *Plasmid* March 27; [Epub ahead of print.] doi:10.1016/j.plasmid.2007.02.002.
46. **Paster, B. J., S. K. Boches, J. L. Galvin, R. E. Ericson, C. N. Lau, V. A. Levanos, A. Sahasrabudhe, and F. E. Dewhirst.** 2001. Bacterial diversity in human subgingival plaque. *J. Bacteriol.* **183:**3770–3783.
47. **Petersen, F. C., G. Fimland, and A. A. Scheie.** 2006. Purification and functional studies of a potent modified quorum-sensing peptide and two-peptide bacteriocin in *Streptococcus mutans*. *Mol. Microbiol.* **61:**1322–1334.
48. **Qazi, S., B. Middleton, S. H. Muharram, A. Cockayne, P. Hill, P. O'Shea, S. R. Chhabra, M. Cámara, and P. Williams.** 2006. *N*-Acylhomoserine lactones antagonize virulence gene expression and quorum sensing in *Staphylococcus aureus*. *Infect. Immum.* **74:**910–919.
49. **Ramachandran, P., P. Boontheung, Y. Xie, M. Sondej, D. T. Wong, and J. A. Loo.** 2006. Identification of N-linked glycoproteins in human saliva by glycoprotein capture and mass spectrometry. *J. Proteome Res.* **5:**1493–1503.
50. **Rasmussen, T. B., and M. Givskov.** 2006. Quorum sensing inhibitors: a bargain of effects. *Microbiology* **152:**895–904.
51. **Ready, D., J. Pratten, A. P. Roberts, R. Bedi, P. Mullany, and M. Wilson.** 2006. Potential role of *Veillonella* spp. as a reservoir of transferable tetracycline resistance in the oral cavity. *Antimicrob. Agents Chemother.* **50:**2866–2868.
52. **Reynolds, E. C.** 1987. The prevention of subsurface demineralization of bovine enamel and change in plaque composition by casein in an intra-oral model. *J. Dent. Res.* **66:**1120–1127.
53. **Rickard, A. H., R. J. Palmer, Jr., D. S. Blehert, S. R. Campagna, M. F. Semmelhack, P. G. Egland, B. L. Bassler, and P. E. Kolenbrander.** 2006. Autoinducer 2: a concentration-dependent signal for mutualistic bacterial biofilm growth. *Mol. Microbiol.* **60:** 1446–1456.
54. **Rosan, B., and R. J. Lamont.** 2000. Dental plaque formation. *Microbes Infect.* **2:**1599–1607.
55. **Ruhl, S., A. L. Sandberg, and J. O. Cisar.** 2004. Salivary receptors for the proline-rich protein-binding and lectin-like adhesins of oral actinomyces and streptococci. *J. Dent. Res.* **83:** 505–510.
56. **Sasaki, H., M. Matsumoto, T. Tanaka, M. Maeda, M. Nakai, S. Hamada, and T. Ooshima.** 2004. Antibacterial activity of polyphenol components in oolong tea extract against *Streptococcus mutans*. *Caries Res.* **38:**2–8.
57. **Scannapieco, F. A.** 2005. Systemic effects of periodontal diseases. *Dent. Clin. N. Am.* **49:**533–550, vi.
58. **Seo, E. S., D. Kim, J. F. Robyt, D. F. Day, D. W. Kim, and H. J. Park.** 2004. Modified oligosaccharides as potential dental plaque control materials. *Biotechnol. Prog.* **5:**1550–1554.
59. **Steinberg, D., M. Feldman, I. Ofek, and E. I. Weiss.** 2005. Cranberry high molecular weight constituents promote *Streptococcus sobrinus* desorption from artificial biofilm. *Int. J. Antimicrob. Agents* **25:**247–251.
60. **Stenudd, C., A. Nordlund, M. Ryberg, I. Johansson, C. Kallestal, and N. Strömberg.** 2001. The association of bacterial adhesion with dental caries. *J. Dent. Res.* **80:**2005–2010.
61. **Van der Ploeg, J. R.** 2005. Regulation of bacteriocin production in *Streptococcus mutans* by the quorum-sensing system required for the development of genetic competence. *J. Bacteriol.* **187:** 3980–3989.

62. **Wescombe, P. A., M. Upton, K. P. Dierksen, N. L. Ragland, S. Sivabalan, R. E. Wirawan, M. A. Inglis, C. J. Moore, G. V. Walker, C. N. Chilcott, H. F. Jenkinson, and J. R. Tagg.** 2006. Production of the lantibiotic salivaricin A and its variants by oral streptococci and use of a specific induction assay to detect their presence in human saliva. *Appl. Environ. Microbiol.* **72:**1459–1466.
63. **Xie, H., G. S. Cook, J. W. Costerton, G. Bruce, T. M. Rose, and R. J. Lamont.** 2000. Intergeneric communication in dental plaque biofilms. *J. Bacteriol.* **182:**7067–7069.
64. **Xie, H., N. L. Rhodus, R. J. Griffin, J. V. Carlis, and T. J. Griffin.** 2005. A catalogue of human saliva proteins identified by free flow electrophoresis-based peptide separation and tandem mass spectrometry. *Mol. Cell. Proteomics* **4:** 1826–1830.
65. **Xiong, X., P. Buekens, W. D. Fraser, J. Beck, and S. Offenbacher.** 2006. Periodontal disease and adverse pregnancy outcomes: a systematic review. *BJOG* **113:**135–143.
66. **Yoshida, A., T. Ansai, T. Takehara, and H. K. Kuramitsu.** 2005. LuxS-based signaling affects *Streptococcus mutans* biofilm formation. *Appl. Environ. Microbiol.* **71:**2372–2380.
67. **Zhang, Y., Y. Lei, A. Khammanivong, and M. C. Herzberg.** 2004. Identification of a novel two-component system in *Streptococcus gordonii* V288 involved in biofilm formation. *Infect. Immun.* **72:**3489–3494.

# THE GUT MICROBIOME: CURRENT UNDERSTANDING AND FUTURE PERSPECTIVES

*Zhongtang Yu and Mark Morrison*

# 2

There has been a recent surge in interest about the gut microbiome and its interactions with diet and host (immuno)physiology. This is driven in part by the recent application of genome sequencing technologies and approaches most commonly referred to as metagenomics. In that context, many countries are now initiating research programs that seek to characterize the human microbiome and its role in health and disease. The Human Microbiome Project, supported by the National Institutes of Health in the United States, MetaHIT, supported by the European Commission, and the International Human Microbiome Consortium are examples of these activities. The International Human Microbiome Consortium seeks to develop a coordinated, multinational approach to the production and sharing of metagenomic data, to support the achievement of meaningful preventive and therapeutic interventions of relevance to all cultures and societies.

The scientific potential arising from these activities has been previously described (123, 131), and recent review articles note how genomic and metagenomic approaches will direct efforts to better understand the pathogenesis of inflammatory bowel diseases (7, 40, 95). Indeed, the knowledge base and impacts arising from individual investigator, national, and international initiatives are rapidly emerging, making it very challenging to produce a review of this topic that is not quickly outdated. In this chapter, we attempt to complement these recent reviews by providing an overview of how the human gut microbiome develops, with a specific emphasis on the colon; some insights into life cycle assessment of gut microbiome structure-function relationships; and description of specific roles that foods might play in modulating these responses.

Where applicable, comparisons with the gut microbiomes of other vertebrate animal species are made. Only limited emphasis is directed towards the applications of metagenomic tools that can be used to better understand "host-pathogen" interactions and host immunophysiological responses to the gut microbiome. These topics are dealt with elsewhere (42, 55, 81, 96), including chapter 18 of this book.

*Zhongtang Yu,* Department of Animal Sciences, The Ohio State University, Columbus, OH 43210. *Mark Morrison,* Department of Animal Sciences, The Ohio State University, Columbus, OH 43210, and CSIRO Livestock Industries, St Lucia, Queensland, Australia.

*Food-Borne Microbes: Shaping the Host Ecosystem,* Edited by L.-A. Jaykus, H. H. Wang, and L. S. Schlesinger, © 2009 ASM Press, Washington, DC

## PHYSIOANATOMICAL ADAPTATIONS OF THE GI TRACT AND HOST-MICROBE ASSOCIATIONS

Detailed reviews of the physioanatomical variations in the gastrointestinal (GI) tracts of animals and humans have been provided by Stevens and Hume (118), Hume (54), and Mackie et al. (79). With respect to nutrient extraction for growth and development, the interactions between the physioanatomical adaptations of the host GI tract and the resident microbiomes can give rise to competitive, cooperative, or combinatorial relationships (77). The GI microbiome in many carnivores is, in effect, in competition with the digestive and absorptive processes of the host animal, though the commensal microflora does serve to protect the host from pathogens, principally by colonization resistance. The tubiform nature of the carnivore GI tract serves to limit this competition by reducing the residence time for luminal microbes. While the GI tract is mainly tubiform in most carnivores, the GI tract takes on greater anatomical differentiation in herbivores, in part to more effectively harbor and retain a microbiome that actively governs nutrient extraction from foods recalcitrant to host digestive processes and/or other endogenous secretions. Animals possessing a forestomach exemplify the cooperative model of host-microbe relationship, whereby the microbiota provides the hydrolytic potential, fermentation products, and microbial biomass that governs the supply of energy- and protein-yielding nutrients for the host animal. The forestomach of all ruminants consists of a reticulum, rumen, and omasum, of which the rumen is the major chamber (15). Due to a high water content and constant contraction of the rumen wall, much of the ruminal content is more fluidic and better mixed than the contents of other gut compartments. Work with exotic ruminant species also clearly demonstrates how physioanatomical adaptations favor variations in nutritional ecology in these herbivores. The browsing ruminants (such as duikers and giraffes) tend to select leafy plant material and possess larger salivary glands and a smaller reticulorumen than their grazing counterparts (cattle, buffalo, and sheep), as well as differences in dentition, palate, and tongue shape. Intermediate feeders (deer and goats) possess some of the characteristics from each of the above-mentioned groups, in support of their more broad selection of plant material. Conversely, the forestomach of some native Australian marsupials (kangaroos and wallabies) is more tubiform and the intestinal contents are poorly mixed. Detailed insights into the physioanatomy of this organ have been provided by Hume (54) and Lentle et al. (68). The combinatorial model of host-microbe association is exemplified by those herbivores which have evolved to possess an enlarged cecum and/or colon. The hippopotamus, koala, and southern hairy-nosed wombat, as well as horses and the Australian brushtail possum, all possess the combinatorial model of host-microbe relationship, which has been described in detail by Mackie et al. (79) and Hume (54).

The structural features of the human GI tract also favor the combinatorial model for host-microbe association and nutrient extraction from foods, with the upper regions of the tract being somewhat more "hostile" to colonization and persistence by a diverse microbiome. Conversely, the ileum and, most notably, the colon function as a facultative afterburner of undigested foods and/or other endogenous materials, which is largely governed by the development and persistence of a microbiome with extraordinary richness. There is also stratification of the microbiome with respect to the colonization of particulate or mucosal surfaces, giving rise to further complexities. The following sections principally focus on these distal regions of the human GI tract.

## MICROBIAL COLONIZATION AND SUCCESSION OF THE LOWER GI TRACT

Virtually all of our knowledge of microbial succession in neonates and infants arises from the examination of stool samples, with our earliest knowledge produced from cultivation-based studies. More recently, studies have almost exclusively employed DNA-based techniques, especially denaturing gradient gel electrophoresis (DGGE) or thermal gradient gel electrophoresis, as well as real-time PCR and fluorescent in situ hybridization, to quantify individual bacterial genera or species. The latter types of studies have provided a deeper understanding of microbial diversity than most cultivation-based studies, and they facilitate the analysis of a greater number of samples with better statistical comparisons, including options for quantification. The collective findings of these studies show that colonization of the lower GI tract and the subsequent development of the microbiome in humans constitute a temporal process (Table 1), with the period postpartum through weaning divided into four phases (88). The first few days postpartum are considered critical to microbiome development. This phase is most often initiated by vaginal delivery, when the bacteria resident within the vaginal canal and perineal area gain access to the previously sterile GI tract of the newborn (reviewed in references 35, 78, and 88). After birth, all neonates will come into contact with microbes other than those colonizing the mother, including those associated with caretakers and the environmental surroundings. These bacteria may also be primary colonizers, especially in neonates delivered by cesarean section and/or those infants subjected to intensive neonatal care and parenteral feeding (35, 78). Although such infants are eventually colonized by maternally derived microbes, their appearance appears to be delayed, and cultivation-based studies show that facultative anaerobes, mainly enterobacteria, streptococci, enterococci, and staphylococci, are the most common bacteria isolated during this phase (35, 59). Recent studies of 100 infants from Göteborg (Sweden), London (United Kingdom), and Rome (Italy) support most of these observations, with a few notable exceptions (2). Most notably, *Escherichia coli* and other *Enterobacteriaceae* appeared to be among the first bacteria detected in infants, but their abundance fluctuated in most children during the first 6 months of life. Similar results were also obtained using real-time PCR analysis: the bacterial counts appeared to be relatively unstable throughout the first week of life, and cesarean section delayed colonization of the GI tract (91). However, another quantitative study (by cultivation-based analysis) showed that *Lactobacillus* colonization was not significantly related to the delivery mode or to the presence of siblings or pets in the household (3).

When newborns begin feeding during the first week of life, facultative anaerobes such as streptococci and enterobacteria rapidly consume the available oxygen, lowering the redox potential below −100 mV. Consequently, the secondary colonizers are more fastidious anaerobes, including *Bacteroides,* clostridia, and bifidobacteria (87). In general terms, the major groups of bacteria that have been isolated from the stools of healthy infants within 1 week of birth include *Bacteroides, Bifidobacterium,* enterobacteria, *Enterococcus, Propionibacterium, Streptococcus, Eubacterium, Fusobacterium, Peptostreptococcus, Veillonella, Bacillus, Corynebacterium, Lactobacillus, Micrococcus, Staphylococcus,* and clostridia. Total bacterial DGGE profiling has also revealed a rather simple microbiome in infants during the first few weeks (especially during the first few days) of life, but rapid successions during the first 2 months led to a more complex GI microbiota (36, 37). In summary, there is an expansion of microbial richness once the prevailing physicochemical conditions in the large intestine favor the growth of obligate anaerobes, but the actual structure of the microbiome is far from stable or consistent among individuals. While the results are largely similar between cultivation-dependent and -independent studies, there are some differences. For instance, the ruminococci, which have not been frequently recovered from infants by cultivation-based studies, are frequently detected by the cultivation-independent approaches such as DGGE.

The impacts of breast-feeding (or otherwise) on the composition of the neonatal in-

**TABLE 1** Numerical abundance of major bacterial groups in infants[a]

| Age | No. of infants | Numerical abundance, log of CFU/g of wet fecal sample (prevalence [%]) | | | | | | Clostridia | Total bacteria | Country |
|---|---|---|---|---|---|---|---|---|---|---|
| | | Enterobacteria | Enterococci | Staphylococci | *Bacteroides* | *Bifidobacterium* | *Lactobacillus* | | | |
| 1 day[b] | 6 | 9.0 (100) | 8.1 (100) | 4.8 (83) | 0 | 0 | 0 | 6 (17) | NA | Japan |
| 1 day[b] | 6 | 8.8 | 4.3 | 4.5 | 4.0 | 5.0 | 0 | NA | NA | Japan |
| 3 days[b] | 6 | 9.5 (100) | 8.9 (100) | 5.2 (83) | 5.5 (33) | 8.6 (100) | 5.9 (33) | 7.1 (17) | NA | Japan |
| 3 days[b] | 6 | 8.5 | 7.2 | 4.8 | 3.3 | 7.0 | 4.7 | ND | ND | Japan |
| 3 days[c] | 13 | 9.1 (54) | 7.3 (85) | 6.2 (62) | 6.5 (54) | 7.0 (54) | 3.5 (15) | 5.3 (23) | NA | Sweden |
| 4 days[c] | 15 | 8.0 (71) | 7.1 (75) | ND | 5.5 (33) | 9.4 (71) | ND | 4.7 (33) | ND | Italy |
| 5 days[c] | 14 | 9.0 (70) | 6.5 (58) | 7.0 (42) | 7.0 (80) | 6.0 (50) | 0 | 0 | NA | Sweden |
| 6 days[b] | 6 | 6.0 | 6.0 | 5.0 | 5.5 | 9.5 | 4.5 | NA | NA | Japan |
| 7 days[b] | 6 | 7.5 (100) | 7.1 (100) | 4.3 (100) | 7.0 (67) | 10.4 (100) | 5.0 (83) | 5.7 (50) | NA | Japan |
| 7 days[c] | 12 | 8.4 (100) | 8.2 (92) | ND | 6.7 (40) | 8.8 (83) | ND | 6.7 (100) | NA | Italy |
| 7 days[c] | 26 | NA | 10.3 (96) | 8.2 (62) | 11 (46) | 10.3 (50) | 5.8 (8) | 5.1 (27) | NA | Estonia, Sweden |
| 10 days[c] | 30 | 8.7 (87) | ND | ND | ND | 10.2 (27) | 7.9 (43) | ND | ND | United Kingdom |
| 16 days[c] | 15 | 8.3 (80) | 6.9 (80) | ND | 8.0 (20) | 9.5 (84) | ND | 5.1 (28) | ND | Italy |
| 3 wk[c] | 12 | 8.0 (92) | 5.0 (75) | 6.0 (50) | 7.0 (90) | 7.0 (8) | 5.0 (22) | 4.5 (20) | ND | Sweden |
| 4 wk[b] | ? | 8.5 | 6.0 | 6.5 | 7.5 | 11 | 6.5 | ND | ND | Japan |
| 4 wk[c] | 12 | 8.0 (75) | 7.5 (100) | ND | 7.3 (33) | 9.2 (100) | ND | 5.3 (75) | ND | Italy |
| 4 wk[c] | 24 | 8.6 (100) | ND | ND | ND | 9.5 (21) | 8.8 (58) | ND | ND | United Kingdom |
| 4 wk[c] | 26 | NA | 10.1 (96) | 8.8 (65) | 9.8 (62) | 10.0 (69) | 8.6 (46) | 6.3 (27) | NA | Estonia, Sweden |
| 3 mo[b] | 6 | 8.5 | 7.0 | 6.5 | 9.0 | 10.5 | 6.5 | NA | NA | Japan |
| 3 mo[c] | 13 | 8.2 (62) | 5.3 (85) | 5.0 (40) | 6.5 (54) | 9.0 (62) | 8.1 (31) | 4.8 (31) | ND | Sweden |
| 3 mo[c] | 12 | 8.2 (100) | 9.0 (100) | ND | 7.5 (50) | 9.5 (100) | ND | 6.7 (100) | ND | Italy |
| 3 mo[c] | 26 | NA | 10.1 (96) | 7.6 (50) | 9.8 (65) | 10.8 (62) | 9.3 (34) | 7.2 (46) | NA | Estonia, Sweden |
| 6 mo[c] | 26 | ND | 10.1 (96) | 6.1 (23) | 10.1 (81) | 10.3 (42) | 8.4 (38) | 7.3 (65) | ND | Estonia, Sweden |
| 12 mo[c] | 26 | ND | 10.3 (96) | 5.4 (27) | 10.6 (96) | 10.6 (69) | 10.2 (38) | 8.5 (54) | ND | Estonia, Sweden |

[a]Data are from references 11 and 35. NA, not analyzed; ND, not detected.
[b]Values in this row are means.
[c]Values in this row are medians.

testinal microbiome have long been recognized, with the World Health Organization recommending that breast-feeding continue for at least 6 months before weaning (48). However, considerable behavioral variations exist in different cultures and social strata. It should also be noted that in modern Western societies, many infants are often fed with both human milk and formula, and the impact(s) of such mixed feeding on the development of GI microbiota in early infancy still requires clarification.

Breast-feeding and human milk are widely recognized to especially favor the proliferation of bifidobacteria, which typically dominate the fecal microbiota in breast-fed infants. In contrast, formula-fed infants harbor, in addition to bifidobacteria, a variety of other bacterial genera, including enterobacteria, streptococci, *Bacteroides,* and clostridia (22, 32, 50), although it should be noted that these differences have not been reported in all studies (75, 103, 106). These differences between the fecal microbiota of breast- and formula-fed infants are often attributed to a variety of bioactive compounds present in human milk, including immunoglobulin A, lysozyme, lactoperoxidase, lactoferrin, oligosaccharides, gangliosides (glycosphingolipids), and cytokines. Indeed, a number of peptides present in human milk appear to stimulate the growth of bifidobacteria (70, 73), and a ganglioside-supplemented infant formula did produce an increase in the abundance of bifidobacteria but decreases in *E. coli* organisms (102). Furthermore, recent findings by Perez et al. (94) suggest that human breast milk contains a variety of "DNA signatures" and perhaps other bacterially derived components which are translocated to the human breast via blood mononuclear cells. These signatures may assist with the programming of the neonatal immune system and its response to ingested microbes.

The feces of breast-fed infants also have a lower pH, presumably due to the production of more short-chain fatty acids (SCFA) (19). Such end products not only serve to negatively affect the growth of the enterobacteria but also give the more acid-tolerant bifidobacteria and lactobacilli a competitive advantage. The low iron level in human milk (20) probably also contributes to the dominance of *Bifidobacterium* and *Lactobacillus* in the GI microbiota of breast-fed infants, because many species of these two genera do not require iron for growth (22). Another, often neglected difference between breast-feeding and formula feeding is the continuous inoculation of breast-fed infants with bacteria present in human milk from the mother. The human milk of healthy lactating women contains $\sim 10^6$ viable bacteria per ml of milk (132). Coagulase-negative staphylococci are most frequently isolated, along with streptococci, corynebacteria, lactobacilli, micrococci, propionibacteria, and bifidobacteria (2, 132). The greater abundance of staphylococci in the feces of breast-fed than in those of formula-fed infants (43, 50, 71) may reflect a continuous inoculation of the GI tract of breast-fed infants. Such a continuous inoculation likely has an impact on the GI microbiota of infants; however, the extent of this contribution has yet to be determined.

Perhaps the most extensively studied aspect differentiating breast-fed from first-generation milk formula-fed infants relates to the types and amounts of oligosaccharides present in human milk (14, 64, 85) and their impact on microbiome structure. A recent annotation of the human milk glycome found approximately 200 different molecular species of oligosaccharides (86). Numerous studies have compared the GI microbiota of breast- and formula-fed infants (60, 93, 103) and specifically examined the bifidogenic/prebiotic effects of fructo-oligosaccharides (FOS) and galacto-oligosaccharides (GOS) when added to infant formulas (13, 33, 50, 62, 100). Brunser et al. (18) recently used fluorescent in situ hybridization in evaluating whether supplementation of milk formulas with FOS or *Lactobacillus johnsonii* La1 could modulate the composition of the fecal microbiota of formula-fed infants. They concluded that after 7 and 15 weeks of feeding, supplementation diminished the previous differences in fecal microbiota between breast- and formula-fed infants. However, only a limited number of bacterial groups (bifidobacteria, lactobacilli, clostridia, *Bacteroides,* and enterococci) were included in the analysis. Although some infant formulas have even been shown to stimulate the autochthonous population of bifidobacteria in infants (93, 120), it is now widely accepted that the inclusion of GOS in particular generally has a bifidogenic effect on the colonic microbiome of formula-fed infants (12). However, the magnitude of the bifidogenic effects of these prebiotics can be quite variable, with several reports of no significant differences in the abundance of bifidobacteria and lactobacilli in comparisons of breast- and formula-fed infants, especially when FOS is used alone (33, 93, 103). Despite these variations in findings, continued refinement of the oligosaccharide content of infant formulas may offer excellent opportunities to diminish differences in gut microbiome development and stability between breast- and formula-fed infants (Q. Xia, T. Premaraj, T. Williams, D. Hustead, P. Price, M. Morrison, and Z. Yu, unpublished data).

As infants are gradually weaned from a diet limited to milk, the GI microbiome is presented with a more diverse and structurally complex range of substrates which further stimulate the growth of a variety of microbes. The resulting microbiome is more complex and diverse and starts to resemble the adult microbiome. Several biochemical parameters, such as SCFA concentrations and pH, also become more comparable with those found in adults. Earlier cultivation-based studies showed that the numbers of enterococci, *Bacteroides* organisms, enterobacteria, clostridia, and anaerobic streptococci substantially increased during this time (43, 117). Relative to those in preweaned infants, more bifidobacteria ($10^{9.7}$ CFU/g of stool) than *Bacteroides* organisms ($10^{8.6}$ CFU/g of stool) and greater levels of enterobacteria ($10^{8.0}$ CFU/g of stool) and enterococci ($10^{7.8}$ CFU/g of stool) were found among weaned infants (47, 112), but both of these investigations showed surprisingly low rates of isolation (12%) of lactobacilli. Ahrne et al. (3) noticed a different trend among 112 Swedish infants, with *Lactobacillus* concentrations peaking at 6 months ($10^{8.8}$ CFU/g of feces) and then dwindling to $10^{5.4}$ CFU/g of feces at 12 months of age. *Lactobacillus* isolation reached a nadir of 17% by 12 months but increased to 31% by 18 months of age. Interestingly, food-related *Lactobacillus* species, such as *Lactobacillus paracasei, L. plantarum, L. acidophilus,* and *L. delbrueckii,* dominated in this period, while *L. rhamnosus* and *L. gasseri* were the most common species before weaning. Nonetheless, while *Lactobacillus* concentrations appear to fluctuate, age and/or diet modifications evidently had little impact on fecal bifidobacterial abundance but increased enterobacterial and enterococcal populations during weaning. One of the first applications of phylogenetic microarrays used to investigate the temporal successions of the gut microbiome in infants was described by Palmer et al. (91). In total, 2,149 nested taxonomic groups and species were detected in these samples, with the vast majority belonging to just 4 of the 22 recognized bacterial phyla: *Bacteroidetes, Proteobacteria, Firmicutes,* and *Actinobacteria.* Consistent with previous studies, the primary colonizers were found to be *Staphylococcus, Streptococcus,* and enterobacteria, whereas the secondary colonizers tended to be *Eubacterium* and *Clostridium.* The initial colonization by *Bacteroides* varied greatly among the infants but was persistent to some degree in nearly all infants by 1 year of age. Each infant showed evidence of an "individualized" stool microbiota within 6 months, and the differences between two infants exceeded the temporal changes in the GI microbiota of each infant. This confirmed the previous conclusions that the GI microbiota in infants is more prone to impacts imposed by their surrounding environment and that each infant acquires and maintains a distinct combination of microbial species in a temporal pattern. Another interesting observation by Palmer et al. (91) was that other organisms, including *Prevotella, Acinetobacter, Desulfovibrio, Veillonella,* and *Clostridium perfringens,* all tended to appear transiently, sometimes appearing and disappearing repeatedly, within the first year of life.

In summary, the GI microbiome of infants takes as long as 24 months to develop the stability and richness seen in adults. Both the mode of delivery (vaginal versus cesarean) and preweaning nutrition (breast milk versus formula) give rise to measurable differences in microbial community structure. More studies, including those with an increased number of subjects and a broader microbiological analysis, are needed to further elucidate these relationships. To this end, the application of new technologies such as the microarray techniques employed by Palmer et al. (91) and the next-generation DNA sequencing methods (e.g., 454 pyrosequencing) offer new opportunities to produce a more mechanistic understanding of this key aspect of gut microbiome development.

## FUNCTIONAL ANALYSIS OF GUT MICROBIOMES BY (META)GENOMIC ANALYSES

The gut microbiome of adult humans is complex and has been investigated principally using stool samples which are subjected to both culturing (29, 108) and DNA-based techniques (87, 138, 139). Only 10 of the recognized bacterial phyla are readily identified in the adult gut microbiome, compared to the 20 or so phyla that can be identified in soils (26). However, while no fewer than 200 bacterial species have already been described, more than 800 species (5) and 142 genera (65) are estimated to reside in this microbiome. Invariably, in all gut microbiomes, the *Bacteroidetes* and *Firmicutes* are numerically the most abundant, with representatives of the *Proteobacteria, Fusobacteria, Spirochaetes, Actinobacteria, Verrucomicrobia, Lentisphaerae,* and *Deinococcus-Thermus* also commonly identified. Table 2 lists the predominant bacterial species reported in the literature. Most of these species were first isolated in studies conducted prior to the advent of cultivation-independent approaches and, in most instances, identified to the genus level. Indeed, since 2000, only 19 new bacterial species have been isolated and described as having originated from the human colon or feces (29). Approximately 20 to 30% of the direct count of bacteria present in fresh fecal samples can be grown using a single nonselective medium (52, 119); however, if the number of species estimated from phylogenetic analysis holds true, then probably 75% of the bacterial species autochthonous to the human GI tract have not yet been cultured (29). Expanding our culture collections with new isolates, especially when combined with the molecular and metagenomic studies, will be crucial to developing a definitive and mechanistic understanding of the impact(s) of the gut microbiome on nutrition, health, and well-being of adult humans.

Phylogenetic studies employing cloning and sequencing of 16S rRNA genes amplified by PCR have greatly expanded our knowledge of the diversity and composition of the adult gut microbiome. Eckburg et al. (31) identified 395 bacterial phylotypes belonging to the phyla *Bacteroides, Proteobacteria, Actinobacteria, Fusobacteria,* and *Verrucomicrobia,* and 1 archaeal phylotype in the human GI tract; many novel species, especially of the *Clostridia* class, were also noted. Although individual stool and mucosal samples produced different microbial profiles, the intersubject variability was greater than the intrasubject variability. Indeed, most of our knowledge of the human gut microbiome has so far been generated from studies of fecal material, which raises questions about the representation and identification of mucosa-associated bacteria in such samples. For instance, *Bacteroides, Clostridium, E. coli, Lactobacillus,* and *Enterococcus* accounted for ~47% of the total bacterial counts in the feces but only 16% in the mucosa (89). Because mucus and gut epithelial cells are constantly shed into the lumen, it is presumed that an equilibrium develops between the growth of bacteria in the mucous layer and their shedding into the lumen. Thus, mucosal bacteria

**TABLE 2** Microbial genera and species isolated from feces of humans[a]

| Genus | Most frequently isolated species |
|---|---|
| *Acidaminococcus* | *Acidaminococcus* sp. strain 2_2_8 |
| *Acidovorax* | *Acidovorax* sp. strain R-24608 |
| *Acinetobacter* | *A. haemolyticus, A. schindleri* |
| *Actinobacterium* | *Actinobacterium* sp. strain M28 |
| *Actinomyces* | *A. coleocanis, A. naeslundii, A. neuii, A. urogenitalis* |
| *Aerococcus* | *A. christensenii* |
| *Aeromicrobium* | *A. marinum* |
| *Akkermansia* | *A. muciniphila* |
| *Alistipes* | *A. putredinis* |
| *Allochromatium* | *A. vinosum* |
| *Anaerococcus* | *A. hydrogenalis, A. lactolyticus, A. tetradius* |
| *Anaerofustis* | *A. stercorihominis* |
| *Anaerostipes* | *A. caccae* |
| *Anaerotruncus* | *A. colihominis* |
| *Aneurinibacillus* | *A. aneurinilyticus* |
| *Arcobacter* | *A. cryaerophilus* |
| *Arthrobacter* | *A. globiformis* |
| *Atopobium* | *A. vaginae* |
| *Bacillus* | *B. subtilis* |
| *Bacteroides* | *B. coprocola, B. coprophilus, B. distasonis, B. dorei, B. eggerthii, B. finegoldii, B. fragilis, B. galacturonicus, B. heparinolyticus, B. intestinalis, B. merdae, B. ovatus, B. pectinophilus, B. plebeius, B. stercoris, B. thetaiotaomicron, B. uniformis, B. vulgatus* |
| *Bdellovibrio* | *Bdellovibrio* sp. strain BEP2 |
| *Bifidobacterium* | *B. adolescentis, B. angulatum, B. bifidum, B. breve, B. catenulatum, B. dentium, B. gallicum, B. globosum, B. infantis, B. longum, B. pseudocatenulatum, B. scardovii, B. thermophilum* |
| *Brachybacterium* | *Brachybacterium* sp. strain C3H-211 |
| *Brevibacterium* | *B. iodinum, B. mcbrellneri, B. paucivorans* |
| *Brevundimonas* | *B. aurantiaca* |
| *Bryantella* | *B. formatexigens* |
| *Butyrivibrio* | *B. crossotus* |
| *Campylobacter* | *C. ureolyticus* |
| *Capnocytophaga* | *C. gingivalis* |
| *Carnobacterium* | *Carnobacterium* sp. strain R-7272 |
| *Catenibacterium* | *C. mitsuokai* |
| *Cedecea* | *C. davisae* |
| *Cetobacterium* | *C. somerae* |
| *Chryseobacterium* | *C. gleum* |
| *Citrobacter* | *C. freundii, C. youngae* |
| *Clostridium* | *C. asparagiforme, C. bartlettii, C. bifermentans, C. bolteae, C. butyricum, C. carnis, C. celatum, C. clostridioforme, C. coccoides, C. difficile, C. eutactus, C. freundii, C. hathewayi, C. hiranonis, C. hylemonae, C. indolis, C. innocuum, C. leptum, C. methylpentosum, C. nexile, C. orbiscindens, C. paraputrificum, C. perfringens, C. ramosum, C. saccharolyticum, C. scatologenes, C. scindens, C. spiroforme, C. sporogenes, C. symbiosum, C. tertium, C. xylanolyticum* |
| *Collinsella* | *C. aerofaciens, C. intestinalis, C. stercoris* |
| *Coprobacillus* | *Coprobacillus* sp. strain 9_1 |
| *Coprococcus* | *C. comes, C. eutactus* |
| *Coriobacterium* | *Coriobacterium* sp. |
| *Corynebacterium* | *C. accolens, C. ammoniagenes, C. aurimucosum, C. curtum, C. efficiens, C. genitalium, C. glutamicum, C. jeikeium, C. lipophiloflavum, C. nigricans, C. pseudogenitalium, C. seminale, C. striatum, C. sundsvallense* |

*(Table continues)*

**TABLE 2** *(continued)*

| Genus | Most frequently isolated species |
|---|---|
| *Delftia acidovorans* | *D. acidovorans* |
| *Desulfomicrobium* | *D. orale* |
| *Desulfovibrio* | *D. fairfieldensis, D. piger* |
| *Dialister* | *D. invisus* |
| *Diaphorobacter* | *D. nitroreducens* |
| *Dorea* | *D. formicigenerans* |
| *Edwardsiella* | *E. tarda* |
| *Eggerthella* | *E. lenta* |
| *Enterobacter* | *E. cancerogenus, E. cloacae* |
| *Enterococcus* | *E. faecalis, E. faecium* |
| *Erysipelothrix* | *E. rhusiopathiae* |
| *Escherichia* | *E. coli* |
| *Eubacterium* | *E. barkeri, E. biforme, E. brachy, E. contortum, E. cylindroides, E. dolichum, E. hadrum, E. hallii, E. limosum, E. moniliforme, E. plautii, E. ramulus, E. rectale, E. siraeum, E. ventriosum* |
| *Faecalibacterium* | *F. prausnitzii* |
| *Falcivibrio* | *F. vaginalis* |
| *Finegoldia* | *F. magna* |
| *Fusobacterium* | *F. gonidiaformans, F. mortiferum, F. nucleatum, F. prausnitzii, F. ulcerans, F. varium* |
| *Gardnerella* | *G. vaginalis* |
| *Gordonia* | *G. sputi* |
| *Grimontia* | *G. hollisae* |
| *Haemophilus* | *H. ducreyi* |
| *Helicobacter* | *H. canadensis, H. cinaedi, H. pullorum, H, pylori, H. rappini, H. winghamensis* |
| *Holdemania* | *H. filiformis* |
| *Hyphomicrobium* | *H. facile* |
| *Janibacter* | *J. melonis* |
| *Janthinobacterium* | *J. lividum* |
| *Klebsiella* | *K. pneumoniae* |
| *Kocuria* | *K. marina* |
| *Lactobacillus* | *L. acidophilus, L. amylolyticus, L. antri, L. brevis, L. casei, L. coleohominis, L. crispatus, L. delbrueckii, L. fermentum, L. gasseri, L. helveticus, L. iners, L. johnsonii, L. lactis, L. paracasei, L. plantarum, L. reuteri, L. rhamnosus, L. sakei, L. salivarius, L. suntoryeus, L. ultunensis* |
| *Leptothrix* | *Leptothrix* sp. strain DhA-711 |
| *Leptotrichia* | *L. amnionii* |
| *Leuconostoc* | *L. argentinum, L. mesenteroides* |
| *Listeria* | *L. grayi* |
| *Megasphaera* | *M. elsdenii, M. micronuciformis* |
| *Methanobrevibacter* | *M. oralis, M. smithii* |
| *Micrococcus* | *M. luteus* |
| *Micromonas* | *M. micros* |
| *Mitsuokella* | *M. multacida* |
| *Mobiluncus* | *M. curtisii, M. mulieris* |
| *Mogibacterium* | *M. timidum* |
| *Moraxella* | *M. osloensis* |
| *Mycobacterium* | *M. parascrofulaceum* |
| *Mycoplasma* | *M. fermentans, M. hominis, M. salivarium, M. spermatophilum* |
| *Paenibacillus* | *Paenibacillus* sp. |
| *Papillibacter* | *Papillibacter* sp. strain 4_1_38B |
| *Pediococcus* | *Pediococcus* sp. strain 7_4A |
| *Pedomicrobium* | *P. australicum* |

*(Table continues)*

**TABLE 2** Microbial genera and species isolated from feces of humans[a] *(continued)*

| Genus | Most frequently isolated species |
|---|---|
| *Peptostreptococcus* | *P. anaerobius, P. micros, P. productus* |
| *Phascolarctobacterium* | *P. faecium* |
| *Porphyromonas* | *P. levii* |
| *Prevotella* | *P. bergensis, P. copri, P. stercorea* |
| *Propionibacterium* | *P. acnes* |
| *Proteus* | *P. mirabilis, P. penneri* |
| *Providencia* | *P. rustigianii, P. stuartii* |
| *Pseudomonas* | *P. aeruginosa* |
| *Rhodococcus* | *R. corynebacterioides, R. equi* |
| *Roseburia* | *R. faecis, R. intestinales, R. inulinivorans* |
| *Roseomonas* | *R. cervicalis* |
| *Rothia* | *R. dentocariosa* |
| *Ruminococcus* | *R. bromii, R. callidus, R. gnavus, R. hansenii, R. hydrogenotrophicus, R. lactaris, R. luti, R. obeum* |
| *Selenomonas* | *S. dianae* |
| *Serratia* | *S. liquefaciens* |
| *Slackia* | *S. exigua* |
| *Sphingobacterium* | *S. spiritivorum* |
| *Sphingobium* | *S. amiense* |
| *Staphylococcus* | *S. aureus, S. capitis, S. epidermidis, S. haemolyticus, S. saprophyticus* |
| *Stenotrophomonas* | *S. maltophilia* |
| *Streptococcus* | *S. agalactiae, S. anginosus, S. equinus, S. infantarius, S. pneumoniae, S. salivarius, S. waineri* |
| *Subdoligranulum* | *S. variabile* |
| *Treponema* | *T. amylovorum* |
| *Tsukamurella* | *T. tyrosinosolvens* |
| *Variovorax* | *V. paradoxus* |
| *Veillonella* | *V. atypica, V. dispar, V. paramesenteroides, V. parvula* |

[a]Compiled from references 50, 90, and 132.

should be represented and detectable in stool samples. In this context, all of the bacterial phyla *(Firmicutes, Bacteroidetes, Proteobacteria, Fusobacteria, Verrucomicrobia,* and *Actinobacteria)* identified previously in colonic mucosal samples were also identified in feces (129). This may help explain the similarity between colonic tissue and fecal communities arising from the same subject, as previously reported by Eckburg et al. (31).

The mucosa-associated bacteria in the small intestine have been less well characterized, largely due to the difficulties in obtaining biopsy samples. In one comparative study, the jejunum was found to have less microbial diversity than the colonic tissue, and the greatest diversity was observed in the ascending colon (129). Frank et al. (41) reported that biopsy samples from the small intestine tended to produce a larger frequency of clones from the class *Bacilli,* principally *Streptococcaceae,* with fewer *Firmicutes* and *Bacteroidetes* clones. Comparative studies have also suggested that differences in bacterial diversity can be detected in *rrs* gene libraries and metagenomic data produced from either stool or tissue samples collected from healthy (control) individuals, compared to persons suffering from inflammatory bowel disease (41, 80) and obesity (69). Using terminal restriction fragment length polymorphism analysis and other techniques to examine the fecal samples from monozygotic twins concordant and discordant for Crohn's disease, Dicksved et al. (25) were able to show that healthy twins possessed fecal microbiomes that were more similar than the microbiomes of twins concordant for Crohn's disease. Furthermore, the differences in fecal microbiome structure were found to be greatest between twins discordant for Crohn's disease, with the afflicted individuals possessing higher relative abundances of *Bacteroides ovatus* and *Bacteroides vulgatus* organisms. A significant gender-specific difference in the *Bacteroides-Prevotella* group in both adults (20 to 50 years of age) and elderly subjects (>60 years of age) has also been observed, with higher levels of *Bacteroides-Prevotella* organisms in males than in females (84). However, even the most comprehensive studies (31, 41) are considered to have only partially captured the biodiversity of the microbial residents of the human gut, and inadequate coverage may underestimate the species richness (~800 species) of the human gut microbiome (67, 136).

As mentioned previously, future studies employing microarray and next-generation sequencing technologies will offer new opportunities to better characterize the spatial and temporal alterations in gut microbiome structure. In our own laboratories, a phylogenetic microarray has recently been developed for use with human stool and gut biopsy tissue samples. The microarray design was first validated for use with fecal DNA samples (80) in collaboration with Joël Doré and colleagues (S. Kang, Z. Yu, M. Morrison, and C. McSweeney, unpublished data). We have also recently validated the utility of the same microarray to examine bacterial DNA recovered from biopsy tissue samples; these studies also suggest that DNA samples in archived biobanks can be analyzed using the microarray. A combination of microarrays and next-generation sequencing technologies are most likely to be the preferred methods to describe microbial diversity in the future, superseding the more cryptic community fingerprinting methods such as DGGE and terminal restriction fragment length polymorphism. Indeed, one of the first applications of 454 pyrosequencing technology with the gut microbiome has very recently been reported by Andersson et al. (4) and is described below. In the future, a greater focus on quantitative measures that monitor specific populations and their fluctuations should also be utilized.

Until recently, the combination of phylogenetic assessments of microbial diversity with metabolic characterization determined biochemically for selected microbial isolates has been used to infer the functional characteristics underlying microbiomes. However, the advent of genome sequencing and the Human Microbiome Project has begun to change the scope and approaches employed to characterize microbiome functions. The findings from all of these studies to date are beyond the scope of this review, so our discussion here is restricted to *Bacteroides thetaiotaomicron, Bifidobacterium,* and *Lactobacillus,* which provide excellent insight into the future opportunities afforded from genome sequence data.

The genome sequence of *B. thetaiotaomicron* was characterized by Xu et al. (134), who revealed an impressive repertoire of genes in this bacterium underlying the hydrolysis, transport, and metabolism of exogenous and endogenous sources of polysaccharides. Additionally, this bacterium was found to have novel "hybrid two-component systems" that coordinate sensing and response to environmental cues (113, 135), and a novel mechanism for the recognition and binding of amylose helices to the cell surface (63). The production and validation of a microarray have supported a whole-genome view of in vivo gene expression of *B. thetaiotaomicron,* in response to its monoassociation with mice (115) or together with either the methanogen *Methanobrevibacter smithii* (105) or *Bifidobacterium longum* (114). At the time of writing, the genomes of five strains of *Bifidobacterium* (*B. adolescentis* ATCC 15703, *B. animalis* subsp. *lactis* Bb-12, *B. bifidum* JCM 1255, *B. breve* UCC2003, and *B. longum* NCC2705) had been sequenced and another nine strains were being sequenced (http://www.genomesonline.org). The sequenced bifidobacterial genomes have a high G+C content, 55 to 67% (10), and the genomes of both *B. longum* and *B. breve* contain all of the genes necessary for the biosynthesis of pyrimidines and purines, 19 amino acids, and some key vitamins (109, 127). The presence of fimbriae and a teichoic acid-linked surface polysaccharide, both of which could mediate attachment to the surface of the colon, were also identified. Clusters of orthologous groups (COG) devoted to cell cycle control and cell division are overrepresented in bifidobacteria compared to other bacteria, and like in *B. thetaiotaomicron,* a large proportion of their genomes appear to be dedicated to transcriptional regulation (109, 127).

Comparative genome analysis of *B. longum* NCC2705 and *B. adolescentis* ATCC 15703 may help explain several of the physiological and ecological traits of this important genus (109, 125). Both genomes contain at least 17 families of glycoside hydrolases (GHs), including several rare or novel GHs acting on nondigestible plant polymers or host-derived glycoproteins. Other bifidobacterial species also have numerous GHs from a wide range of families (125). Interestingly, *B. longum* contains more GH 51 enzymes (presumably arabinofuranosidases), whereas *B. adolescentis* contains more GH 13 enzymes, which are probably involved in the degradation of a much broader range of poly- and oligosaccharides containing α-glucosidic bonds. Other than the studies of Sonnenburg et al. (114), we are unaware of similar applications of microarrays to examine the transcriptome of bifidobacteria in animal studies, although Gonzalez et al. (46) have examined how the transcriptome of *B. longum* responds to in vitro cultivation in the presence of breast milk, infant formula, or GOS.

No fewer than 12 genomes from the lactobacilli have been sequenced *(Lactobacillus acidophilus, L. brevis, L. casei, L. delbrueckii, L. gasseri, L. helveticus, L. johnsonii, L. plantarum, L. reuteri, L. rhamnosus, L. salvarius,* and *L. sakei).* Analysis of the *L. acidophilus, L. casei, L. gasseri, L. johnsonii,* and *L. plantarum* genomes revealed many transporters, especially phosphotransferase system (PTS) transporters, corroborating the organism's broad capacity to metabolize different carbohydrates found in the GI tract (58). Most of the sequenced genomes of lactobacilli are relatively small (1.9 to 3.3 Mbp), and many of them lack the full genetic complement required for amino acid biosynthesis. The genomes do encode numerous peptidases, amino acid permeases, and multiple oligopeptide transporters that support efficient processing and uptake of amino acids from environments rich in nutrients. Despite these similarities, there are also some striking differences in gene content and synteny among these five genomes. Further comparison of similarities and differences is expected to provide important insights about gene content, organization, and regulation that contribute to both gut and probiotic functionality (24). To this end, transcriptomic analysis of the lactobacilli using microarrays is well under way, reflected in the relatively large number of publications in the last several years, providing important insights into carbohydrate utilization (8, 107) and stress responses (17, 130).

Building on these studies of individual bacterial genomes, metagenomics provides insight into the genetic potential of complex microbial communities, including previously uncultured species (49). Metagenomic studies of the rumen (38), mouse colon (128), and human feces (45, 80) have revealed the rich gene content of these microbiomes. From humans, Gill et al. (45) produced metagenomic data from DNA extracted from fecal samples of two individuals. They identified functional genes coding for the metabolism of glycans, amino acids, and xenobiotics; methanogenesis; and 2-methyl-D-erythritol 4-phosphate pathway-mediated biosynthesis of vitamins and isoprenoids. The data set suggests that the human gut microbiome contains a great abundance of genes involved in the transport and metabolism of various carbohydrates. No fewer than 81 different GH families were represented in the data set, with many showing limited homology with GH genes encountered in other sequenced bacterial genomes. Key genes underpinning the formation of acetate, butyrate, lactate, and succinate are also relatively abundant in the gut metagenomic data, compared with all microbial genomes in the COG database (45). The most abundant COG was related to butyrate kinase, an enzyme that underscores the relative importance of this SCFA in the colon. Unfortunately, the metagenomic data from this study appear to provide only a limited representation of the members of *Bacteroidetes,* a bacterial phylum that was presumed to be abundant in both microbiomes. While this might be compensated in part with genome sequence data from individual isolates, the metabolic potential revealed from this study is probably far from complete.

A comparative metagenomics study on 13 Japanese human subjects has provided some interesting additional insights (65). These investigators identified 237 gene families commonly enriched in both adult and juvenile gut microbiomes; only 136 gene families were detected from infant (preweaned) microbiomes, and there was only a small amount of overlap between infant and adult microbiomes. There is also little overlap between the gene families identified in the human gut microbiomes and those from other environments (65). Although there were interindividual differences, this study seems to support the presence of a "functional core" within the human gut microbiome, encoded probably by "adult or infant gut commonly enriched genes." Such findings provide a foundation for more comparative studies with persons of different racial, ethnic, and/or geographical origins to reveal how similar, or different, we are in terms of gut microbiome structure and function.

As mentioned above, high-throughput DNA sequencing technologies, such as 454 pyrosequencing, now offer exciting opportunities to better characterize the spatial and temporal alterations in the gut microbiome. For instance, Turnbaugh et al. (124) reported comparative metagenomic analysis of the microbiomes recovered from genetically obese mice and lean littermates. They not only showed an increase in the relative abundance of the *Firmicutes* in obese mice (at the expense of *Bacteroidetes*) but also revealed a greater abundance of sequences annotated as encoding specific families of GHs, as well as functions required for the assimilation and fermentation of sugars to acetate and butyrate (which are typical fermentation end products of the *Firmicutes*). One of the first applications of 454 pyrosequencing technology to the human gut microbiome was very recently reported by Andersson et al. (4). In that study, bacterial community DNA was recovered from stomach tissue biopsy samples collected from six elderly, healthy individuals; throat and stool samples were also recovered from three individuals diagnosed with duodenal ulcer and three more individuals as dyspeptic controls. In brief, a neighbor-joining phylogenetic tree and hierarchical clustering of the sequences showed that the samples could be clustered according to site, suggesting that each site bears its own distinct microbiome. Interestingly, all the fecal samples were found to possess more *Actinobacteria* sequences than *Bacteroidetes* sequences. However, the authors noted that the reasons for this observation might be multifactorial, including a possible PCR bias arising from primer design or the fact that the stool samples were obtained only from elderly subjects. Despite these limitations, this study clearly demonstrates the potential of comparative metagenomic analysis, especially when it is applied in conjunction with well-designed clinical and/or animal trials.

## WHERE TO, FROM HERE?

Many conceptual breakthroughs in the life sciences arose because of the development of novel techniques and instrumentation to investigate cellular components and/or physiological processes. For instance, the development of culturing techniques for obligate anaerobes in the 1950s and 1960s rapidly advanced the field of gut microbiology research. The ensuing decades provided many of the bacterial isolates in culture collections, much of our current understanding of the biochemistry and metabolism of gut archaea and bacteria, and, indeed, much of the available genome sequence data. During the middle to late 1980s, the utility of *rrs* gene sequencing was established (66) and cultivation-independent techniques superseded cultivation-based studies; they continue to do so. Since the production of whole-genome sequence data for the first bacterium a little more than 10 years ago (39), microbial biology has advanced from the sequencing of individual genomes to comprehensive assessments of microbial diversity and the genetic potential residing within entire microbial communities. This has been driven largely by (i) the widespread appreciation that culturable microbes represent only a small percentage of the total microbial world, (ii) the development of cloning vectors that support the stable maintenance of large DNA inserts in *E. coli* and other bacterial hosts, and (iii) advances in high-throughput DNA sequencing and in silico analytical technologies.

So, where do we go from here? In their recent review, Peterson et al. (95) described a "wish list for the future" of metagenomics analyses of gut microbiomes, which emphasized the use of animal models (principally the mouse) as well as human trials. Rather than reiterate all their suggestions in detail here, we present only a brief description of some of their major points, as well as possible research directions and approaches for the future, with a specific focus on the role of foods in such studies.

Technically, much of the research output to date is at the DNA level of investigation. As noted by Peterson et al. (95), advances in our abilities to examine gut microbiomes at the RNA, protein, and metabolite levels of investigation will further our mechanistic understanding of microbiome function. While efforts are already under way to achieve that (57), advances will be supported by a more complete functional annotation of (meta)-genome sequences. To that end, the research field will be well served by a renewed commitment to isolating "novel" gut microbes, as well as the development of techniques in bacterial genetics with a wider range of gut microbes that will support mutational analysis of bacterial physiology. Experimentally, Peterson et al. (95) also emphasized the value of working with mouse models as well as mono- and dizygotic twins in future gut microbiome research. Since the insightful studies by Zoetendal et al. (137), both concordant and discordant sets of twins are recognized as offering some valuable opportunities to better understand gut microbiome development and maturation, as well as the role of the gut microbiome in the etiology, potentiation, and/or progression of disease (25). Similarly, animal models such as the mouse have already been found to be a powerful tool in gut microbiome research, especially with the capability to undertake studies using germfree and gnotobiotic mice.

The pig is another potentially useful animal model. These animals can be delivered and reared in either a germfree or gnotobiotic state and can be colonized with a human-like microbiome (92). They also have a relatively large litter size and can be maintained by total parenteral nutrition. The systemic physiology of the neonatal piglet is already widely recognized as an excellent model for human neonatal nutrition and metabolism research (98, 99) as well as obesity research (116). In addition, the porcine genome has been sequenced and can also be manipulated using transgenic technologies (21), providing a suite of useful tools to examine host responses to microbiome perturbations. Finally, we contend that the pig would be an excellent animal model for life cycle assessment of microbiome structure-function relationships, from pediatric colonization events through geriatric conditions. For these reasons, an expansion of research with large-animal models, and especially swine, offers the opportunity to contribute novel findings that can truly augment the knowledge base that can be produced using rodent models.

From a more pragmatic perspective, hydrogen transactions in the gut probably serve as a central underpinning of microbiome structure-function relationships and warrant further attention. More than 10% of the diet ingested reaches the colon as fiber, resistant starch, and proteins (87). The colonic microbiota efficiently converts the undigested foods to mainly SCFA and gases ($CH_4$, $CO_2$, and $H_2$) (101). Between 300 and 400 mmol of SCFA can be produced by an adult per day, and these SCFA, especially butyrate, provide the preferred energy substrate for colonocytes (61, 83, 101). The SCFA can also stimulate the growth of probiotic lactobacilli and acid-tolerant bifidobacteria (82), and probiotic bacteria may function via stimulating the production of SCFA (104). Much of the SCFA is absorbed by the host and can provide 2 to 4% of the host daily energy requirement (122). Carbohydrates are fermented by saccharolytic bacteria primarily in the proximal colon, producing linear forms of SCFA, while fermentation of proteins and amino acids by proteolytic bacteria yields both linear and branched SCFA (76, 133).

Of the SCFA, the production of lactate and butyrate has drawn considerable attention. Lactate is normally present at <3 mM in stool samples from healthy adults, but concentrations of up to 100 mM have been reported in gut disorders such as ulcerative colitis (53). Lactate can be rapidly converted to acetate, butyrate, and propionate within the healthy gut (16). *Anaerostipes caccae, Clostridium homopropionicum, Eubacterium hallii, Propionibacterium freudenreichii,* and *Veillonella parvula* are all species shown to utilize lactate in humans, and *Veillonella* spp. and *Megasphaera elsdenii* perform a similar function in the gut of animals (30, 51, 111). Butyrate is perceived as the most important SCFA, not only as a preferred energy source but also as an anti-inflammatory and chemopreventive compound in colorectal disease (83, 110, 133). Carbohydrates, especially resistant starch, inulin, raftilose, and FOS, are the primary substrates for direct butyrate production (56, 72, 126). At least 42 species of butyrate-producing bacteria have been isolated from human colon, but more butyrate producers remain to be identified (97). *Anaerostipes caccae, Anaerotruncus colihominis, Butyrivibrio fibrisolvens, Clostridium nexile, Clostridium saccharolyticum, Clostridium indolis, Coprococcus* spp., *Eubacterium cylindroides, Eubacterium hominis, Eubacterium rectale, Eubacterium ramulus, Eubacterium ruminantium, Faecalibacterium prausnitzii, Roseburia cecicola, Roseburia intestinalis, Roseburia inulinivorans,* and *Ruminococcus obeum* are among the common colonic butyrate producers (6, 27, 30, 34, 74, 97).

Australian researchers have further established how the inclusion of resistant starches in human diets might serve to promote butyrate production in the colon. In addition to resistant starches promoting fecal bulking and decreasing transit time, there is growing evidence that resistant starches may contribute to a decreased risk of colorectal disease (121). For instance, Clarke et al. (23) reported that a butyrylated form of resistant starch resulted in higher luminal butyrate concentrations in rats, and this was associated with reduced tumor incidence, number, and size. This follows from earlier rat studies showing that resistant starch was protective against colonocyte DNA damage (121). In a human trial, Abell et al. (1) noted that the inclusion of resistant starch in the diet significantly increased butyrate and acetate concentrations (by approximately 22 and 10%, respectively). The community fingerprinting analyses of the fecal samples in this study showed that bacteria closely related to *Ruminococcus bromii* were stimulated by the inclusion of resistant starch in these diets, which was confirmed by real-time PCR assays. Considering that *R. bromii* is not recognized as a butyrate-producing bacterium, these findings suggest that butyrate production may arise from cross-feeding, with other colonic bacteria using the acetate produced by the *R. bromii*-like bacteria for butyrate production. This is perhaps not surprising, given the findings of Flint and colleagues in terms of a robust population of microbes that use acetate (and other fermentation end products) to produce butyrate (9, 28).

In summary, dietary interventions and the inclusion of resistant starches in human diets offer great opportunities to better understand the regulation of butyrate production in the human colon. More physiological and ecological studies on the different hydrogen-utilizing bacteria should help develop dietary strategies (e.g., prebiotics and probiotics) that enhance butyrate production in the colon and maximize the preventive effects associated with the microbiome favored under these conditions, in terms of colorectal disease.

## CONCLUDING REMARKS

The recent advances in genomics and metagenomics provide the opportunity to develop a better understanding of human gut microbiome structure-function relationships and their role in human health and disease, including modulating effects. Consequently, these techniques permit the assessment of interactions among different gut bacteria and the relationships among the host, the foods we eat, and the entire gut microbiome, rather than selecting individual bacterial populations perceived to be important. However, it should be noted that although (meta)genomic studies can reveal the genetic potential present within the gut microbiome, the interactions and processes actually happening in the gut can only be determined through (meta)transcriptomic and (meta)proteomic analyses. Transcriptomic studies have been reported for individual gut bacteria *(Bacteroides, Bifidobacterium,* and *Lactobacillus)*, but metatranscriptomic studies have not yet been reported for the gut microbiome. Very recently, a successful metatranscriptomic study has been reported in the analysis of genes expressed in marine microbial communities (44), which are more complex and diverse than the gut microbiome. Such studies with gut microbiomes should be forthcoming in the near future.

The widespread application of cultivation-independent studies in microbial community analyses should not be used as a replacement for culture methods. In fact, bacterial culture is equally important, if not more so than ever before. Recognizing that it is not necessary or practical to isolate all of the bacteria resident in the gut, more effort should be directed toward isolating representative strains of the important species and toward the recovery of previously nonculturable bacteria. Coupled with an increased number of isolates, future (meta)genomic, (meta)transcriptomic, and (meta)proteomic studies can eventually advance our understanding of the gut microbiome, with the goal of contributing to new approaches to either prevent or treat gut diseases. The use of alternative animal models (such as pigs) will also offer new opportunities for research, especially in relation to interdisciplinary projects involving food, nutritional, and clinical sciences.

## REFERENCES

1. **Abell, G. C., C. M. Cooke, C. N. Bennett, M. A. Conlon, and A. L. McOrist.** 2008. Phylotypes related to *Ruminococcus bromii* are abundant in the large bowel of humans and increase in response to a diet high in resistant starch. *FEMS Microbiol. Ecol.* **63:**505–515.
2. **Adlerberth, I., D. P. Strachan, P. M. Matricardi, S. Ahrne, L. Orfei, N. Aberg, M. R. Perkin, S. Tripodi, B. Hesselmar, R. Saalman, A. R. Coates, C. L. Bonanno, V. Panetta, and A. E. Wold.** 2007. Gut microbiota and development of atopic eczema in 3 European birth cohorts. *J. Allergy Clin. Immunol.* **120:**343–350.
3. **Ahrne, S., E. Lonnermark, A. E. Wold, N. Aberg, B. Hesselmar, R. Saalman, I. L. Strannegard, G. Molin, and I. Adlerberth.** 2005. Lactobacilli in the intestinal microbiota of Swedish infants. *Microbes Infect.* **7:**1256–1262.
4. **Andersson, A. F., M. Lindberg, H. Jakobsson, F. Backhed, P. Nyren, and L. Engstrand.** 2008. Comparative analysis of human gut microbiota by barcoded pyrosequencing. *PLoS ONE* **3:**e2836.
5. **Bäckhed, F., R. E. Ley, J. L. Sonnenburg, D. A. Peterson, and J. I. Gordon.** 2005. Host-bacterial mutualism in the human intestine. *Science* **307:**1915–1920.
6. **Barcenilla, A., S. E. Pryde, J. C. Martin, S. H. Duncan, C. S. Stewart, C. Henderson, and H. J. Flint.** 2000. Phylogenetic relationships of butyrate-producing bacteria from the human gut. *Appl. Environ. Microbiol.* **66:**1654–1661.
7. **Barclay, A. R., D. J. Morrison, and L. T. Weaver.** 2008. What is the role of the metabolic activity of the gut microbiota in inflammatory bowel disease? Probing for answers with stable

isotopes. *J. Pediatr. Gastroenterol. Nutr.* **46:**486–495.

8. **Barrangou, R., M. A. Azcarate-Peril, T. Duong, S. B. Conners, R. M. Kelly, and T. R. Klaenhammer.** 2006. Global analysis of carbohydrate utilization by *Lactobacillus acidophilus* using cDNA microarrays. *Proc. Natl. Acad. Sci. USA* **103:**3816–3821.
9. **Belenguer, A., S. H. Duncan, G. Holtrop, S. E. Anderson, G. E. Lobley, and H. J. Flint.** 2007. Impact of pH on lactate formation and utilization by human fecal microbial communities. *Appl. Environ. Microbiol.* **73:**6526–6533.
10. **Biavati, B., M. Vescovo, S. Torriani, and V. Bottazzi.** 2000. Bifidobacteria: history, ecology, physiology and applications. *Ann. Microbiol.* **50:** 117–131.
11. **Bjorksten, B., E. Sepp, K. Julge, T. Voor, and M. Mikelsaar.** 2001. Allergy development and the intestinal microflora during the first year of life. *J. Allergy Clin. Immunol.* **108:**516–520.
12. **Boehm, G., J. Jelinek, B. Stahl, K. van Laere, J. Knol, S. Fanaro, G. Moro, and V. Vigi.** 2004. Prebiotics in infant formulas. *J. Clin. Gastroenterol.* **38:**S76–S79.
13. **Boehm, G., M. Lidestri, P. Casetta, J. Jelinek, F. Negretti, B. Stahl, and A. Marini.** 2002. Supplementation of a bovine milk formula with an oligosaccharide mixture increases counts of faecal bifidobacteria in preterm infants. *Arch. Dis. Child. Fet. Neonat. Ed.* **86:**F178–F181.
14. **Boehm, G., and B. Stahl.** 2003. Oligosaccharides, p. 203–243. *In* T. Mattila-Sandholm (ed.), *Functional Dairy Products*. Woodhead Publishing Limited, Cambridge, United Kingdom.
15. **Bomba, A., Z. Jonecova, S. Gancarcikova, and R. Nemcova.** 2006. The gastrointestinal microbiota of farm animals, p. 381–400. *In* A. C. Ouwehan and E. E. Vaughan (ed.), *Gastrointestinal Microbiology*. Taylor & Francis Group, New York, NY.
16. **Bourriaud, C., R. J. Robins, L. Martin, F. Kozlowski, E. Tenailleau, C. Cherbut, and C. Michel.** 2005. Lactate is mainly fermented to butyrate by human intestinal microfloras but inter-individual variation is evident. *J. Appl. Microbiol.* **99:**201–212.
17. **Bron, P. A., D. Molenaar, W. M. de Vos, and M. Kleerebezem.** 2006. DNA microarray-based identification of bile-responsive genes in *Lactobacillus plantarum*. *J. Appl. Microbiol.* **100:** 728–738.
18. **Brunser, O., G. Figueroa, M. Gotteland, E. Haschke-Becher, C. Magliola, F. Rochat, S. Cruchet, R. Palframan, G. Gibson, F. Chauffard, and F. Haschke.** 2006. Effects of probiotic or prebiotic supplemented milk formulas on fecal microbiota composition of infants. *Asia Pac. J. Clin. Nutr.* **15:**368–376.
19. **Bullen, C. L., and A. T. Willis.** 1971. Resistance of the breast-fed infant to gastroenteritis. *Br. Med. J.* **3:**338–343.
20. **Bullen, J. J., H. J. Rogers, and L. Leigh.** 1972. Iron-binding proteins in milk and resistance to *Escherichia coli* infection in infants. *Br. Med. J.* **1:**69–75.
21. **Chen, K., T. Baxter, W. M. Muir, M. A. Groenen, and L. B. Schook.** 2007. Genetic resources, genome mapping and evolutionary genomics of the pig *(Sus scrofa)*. *Int. J. Biol. Sci.* **3:**153–165.
22. **Chierici, R., S. Fanaro, D. Saccomandi, and V. Vigi.** 2003. Advances in the modulation of the microbial ecology of the gut in early infancy. *Acta Paediatr.* **92:**56–63.
23. **Clarke, J. M., D. L. Topping, A. R. Bird, G. P. Young, and L. Cobiac.** 2008. Effects of high-amylose maize starch and butyrylated high-amylose maize starch on azoxymethane-induced intestinal cancer in rats. *Carcinogenesis* **29:**2190–2194.
24. **de Vos, W. M., P. A. Bron, and M. Kleerebezem.** 2004. Post-genomics of lactic acid bacteria and other food-grade bacteria to discover gut functionality. *Curr. Opin. Biotechnol.* **15:**86–93.
25. **Dicksved, J., J. Halfvarson, M. Rosenquist, G. Järnerot, C. Tysk, J. Apajalahti, L. Engstrand, and J. K. Jansson.** 2008. Molecular analysis of the gut microbiota of identical twins with Crohn's disease. *ISME J.* **2:**716–727.
26. **Dunbar, J., S. M. Barns, L. O. Ticknor, and C. R. Kuske.** 2002. Empirical and theoretical bacterial diversity in four Arizona soils. *Appl. Environ. Microbiol.* **68:**3035–3045.
27. **Duncan, S. H., A. Barcenilla, C. S. Stewart, S. E. Pryde, and H. J. Flint.** 2002. Acetate utilization and butyryl coenzyme A (CoA): acetate-CoA transferase in butyrate-producing bacteria from the human large intestine. *Appl. Environ. Microbiol.* **68:**5186–5190.
28. **Duncan, S. H., G. Holtrop, G. E. Lobley, A. G. Calder, C. S. Stewart, and H. J. Flint.** 2004. Contribution of acetate to butyrate formation by human faecal bacteria. *Br. J. Nutr.* **91:** 915–923.
29. **Duncan, S. H., P. Louis, and H. J. Flint.** 2007. Cultivable bacterial diversity from the human colon. *Lett. Appl. Microbiol.* **44:**343–350.
30. **Duncan, S. H., P. Louis, and H. J. Flint.** 2004. Lactate-utilizing bacteria, isolated from human feces, that produce butyrate as a major fermentation product. *Appl. Environ. Microbiol.* **70:** 5810–5817.

31. **Eckburg, P. B., E. M. Bik, C. N. Bernstein, E. Purdom, L. Dethlefsen, M. Sargent, S. R. Gill, K. E. Nelson, and D. A. Relman.** 2005. Diversity of the human intestinal microbial flora. *Science* **308:**1635–1638.
32. **Edwards, C. A., and A. M. Parrett.** 2002. Intestinal flora during the first months of life: new perspectives. *Br. J. Nutr.* **88**(Suppl. 1)**:**S11–S18.
33. **Euler, A. R., D. K. Mitchell, R. Kline, and L. K. Pickering.** 2005. Prebiotic effect of fructo-oligosaccharide supplemented term infant formula at two concentrations compared with unsupplemented formula and human milk. *J. Pediatr. Gastroenterol. Nutr.* **40:**157–164.
34. **Falony, G., A. Vlachou, K. Verbrugghe, and L. De Vuyst.** 2006. Cross-feeding between *Bifidobacterium longum* BB536 and acetate-converting, butyrate-producing colon bacteria during growth on oligofructose. *Appl. Environ. Microbiol.* **72:**7835–7841.
35. **Fanaro, S., R. Chierici, P. Guenini, and V. Vigi.** 2003. Intestinal microflora in early infancy: composition and development. *Acta Paediatr.* **92:** S48–S55.
36. **Favier, C. F., W. M. de Vos, and A. D. L. Akkermans.** 2003. Development of bacterial and bifidobacterial communities in feces of newborn babies. *Anaerobe* **9:**219–229.
37. **Favier, C. F., E. E. Vaughan, W. M. De Vos, and A. D. L. Akkermans.** 2002. Molecular monitoring of succession of bacterial communities in human neonates. *Appl. Environ. Microbiol.* **68:**219–226.
38. **Ferrer, M., O. V. Golyshina, T. N. Chernikova, A. N. Khachane, D. Reyes-Duarte, V. A. Santos, C. Strompl, K. Elborough, G. Jarvis, A. Neef, M. M. Yakimov, K. N. Timmis, and P. N. Golyshin.** 2005. Novel hydrolase diversity retrieved from a metagenome library of bovine rumen microflora. *Environ. Microbiol.* **7:**1996–2010.
39. **Fleischmann, R. D., M. D. Adams, O. White, R. A. Clayton, E. F. Kirkness, A. R. Kerlavage, C. J. Bult, J. F. Tomb, B. A. Dougherty, J. M. Merrick, et al.** 1995. Whole-genome random sequencing and assembly of *Haemophilus influenzae* Rd. *Science* **269:**496–512.
40. **Frank, D. N., and N. R. Pace.** 2008. Gastrointestinal microbiology enters the metagenomics era. *Curr. Opin. Gastroenterol.* **24:**4–10.
41. **Frank, D. N., A. L. St Amand, R. A. Feldman, E. C. Boedeker, N. Harpaz, and N. R. Pace.** 2007. Molecular-phylogenetic characterization of microbial community imbalances in human inflammatory bowel diseases. *Proc. Natl. Acad. Sci. USA* **104:**13780–13785.
42. **Gaskins, H. R., J. A. Croix, N. Nakamura, and G. M. Nava.** 2008. Impact of the intestinal microbiota on the development of mucosal defense. *Clin. Infect. Dis.* **46**(Suppl. 2)**:**S80–S86; discussion, S144–S151.
43. **George, M., K. E. Nord, G. Ronquist, G. Hedenstierna, and L. Wiklund.** 1996. Faecal microflora and urease activity during the first six months of infancy. *Upsala J. Med. Sci.* **101:**233–250.
44. **Gilbert, J. A., D. Field, Y. Huang, R. Edwards, W. Li, P. Gilna, and I. Joint.** 2008. Detection of large numbers of novel sequences in the metatranscriptomes of complex marine microbial communities. *PLoS ONE* **3:**e3042.
45. **Gill, S. R., M. Pop, R. T. Deboy, P. B. Eckburg, P. J. Turnbaugh, B. S. Samuel, J. I. Gordon, D. A. Relman, C. M. Fraser-Liggett, and K. E. Nelson.** 2006. Metagenomic analysis of the human distal gut microbiome. *Science* **312:**1355–1359.
46. **Gonzalez, R., E. S. Klaassens, E. Malinen, W. M. de Vos, and E. E. Vaughan.** 2008. Differential transcriptional response of *Bifidobacterium longum* to human milk, formula milk, and galactooligosaccharide. *Appl. Environ. Microbiol.* **74:**4686–4694.
47. **Guerin-Danan, C., C. Andrieux, F. Popot, A. Charpilienne, P. Vaissade, C. Gaudichon, C. Pedone, C. Bouley, and O. Szylit.** 1997. Pattern of metabolism and composition of the fecal microflora in infants 10 to 18 months old from day care centers. *J. Pediatr. Gastroenterol. Nutr.* **25:**281–289.
48. **Gupta, A., G. P. Mathur, and J. C. Sobti.** 2002. World Health Assembly recommends exclusive breastfeeding for first six months. *J. Indian Med. Assoc.* **100:**510–511, 515.
49. **Handelsman, J.** 2004. Metagenomics: application of genomics to uncultured microorganisms. *Microbiol. Mol. Biol. Rev.* **68:**669–685.
50. **Harmsen, H. J. M., A. C. M. Wildeboer-Veloo, G. C. Raangs, A. A. Wagendorp, N. Klijn, J. G. Bindels, and G. W. Welling.** 2000. Analysis of intestinal flora development in breast-fed and formula-fed infants by using molecular identification and detection methods. *J. Pediatr. Gastroenterol. Nutr.* **30:**61–67.
51. **Hashizume, K., T. Tsukahara, K. Yamada, H. Koyama, and K. Ushida.** 2003. *Megasphaera elsdenii* JCM1772T normalizes hyperlactate production in the large intestine of fructooligosaccharide-fed rats by stimulating butyrate production. *J. Nutr.* **133:**3187–3190.
52. **Hayashi, H., M. Sakamoto, and Y. Benno.** 2002. Phylogenetic analysis of the human gut microbiota using 16S rDNA clone libraries and

strictly anaerobic culture-based methods. *Microbiol. Immunol.* **46:**535–548.

53. **Hove, H., I. Nordgaard-Andersen, and P. B. Mortensen.** 1994. Faecal DL-lactate concentration in 100 gastrointestinal patients. *Scand. J. Gastroenterol.* **29:**255–259.

54. **Hume, I. D.** 1999. *Marsupial Nutrition.* Cambridge University Press, Cambridge, United Kingdom.

55. **Janssen, R., K. A. Krogfelt, S. A. Cawthraw, W. van Pelt, J. A. Wagenaar, and R. J. Owen.** 2008. Host-pathogen interactions in *Campylobacter* infections: the host perspective. *Clin. Microbiol. Rev.* **21:**505–518.

56. **Khan, K. M., and C. A. Edwards.** 2005. In vitro fermentation characteristics of a mixture of raftilose and guar gum by human faecal bacteria. *Eur. J. Nutr.* **44:**371–376.

57. **Klaassens, E. S., W. M. de Vos, and E. E. Vaughan.** 2007. Metaproteomics approach to study the functionality of the microbiota in the human infant gastrointestinal tract. *Appl. Environ. Microbiol.* **73:**1388–1392.

58. **Klaenhammer, T. R., R. Barrangou, B. L. Buck, M. A. Azcarate-Peril, and E. Altermann.** 2005. Genomic features of lactic acid bacteria effecting bioprocessing and health. *FEMS Microbiol. Rev.* **29:**393–409.

59. **Kleessen, B., E. Bezirtzoglou, and J. Matto.** 2000. Culture-based knowledge on biodiversity, development and stability of human gastrointestinal microflora. *Microb. Ecol. Health Dis.* **12**(Suppl. 2):53–63.

60. **Kleessen, B., H. Bunke, K. Tovar, J. Noack, and G. Sawatzki.** 1995. Influence of two infant formulas and human milk on the development of faecal flora in newborn infants. *Acta Paediatr.* **84:**1347–1356.

61. **Kles, K. A., and E. B. Chang.** 2006. Short-chain fatty acids impact on intestinal adaptation, inflammation, carcinoma, and failure. *Gastroenterology* **130:**S100–S105.

62. **Knol, J., P. Scholtens, C. Kafka, J. Steenbakkers, S. Gro, K. Helm, M. Klarczyk, H. Schöpfer, H.-M. Böckler, and J. Wells.** 2005. Colon microflora in infants fed formula with galacto- and fructo-oligosaccharides: more like breast-fed infants. *J. Pediatr. Gastroenterol. Nutr.* **40:**36–42.

63. **Koropatkin, N. M., E. C. Martens, J. I. Gordon, and T. J. Smith.** 2008. Starch catabolism by a prominent human gut symbiont is directed by the recognition of amylose helices. *Structure* **16:**1105–1115.

64. **Kunz, C., and S. Rudloff.** 1993. Biological functions of oligosaccharides in human milk. *Acta Paediatr.* **82:**903–912.

65. **Kurokawa, K., T. Itoh, T. Kuwahara, K. Oshima, H. Toh, A. Toyoda, H. Takami, H. Morita, V. K. Sharma, T. P. Srivastava, T. D. Taylor, H. Noguchi, H. Mori, Y. Ogura, D. S. Ehrlich, K. Itoh, T. Takagi, Y. Sakaki, T. Hayashi, and M. Hattori.** 2007. Comparative metagenomics revealed commonly enriched gene sets in human gut microbiomes. *DNA Res.* **14:**169–181.

66. **Lane, D. J., B. Pace, G. J. Olsen, D. A. Stahl, M. L. Sogin, and N. R. Pace.** 1985. Rapid determination of 16S ribosomal RNA sequences for phylogenetic analyses. *Proc. Natl. Acad. Sci. USA* **82:**6955–6959.

67. **Larue, R., Z. Yu, V. A. Parisi, A. R. Egan, and M. Morrison.** 2005. Novel microbial diversity adherent to plant biomass in the herbivore gastrointestinal tract, as revealed by ribosomal intergenic spacer analysis and *rrs* gene sequencing. *Environ. Microbiol.* **7:**530–543.

68. **Lentle, R. G., K. J. Stafford, M. S. Kennedy, and S. J. Haslett.** 2002. Rheological properties of digesta suggest little radial or axial mixing in the forestomach of the tammar *(Macropus eugenii)* and the parma *(Macropus parma)* wallaby. *Physiol. Biochem. Zool.* **75:**572–582.

69. **Ley, R. E., P. J. Turnbaugh, S. Klein, and J. I. Gordon.** 2006. Microbial ecology: human gut microbes associated with obesity. *Nature* **444:**1022–1023.

70. **Liepke, C., K. Adermann, M. Raida, H. J. Mägert, W. G. Forssmann, and H. D. Zucht.** 2002. Human milk provides peptides highly stimulating the growth of bifidobacteria. *Eur. J. Biochem.* **269:**712–718.

71. **Lindberg, E., I. Adlerberth, B. Hesselmar, R. Saalman, I. L. Strannegard, N. Aberg, and A. E. Wold.** 2004. High rate of transfer of *Staphylococcus aureus* from parental skin to infant gut flora. *J. Clin. Microbiol.* **42:**530–534.

72. **Loh, G., M. Eberhard, R. M. Brunner, U. Hennig, S. Kuhla, B. Kleessen, and C. C. Metges.** 2006. Inulin alters the intestinal microbiota and short-chain fatty acid concentrations in growing pigs regardless of their basal diet. *J. Nutr.* **136:**1198–1202.

73. **Lonnerdal, B.** 2003. Nutritional and physiologic significance of human milk proteins. *Am. J. Clin. Nutr.* **77:**1537S–1543S.

74. **Louis, P., S. H. Duncan, S. I. McCrae, J. Millar, M. S. Jackson, and H. J. Flint.** 2004. Restricted distribution of the butyrate kinase pathway among butyrate-producing bacteria from the human colon. *J. Bacteriol.* **186:**2099–2106.

75. **Lundequist, B., C. E. Nord, and J. Winberg.** 1985. The composition of the faecal microflora

in breastfed and bottle fed infants from birth to eight weeks. *Acta Paediatr. Scand.* **74:**45–51.

**76. Macfarlane, S., and G. T. Macfarlane.** 2003. Regulation of short-chain fatty acid production. *Proc. Nutr. Soc.* **62:**67–72.

**77. Mackie, R. I.** 2002. Mutualistic fermentative digestion in the gastrointestinal tract: diversity and evolution. *Integr. Comp. Biol.* **42:**319–326.

**78. Mackie, R. I., A. Sghir, and H. R. Gaskins.** 1999. Developmental microbial ecology of the neonatal gastrointestinal tract. *Am. J. Clin. Nutr.* **69:**1035S–1045S.

**79. Mackie, R. I., B. A. White, and R. E. Isaacson.** 1997. *Gastrointestinal Microbiology.* Chapman & Hall, New York, NY.

**80. Manichanh, C., L. Rigottier-Gois, E. Bonnaud, K. Gloux, E. Pelletier, L. Frangeul, R. Nalin, C. Jarrin, P. Chardon, P. Marteau, J. Roca, and J. Dore.** 2006. Reduced diversity of faecal microbiota in Crohn's disease revealed by a metagenomic approach. *Gut* **55:**205–211.

**81. Mason, K. L., G. B. Huffnagle, M. C. Noverr, and J. Y. Kao.** 2008. Overview of gut immunology. *Adv. Exp. Med. Biol.* **635:**1–14.

**82. McGarr, S. E., J. M. Ridlon, and P. B. Hylemon.** 2005. Diet, anaerobic bacterial metabolism, and colon cancer: a review of the literature. *J. Clin. Gastroenterol.* **39:**98–109.

**83. Miller, S. J.** 2004. Cellular and physiological effects of short-chain fatty acids. *Mini Rev. Med. Chem.* **4:**839–845.

**84. Mueller, S., K. Saunier, C. Hanisch, E. Norin, L. Alm, T. Midtvedt, A. Cresci, S. Silvi, C. Orpianesi, M. C. Verdenelli, T. Clavel, C. Koebnick, H.-J. F. Zunft, J. Dore, and M. Blaut.** 2006. Differences in fecal microbiota in different European study populations in relation to age, gender, and country: a cross-sectional study. *Appl. Environ. Microbiol.* **72:**1027–1033.

**85. Newburg, D. S.** 2000. Oligosaccharides in human milk and bacterial colonization. *J. Pediatr. Gastroenterol. Nutr.* **30:**S8–S17.

**86. Ninonuevo, M. R., Y. Park, H. Yin, J. Zhang, R. E. Ward, B. H. Clowers, J. B. German, S. L. Freeman, K. Killeen, R. Grimm, and C. B. Lebrilla.** 2006. A strategy for annotating the human milk glycome. *J. Agric. Food Chem.* **54:**7471–7480.

**87. O'Keefe, S. J.** 2008. Nutrition and colonic health: the critical role of the microbiota. *Curr. Opin. Gastroenterol.* **24:**51–58.

**88. Orrhage, K., and C. E. Nord.** 1999. Factors controlling the bacterial colonization of the intestine in breastfed infants. *Acta Paediatr.* Suppl. **430:**47–57.

**89. Ouwehand, A. C., S. Salminen, T. Arvola, T. Ruuska, and E. Isolauri.** 2004. Microbiota composition of the intestinal mucosa: association with fecal microbiota? *Microbiol. Immunol.* **48:** 497–500.

**90. Ouwehand, A. C., and E. E. Vaughan.** 2006. *Gastrointestinal Microbiology.* Taylor & Francis Group, New York, NY.

**91. Palmer, C., E. M. Bik, D. B. Digiulio, D. A. Relman, and P. O. Brown.** 2007. Development of the human infant intestinal microbiota. *PLoS Biol.* **5:**e177.

**92. Pang, X., X. Hua, Q. Yang, D. Ding, C. Che, L. Cui, W. Jia, P. Bucheli, and L. Zhao.** 2007. Inter-species transplantation of gut microbiota from human to pigs. *ISME J.* **1:**156–162.

**93. Penders, J., C. Vink, C. Driessen, N. London, C. Thijs, and E. E. Stobberingh.** 2005. Quantification of *Bifidobacterium* spp., *Escherichia coli* and *Clostridium difficile* in faecal samples of breast-fed and formula-fed infants by real-time PCR. *FEMS Microbiol. Lett.* **243:**141–147.

**94. Perez, P. F., J. Doré, M. Leclerc, F. Levenez, J. Benyacoub, P. Serrant, I. Segura-Roggero, E. J. Schiffrin, and A. Donnet-Hughes.** 2007. Bacterial imprinting of the neonatal immune system: lessons from maternal cells? *Pediatrics* **119:**e724–e732.

**95. Peterson, D. A., D. N. Frank, N. R. Pace, and J. I. Gordon.** 2008. Metagenomic approaches for defining the pathogenesis of inflammatory bowel diseases. *Cell Host Microbe* **3:** 417–427.

**96. Pillinger, M. H., and M. J. Blaser.** 2007. The language used by *Helicobacter pylori* to regulate human cells. *J. Infect. Dis.* **196:**6–9.

**97. Pryde, S. E., S. H. Duncan, G. L. Hold, C. S. Stewart, and H. J. Flint.** 2002. The microbiology of butyrate formation in the human colon. *FEMS Microbiol. Lett.* **217:**133–139.

**98. Puiman, P., and B. Stoll.** 2008. Animal models to study neonatal nutrition in humans. *Curr. Opin. Clin. Nutr. Metabolic Care* **11:**601–606.

**99. Riedijk, M. A., B. Stoll, S. Chacko, H. Schierbeek, A. L. Sunehag, J. B. van Goudoever, and D. G. Burrin.** 2007. Methionine transmethylation and transsulfuration in the piglet gastrointestinal tract. *Proc. Natl. Acad. Sci. USA* **104:**3408–3413.

**100. Rinne, M. M., M. Gueimonde, M. Kalliomäki, U. Hoppu, S. J. Salminen, and E. Isolauri.** 2005. Similar bifidogenic effects of prebiotic-supplemented partially hydrolyzed infant formula and breastfeeding on infant gut microbiota. *FEMS Immunol. Med. Microbiol.* **43:** 59–65.

101. **Roy, C. C., C. L. Kien, L. Bouthillier, and E. Levy.** 2006. Short-chain fatty acids: ready for prime time? *Nutr. Clin. Practice* **21:**351–366.
102. **Rueda, R., J. Maldonado, E. Narbona, and A. Gil.** 1998. Neonatal dietary gangliosides. *Early Hum. Dev.* **53**(Suppl.)**:**S135–S147.
103. **Sakata, S., T. Tonooka, S. Ishizeki, M. Takada, M. Sakamoto, M. Fukuyama, and Y. Benno.** 2005. Culture-independent analysis of fecal microbiota in infants, with special reference to *Bifidobacterium* species. *FEMS Microbiol. Lett.* **243:**417–423.
104. **Sakata, T., T. Kojima, M. Fujieda, M. Takahashi, and T. Michibata.** 2003. Influences of probiotic bacteria on organic acid production by pig caecal bacteria in vitro. *Proc. Nutr. Soc.* **62:**73–80.
105. **Samuel, B. S., and J. I. Gordon.** 2006. A humanized gnotobiotic mouse model of host-archaeal-bacterial mutualism. *Proc. Natl. Acad. Sci. USA* **103:**10011–10016.
106. **Satokari, R. M., E. E. Vaughan, C. F. Favier, J. Dore, C. Edwards, and W. M. de Vos.** 2002. Diversity of *Bifidobacterium* and *Lactobacillus* spp. in breast-fed and formula-fed infants as assessed by 16S rDNA sequence differences. *Microb. Ecol. Health Dis.* **14:**97–105.
107. **Saulnier, D. M., D. Molenaar, W. M. de Vos, G. R. Gibson, and S. Kolida.** 2007. Identification of prebiotic fructooligosaccharide metabolism in *Lactobacillus plantarum* WCFS1 through microarrays. *Appl. Environ. Microbiol.* **73:**1753–1765.
108. **Savage, D. C.** 1977. Microbial ecology of the gastrointestinal tract. *Annu. Rev. Microbiol.* **31:** 107–133.
109. **Schell, M. A., M. Karmirantzou, B. Snel, D. Vilanova, B. Berger, G. Pessi, M. C. Zwahlen, F. Desiere, P. Bork, M. Delley, R. D. Pridmore, and F. Arigoni.** 2002. The genome sequence of *Bifidobacterium longum* reflects its adaptation to the human gastrointestinal tract. *Proc. Natl. Acad. Sci. USA* **99:**14422–14427.
110. **Scheppach, W., and F. Weiler.** 2004. The butyrate story: old wine in new bottles? *Curr. Opin. Clin. Nutr. Metabolic Care* **7:**563–567.
111. **Seeliger, S., P. H. Janssen, and B. Schink.** 2002. Energetics and kinetics of lactate fermentation to acetate and propionate via methylmalonyl-CoA or acrylyl-CoA. *FEMS Microbiol. Lett.* **211:**65–70.
112. **Sepp, E., K. Julge, M. Vasar, P. Naaber, B. Bjorksten, and M. Mikelsaar.** 1997. Intestinal microflora of Estonian and Swedish infants. *Acta Paediatr.* **86:**956–961.
113. **Sonnenburg, E. D., J. L. Sonnenburg, J. K. Manchester, E. E. Hansen, H. C. Chiang, and J. I. Gordon.** 2006. A hybrid two-component system protein of a prominent human gut symbiont couples glycan sensing in vivo to carbohydrate metabolism. *Proc. Natl. Acad. Sci. USA* **103:**8834–8839.
114. **Sonnenburg, J. L., C. T. Chen, and J. I. Gordon.** 2006. Genomic and metabolic studies of the impact of probiotics on a model gut symbiont and host. *PLoS Biol.* **4:**e413.
115. **Sonnenburg, J. L., J. Xu, D. D. Leip, C. H. Chen, B. P. Westover, J. Weatherford, J. D. Buhler, and J. I. Gordon.** 2005. Glycan foraging in vivo by an intestine-adapted bacterial symbiont. *Science* **307:**1955–1959.
116. **Spurlock, M. E., and N. K. Gabler.** 2008. The development of porcine models of obesity and the metabolic syndrome. *J. Nutr.* **138:**397–402.
117. **Stark, P. L., and A. Lee.** 1982. The microbial ecology of the large bowel of breast-fed and formula-fed infants during the first year of life. *J. Med. Microbiol.* **15:**189–203.
118. **Stevens, C. E., and I. D. Hume.** 1998. Contributions of microbes in vertebrate gastrointestinal tract to production and conservation of nutrients. *Physiol. Rev.* **78:**393–427.
119. **Suau, A., R. Bonnet, M. Sutren, J.-J. Godon, G. R. Gibson, M. D. Collins, and J. Dore.** 1999. Direct analysis of genes encoding 16S rRNA from complex communities reveals many novel molecular species within the human gut. *Appl. Environ. Microbiol.* **65:**4799–4807.
120. **Tannock, G. W.** 2002. Probiotics and prebiotics: where are we going? p. 1–39. *In* G. W. Tannock (ed.), *Probiotics and Prebiotics.* Caister Academic Press, Wymondham, United Kingdom.
121. **Toden, S., A. R. Bird, D. L. Topping, and M. A. Conlon.** 2007. High red meat diets induce greater numbers of colonic DNA double-strand breaks than white meat in rats: attenuation by high-amylose maize starch. *Carcinogenesis* **28:**2355–2362.
122. **Topping, D. L., and P. M. Clifton.** 2001. Short-chain fatty acids and human colonic function: roles of resistant starch and nonstarch polysaccharides. *Physiol. Rev.* **81:**1031–1064.
123. **Turnbaugh, P. J., R. E. Ley, M. Hamady, C. M. Fraser-Liggett, R. Knight, and J. I. Gordon.** 2007. The human microbiome project. *Nature* **449:**804–810.
124. **Turnbaugh, P. J., R. E. Ley, M. A. Mahowald, V. Magrini, E. R. Mardis, and J. I. Gordon.** 2006. An obesity-associated gut mi-

crobiome with increased capacity for energy harvest. *Nature* **444:**1027–1031.

125. **van den Broek, L. A., S. W. Hinz, G. Beldman, J. P. Vincken, and A. G. Voragen.** 2008. Bifidobacterium carbohydrases—their role in breakdown and synthesis of (potential) prebiotics. *Mol. Nutr. Food Res.* **52:**146–163.
126. **van de Wiele, T., N. Boon, S. Possemiers, H. Jacobs, and W. Verstraete.** 2007. Inulin-type fructans of longer degree of polymerization exert more pronounced in vitro prebiotic effects. *J. Appl. Microbiol.* **102:**452–460.
127. **Ventura, M., C. Canchaya, G. F. Fitzgerald, R. S. Gupta, and D. van Sinderen.** 2007. Genomics as a means to understand bacterial phylogeny and ecological adaptation: the case of bifidobacteria. *Antonie van Leeuwenhoek* **91:**351–372.
128. **Walter, J., M. Mangold, and G. W. Tannock.** 2005. Construction, analysis, and β-glucanase screening of a bacterial artificial chromosome library from the large-bowel microbiota of mice. *Appl. Environ. Microbiol.* **71:** 2347–2354.
129. **Wang, M., S. Ahrné, B. Jeppsson, and G. Molin.** 2005. Comparison of bacterial diversity along the human intestinal tract by direct cloning and sequencing of 16S rRNA genes. *FEMS Microbiol. Ecol.* **54:**219–231.
130. **Whitehead, K., J. Versalovic, S. Roos, and R. A. Britton.** 2008. Genomic and genetic characterization of the bile stress response of probiotic *Lactobacillus reuteri* ATCC 55730. *Appl. Environ. Microbiol.* **74:**1812–1819.
131. **Whitlock, D. R.** 2008. Human microbiome: hype or false modesty? *Nature* **454:**690.
132. **Wilson, M.** 2005. The gastrointestinal tract and its indigenous microbiota, p. 251–317. *In* M. Wilson (ed.), *Microbial Inhabitants of Humans.* Cambridge University Press, Cambridge, United Kingdom.
133. **Wong, J. M., R. de Souza, C. W. Kendall, A. Emam, and D. J. Jenkins.** 2006. Colonic health: fermentation and short chain fatty acids. *J. Clin. Gastroenterol.* **40:**235–243.
134. **Xu, J., M. K. Bjursell, J. Himrod, S. Deng, L. K. Carmichael, H. C. Chiang, L. V. Hooper, and J. I. Gordon.** 2003. A genomic view of the human-*Bacteroides thetaiotaomicron* symbiosis. *Science* **299:**2074–2076.
135. **Xu, J., and J. I. Gordon.** 2003. Honor thy symbionts. *Proc. Natl. Acad. Sci. USA* **100:** 10452–10459.
136. **Yu, Z., M. Yu, and M. Morrison.** 2006. Improved serial analysis of V1 ribosomal sequence tags (SARST-V1) provides a rapid, comprehensive, sequence-based characterization of bacterial diversity and community composition. *Environ. Microbiol.* **8:**603–611.
137. **Zoetendal, E. G., A. D. L. Akkermans, W. M. Akkermans-van Vliet, J. A. G. M. de Visser, and W. M. de Vos.** 2001. The host genotype affects the bacterial community in the human gastrointestinal tract. *Microb. Ecol. Health Dis.* **13:**129–134.
138. **Zoetendal, E. G., B. Cheng, S. Koike, and R. I. Mackie.** 2004. Molecular microbial ecology of the gastrointestinal tract: from phylogeny to function. *Curr. Issues Intest. Microbiol.* **5:**31–47.
139. **Zoetendal, E. G., C. T. Collier, S. Koike, R. I. Mackie, and H. R. Gaskins.** 2004. Molecular ecological analysis of the gastrointestinal microbiota: a review. *J. Nutr.* **134:**465–472.

# NATURAL MICROBIAL ECOSYSTEMS AND THEIR PROGRESSION IN FRESH FOODS

*James M. Jay*

# 3

It is not an easy task to cover microbial ecosystems of fresh foods in a single chapter or even in several chapters. Much of this information has been published but is widely scattered throughout the microbiological and food science and technology literature. As a means of providing substance and the current status of research in this broad area, only fresh foods are covered.

## MICROORGANISMS IN FRESH FOODS

In general, the chemoheterotrophic microorganisms that are found in fresh meats, poultry, seafoods, and plant products are those that occur in the pre- and postharvest environments. Neither chemolithotrophs nor cyanobacteria are of any consequence on food products. Although high numbers of indigenous or autochthonous bacteria, such as streptomycetes and mycobacteria, exist in farm soils, they as well as most protozoa do not attach firmly to plant products, and thus, they are generally removed by washing. The bacterial genera most often identified in foods are listed in Table 1. Due to the many changes in bacterial taxonomy over the past 8 to 10 years, many changes in the list of food-borne genera have already been made and more are likely to occur. For example, at least 7 of the well-known food-borne genera prior to the 1980s have undergone reclassifications resulting in the establishment of 38 new genera from the previous 7 with the genus *Pseudomonas* being delimited by the creation of at least 12 new genera (41).

In Table 1, only one genus of *Archaea (Halobacterium)* is listed, and it has been found in some salted fish/seafood products. Among the *Bacteria,* the genera of *Alphaproteobacteria* listed are found on fruits and vegetables, and several oxidize ethanol to acetic acid. The listed genera of *Betaproteobacteria* are common in soils and on plant matter. The class *Gammaproteobacteria* constitutes the largest broad taxonomic group and includes the *Enterobacteriaceae* and *Pseudomonadaceae* families. As common gastrointestinal tract inhabitants, they are constantly added to soils and waters from farm animals and human wastes. Some of the genera that contain chromogenic species, such as *Chromobacterium, Citrobacter, Erwinia,* and *Serratia,* are commonly associated with fresh plant matter. The *Deltaproteobacteria* are rarely, if ever, identified from foods, while two genera of *Epsilonproteobacteria* are commonly identified, with *Helicobacter* spp. suspected because of

*James M. Jay (deceased),* Department of Biological Sciences, Wayne State University, Detroit, MI 48202.

*Food-Borne Microbes: Shaping the Host Ecosystem,* Edited by L.-A. Jaykus, H. H. Wang, and L. S. Schlesinger, © 2009 ASM Press, Washington, DC

**TABLE 1** The most common genera of bacteria found in foods, beverages, and food processing environments

| Classification[a] | Food sources |
|---|---|
| **Domain: *Archaea*** | |
| Phylum: *Euryarchaeota* | |
| Class: *Halobacteria* | |
| Genus: *Halobacterium* | Salted fish |
| **Domain: *Bacteria*** | |
| Phylum: *"Deinococcus-Thermus"* | |
| Class: *Deinococci* | |
| Genus: *Deinococcus* | Water, fresh meats |
| Phylum: *Proteobacteria* | All are gram negative |
| Class: *Alphaproteobacteria* | |
| Genus: *Acetobacter* | Fruits, cider |
| *Acidomonas* | Plants, fruits |
| *Asaia* | Plants |
| *Brevundimonas* | Spoiled seafoods |
| *Devosia* | Formerly *Pseudomonas* |
| *Gluconacetobacter* | Plants, fruits |
| *Gluconobacter* | Plants |
| *Kozakia* | See *Acetobacter* |
| *Roseobacter* | Plants |
| *Saccharibacter* | Spoiled wine, fruits |
| *Sphingomonas* | Water, vegetables; formerly *Pseudomonas* |
| *Swaminathania* | Spoiled wine, fruits |
| *Zymomonas* | Beer, cider spoilage |
| Class: *Betaproteobacteria* | |
| Genus: *Acidovorax* | Lamb's lettuce |
| *Alcaligenes* | Water, meats, vegetables |
| *Burkholderia* | Plants, raw milk, flowers |
| *Chromobacterium* | Plants, raw milk |
| *Comamonas* | Water, soil |
| *Hydrogenophaga* | Formerly *Pseudomonas* |
| *Janthinobacterium* | Vegetable soft rots |
| *Neisseria* | Mucous membranes of humans, animals |
| *Ralstonia* | Water, soil, plants |
| *Telluria* | Formerly *Pseudomonas* |
| *Wautersia* | Formerly *Ralstonia* |
| Class: *Gammaproteobacteria* | |
| Genus: *Acinetobacter* | Water, plants, meats |
| *Aeromonas* | Water, meats, vegetables |
| *Alteromonas* | Water, meats, plants |
| *Brenneria* | Formerly *Erwinia* |
| *Cedecea* | Water, plants, meats |
| *Citrobacter*[b] | Vegetables, meats |
| *Coxiella* | Raw milk |
| *Dickeya* | Plant rots, wilts |
| *Enterobacter*[b] | Water, plants, meats |
| *Erwinia* | Rooty vegetables, soil |
| *Escherichia*[b] | Water, meats, vegetables |
| *Hafnia* | Water, seafoods, meats |
| *Klebsiella*[b] | Water, seafoods, meats |
| *Moraxella* | Plants, fresh meats |
| *Morganella* | Water, seafoods |

*(Table continues)*

**TABLE 1** *(continued)*

| Classification[a] | Food sources |
|---|---|
| *Obesumbacterium* | Spoiled beer |
| *Pantoea* | Plants, meats, water |
| *Pectobacterium* | Formerly *Erwinia* |
| *Photobacterium* | Spoiled marine fish |
| *Plesiomonas* | River, lake waters |
| *Proteus* | Meats, vegetables |
| *Pseudoalteromonas* | Shellfish |
| *Pseudomonas* | Ubiquitous; soils, water |
| *Psychrobacter* | Refrigerated meats |
| *Raoultella* | Formerly *Klebsiella* |
| *Salmonella* | Farm animals, water |
| *Serratia* | Plants, refrigerated meats |
| *Shewanella* | Meats, vegetables |
| *Shigella* | Mainly polluted waters |
| *Stenotrophomonas* | Formerly *Pseudomonas* |
| *Vibrio* | Seawater, seafoods |
| *Xanthomonas* | Vegetables, other plants |
| *Yersinia* | Pork, other meats |
| Class: *Epsilonproteobacteria* | |
| Genus: *Arcobacter* | Poultry, raw milk, water |
| *Campylobacter* | Raw poultry |
| *Helicobacter* | River, groundwaters |
| Phylum: *Firmicutes* | |
| Class: "Clostridia" | |
| Genus: *Clostridium* | Soils, waters, meats |
| *Desulfotomaculum* | Spoiled canned foods |
| *Pectinatus* | Spoiled beers |
| *Thermoanaerobacterium* | Some canned foods |
| Class: *Mollicutes* | |
| Genus: *Erysipelothrix* | Swine erysipelas, pork |
| Class: "Bacilli" | |
| Genus: *Alicyclobacillus* | Heat-processed foods |
| *Amphibacillus* | Plants, plant wastes |
| *Aneurinibacillus* | Formerly *Bacillus* |
| *Bacillus* | Ubiquitous; soils, dust |
| *Brevibacillus* | Formerly *Bacillus* |
| *Brochothrix* | Spoiled refrigerated meats |
| *Carnobacterium* | Refrigerated meats |
| *Enterococcus*[c] | Polluted waters, meats |
| *Geobacillus* | Some canned foods |
| *Gracilibacillus* | Formerly *Bacillus* |
| *Jeotgalibacillus* | Some fermented foods |
| *Jeotgalicoccus* | Korean fermented food |
| *Kurthia* | Refrigerated meats, raw milk |
| *Lactobacillus*[c] | Dairy starters, meats |
| *Lactococcus*[c] | Raw milk, dairy starters |
| *Lactosphaera*[c] | Plants |
| *Leuconostoc*[c] | Dairy products, plants |
| *Listeria* | Plants, meats |
| *Macrococcus* | Human, animal skin |
| *Megasphaera* | Spoiled beer |
| *Oenococcus*[c] | Grapes, other fruits |
| *Paenibacillus* | Formerly *Bacillus* |
| *Paralactobacillus*[c] | Plants |

*(Table continues)*

**TABLE 1** The most common genera of bacteria found in foods, beverages, and food processing environments *(continued)*

| Classification[a] | Food sources |
|---|---|
| *Pediococcus*[c] | Fermented foods |
| *Sporolactobacillus* | Plants, fermented drinks |
| *Staphylococcus* | Human, animal skin |
| *Streptococcus*[c] | Groups D and E, dairy |
| *Tetragenococcus*[c] | Fermented meats |
| *Vagococcus*[c] | Meats, vegetables |
| *Virgibacillus* | Formerly *Bacillus* |
| *Weissella*[c] | Green wieners |
| Phylum: *Bacteroidetes* | |
| Class: *"Bacteroides"* | |
| Genus: *Bacteroides* | Fecal matter, waters |
| Phylum: *Actinobacteria* | |
| Class: *Actinobacteria* | |
| Genus: *Bifidobacterium* | Some probiotic products |
| *Brevibacterium* | Dairy products, plants |
| *Chryseobacterium* | Formerly *Flavobacterium* |
| *Clavibacter* | Plant pathogens |
| *Corynebacterium* | Vegetable spoilage |
| *Curtobacterium* | Plants, plant pathogens |
| *Frigoribacterium* | Hay |
| *Janibacter* | Raw poultry |
| *Jonesia* | Formerly *Listeria* |
| *Kocuria* | Formerly *Micrococcus* |
| *Kytococcus* | Formerly *Micrococcus* |
| *Microbacterium* | Dairy products, plants |
| *Micrococcus* | Dust, animals, vegetables |
| *Mycobacterium* | Soils, raw milk |
| *Nesterenkonia* | Formerly *Micrococcus* |
| *Propionibacterium* | Dairy products, plants |
| *Rathayibacter* | Plant pathogens |
| *Streptomyces* | Potato scab |
| Class: *"Flavobacteria"* | |
| Genus: *Chryseobacterium* | Formerly *Flavobacterium* |
| *Empedobacter* | Formerly *Flavobacterium breve* |
| *Flavobacterium* | Plants, fresh meats |
| *Myroides* | Formerly *Flavobacterium odoratum* |
| Class: *"Sphingobacteria"* | |
| Genus: *Sphingobacterium* | Formerly *Flavobacterium* |

[a] Taxonomy was extracted from references 15a, 20a, and other published sources.
[b] Member of the coliform group.
[c] Lactic acid bacteria.

their demonstrated occurrence in groundwaters.

Most of the commonly found gram-positive bacteria belong to the phylum *Firmicutes,* and they include all genera of the lactic acid bacteria. Although some of these catalase-negative, lactic acid-producing bacteria are commonly found in human and animal wastes, they are very commonly associated with plants, in which catalase and simple nutrients in plant exudates aid in their growth and survival.

It should be noted that not all recognized species of the bacterial genera noted have been found in foods, and the same applies to the fungal genera listed in Table 2.

**TABLE 2** Commonly reported genera of yeasts and molds in most fresh, preserved, and fermented foods[a]

| Genus | Food sources |
|---|---|
| **Yeasts** | |
| *Brettanomyces/Dekkera* | Spoiled soft drinks, fermenting cacao beans |
| *Candida* | Ubiquitous; fruits, meats, pickles, olives |
| *Cryptococcus* | Poultry, shrimp, plants, meats |
| *Debaryomyces* | Slimy wieners, fruit juices, meat brines |
| *Hanseniaspora* | Citrus fruits, cacao bean fermentation |
| *Hansenula* | Fruit juices, meat brines |
| *Issatchenkia* | Many foods: meats, vegetables |
| *Kluyveromyces* | Many fruits, cheese, yogurt spoilage |
| *Lodderomyces* | Soft-drink spoilage |
| *Metschnikowia* | Wines |
| *Pichia* | Citrus fruits, seafoods, spoiled kraut |
| *Rhodotorula* | Fresh meats, bacon, seafoods |
| *Saccharomyces* | Most or all fruits, baker's and brewer's yeast |
| *Saccharomycodes* | Wine spoilage |
| *Saccharomycopsis* | Fermenting fruits, cacao bean fermentation |
| *Schizosaccharomyces* | Raw sugar, molasses, syrup, cacao beans |
| *Torulaspora* | Cucumbers, pickling brines |
| *Trichosporon* | Fresh meats, shrimp |
| *Yarrowia* | Dairy products, mayonnaise spoilage |
| *Zygosaccharomyces* | Fermenting soy beans, miso, vegetables |
| **Molds** | |
| *Alternaria* | Field fungus, stem-end rot of fruits |
| *Aspergillus* | Ubiquitous storage fungus, cured hams |
| *Basipetospora* | Xerophile; dried foods |
| *Botrytis* | Gray mold rot of many plant products |
| *Byssochlamys* | Black rot spoilage of canned fruits |
| *Cephalosporium* | Frozen meats and vegetables |
| *Cladosporium* | Rot of stone fruits, oleomargarine spoilage |
| *Colletotrichum* | Anthracnose of citrus fruits, mangoes, others |
| *Diplodia* | Stem-end rot of citrus fruits, melons |
| *Eremascus* | Xerophile; dried foods |
| *Fusarium* | Field fungus, vegetables, grains |
| *Geotrichum* | Fruit rots, meats, the "dairy mold" |
| *Gloeosporium* | Black rot of bananas, other fruits |
| *Helminthosporium* | Vegetable spoilage |
| *Monilia* | Beef, bacon, brown rot of stone fruits |
| *Mucor* | Citrus fruits, vegetables, beef "whiskers" |
| *Neosartorya* | Spoiled canned fruits |
| *Neurospora* | "Red bread mold," ontjom |
| *Penicillium* | Ubiquitous; fruit rots, blue cheese |
| *Pezicula* | Apple spoilage |
| *Phytophthora* | White potato tubers |
| *Rhizoctonia* | Carrot spoilage |
| *Rhizopus* | Ubiquitous; black rot of beef, yam rot |
| *Sclerotinia* | Rots of lemons |
| *Sporotrichum* | "White spot" of beef |
| *Trichothecium* | Vegetable soft rots, fruit spoilage |
| *Wallemia* | Spoiled dried and salted fish |
| *Xeromyces* | Xerophile; grows at an $a_w$ value of 0.61 |

[a]Extracted from reference 1 and other published sources.

Most of the fungi that are known to occur in fresh foods are listed in Table 2. The genera *Candida* and *Saccharomyces* are the most ubiquitous among yeasts, and plants, especially fruits, are their most common habitats. The presence of B vitamins and simple carbohydrates in plant exudates is beneficial to this group. The most ubiquitous mold genera are *Aspergillus, Penicillium,* and *Rhizopus.* Their capacity to subsist in locations in which neither bacteria nor yeasts can grow accounts in large part for their presence in various numbers on all foods. The field fungi are the molds that invade and damage preharvest seeds such as barley and wheat, and the genera *Alternaria, Fusarium,* and *Cladosporium* are typical of this group. Storage fungi grow on postharvest grain crops, and species of the genus *Aspergillus* are very common. In general, storage fungi can grow at lower water activity ($a_w$) values than the field fungi.

Overall, the microbial ecosystem of fresh foods is not as complex as that of the corresponding preharvest products, and this is due in large part to the removal of many microbes from meats by animal carcass washing in which hot water, steam, and/or organic acids are applied. The washing of vegetables leads to the removal of most transients and some of the indigenous types as noted above. Finally, microbes are added to finished and near-finished products from the hands of handlers, processing room air, and holding and packaging containers. How this altered microbial ecosystem behaves when these products are stored at refrigerator temperatures (40°F, 5 to 7°C) is outlined and discussed below. More detailed information on food microbiota can be found in the works of Doyle and Beuchat (14), Jay et al. (43), and Matthews (53).

## GROWTH PARAMETERS AND FACTORS THAT AFFECT MICROORGANISMS IN FOODS

Assuming that fresh foods are harvested, processed, packaged, and stored according to acceptable procedures, the fate of each group of microbes can be predicted if one knows how the foods are packaged along with the time and temperature of storage. This may be achieved if the other growth parameters are known for given groups of food-borne microorganisms. The key parameters that affect microbial growth are summarized below, and it should be noted that most apply to the growth of individual microbes in laboratory culture media. See the work of Boddy and Wimpenny (2) or Jay et al. (43) for more detailed information on growth parameters of food-borne microbes.

### pH of Food

Fruits such as apples, oranges, and grapes have pH values of $<4.5$, which are inhibitory to gram-negative bacteria and most gram-positive bacteria. Yeasts, molds, and some lactic acid bacteria grow well in or on such products. Since the pH of many vegetables and all meats, poultry, and seafoods is $>5.0$, the growth of gram-negative and many gram-positive bacteria is favored over that of fungi. However, the growth of individual strains in association with others is not as easy to predict, in part because of competition, which may result in one being inhibited by another. See the section below for more information on microbial interference.

### Available Moisture ($a_w$)

Yeasts and molds grow at lower $a_w$ values than all bacteria, with some xerophilic molds and osmophilic yeasts being able to grow at $a_w$ values as low as 0.61, albeit very slowly. Gram-negative bacteria are the most sensitive to low $a_w$ values, with most food spoilage types requiring values of $>0.90$ and pathogens requiring values of $>0.94$. Gram-positive bacteria grow at lower $a_w$ values than gram-negative bacteria but not as low as for fungi. One of the most sensitive is *Clostridium botulinum,* which does not grow below 0.94. The lowest $a_w$ value at which a bacterial pathogen can grow is 0.86, and the bacterium which grows at this $a_w$ is *Staphylococcus aureus.* The food additives NaCl and sugars such as sucrose affect microbial growth by their depression of the $a_w$. Ten percent NaCl depresses the $a_w$ value to ca. 0.94, and about five times more sucrose is required to achieve the same inhibitory values. An $a_w$ value of 0.86 requires ca. 22% NaCl.

### Oxidation-Reduction Potential

In general, yeasts and molds require higher oxidation-reduction potential ($E_h$) values than most bacteria, generally 200 to 400 mV. The strictly aerobic gram-positive bacteria, such as *Bacillus* and *Paenibacillus* spp., grow best under these conditions, as do the strict aerobes, such as the pseudomonads. The facultative anaerobes (most gram-negative bacteria) can grow at $E_h$ values of −100 to −300 mV, about the same as for strict anaerobes such as *Clostridium botulinum*. The catalase-negative, gram-positive bacteria such as lactobacilli, enterococci, and streptococci also grow under these low-$E_h$ conditions. In general, fruit juices and vegetables have much higher $E_h$ values than meats and processed cheeses (such as cheddar), as well as meat cuts such as roasts, all of which possess $E_h$ values in the anaerobic range. The $E_h$ of meat animals such as beef steers and hogs is ca. 200 to 350 mV immediately after slaughter, but it decreases to ca. −250 mV in large cuts after rigor mortis sets in. At the same time, pH decreases from an initial value of ca. 7.35 to ca. 5.5 to 5.8 in the meat of well-rested animals in which a maximum level of muscle glycogen is present.

### Nature and State of Nutrients

Since all food spoilage and food poisoning microbes are chemoheterotrophs, they require organic sources of carbon and nitrogen in utilizable forms. The simpler the C and N sources, the faster the growth of all food-borne microorganisms. When the primary C sources are starches and other polysaccharides, only those that produce enzymes that degrade these polymers can grow, and these include most molds and some gram-negative and gram-positive bacteria. The same principle applies to proteins as N sources. Molds and a number of bacteria produce proteases that degrade structural proteins to amino acids. Food-borne yeasts do not produce amylase, and proteases are produced only weakly. The microbes that cannot degrade macromolecules are dependent upon those that do.

### Naturally Occurring Antimicrobial Constituents

Foods such as raw eggs and milk contain lysozyme, which is inhibitory to gram-positive bacteria. Milk also contains lactoferrin, conglutinin, and the lactoperoxidase system, all of which possess antimicrobial properties. The essential oils that exist in many spices are effective against a number of food-borne microbes. The most effective of these are eugenol in allspice and cloves, cinnamic aldehyde in cinnamon, vanillin in vanilla beans, thymol in thyme and oregano, and the like. In general, gram-negative bacteria are more sensitive than gram-positive bacteria to spice oils, and the latter are more sensitive than the lactic acid bacteria. Yeasts and molds are intermediate in their sensitivity to these compounds. A large number of spices and spice oils have been shown to be inhibitory to certain food-borne bacteria, but the needed concentrations adversely affect food flavor.

### Natural Coverings

The natural covering of fresh fruits, vegetables, and whole eggs is very effective in preventing the entry of most microorganisms to the inside of these products. The same is true for the skin or hide on whole meat cuts. In addition to being physical barriers, the chemical composition of these structures is too complex for most microbes to degrade. Among the exceptions are some vegetables such as whole carrots, for which the outer covering is specifically broken down by soil bacteria such as some *Erwinia* and *Pectobacterium* spp. Because of the complex chemical composition of natural food coverings, they tend to possess surface $a_w$ values that prevent the growth of most bacteria. This alone explains why whole fruits and many vegetables support the growth of molds over bacteria.

In the case of whole eggs, both the inner membrane and outer shell are very effective in preventing the entry of bacteria.

### Temperature of Storage

The growth temperature groups that apply to food-borne microorganisms are psychrotrophs (grow at 0 or 4°C up to ca. 30°C), mesophiles (grow best from 20°C up to ca. 45°C), and thermophiles (grow at >45°C). True psychrophiles do not grow above ca. 15°C, and they are encountered only in certain seafoods. A practical way of thinking about psychrotrophs is that they are mesophiles that can grow at refrigerator temperatures. Yeasts, molds, and many gram-negative bacteria can grow from 0°C to ca. 45°C. With the exception of many gram-positive lactic acid bacteria and some sporeformers, food-borne gram-positive bacteria can grow between ca. 10°C and ca. 80°C. With a few exceptions, food-borne human pathogens do not grow below ca. 5°C or above ca. 50°C. If three samples of a fresh food from the same batch are stored at 5 to 7, 35, and 50°C, the microbes that dominate each will be different based solely on the selective effect of growth temperature.

### RH of the Storage Environment

The relative humidity (RH) of the food storage environment is especially important to the surface growth of microorganisms on foods that are stored in gas-permeable packages. A low RH favors low moisture on food surfaces, and this retards the growth of surface microbes, while a high RH favors moisture uptake that leads to an increased surface $a_w$, which permits surface mold growth. High humidity is necessary to maintain vegetables in a desirable state, and this allows for the surface growth of some bacteria.

### Presence and Concentration of Gases (Vacuum and MAP)

The two gases of interest are ozone ($O_3$) and $CO_2$. Ozone is well known for its surface antimicrobial activity. However, because of its strong oxiding properties, its use in food packages is limited even though it has been shown to kill a number of food-borne organisms. Carbon dioxide is the gas of greatest importance in the food storage environment, and it comes from two sources: vacuum packaging and modified-atmosphere packaging (MAP). Vacuum packaging is accomplished by the physical removal of $O_2$ from food that is stored in a gas-impermeable bag or pouch followed by sealing. In the case of fresh meats and vegetables, the normally occurring aerobes continue to grow and produce $CO_2$, and up to ca. 20% $CO_2$ may be found in such packaged products following prolonged storage. MAP is achieved by the physical removal of atmospheric air, which is replaced by a known mixture of gases such as 10% $O_2$ and 30% $CO_2$, with the remainder being $N_2$. The increased level of $CO_2$ slows and prevents growth of aerobic microbes and extends product shelf life. The microorganisms most sensitive to high $CO_2$ levels are the fungi and the strictly aerobic gram-negative bacteria. Most resistant are the catalase-positive, gram-positive bacteria, followed by the catalase-negative bacteria consisting of lactic acid bacteria such as the lactobacilli. Growth of the latter organisms is accompanied by the production of lactic acid from simple carbohydrates, which leads to a lowering of product pH to a level below that which can be tolerated by the gram-negative bacteria. The refrigerated shelf life of MAP meats may be extended from one to several weeks, and the biota is dominated by lactic acid bacteria.

### Competition among Members of the Normally Occurring Food-Borne Microbiota

All fresh foods contain a highly varied microbiota including some organisms that produce antimicrobial substances such as diacetyl, $H_2O_2$, organic acids, bacteriocins, and the like. As noted above, some grow faster than others, and the growth of some is highly dependent upon nutrients produced by others in the consortium. While some are inhibited by specific toxic products of others, some are inhibited by as-yet-unknown means, and this is discussed further in the section below on nonspecific microbial interference. Just as the normal gastrointestinal microbiota protects us from many pathogens that enter through food and water, the growth of many or most foodborne pathogens is adversely affected by the normally occurring higher numbers of nonpathogens.

## SYNOPSIS OF MICROBIAL PROGRESSION IN SOME FOODS FROM FRESHNESS TO SPOILAGE

When fresh foods that contain a typical microbiota undergo refrigerator spoilage, some or all of the microbial growth parameters noted above come into play. While pH and $a_w$ are the principal controlling factors in the spoilage of citrus fruits, most of the other factors noted above come into play when products such as fresh meats, raw milk, and some vegetables undergo refrigerator spoilage. Details on the spoilage mechanisms of a number of refrigerator-stored foods can be found in reference 43. Brief summaries of microbial progression in three products are presented below.

### Red Meat

To predict microbial progression in retail ground beef stored in air (gas-permeable packages) at 5°C, the microbial growth parameters of the product need to be assessed. Typical findings for the relevant parameters are these: pH, 5.7; $a_w$ value, ca. 0.99; $E_h$, 50 to 100 mV; abundant supply of simple nutrients; no antimicrobial substances; no $O_3$ or surplus $CO_2$; and an initial aerobic plate count (APC) of $10^4$ CFU/g in which most organisms are psychrotrophs. Natural covering is not relevant since this is a comminuted product, and refrigeration RH is generally high. This product should support growth of large numbers of the genera in Tables 1 and 2 that contain psychrotrophic species.

Fresh red meats such as beef contain the following: water, 75%; protein, 18% (about 10% of the proteins are structural [such as actin], and the remainder are sacroplasmic [such as enzymes]); fats, ca. 3%; soluble nonprotein nitrogen compounds, ca. 1.5%; carbohydrates (lactic acid, glucose, etc.), ca. 1.1%; and ash, ca. 1.0%. In aerobically stored beef, the simple carbohydrates can support growth of bacteria to $10^8$ to $10^9$/g or $cm^2$ and the nonprotein nitrogen constituents are all utilized when the bacterial population reaches ca. $10^9$/g or $cm^2$ (23).

First to grow in refrigerated ground beef are the pseudomonads, and at 5°C, their generation time has been found to be 5.1 to 5.4 h (10). *Brochothrix thermosphacta* had a generation time of ca. 7.3 h, *Enterobacter* spp. ca. 7.8 h, and *Acinetobacter* spp. ca. 8.9 h. The pseudomonads consist of both fluorescent and nonfluorescent types, and in one study 91 and 100% of those on fresh meats and fresh fish, respectively, were fluorescent, with *Pseudomonas fluorescens* being the single most common species (21). After spoilage of these two products, the fluorescent organisms represented 100% of the pseudomonads.

As noted above, the sugars and lactic acid are the first energy sources attacked by the fresh meat microbiota, which is dominated by *Pseudomonas* and *Acinetobacter* spp. Once this source is exhausted, free amino acids are utilized as energy sources, and this is the source of off-odor compounds such as ammonia, indole, and $H_2S$. Assuming that the initial APC of this meat was $10^4$ CFU/g, the number is now around $10^7$ to $10^8$ CFU/g, with the appearance of off-odors and surface slime to produce a biofilm, a pH of ca. 8.0 to 8.5, and a lower $E_h$. At this point, the structural proteins, such as actomyosin, have not been attacked because of the availability of simpler nutrients. When the APC reaches ca. $10^9$ CFU/g or $cm^2$, the breakdown of the simpler proteins of the sarcoplasm can be detected (11). Even if this product is held in the refrigerator for 12 months, it does not liquefy.

From a study of a fresh ground beef sample held in air at 5 to 7°C for up to 21 days, the initial APC of 7.41 $log_{10}$ CFU/g increased to 10.78 $log_{10}$/g after 21 days (42). Numbers of identified genera decreased from 21 to 7, and the percentage of gram-negative bacteria increased from 84 to 100 after 21 days. The six identified genera after 21 days were *Pseudomonas, Enterococcus, Morganella, Providentia, Psychrobacter,* and *Pantoea,* with one unidentified genus (42). Although yeasts and molds are common in ground beef, they were not detected in this sample since selective culture media were not employed.

As noted above, yeasts are common in comminuted red meats, and numbers are typically 3.0 to 6.0 $\log_{10}$ CFU/g (34). From meat samples held to spoilage in air at 5°C for up to 18 days, numbers increased from the initial count to ca. 10.0 $\log_{10}$ CFU/g (32). The single most common genus was *Candida,* which represented 82% of the 79 yeast isolates, with *Rhodotorula, Torulopsis, Trichosporon,* and *Cryptococcus* present in lower numbers. Dillon and Board (13) also found *Candida* spp. to be the most common yeasts on red meats.

The microbiota of vacuum-packaged and MAP ground beef would differ significantly from that of aerobically stored samples. The most dramatic difference would be slower growth of the strictly aerobic pseudomonads because of reduced $O_2$. The dominant microbiota would be lactic acid bacteria, which utilize the simple carbohydrates and produce lactic acid, which depresses pH to a level that slows or inhibits growth of gram-negative bacteria. The most commonly occurring lactic acid bacteria in refrigerated meats are *Lactobacillus, Carnobacterium,* and *Leuconostoc* spp. (10).

Beef cuts such as roasts are quite different from comminuted meats relative to the progression of the microbiota. First, the outer skin portions cannot be penetrated by bacteria. The low surface $a_w$ favors molds over yeasts, since the latter have a higher requirement for simple constituents. The cut surfaces may support growth of some gram-negative bacteria as well as some lactic acid bacteria, depending on the RH of the storage environment. In the case of beef quarters that are shipped long distances in refrigerated containers without increased $CO_2$, surface mold growth often occurs in such a manner as to produce long chains of mycelia that hang from carcasses to produce what is referred to as "whiskers." The effect of the intact hide of lamb carcasses stored at 5°C on microbial progression can be seen from a study by Prieto et al. (63). At day 0, most of the bacteria were gram-negative organisms dominated by pseudomonads, followed by two groups of gram-positive bacteria (Table 3). After 15 days, the pseudomonads were still the most predominant but the percentage of gram-positive bacteria had increased. However, after detectable signs of microbial spoilage appeared, the gram-positive bacteria surpassed the gram-negative organisms, and the most prominent were *Staphylococcus* and *Brochothrix* spp. (Table 3). The dominance of the gram-positive bacteria is a reflection of the drier state of the carcasses over time, since these organisms can grow at lower $a_w$ values than the gram-negative ones. Also, staphylococci are indigenous to animal skin, and they can grow at lower $a_w$ values than *Brochothrix* spp. Molds would be expected to replace all bacteria over a holding period of months.

## Poultry

The progression of microbes on refrigerated poultry is quite similar to that for ground beef, with the most significant difference being the higher initial pH of 6.2 to 6.4. Since the tissues of healthy chickens are free of microbes, for refrigerated chicken meat such as wings,

**TABLE 3** Predominant microbial groups on intact lamb carcasses held at 5°C in air[a]

| |
|---|
| Day 0: gram-negative rods, 34.75%, of which 23.5% were pseudomonads, 24.25% *Brochothrix,* and 22.5% staphylococci |
| After 15 days: pseudomonads, 30.86%; *Brochothrix,* 26.27%; staphylococci, 19.60% |
| After spoilage (appearance of slime and/or off-odors): staphylococci, 39.9%; *Brochothrix,* 23.8%; pseudomonads, 14.2% |

[a]Taken from reference 63. *Brochothrix* is *B. thermosphacta,* and the dominant *Staphylococcus* species is *Staphylococcus xylosus.*

thighs, and breasts, microbial progression and subsequent spoilage occur on the cut and skin surfaces since microbes do not pass through the intact skin. The fresh poultry microbiota is fairly uniform because of the commonly used mechanized slaughter and washing procedures, and it consists of psychrotrophic gram-negative and gram-positive bacteria along with yeasts and molds. Because cut surfaces possess a high $E_h$, a high $a_w$, and simple nutrients, gram-negative psychrotrophs outgrow their competitors. When numbers/$cm^2$ reach ca. 7.0 $log_{10}$, off-odors may be detected. Surface growth becomes more or less confluent when bacterial numbers reach ca 8.0 $log_{10}$ CFU/$cm^2$ or higher, and the product is slimy. Neither lactic acid bacteria nor yeasts and molds are prominent because of the much faster growth of gram-negative rods such as *Pseudomonas* spp.

In the case of eviscerated whole chickens, the first signs of off-odors appear on the inside cavity, where growth is favored over that on surfaces. More psychrotrophic enteric bacteria may be found here than on surfaces. More on the microbial progression and spoilage of refrigerated meats can be obtained from references 22 and 41.

### Vegetables

The composition of fresh vegetables differs from that of meats in several ways. First, the water content varies from 74 to 94%, with a mean of ca. 85% from 22 vegetables (43). The carbohydrate content varies from ca. 3% for spinach to ca. 21% for corn; the protein content varies from ca. 1.0% for radishes to 6.7% for peas (41). Carbohydrates in vegetables are largely in the form of starch, which can be utilized only by amylase producers. Although not as abundant as in meats, the supply of simple carbohydrates and amino acids is sufficient to support early and rapid growth of many bacteria.

The pHs of market vegetables differ widely, with values of 6.0 to 6.6 being common. In the case of uncut vegetables such as string beans and celery, the surface $a_w$ value is ca. 0.95 or lower in low-RH environments, and the surface $E_h$ is >50 mV. The initial numbers of microorganisms on fresh vegetables vary widely, with 5.0 to 6.0 $log_{10}$ CFU/g being typical for some and as high as 8.0 $log_{10}$ CFU/g for seed sprouts. Those that are grown in soil may be expected to contain larger numbers of the organisms in Tables 1 and 2. Lund (52) has noted that leafy vegetables contain $2 \times 10^6$ mostly gram-negative bacteria per g, with $4 \times 10^3$ lactic acid bacteria per g. Among the latter, *Leuconostoc* spp. predominated on fresh vegetables, with numbers of ca. $2.5 \times 10^4$/g for green beans, okra, and peas (52). Among gram-negative bacteria, *Enterobacter cloacae, Citrobacter freundii,* and *Pantoea agglomerans* are common.

Following the growth of the lactic acid bacteria and other gram-positive and gram-negative bacteria such as those that initiate early growth, the supply of simple nutrients is not sufficient to sustain them for long, and this is where the molds enter. The low surface $a_w$, the high $E_h$, and a declining population of bacteria make these products very susceptible to mold growth. The genera *Botrytis, Penicillium,* and *Rhizopus* grow well and produce various rot conditions.

## MICROBIAL INTERFERENCE IN FOODS

The phenomenon in which one or more different microorganisms inhibit or kill others is variously referred to as microbial antagonism, antibiosis, biocontrol, competitive exclusion, lactic antagonism, etc., but it is referred to as microbial interference in this chapter. Some of the general principles of this phenomenon have been discussed by Fredrickson and Stephanopoulos (19).

Microbial interference was noted as early as 1877 by Pasteur and Joubert, who found that the anthrax bacterium showed little or no growth and later died out when it was inoculated in urine along with a "common bacterium" (cited by Florey [18]). Prior to the 1950s, microbial interference was used by medical researchers before the general availability of antibiotics. The history of the medical applications of this phenomenon has been reviewed by Howard Florey (18). When *Staphylococcus aureus* phage type 80/81 became resistant to all clinically useful antibiotics in the mid-1950s, a number of clinical studies showed that if the nasal cavities of newborn infants were swabbed with an avirulent strain of *S. aureus,* it prevented colonization by the more virulent phage type 80/81 strain (5, 50, 68). This procedure was fairly widely used until new antibiotics became available. Summaries of some of the more specialized forms of microbial interference are presented below.

### Competitive Exclusion

Competitive exclusion (CE) is a phenomenon that is highly similar to that noted above for the exclusion of *Staphylococcus aureus* from the nasal cavities of newborn infants, but this terminology is more or less restricted to the phenomenon in food animals, especially poultry. CE is employed in the rearing of poultry to prevent intestinal and cecal colonization by salmonellae and campylobacters, and it is often referred to as the Nurmi concept when used to prevent salmonella colonization (61). When young chicks become colonized by these pathogens, their fecal droppings serve to spread the pathogens to other young chicks, which acquire them by picking through the droppings of other chicks. When the chicken gut microbiota is obtained from older salmonella-free birds and administered to the hatchlings via water or food, they occupy the ceca or intestinal sites required by the pathogens noted, and thus, they are excluded. After the protective microbiota is established in young chicks, it remains for the life of the bird. More detailed information may be obtained from references 17, 44, 67, and 75.

### Replacement Therapy

Replacement therapy, which is quite similar to CE, was devised by Hillman et al. (30), who employed it to prevent the attachment to teeth of an acid-producing strain of *Streptococcus mutans,* which causes dental caries. The genetically engineered strain does not produce lactic acid, but it colonized the teeth of laboratory rats and served as an effector strain to prevent dental caries.

### Lactic Antagonism

The expression "lactic antagonism" was used in the early 1900s for foods that contained high numbers of actively growing lactic acid bacteria, which retarded the growth of both spoilage and food poisoning bacteria. Not only is it another example of microbial interference but also it is the most widely employed. In addition to their roles in the production of fermented foods such as sauerkraut and pickles, lactic acid bacteria produce bacteriocins, organic acids, and other products that prevent the growth of human pathogens and spoilage bacteria. The early history of lactic antagonism and bacteriocins in foods has been reviewed (33, 34, 36), and some of their specific activities in foods have been reviewed (24, 45, 55, 66, 71). A thorough review of the genetics and other aspects of lactic acid bacteriocins has been presented by Klaenhammer (46).

The extended shelf life of vacuum-packaged and MAP fresh foods is due not only to growth inhibition of strict aerobic microbes but also significantly to interference of aerobic spoilage microbes by the production of organic acids such as lactic acid, which depresses product pH below that required for growth by the spoilers. A large number of studies have been published in which the presence and growth of various lactic acid bacteria repressed or killed a variety of bacterial food-borne pathogens.

In general, microbial interference by the lactic acid bacteria results from their production of defined products, such as those noted in Table 4. Agar plate diffusion assays can be used to isolate effector strains. However, not all effector organisms can be identified in this way, and the section below deals with the nonspecific inhibition of some food-borne organisms.

**TABLE 4** Demonstrated and suggested mechanisms of microbial interference in foods

| Inhibitory or causal factor | Producing biota, status |
|---|---|
| Bacteriocins | Lactic acid bacteria, some micrococci and staphylococci |
| $H_2O_2$ | Some lactic acid bacteria |
| Diacetyl (2,3-butanedione) | Lactic acid bacteria, especially *Lactococcus* spp. |
| Iron sequestration | Siderophore-producing gram-negative bacteria |
| Free fatty acids | Some gram-negative bacteria |
| Competition for nutrients | Gram-positive bacteria, yeasts |
| Competition for adhesion sites | Possibly many different microbes |
| Polypeptide antibiotics | Some gram-positive bacteria |
| Bacteriolytic enzymes | Some gram-positive bacteria |
| Change in acid tolerance | Gram-negative biota |
| Biofilm formation | Suggested |
| Quorum sensing | Suggested |
| Induction of viable but nonculturable state | Suggested |

## NONSPECIFIC INTERFERENCE RELATIVE TO THE CONTROL OF FOOD-BORNE PATHOGENS AND SPOILAGE

As noted above, microbial interference within the naturally occurring food-borne ecosystem is not always defined by the production of specific inhibitory substances, and this is described here as nonspecific interference relative to some food-borne diseases and spoilage organisms.

### *Staphylococcus aureus*

In the 1950s and 1960s, staphylococcal gastroenteritis was regarded as the leading food-borne illness in the United States and some other countries. It was associated in large part with changes in eating habits that included a wider variety of convenience or ready-to-eat foods, many of which were not heated between preparation and subsequent eating. Many of these products were vehicles for enterotoxigenic *S. aureus*. In addition to sandwiches and cream-filled pastries, frozen meat and vegetable pot pies and frozen TV dinners were common vehicles of disease. Although the frozen products had to be heated or cooked before consumption, it was demonstrated that *S. aureus* could grow and produce heat-resistant enterotoxins if the frozen products were temperature abused. Synopses of some of this research are presented below.

Straka and Combs (70) were the first to show that *S. aureus* did not grow when inoculated into creamed chicken because of overgrowth by the background microbiota. The background organisms of ground pork at 15°C were shown to inhibit *S. aureus* such that numbers needed to cause food poisoning (estimated to be ca. $10^7$/g) were never attained (56).

In research carried out at the University of Chicago, Dack and Lippitz (8) inoculated *S. aureus*, *Escherichia coli*, and *Salmonella enterica* serovar Typhimurium into slurries of frozen pot pies and incubated them at 35°C for 18 h. Both *S. aureus* and serovar Typhimurium were more inhibited than *E. coli*, and pH was only partly responsible. A number of studies on the fate of *S. aureus* in chicken pot pies, in macaroni and cheese dinners, and in laboratory culture media were published by researchers at the Campbell Soup Co. in the 1960s (60). Overall, this research demonstrated the inability of this pathogen to compete with naturally occurring organisms under various conditions. Overall, *S. aureus* growth was effectively inhibited at 20°C and below by the background microbiota, with initial numbers at 6.0 $\log_{10}$ CFU/g. The absence of significant changes in pH during the 14-day incubation suggested that inhibition was not due to the growth of lactic acid bacteria. In precooked frozen foods, it was not possible to promote the growth of appreciable numbers of staphylococci under any conditions of defrost (60). The inhibitory effects were about the same at 5°C for 16 days as at 37°C for 73 h even though the staphylococci grew to higher numbers at the higher temperature. The inhibition of *S. aureus* by the harmless background microbiota has been demonstrated by a number of investigators (9, 12, 25, 35, 72). As to the possible basis for *S. aureus* inhibition by non-lactic acid food-borne microbes, Troller and Frazier (73) suggested that it was due to amino acid depletion by the background microbiota.

### *Escherichia coli* O157:H7

The 1993 U.S. Pacific Northwest outbreak of hemorrhagic colitis and hemolytic uremic syndrome caused by *E. coli* O157:H7 traced to ground beef was the single largest food-borne outbreak from this product ever recorded in the United States, and there were ca. 732 cases and three deaths. There has been much speculation as to why such a large outbreak occurred at this time and location. The one certainty is that it would not have occurred had the beef been cooked to 160°F. Prior to 1993, ground beef was the vehicle for only three published bacterial food-borne outbreaks. After comparing published APCs of retail ground beef throughout the United States between 1914 and 1977, mean numbers were found to range from $10^5$ to $10^7$/g, with numbers as high as $10^8$/g from four of the nine studies (39). Oregon is the only state in America ever to have a law on the numbers of bacteria for ground beef, and the maximum acceptable APC was $5 \times 10^6$/g. On the other hand, the average APCs for 563 samples of ground beef collected throughout the United States by the U.S. Department of Agriculture averaged 3.90 $\log_{10}$ CFU/g (74). In 1976, the maximum limit suggested by the International Commission on Microbiological Specifications for Foods was $10^7$ CFU/g (39). With this information in hand, it was postulated in 1995 (38) that the ground beef in the Pacific Northwest outbreak probably contained too few background organisms to inhibit low numbers of *E. coli* O157:H7, and the first published data in support of this hypothesis appeared in 1997 (40). In that limited study, coliform-free retail ground beef with a background microbiota of 5.97 $\log_{10}$ CFU/g was inoculated with 4.28 $\log_{10}$ of a strain of *E. coli* O157:H7 per g and incubated at 3 to 5°C for 11 days. The pathogen could not be detected when the APC was 7.05 $\log_{10}$/g. In a study in which four *E. coli* O157:H7 strains were inoculated into ground beef and incubated at 8°C, their growth was inhibited by the high background microbiota, although *E. coli* O157:H7 remained viable (59).

In a more extensive study in Norway, Vold et al. (78) collected background microbiota from eight commercial ground beef samples, added these organisms to ground beef along with *E. coli* O157:H7, and stored the meat at 12°C for 14 days. The pathogen was added at a level of ca. $10^3$/g, and the background biota was $10^5$ to $10^6$/g. The pathogen was inhibited by the background biota under both aerobic and anaerobic conditions of packaging, with inhibition being more pronounced under anaerobic conditions. Under anaerobic conditions, the background biota consisted largely of lactic acid bacteria, with ca. 80% identified as *Lactobacillus sakei*.

In a study on the effect of epiphytic bacteria from lettuce and thale cress on *E. coli* O157:H7, *Enterobacter asburiae* decreased the pathogen 20- to 30-fold on foliage, and inhibition also occurred in the rhizosphere and in plant exudates (7). On the other hand, the epiphyte *Wautersia paucula* enhanced survival of the pathogen sixfold on lettuce foliage grown from coinoculated lettuce seed.

The relative growth of *E. coli* O157:H7 on beef that was decontaminated (steam vacuuming followed by a lactic acid spray) was compared to that on untreated beef by Nissen et al. (57). On decontaminated beef the pathogen increased ca. 3 log cycles after 5 days in air at 10°C, while on untreated beef it increased by only 1 log cycle. When stored in vacuum for up to 21 days, inhibition of the pathogen was more significant than by aerobic storage. Although the normal microbiota of the untreated beef reduced the pathogen in air, inhibition was greater under anaerobic conditions, most likely due to the growth of lactic acid bacteria.

When laboratory culture media were inoculated with a fresh beef biota and ca. 2 $log_{10}$ CFU of *E. coli* O157:H7 per ml, growth of the pathogen was halted when the maximal beef microbiota was reached at 37°C (77). A study on the growth kinetics of *E. coli* O157: H7 in association with nonpathogenic competitors at 15 and 37°C was reported by Duffy et al. (15). At 37°C and pH 7.0, *Hafnia alvei* significantly inhibited and outgrew the pathogen. At 15°C, *Pseudomonas fragi* at a level of ca. 4.0 $log_{10}$ CFU/ml extended the pathogen's lag phase by ca. 3 h when inoculated at a level of ca. 3.0 $log_{10}$ CFU/ml (15).

## Other Pathogens and Spoilage Microorganisms

Among the large number of published studies on the nonspecific inhibition of food-borne pathogens and spoilage microbes are those below, which cover a number of pathogens. In a study of the natural biota of ground beef, Mattila-Sandholm and Skyttä (54) found that five food-borne pathogens decreased significantly after 3 days at 6°C when the initial ground beef microbiota was ca. $10^6$ CFU/g. Over the incubation period, the initial number increased to $10^{10}$ CFU/g after 4 days in the absence of significant changes in meat pH. Specifically, *S. aureus* and *Yersinia enterocolitica* were inhibited in samples inoculated with *Pseudomonas fluorescens* and *P. fragi,* and *Listeria monocytogenes* and *Salmonella enterica* serovar Infantis were inhibited by *Pseudomonas* spp.; however, inhibition of *Bacillus cereus* was only minimal. These authors suggested nutrient depletion, especially of iron, as the possible basis for the inhibition.

In a study of the possible inhibitory capacity of 209 *Pseudomonas* isolates from fresh and spoiled fish, 67 showed inhibition of one or more of six targeted bacteria in an agar plate assay using modified iron agar incubated at 25°C for 48 h (27). Overall, 53 strains inhibited *Staphylococcus aureus,* 45 *Aeromonas sobria,* and 35 *Shewanella putrefaciens,* but only 8 inhibited *L. monocytogenes.* The mode of inhibition was associated with siderophore production by effector strains: 55 siderophore-positive strains were inhibitory, while another 20 siderophore producers were uninhibitory. The addition of iron to the modified iron agar medium eliminated the inhibitory activity. Among the targeted bacteria, *A. sobria* and *S. aureus* were the most sensitive, followed by *Shewanella putrefaciens* and *Escherichia coli* (27).

Growth of *L. monocytogenes* and *Salmonella enterica* serovar Poona on fresh-cut apple tissues was inhibited by four microorganisms isolated from apple surfaces (49). The four effectors were the bacterium *Gluconobacter asaii,* two yeasts (a *Candida* sp. and *Metschnikowia pulcherrina*), and the mold *Discosphaerina fagi.* The inoculated apple tissues were held for up to 7 days at 10 or 25°C, and the inhibition was highest at the latter temperature. Effector cultures were applied at levels of ca. $10^4$ to $10^5$ CFU/ml, and they became active only after 2 or 5 days. In another study, *L. monocytogenes* and *Listeria innocua* were shown to be inhibited in smear cheese by an undefined microbiota used to ripen cheese of this type over and above the inhibition ascribed to bacteriocin-producing lactic acid bacteria (16). Although the undefined organisms were not characterized, they were recovered from smears of industrially produced cheeses. The overall effectiveness of identified bacteria in preventing growth of pathogens on smear cheeses has been reviewed previously (16).

In a study of 304 yeast isolates from smear-ripened cheeses, about 4% were clearly inhibitory to *L. monocytogenes* growth on smear cheese surfaces (26). One *Candida intermedia* isolate reduced *L. monocytogenes* counts by 4 $log_{10,}$ and four other yeasts reduced counts of the pathogen by 3 $log_{10}$. The inhibition was clearly pronounced in coculture, and there was no evidence from agar diffusion assays for inhibitory substances.

Epiphytic yeasts from peaches *(Candida maltosa, Pichia fermentans,* and *Pichia kluyveri)* inhibited the growth of two peach spoilage molds (*Penicillium crustosum* and *Mucor circinelloides)* when sprayed on wounded peaches incubated at 4°C and held for up to 30 days (64). The effectors and spoilage molds were prepared with ca. $10^8$ CFU/ml. While *C. maltosa* and *P. fermentans* were the most effective, all three yeasts were effective inhibitors and controlled the pathogens on artificially wounded peaches when stored at 20°C. Neither enzymes nor antimetabolites were detected. Direct competition for nutrients or space was proposed as the inhibitory mechanism.

The postharvest disease or spoilage of Golden Delicious apples is caused by three molds: *Penicillium expansum, Botrytis cinerea,* and *Rhizopus stolonifer.* When *Pantoea agglomerans* was applied to apples stored in air or a low-$O_2$ atmosphere, it effected an 80 to 100% reduction in the spoilage molds (58). When placed inside apples and held at 1°C, numbers of *P. agglomerans* increased 14-fold from the initial number of ca. $8 \times 10^7$ CFU/ml. The exact mechanism of inhibition was not clear, but the authors noted that the following possibilities have been suggested by others: acid production, induction of plant defense, parasitism, and the production of direct antimicrobials (58).

With regard to the interference or inhibition of *L. monocytogenes* and other food-borne pathogens by nonpathogens in laboratory culture media, a large number of studies have been published. One is a study of the effect of *Pseudomonas fluorescens* on *L. monocytogenes* in brain heart infusion broth by Buchanan and Bagi (6) in which the effects of incubation temperature (4, 12, and 19°C), pH (5.0, 6.0, and 7.0), and NaCl (0.5, 2.5, and 4.5%) were assessed. The pathogen was suppressed in coculture at 4°C with 0.5 and 2.5% NaCl at an inoculum level of ca. $10^3$ CFU/ml. When *Lactobacillus lactis* was inoculated into raw chicken breast meat along with *E. coli* O157: H7 and stored at 5°C, it effected a reduction in numbers of the pathogen, although the pathogen did not decline when inoculated alone (3). The production of $H_2O_2$ was thought to be the primary inhibitory factor.

The lack of growth of *Clostridium botulinum* in foods that contain a natural background microbiota of $10^8$ to $10^9$ organisms per g or ml is well established. This is especially true with refrigerated meats in which the $E_h$ is within the range required for this pathogen. Although antibotulinal substances from some soil-borne bacteria are not unknown, the more rapid growth of spoilage bacteria may be the most significant factor, along with the inability of the proteolytic *C. botulinum* strains to grow at $<10°C$. In a study of soil bacteria, three *Clostridium perfringens* strains inhibited 11 type A, 7 type F, 1 nonproteolytic type B, 9 type E, and 7 type F *C. botulinum* strains (69). An agar diffusion assay was used, and both growth and toxin production were inhibited. In an earlier study, *C. botulinum* type F was shown to be inhibited by *Bacillus licheniformis* (79).

## PROTECTIVE CULTURES AND THE HURDLE TECHNOLOGY CONCEPT

The protective-culture concept was advanced by Holzapfel et al. in Germany; it entails the addition of certain microbes and/or their products to perishable processed foods to inhibit harmful and/or spoilage microbes. The lactic acid bacteria which produce bacteriocins and other demonstrated inhibitory substances, along with probiotic bacteria, are the major protective cultures used. For a review of this concept, see the works of Holzapfel et al. (31) and Lücke (51). The addition to sanitized beef carcasses of a competitive microbiota consisting of naturally occurring gram-positive and gram-negative bacteria has been suggested as one possible means to limit growth of low numbers of pathogens such as salmonellae and enterohemorrhagic *E. coli* (39). A specific example of the use of protective cultures to control *Listeria monocytogenes* on ham and sausage is that reported by Bredholt et al. (4). Cooked vacuum-packaged ham and cervelat sausages were inoculated with $10^3$ CFU of *L. monocy-*

*togenes* per g and stored at 8°C for 4 weeks. No *L. monocytogenes* organisms were found where the lactic acid bacteria grew, and both *L. monocytogenes* and *E. coli* O157:H7 were inhibited, but not *Yersinia enterocolitica* (4).

The hurdle technology concept was advanced in the 1980s by L. Leistner et al. in Germany, and it entails the employment of multiple factors to control targeted food-borne microorganisms (47, 48). Although the concept is generally applied to perishable processed foods, multiple barriers exist naturally in some food products, such as the protective coating on citrus fruits and the low pH and generally high $E_h$, each of which must be "hurdled" before anaerobic bacteria can grow in such products. A relatively high background microbiota on fresh meats and vegetables fits the definition of a hurdle relative to the growth of pathogens on products of this type.

## SUMMARY AND MECHANISMS OF MICROBIAL INTERFERENCE

The microbial ecosystem of fresh foods is composed of a large number of bacterial genera and species, with most consisting of *Gammaproteobacteria* and *Firmicutes*. Among the fungi, fewer genera of yeasts than molds are found. Although the progression of the food-borne microbiota is fairly well understood for some refrigerated meats, vegetables, and fruits, it can only be surmised for a number of other food products by assessing the state and activity of the growth parameters of each food product and fitting these to the requirements of the existing microbiota.

Much attention has been devoted to microbial interference in foods relative to the control of both food-borne pathogens and spoilage organisms. The most studied and best understood of the interference phenomena is the demonstrated production of antimicrobial factors by the lactic acid-producing bacteria and some other gram-positive bacteria. Most notable of these inhibitory factors are bacteriocins, diacetyl, and organic acids such as lactic acid. Less noted among the inhibitory factors is $H_2O_2$, which is produced by some lactic acid bateria, especially the lactobacilli. The role of this compound in lactic acid baterial interference has been reported by a number of investigators (3, 9, 62). The antimicrobial activity of diacetyl is well established (37).

Less well understood is indirect interference, in which specific inhibitory substances are rarely found. It is postulated that this form of interference may well be more significant than that produced by the lactic acid bacteria in the progression of individual components of the food microbiota. The gist of this hypothesis is the much wider occurrence of gram-negative bacteria in fresh foods than lactic acid bacteria, and their faster growth. Competition for iron is associated with some fluorescent pseudomonads, and its production by food-borne gram-negative bacteria has been noted (20, 27–29).

Nutrient limitation or depletion has been demonstrated to be a mechanism by which gram-positive bacteria such as *Staphylococcus aureus* may be inhibited in some foods, but probably not in fresh meats and raw milk, in which there is an overabundance of simple nutrients. This effect has been noted by a number of investigators (55, 64, 73). Among other inhibitory mechanisms is the production of an antibiotic by *Erwinia herbicola* Eh252, which inhibited the fireblight pathogen, *Erwinia amylovora*. The antibiotic was inactivated by histidine and proteolytic enzymes but not by iron (76). In a study of the acid tolerance response of *L. monocytogenes* in fresh meat decontamination fluids, the natural gram-negative meat biota appeared to increase the acid sensitivity of the pathogen and thus increase its acid stress (65).

As to other possible mechanisms of indirect interference, it is not inconceivable that gram-negative bacteria, such as the pseudomonads, incorporate autoinducer substances, such as acylhomoserine lactones, produced by other gram-negative bacteria and thus multiply at a higher rate than their gram-negative competitors, but clear-cut evidence is wanting. Also, since nonindigenous bacteria on meats, such as food-borne pathogens, are generally inhibited when the indigenous microbiota is ca. $10^7$ to $10^8$ CFU/g, biofilm formation could be a means of exclusion. It is well demonstrated that the growth of pathogens does not occur on fresh refrigerated ground beef when the background microbiota forms a discernible biofilm on which numbers reach ca. $10^8$ cells/g. The consistent dominance of pseudomonads in such products has not been explained (42).

Whatever the mechanism of microbial interference, both the direct and demonstrated mechanisms, along with the less well-defined mechanisms, interact to affect microbial progression in fresh foods. More research is needed to explain the indirect or unknown phenomena by use of food substrates and food isolates, since the use of stock cultures and laboratory culture media does not always produce results that correspond to those obtained from food products and their native environments.

## REFERENCES


1. **Beuchat, L. R. (ed.).** 1987. *Food and Beverage Mycology,* 2nd ed. Springer, New York, NY.
2. **Boddy, L., and J. W. T. Wimpenny.** 1992. Ecological concepts in food microbiology. *Soc. Appl. Bacteriol. Symp. Ser.* **21:**23S–38S.
3. **Brashears, M. M., S. S. Reilly, and S. E. Gilliland.** 1998. Antagonistic action of cells of *Lactobacillus lactis* toward *Escherichia coli* O157:H7 on refrigerated raw chicken meat. *J. Food Protect.* **61:** 166–170.
4. **Bredholt, S., T. Nesbakken, and A. Holek.** 1999. Protective cultures inhibit growth of *Listeria monocytogenes* and *Escherichia coli* O157:H7 in cooked, sliced, vacuum- and gas-packaged meat. *Int. J. Food Microbiol.* **53:**43–52.
5. **Brook, I.** 1999. Bacterial interference. *Crit. Rev. Microbiol.* **25:**155–172.
6. **Buchanan, R. L., and L. K. Bagi.** 1999. Microbial competition: effect of *Pseudomonas fluorescens* on the growth of *Listeria monocytogenes*. *Food Microbiol.* **16:**523–529.
7. **Cooley, M. B., D. Chao, and R. E. Mandrell.** 2006. *Escherichia coli* O157:H7 survival and growth on lettuce is altered by the presence of epiphytic bacteria. *J. Food Prot.* **69:**2329–2335.
8. **Dack, G. M., and G. Lippitz.** 1962. Fate of staphylococci and enteric microorganisms introduced into slurry of frozen pot pies. *Appl. Microbiol.* **10:** 472–479.
9. **Dahiya, R. S., and M. L. Speck.** 1968. Hydrogen peroxide formation by lactobacilli and its effect on *Staphylococcus aureus*. *J. Dairy Sci.* **51:** 1568–1572.
10. **Dainty, R. H., and B. M. Mackey.** 1992. The relationship between the phenotypic properties of bacteria from chill-stored meat and spoilage processes. *J. Appl. Bacteriol.* **73:**1035–1045.
11. **Dainty, R. H., B. G. Shaw, K. A. DeBoer, and E. S. J. Scheps.** 1975. Protein changes caused by bacterial growth on beef. *J. Appl. Bacteriol.* **39:**73–81.
12. **Daly, C., W. E. Sandine, and P. R. Elliker.** 1972. Interactions of food starter cultures and food-borne pathogens: *Streptococcus diacetilactis* versus food pathogens. *J. Milk Food Technol.* **35:** 349–357.
13. **Dillon, V. M., and R. G. Board.** 1991. Yeasts associated with red meats. *J. Appl. Bacteriol.* **71:** 93–108.
14. **Doyle, M. P., and L. R. Beuchat (ed.).** 2007. *Food Microbiology—Fundamentals and Frontiers,* 3rd ed. ASM Press, Washington, DC.
15. **Duffy, G., R. C. Whiting, and J. J. Sheridan.** 1999. The effect of a competitive microflora, pH and temperature on the growth kinetics of *Escherichia coli* O157:H7. *Food Microbiol.* **16:**299–307.

15a. **Dworkin, M., S. Falkow, E. Rosenberg, K.-H. Schleifer, and E. Stackebrandt (ed.).** 2007. *The Prokaryotes,* 3rd ed. Springer, New York, NY.

16. **Eppert, I., N. Valdés-Stauber, H. Götz, M. Busse, and S. Scherer.** 1997. Growth reduction of *Listeria* spp. caused by undefined industrial red smear cheese cultures and bacteteriocin-producing *Brevibacterium linens* as evaluated in situ on soft cheese. *Appl. Environ. Microbiol.* **63:**4812–4817.
17. **Fedorka-Cray, P. J., J. S. Bailey, N. J. Stern, N. A. Cox, S. R. Ladely, and M. Musgrove.** 1999. Mucosal competitive exclusion to reduce *Salmonella* in swine. *J. Food Prot.* **62:**1376–1380.
18. **Florey, H. W.** 1946. The use of microorganisms for therapeutic purposes. *Yale J. Biol. Med.* **19:**101–118.
19. **Fredrickson, A. G., and G. Stephanopoulos.** 1981. Microbial competition. *Science* **213:**972–979.
20. **Freedman, D. J., J. K. Kondo, and D. L. Willrett.** 1989. Antagonism of foodborne bacteria by *Pseudomonas* spp.: a possible role for iron. *J. Food Prot.* **52:**484–489.

20a. **Garrity, G. M.** 2001. *Bergey's Manual of Systematic Bacteriology,* 2nd ed. Springer, New York, NY.

21. **Gennari, M., and F. Dragotto.** 1992. A study of the incidence of different fluorescent *Pseudomonas* species and biovars in the microflora of fresh and spoiled meat and fish, raw milk, cheese, soil and water. *J. Appl. Bacteriol.* **72:**281–288.

22. **Gill, C. O.** 1983. Meat spoilage and evaluation of the potential storage life of fresh meat. *J. Food Prot.* **46:**444–452.

23. **Gill, C. O.** 1976. Substrate limitation of bacterial growth at meat surfaces. *J. Appl. Bacteriol.* **41:** 401–410.

24. **Gilliland, S. E., and M. L. Speck.** 1975. Inhibition of psychrotrophic bacteria by lactobacilli and pediococci in nonfermented refrigerated foods. *J. Food Sci.* **40:**903–905.

25. **Goepfert, J. M., and H. U. Kim.** 1975. Behavior of selected food-borne pathogens in raw ground beef. *J. Milk Food Technol.* **38:**449–452.

26. **Goerges, S., U. Aigner, B. Silakowski, and S. Scherer.** 2006. Inhibition of *Listeria monocytogenes* by food-borne yeasts. *Appl. Environ. Microbiol.* **72:**313–318.

27. **Gram, L.** 1993. Inhibitory effect against pathogenic and spoilage bacteria of *Pseudomonas* strains isolated from spoiled and fresh fish. *Appl. Environ. Microbiol.* **59:**2197–2203.

28. **Hamdan, H., D. M. Weller, and L. S. Thomashow.** 1991. Relative importance of fluorescent siderophores and other factors in biological control of *Gaeumannomyces graminis* var. *tritici* by *Pseudomonas fluorescens* 2-79 and M4-80R. *Appl. Environ. Microbiol.* **57:**3270–3277.

29. **Henry, M. B., J. M. Lynch, and T. R. Fermor.** 1991. Role of siderophores in the biocontrol of *Pseudomonas tolaasii* by fluorescent pseudomonad antagonists. *J. Appl. Bacteriol.* **70:**104–106.

30. **Hillman, J. D., T. A. Brooks, S. M. Michalek, C. C. Harmon, J. L. Snoep, and C. C. van der Weijden.** 2000. Construction and characterization of an effector strain of *Streptococcus mutans* for replacement therapy of dental caries. *Infect. Immun.* **68:**543–549.

31. **Holzapfel, W. H., R. Geisen, and U. Schillinger.** 1995. Biological preservation of foods with reference to protective cultures, bacteriocins and food-grade enzymes. *Int. J. Food Microbiol.* **24:**343–362.

32. **Hsieh, D. Y., and J. M. Jay.** 1984. Characterization and identification of yeasts from fresh and spoiled ground beef. *Int. J. Food Microbiol.* **1:**141–147.

33. **Hurst, A.** 1973. Microbial antagonism in foods. *Can. Inst. Food Sci. Technol. J.* **6:**80–90.

34. **Hurst, A.** 1978. Nisin: its preservative effect and function in the growth cycle of the producer organism, p. 297–313. *In* F. A. Skinner and L. B. Quesnel (ed.), *Streptococci.* Academic Press, London, United Kingdom.

35. **Iandolo, J. J., C. W. Clark, L. Bluhm, and Z. J. Ordal.** 1965. Repression of *Staphylococcus aureus* in associative culture. *Appl. Microbiol.* **13:** 646–649.

36. **Jack, R. W., J. R. Tagg, and B. Ray.** 1995. Bacteriocins of gram-positive bacteria. *Microbiol. Rev.* **59:**171–200.

37. **Jay, J. M.** 1982. Antimicrobial properties of diacetyl. *Appl. Environ. Microbiol.* **44:**525–532.

38. **Jay, J. M.** 1995. Foods with low numbers of microorganisms may not be the safest foods or, why did human listeriosis and hemorrhagic colitis become foodborne diseases? *Dairy Food Environ. Sanit.* **15:**674–677.

39. **Jay, J. M.** 1996. Microorganisms in fresh ground meats: the relative safety of products with low versus high numbers. *Meat Sci.* **43:**S59–S66.

40. **Jay, J. M.** 1997. Do background microorganisms play a role in the safety of fresh foods? *Trends Food Sci. Technol.* **8:**421–424.

41. **Jay, J. M.** 2003. A review of recent taxonomic changes in seven genera of bacteria commonly found in foods. *J. Food Prot.* **66:**1304–1309.

42. **Jay, J. M., J. P. Vilai, and M. E. Hughes.** 2003. Profile and activity of the bacterial biota of ground beef held from freshness to spoilage at 5–7°C. *Int. J. Food Microbiol.* **81:**105–111.

43. **Jay, J. M., M. J. Loessner, and D. A. Golden.** 2005. *Modern Food Microbiology,* 7th ed. Springer, New York, NY.

44. **Juven, B. J., R. J. Meinersmann, and N. J. Stern.** 1991. Antagonistic effects of lactobacilli and pediococci to control intestinal colonization by human enteropathogens in live poultry. *J. Appl. Bacteriol.* **70:**95–103.

45. **Klaenhammer, T. R.** 1988. Bacteriocins of lactic acid bacteria. *Biochimie* **70:**337–349.

46. **Klaenhammer, T. R.** 1993. Genetics of bacteriocins produced by lactic acid bacteria. *FEMS Microbiol. Rev.* **12:**39–86.

47. **Leistner, L., and L. G. M. Gorris.** 1995. Food preservation by hurdle technology. *Trends Food Sci. Technol.* **6:**41–46.

48. **Leistner, L., and G. Gould.** 2002. *Hurdle Technology—Combination Treatments for Food Stability, Safety and Quality.* Springer, New York NY.

49. **Leverentz, B., W. S. Conway, W. Janisiewicz, M. Abadias, C. P. Kurtman, and M. J. Camp.** 2006. Biocontrol of the food-borne pathogens *Listeria monocytogenes* and *Salmonella enterica* serovar Poona on fresh-cut apples with naturally occurring bacterial and yeast antagonists. *Appl. Environ. Microbiol.* **72:**1135–1140.

50. **Light, I. J., R. L. Walton, J. M. Sutherland, H. R. Shinefield, and V. Brackvogel.** 1967. Use of bacterial interference to control a staphylococcal nursery outbreak. *Am. J. Dis. Child.* **113:**291–299.
51. **Lücke, F.-K.** 2000. Utilization of microbes to process and preserve meat. *Meat Sci.* **56:**105–115.
52. **Lund, B. M.** 1992. Ecosystems in vegetable foods. *J. Appl. Bacteriol.* **73:**S115–S126.
53. **Matthews, K. R. (ed.).** 2006. *Microbiology of Fresh Produce.* ASM Press, Washington, DC.
54. **Mattila-Sandholm, T., and E. Skyttä.** 1991. The effect of spoilage flora on the growth of food pathogens in minced meat stored at chilled temperature. *Lebensm.-Wiss. Technol.* **24:**116–120.
55. **Mattila-Sandholm, T., A. Haikara, and E. Skyttä.** 1991. The effect of *Pediococcus damnosus* and *Pediococcus pentosaceus* on the growth of pathogens in minced meat. *Int. J. Food Microbiol.* **13:** 87–94.
56. **Miller, W. A.** 1955. Effect of freezing ground pork and subsequent storing above 32°F upon the bacterial flora. *Food Technol.* **9:**332–334.
57. **Nissen, H., T. Maugesten, and P. Lea.** 2001. Survival and growth of *Escherichia coli* O157:H7, *Yersinia enterocolitica* and *Salmonella enteritidis* on decontaminated and untreated meat. *Meat Sci.* **57:** 291–298.
58. **Nunes, C., J. Usall, N. Teixidó, E. Fons, and I. Viñas.** 2002. Post-harvest biological control by *Pantoea agglomerans* (CPA-2) on Golden Delicious apples. *J. Appl. Microbiol.* **92:**247–255.
59. **Palumbo, S. A., A. Pickard, and J. E. Call.** 1997. Population changes and verotoxin production of enterohemorrhagic *Escherichia coli* strains inoculated in milk and ground beef held at low temperatures. *J. Food Prot.* **60:**746–750.
60. **Peterson, A. C., J. J. Black, and M. F. Gunderson.** 1962. Staphylococci in competition. II. Effect of total numbers and proportion of staphylococci in mixed cultures on growth in artificial culture medium. *Appl. Microbiol.* **10:**23–30.
61. **Pivnick, H., and E. Nurmi.** 1982. The Nurmi concept and its role in the control of salmonellae in poultry, p. 41–70. *In* R. Davies (ed.), *Developments in Food Microbiology.* Applied Science Publishers, London, United Kingdom.
62. **Price, R. J., and J. S. Lee.** 1970. Inhibition of *Pseudomonas* species by hydrogen peroxide producing lactobacilli. *J. Milk Food Technol.* **33:**13–18.
63. **Prieto, M., M. L. Garcia-Lópes, T. M. López, and B. Moreno.** 1993. Factors affecting spoilage microflora succession on lamb carcasses at refrigerator temperatures. *J. Appl. Bacteriol.* **74:** 521–525.
64. **Restuccia, C., F. Giusino, F. Licciardello, C. Randazzo, C. Caggia, and G. Muratore.** 2006. Biological control of peach fungal pathogens by commercial products and indigenous yeasts. *J. Food Prot.* **69:**2465–2470.
65. **Samelis, J., J. N. Sofos, P. A. Kendall, and G. C. Smith.** 2001. Influence of the natural microbial flora on acid tolerance response of *Listeria monocytogenes* in a model system of fresh meat decontamination fluids. *Appl. Environ. Microbiol.* **67:** 2410–2420.
66. **Schillinger, U., R. Geisen, and W. H. Holzapfel.** 1996. Potential of antagonistic microorganisms and bacteriocins for the biological preservation of foods. *Trends Food Sci. Technol.* **7:** 158–164.
67. **Schoeni, J. L., and A. C. L. Wong.** 1994. Inhibition of *Campylobacter jejuni* colonization in chicks by defined competitive exclusion bacteria. *Appl. Environ. Microbiol.* **60:**1191–1197.
68. **Shinefield, H. R., J. C. Ribble, and M. Boris.** 1971. Bacterial interference between strains of *Staphylococcus aureus,* 1960 to 1970. *Am. J. Dis. Child.* **121:**148–153.
69. **Smith, L. D.** 1975. Inhibition of *Clostridium botulinum* by strains of *Clostridium perfringens* isolated from soil. *Appl. Microbiol.* **30:**319–323.
70. **Straka, R. P., and F. M. Combs.** 1952. Survival and multiplication of *Micrococcus pyogenes* var. *aureus* in creamed chicken under various holding, storage and defrosting conditions. *Food Res.* **17:**448–455.
71. **Tagg, J. R., A. S. Dajani, and L. W. Wannamaker.** 1976. Bacteriocins of gram-positive bacteria. *Bacteriol. Rev.* **40:**722–756.
72. **Troller, J. A., and W. C. Frazier.** 1963. Repression of *Staphylococcus aureus* by food bacteria. I. Effect of environmental factors on inhibition. *Appl. Microbiol.* **11:**11–14.
73. **Troller, J. A., and W. C. Frazier.** 1963. Repression of *Staphylococcus aureus* by food bacteria. II. Causes of inhibition. *Appl. Microbiol.* **11:**163–165.
74. **U.S. Department of Agriculture.** 1996. *Nationwide Federal Plant Raw Ground Beef Microbiological Survey, August 1993–March 1994.* U.S. Department of Agriculture, Washingon, DC.
75. **van der Wielen, W. J. J., L. J. A. Lipman, F. van Knapen, and S. Biesterveld.** 2002. Competitive exclusion of *Salmonella enterica* serovar Enteritidis by *Lactobacillus crispatus* and *Clostridium lactatifermentans* in a sequencing fed-batch culture. *Appl. Environ. Microbiol.* **68:**555–559.
76. **Vanneste, J. L., J. Yu, and S. V. Beer.** 1992. Role of antibiotic production by *Erwinia herbicola* Eh252 in biological control of *Erwinia amylovora. J. Bacteriol.* **174:**2785–2796.
77. **Vimont, A., C. Vernozy-Rozand, M. P. Montet, C. Laizzera, C. Bavai, and M.-L. Delignette-Muller.** 2006. Modeling and predicting the simultaneous growth of *Escherichia coli*

O157:H7 and ground beef background microflora for various enrichment protocols. *Appl. Environ. Microbiol.* **72:**261–268.

78. **Vold, L., A. Holck, Y. Wasteson, and H. Nissen.** 2000. High levels of background flora inhibits growth of *Escherichia coli* O157:H7 in ground beef. *Int. J. Food Microbiol.* **56:**219–225.

79. **Wentz, M., H. Scott, and J. Vennes.** 1967. *Clostridium botulinum* type F: seasonal inhibition by *Bacillus licheniformis*. *Science* **155:**89–90.

# MICROBIAL SUCCESSION AND GUT HEALTH: PROBIOTICS

*Gerald W. Tannock*

# 4

The human bowel is home to a large and diverse community of bacterial species. A complete catalogue of the members of the community is lacking, but nucleic acid-based analyses of bowel communities of adult humans have shown that, qualitatively and quantitatively, there is impressive compositional stability. In contrast, during the first year of life, the composition of the bowel community of human infants is relatively simple and undergoes major qualitative shifts. Genomic and biochemical investigations have demonstrated that the bacterial community has a huge capacity to degrade and ferment complex carbohydrates, proteins, and lipids that pass undigested to the large bowel. It is clear from experimental animal studies that the bowel community has major impacts on the physiology of the host.

Therefore, there is considerable interest in the influence of the bowel community on the well-being of the human host and whether the composition or activities of the bacterial collection can be engineered to promote health. Of particular note are efforts to understand the protective role of bowel bacteria against allergies and to understand inflammatory bowel diseases, including pouchitis. Probiotics are widely promoted by the self-care health industry but have not yet attained general acceptance as therapeutic or prophylactic agents in medical practice. There is a need to develop probiotic products that will target specific disease conditions, rather than to assume that one probiotic constitutes an elixir of life. The derivation of these novel probiotics will require mechanistic explanations of disease processes in which bowel communities of humans are intimately involved. This chapter considers all of these aspects and suggests ways in which knowledge of the microbial ecology of the human bowel can be obtained using modern technologies.

## PROBIOTICS

The origin of the "probiotic" notion rests with Elie Metchnikoff, director of the Pasteur Institute, Paris, France, during the early years of the 20th century. Through his research concerning resistance to diseases, Metchnikoff developed an interest in the aging process and published his views on the large bowel as a source of toxic substances that damaged the nervous and vascular systems when absorbed from the gut and circulated in the blood-

*Gerald W. Tannock,* Department of Microbiology and Immunology, University of Otago, Dunedin, New Zealand.

*Food-Borne Microbes: Shaping the Host Ecosystem,* Edited by L.-A. Jaykus, H. H. Wang, and L. S. Schlesinger, © 2009 ASM Press, Washington, DC

stream. Microbes inhabiting the bowel produced these toxic substances (amines and ammonia) and were thus responsible for "autointoxication." The microbial products, it was explained, originated from the digestion of proteins by "putrefactive bacteria." Metchnikoff proposed that humans would benefit by encouraging the correct balance of microbial types in the large bowel, especially from the reduction of putrefactive activity. The implantation of lactic acid-producing bacteria in the gut would inhibit the putrefactive microbes, just as happened in milk in which a lactic fermentation had occurred (66, 67). Metchnikoff's fantasies on the relationship among intestinal putrefaction, degenerative diseases, and longevity had no basis in fact, but as a result of his philosophical concepts ("orthobiosis"), fermented milk products acquired their healthy image, which persists today under the guise of probiotics, an industry estimated to be worth many millions of dollars per year (95).

Probiotics are defined as "live microorganisms, which when administered in adequate amounts confer a health benefit on the host" (86). Members of the genus *Lactobacillus* are the most common constituents of probiotic products, although *Bifidobacterium* species, *Escherichia coli, Enterococcus faecium,* and *Saccharomyces* species may also be encountered (96). Although probiotics aimed at altering the composition of oral and vaginal ecosystems have been developed or are under development, most products target bowel health (38, 80, 84, 85). There is little doubt that lactobacilli transit the digestive tract following consumption of probiotic products (2, 23, 89, 98, 106, 114, 131), but in order to evaluate the probiotics phenomenon, one must consider the microbial ecology of the human gut.

## THE BOWEL COMMUNITY

The remnants of microbial cells, particularly bacterial DNA, can be detected along the entire length of the gastrointestinal tract (9, 10, 24, 37, 126). This does not mean, however, that all regions of the gut are colonized by bacterial communities. Not all parts of the gut are suitable for microbial persistence; the acid secreted in the stomach and the swift flow of contents in the duodenum and jejunum ensure that the more proximal regions contain only transient microbial cells in the healthy human host (99). *Helicobacter pylori* associates with the epithelial surface of the gastric and duodenal mucosa, but since the presence of this organism is associated with inflammation of the tissue, it is probably best for the present to consider it a pathogen and for its presence to be termed an infection (53). The terminal ileum and the large bowel are hospitable places for microbial proliferation because bowel motility is slower, and microbial communities (mostly bacterial species) reside in these sites. The surfaces of undigested plant fragments in the digesta have adherent bacterial associates that are involved in the breakdown of complex carbohydrate molecules, just as occurs in the rumen of sheep and cows (58). The existence of a discrete bacterial community associated with the mucosal surface of the human bowel has been postulated by researchers drawing on knowledge of the distribution of bacteria in the proximal colon of mice (100). There is not yet convincing evidence, however, that such a mucosal community exists in humans, but studies aimed at defining the composition of the mucosa-associated community have been reported (15, 50, 54, 57, 61, 73, 78, 83, 118).

Most of this work is related to the bacteriology of inflammatory bowel diseases. Accurate definition of the mucosa-associated community, if indeed it exists, will be difficult because mucosal biopsy specimens must be collected; such collection is invasive, and collection procedures are ill defined from study to study. Prior to colonoscopy and biopsy, the patient is purged to remove bowel contents. The effect of this preparative treatment on the composition of the bowel community is unknown. Further, the bowel is not completely decontaminated by this procedure and a fecal fluid continues to be present in the bowel and to bathe mucosal surfaces. Therefore, it is not clear what is being sampled: the mucosal surface contaminated with luminal bacteria or true mucosal inhabitants. Additionally, the extent of contamination of the colonoscope with bacteria from the fecal fluid has never been determined. Therefore, reports about the mucosa-associated community of humans must be viewed with caution. At the least, these studies have relieved the previous focus on the fecal bacterial community; they now seek to define the microbiology of specific regions of the bowel.

Large bowel contents and feces contain about $10^{11}$ bacterial cells per gram (wet weight), and bacterial cells comprise about 50% of fecal mass (110). Four bacterial phyla are represented *(Firmicutes, Bacteroidetes, Actinobacteria,* and *Proteobacteria),* and three phyogenetically broad groups, each containing many genera and species, are numerically dominant in the feces of healthy humans (*Clostridium coccoides* group, *Clostridium leptum* subgroup, and *Bacteroides-Prevotella* group) (27, 52). Bacteria dominate the bowel community, but fungi and *Archaea* may also be resident, comprising less than 0.05 and 1% of the total inhabitants, respectively (68, 105).

Much of the information pertaining to the residents of the bowel has been generated through the application of nucleic acid-based methodologies, most of which target the nucleotide base sequence of the small ribosomal subunit RNA (16S rRNA in the case of bacteria), which provides a cornerstone of microbial taxonomy. An earlier estimate of the number of bacterial species that might be resident in the human large bowel was based on bacteriological culture. Four hundred species seemed a likely number by extrapolation from what had already been cultured (25, 70). Nucleic acid-based methods of detection suggest that about 50% of the bacterial cells seen microscopically in feces cannot yet be cultured in the laboratory, even when accounting for the fact that some of them are dead (8, 114). "Operational taxonomic units" (molecular species) never encountered in culture-based bacteriology are detectable by molecular analytical methods. Estimates of biodiversity now seem to continually inflate, probably because many of the 16S rRNA sequences in data banks differ by one nucleotide base. These are likely to be sequences representing the same operational taxonomic units but containing sequencing errors (5, 6). Curiously, therefore, despite the application of state-of-the-art technology, we still do not really know, with any accuracy, the composition of the bowel bacterial community.

We do know that the bacterial community is highly metabolically active. A chemostat is a culture apparatus that is used to maintain the growth of bacterial cultures continuously and at a constant rate under laboratory conditions. Culture medium is fed into a culture vessel containing the bacterial cells at a rate controlled by a flow regulator. The culture volume is held constant by means of an overflow tube that enables spent medium (effluent) to pass from the culture vessel. The human large bowel is the equivalent of a chemostat. It is fed with culture medium derived from the undigested components of the diet that pass from the small bowel, as well as substances produced endogenously by the human host. Hence, this culture medium is particularly rich in complex carbohydrates derived from plant cell walls and complex glycoproteins from mucus (17, 88). The ileocecal valve regulates the flow of culture medium into the large bowel; feces are the effluent. Under these circumstances, a continuous fermentation of exogenous and endogenous substrates by consortia of bacteria proceeds, resulting in the formation of short-chain fatty acids (mainly acetic, propionic, and butyric acids), amines, phenols, indoles, and gases as the major products (17).

The absorption of the fatty acids from the bowel lumen once probably provided important caloric and carbon sources for prehistoric humans, just as it does for other animal species even today (64), and nutritionists still invoke butyric acid as an important fuel for colonocytes (90). One would imagine that the bacterial culture responsible for the fermentation would be the same from human to human, since bowel biochemistry follows the same pattern from one person to the next (112). Yet, again on the basis of molecular methods of analysis, it appears that bowel communities are about 30% dissimilar among humans. Though a surprising fact when first encountered, the uniqueness of bowel communities is likely to be due to extensive functional redundancy in the bacterial world: more than one bacterial species can carry out a particular catabolic process. Thus, just as there is a normal range of blood chemistry values in the human population, there is also a range of inhabitants that can fill the ecological niches responsible for the overall bowel fermentation. What factors mediate which organism fills a particular niche in an individual human are unknown but may reflect physiological idiosyncrasies, as well as "windows of opportunity" with respect to specific microbial exposure. Moreover, the composition of individual bowel communities has impressive temporal stability (108, 133).

## AUTOCHTHONY AND ALLOCHTHONY

The human bowel is daily exposed to bacterial cells present in food and water. Some of these bacteria survive and transit the digestive tract, possibly continue to metabolize (77), and are detectable in the feces (124). The same applies to probiotic bacteria that are intentionally consumed in self-care health products, including yogurts (111). These "allochthonous" bacteria do not become part of the bowel community, because they are no longer detected in feces once the source of their ingestion is removed. This can be explained in terms of the niche exclusion principle, which states that a single ecological niche can support no more than one type of organism, either genotype or species (34). Hence, in the large bowel, where heterogeneous environments exist in relation to bacterial nutrients and spatial arrangements, hundreds of niches must exist, because there is a huge diversity of bacterial types. Different types of bacteria are favored in each niche. Lactobacilli are not particularly favored in the human bowel, since they represent much less than 1% of the total membership of the community (114). They are not genetically endowed with the ability to degrade complex polysaccharide molecules and hence lack the means to fulfill more than an oligosaccharide-scavenging role (3). There is likely to be fierce competition from obligately anaerobic bacteria for this role, and the less favored bacteria are competitively excluded from the ecosystem (26, 101, 129).

"Autochthonous" bacteria have a habitat in the bowel. They have long-term associations with the human host, form stable populations of characteristic size, and have demonstrable means of earning a living (an ecological function or niche) in the distal bowel (113). Therefore, snapshots of the composition of bowel communities generated by 16S rRNA gene clone library catalogues or denaturing gradient gel electrophoresis are only the beginning of ecological investigations, and DNA-based methods reveal only "who might have been there." Temporal (quantitative sequential measurements) and function-based (transcriptomics and proteomics) observations are necessary to properly define the composition and metabolism of bacterial communities.

## PROBIOTICS AND THE SELF-CARE HEALTH MARKET

The probiotic concept has a long history, but progress in the scientific and medical evaluation and validation of these products has been slow. Even today, adequate information by which the consumer and health professional can judge the efficacy and safety of retailed probiotics is lacking (111). Murch (71) has pointed out that there are more than 1.25 million Internet pages concerning probiotics, leading to his conclusion that "no other therapeutic modality spans the divide between internet voodoo and cutting edge high tech in this way." Probiotics have, in general, not been subjected to the large-scale trials of efficacy that are used in the pharmaceutical industry. Without these trials and subsequent approval by fastidious regulatory agencies such as the FDA (United States), probiotics continue to be pertinent to the self-care health market rather than clinical medicine. As pointed out by Katz (47), more than half of the "probiotic papers" recorded in the PubMed database (NCBI) are reviews, not reports of the results of experimental science.

There is a considerable list of health-promoting activities of probiotics touted in the literature (Table 1), and amazement may be expressed that these products can target such a wide range of diseases. Perhaps a small number of phenomena shared in common may be involved? The mechanistic explanations are broad and therefore somewhat vague, encompassing production of antimicrobial agents, blocking adhesion of pathogens and toxins to epithelial cells, and modulation of the immune response. Overall, it must be concluded that probiotics continues to be long on speculation and hope but short on medical validation. Optimistically, a new generation of probiotics will emerge before long, in which the products will be designed and derived for the support of patients suffering from specific diseases and for which mechanistic explanations of efficacy will be provided. These products will have been demonstrated in large clinical trials to be efficacious, safe to administer (even to immunologically dysfunctional patients), and available by medical prescription (111). Nevertheless, some members of the medical profession, as well as the laity, have greeted with enthusiasm the use of current probiotic products as prophylaxis for atopic diseases (allergies), inflammatory bowel diseases (Crohn's disease [CD] and ulcerative colitis [UC]), and pouchitis.

**TABLE 1** Probiotic health claims

| Disease target | Reference(s) |
|---|---|
| Allergies (atopic eczema, milk allergy) | 42, 63 |
| Cancer | 128 |
| Hypercholesterolemia | 18 |
| Diarrhea (antibiotic associated, rotavirus infection, traveler's diarrhea) | 62 |
| Endotoxemia (liver disease) | 74 |
| Gastritis *(Helicobacter pylori)* | 92 |
| Hypertension | 36 |
| Immunomodulation | 43 |
| Irritable bowel syndrome | 59 |
| Urolithiasis (urinary tract stones) | 14 |
| Lactose intolerance | 19 |
| Contaminated small bowel syndrome | 22 |
| Inflammatory bowel diseases (CD, UC) | 103 |

## PROBIOTICS AND ALLERGIES

The gut of newborn human infants resembles that of the germfree animal because it does not yet harbor a bacterial or other microbial community. This germfree state is short-lived because within minutes of birth, the baby is exposed to bacteria in the feces that have been involuntarily expelled by the mother during labor, as well as to environmental microbes. Suckling, kissing, and caressing the infant after birth provide additional assurance that maternal microbes are transmitted from one human generation to another. Regulatory mechanisms generated within the ecosystem (autogenic factors) and by external forces (allogenic factors) permit the episodic persistence of some bacterial populations but the elimination of others in a classical biological succession (16). It is likely to take several years to produce a climax community resembling that of adults (76).

Members of the bacterial genus *Bifidobacterium* are numerically predominant in the gut of infants during the first months of life. Nucleic acid-based methods of analysis show that bifidobacteria form between 60 and 91% of the total bacterial community in the feces of breast-fed babies and 28 to 75% (average, 50%) in formula-fed infants, whereas they comprise only a small percentage of the fecal community of adult humans (35). The infant during early life is therefore almost a monoassociated gnotobiote, and bifidobacterial antigens may be important instigators of immunological development. By reference to the results of recent experiments with gnotobiotic animals (7, 39, 40, 55, 91, 107), it can be concluded that the bowel bacteria have a substantial impact, at least in the short term, on the physiology of the animal host. The environmental conditions under which babies are born and nurtured may affect which microbes they are exposed to and subsequently influence the composition of their bowel community. Differences in neonatal bowel communities might occur due to the common occurrence of hospital deliveries, cesarean sections, special-care baby unit admissions, smaller family size, widespread use of antibiotics, good hygiene, and the nature of the maternal diet, which is likely to differ according to ethnic, and therefore cultural, preferences (12). Either the lack of exposure of babies to particular bifidobacterial species or elimination of bifidobacterial species from the bowel through the use of antibiotics might reduce the exposure of children in early life to important bacterial antigens at a critical time in the maturation of the immune system, for example, in removing (immune deviation) the T helper 2 (Th2) skew apparently characteristic of the newborn (82, 130).

In recent decades, many affluent countries have experienced an increase in the prevalence of atopic diseases, including asthma (41). Several aspects of lifestyle have changed in these countries over the same period, and theories have been advanced to explain the altered prevalence of allergies. The "hygiene hypothesis" proposes that atopic diseases could be prevented by infections in early childhood because the neonatal immune system would be driven towards a T helper 1 (Th1) response, but a specific "infectious protective factor" has never been identified (72, 109). Attention has thus turned to the bacterial community of the bowel and the possibility that colonization of the gut by specific bacterial species might be more important than the impact of sporadic infections in programming the mucosal immune system in early life. For example, the composition of the bowel community has been reported to differ in the numbers of lactobacilli and clostridia in the feces of Estonian and Swedish children (102). Atopic diseases were less prevalent in the Estonian than in the Swedish population. Ouwehand and colleagues (79) have reported that the rates of prevalence of *Bifidobacterium adolescentis* differed in the feces of healthy and allergic Finnish children aged 2 to 7 months. Specifically, six of seven allergic children harbored *B. adolescentis* in the feces, whereas this species was not detected in six healthy children.

Young and colleagues (132) reported the results of a study in which the fecal populations of bifidobacteria from children aged 25 to 35 days in Ghana (low prevalence of atopy) and in New Zealand and the United Kingdom (high-prevalence countries) were compared. Natal origin influenced the detection of bifidobacterial species, because fecal samples from Ghana almost exclusively contained *Bifidobacterium infantis,* whereas those of the other children did not. Choosing species on the basis of the bacteriological results, bifidobacterial preparations were tested for their effect on cell surface markers of dendritic cells harvested from cord blood. Bifidobacterial species-specific effects on dendritic cell activation were observed in that CD83 expression was increased by *Bifidobacterium bifidum, Bifidobacterium longum,* and *Bifidobacterium pseudocatenulatum.* One or more of these species were detected in the feces of 40 of 46 children from New Zealand and the United Kingdom but in only a few (*B. longum:* 2 of 32) of the samples from infants living in Ghana. *B. infantis,* common in the feces of babies from Ghana, failed to produce this effect. Further investigations of the molecular interplay among bifidobacteria, human dendritic cells, T cells, and allergens are clearly required and may provide the first clear evidence of "immunological Freudianism" in relation to human diseases.

Finnish researchers have been at the forefront in the investigation of probiotic prophylaxis of atopic dermatitis (atopic eczema). *Lactobacillus rhamnosus* strain GG has mainly been used in these studies, beneficial influences have been reported, and mechanistic speculations have been made (Table 2). The use of

**TABLE 2** *Lactobacillus rhamnosus* GG and atopic eczema / dermatitis

| Details | Outcome | Reference(s) |
|---|---|---|
| A double-blind, randomized placebo-controlled trial in which *L. rhamnosus* GG was administered prenatally to mothers who had at least one first-degree relative (or partner) with atopic eczema, allergic rhinitis, or asthma, and postnatally for 6 mo to their infants. Chronic recurring atopic eczema was the primary end point. | At 2 yr of age, the frequency of atopic eczema in the probiotic group was half that in the placebo group (15 of 64 [23%], vs 31 of 68 [46%]). There was no difference between probiotic and placebo groups with respect to total concentrations of IgE,[a] concentrations of antigen-specific IgE, and positive reactions to skin prick tests. In a follow-up study of the infants, the cumulative risk for developing eczema during the first 7 yr of life was lower in the probiotic group than in the placebo group (42.6%, versus 66.1%). Frequencies of atopic sensitization (IgE, skin pricks) remained similar between the two groups, suggesting that the preventive effect was not IgE mediated. | 44, 45 |
| A randomized, double-blind study of infants with suspected cow's milk allergy was conducted. The infants received probiotic or placebo for 4 wk. Severity of atopic eczema / dermatitis syndrome was measured using the SCORAD[b] index. The infants received extensively hydrolyzed whey formula and skin treatment. In a further study, markers of immune responses were measured in plasma collected from the infants. | There was a greater reduction in the SCORAD (4 wk after treatment) in the infants that were atopically sensitized (IgE concentrations, skin pricks) and fed *Lactobacillus rhamnosus* GG. In these infants, C-reactive protein levels were higher in the probiotic group (0.83 μg / ml) than in the placebo group (0.42; $P = 0.021$). Interleukin-6 and soluble E-selectin concentrations were higher after treatment with the probiotic than with the placebo ($P = 0.023$). The immunological results might reflect a systemic, low-grade inflammation produced by administration of *L. rhamnosus* GG. | 122, 123 |
| A study of 132 infants whose mothers were administered *L. rhamnosus* GG or a placebo prenatally, and the infants for 6 mo postnatally. The compositions of the fecal bacterial communities of the two groups of infants were compared during the first 2 yr of life. | The bacterial succession in the feces did not differ between the two groups of infants. Therefore, probiotic administration did not alter the composition of the bowel community. | 87 |
| A randomized, double-blind study of infants less than 5 mo old with atopic dermatitis (Hanifin criteria) and suspected of having cow's milk allergy. The infants received a hydrolyzed whey-based diet (placebo) or the diet supplemented with *L. rhamnosus* GG (probiotic) for 3 mo. The SCORAD index was used to evaluate the clinical condition of the infants, total IgE and antigen-specific IgE were measured, and skin prick tests were carried out. Inflammatory parameters were measured in blood, urine, and fecal samples. | Probiotic administration did not affect SCORAD, sensitization, or immunological measurements. Therefore, probiotic administration did not have a beneficial impact on infantile atopic dermatitis. | 13 |

[a] IgE, immunoglobulin E.
[b] SCORAD, Severity Scoring of Atopic Dermatitis.

lactobacilli, however, does not really fit with the microbial ecology of the infant gut, and better effects might be obtained through the use of bifidobacteria with antigenic properties relevant to the generation of immune deviation.

## PROBIOTICS AND INFLAMMATORY BOWEL DISEASES

CD and UC are chronic immune inflammatory conditions of the alimentary tract referred to collectively as inflammatory bowel diseases. CD lesions can occur even in upper regions of the tract but are usually located where there are resident bowel bacteria (ileum and colon), whereas UC is limited to the large bowel. The diseases are commonly diagnosed in adolescents and young adults, are lifelong, and have a marked impact on lifestyle. Surgical intervention may be required to alleviate symptoms and to save life (11, 20, 81). Modern drug treatments are anti-inflammatory and largely effective, but better treatments might be derived if the events that initiate, and the factors that drive, the continuing inflammation were known. Experimental animal models of colitis do not mimic exactly CD or UC but can be used to examine the role of specific bacteria in the etiology of enterocolitis in general terms. It is clear from the results of this work that bacteria resident in the bowel of the animals have an essential role in the pathogenesis of colitis because, when maintained germfree, the animals do not develop disease (97). Current interest therefore is focused on the bowel community as the source of antigens that fuel the chronic inflammation seen in inflammatory bowel diseases.

Macdonald and Monteleone (56) proposed an attractive explanation of CD in which bacterial cells and/or their antigens pass via M cells into Peyer's patches, where CD4 T cells are activated, and migrate to the lamina propria. In healthy humans, because there is not further antigenic challenge, the T cells die there by apoptosis. Increased epithelial permeability in CD patients, compounded by a dysfunctional immune system, allows sufficient bacterial antigens to enter the lamina propria from the bowel lumen to trigger T-cell activation, breaking tolerance mediated by immunosuppressive cytokines and T regulatory cells. The immune cells pursue their normal activity of eradicating threatening bacteria and their components, but this is impossible because the antigenic sources are bacteria residing in the bowel. Hence, a chronic inflammation occurs because the antigenic fuel is supplied continuously from the bowel lumen through the defective epithelial barrier.

Currently, good evidence for beneficial probiotic therapy in the treatment of CD is lacking. UC has been considered to be more amenable to the effects of probiotics, since *Escherichia coli* Nissle 1917 appeared to assist in the maintenance of remission of this disease (51). However, in a relatively large (157 subjects) randomized, double-blind, placebo-controlled trial, probiotic products containing *Lactobacillus salivarius* UCC118 or *Bifidobacterium infantis* 35624 consumed daily for 1 year did not provide a significant benefit over placebo in the maintenance of steroid-induced remission of UC (104). These probiotic products had been found to attenuate disease in experimental animal colitis (65). The outcome of this useful study reinforced the need to validate, by conducting powerful clinical trials, probiotic phenomena observed in vitro or in experimental animals.

## PROBIOTICS AND POUCHITIS

The best evidence for the efficacy of probiotics in inflammatory conditions of the bowel comes from studies of the maintenance of remission of pouchitis (31). Chronic or recurrent pouchitis is the most common cause of troublesome, long-term functional disturbance for patients, with pouches created by ileoanal anastomosis following removal of the large bowel to cure UC (94). Pelvic pouches show various degrees of chronic mucosal inflammation and other alterations described as "colonic phenotype change" (28, 116, 117). In some patients, however, acute inflammation and clinical pouchitis develop by a process that possibly parallels UC (115). Pouchitis is less common in pouches formed in patients because of familial adenomatous polyposis than in patients with UC (1, 4). It appears, therefore, that the dysfunctional immune system of UC patients reacts with as-yet-unidentified factors associated with the pouch contents. Empirical success in the treatment of pouchitis with antibiotics points, however, to bacteria as the likely factors with which the immune system reacts (60).

VSL#3 is a probiotic product consisting of a mixture of eight bacterial species that belong to the lactic acid bacterial consortium. In addition to minimizing flare-ups of pouchitis, the probiotic has been used as prophylaxis against pouchitis and in the maintenance of remission of UC (Table 3). The probiotic bacteria are presumed to have anti-inflammatory activities in the bowel. These promising outcomes need to be confirmed in multinational trials with improved statistical power.

## DRAWING BOWEL BLUEPRINTS

Bowel bacteria are capable of influencing gene expression in enterocytes lining the bowel. Experiments utilizing monoassociated ex-germfree mice, although relatively short-term in nature, have demonstrated both up- and down-regulatory effects on murine gene transcription in the ileum, one of which has been shown to have systemic consequences (fat deposition in epididymal pads) (7, 39, 107). Moreover, the incalculable antigenic load as-

**TABLE 3** Clinical studies with VSL#3

| Details | Outcome | Reference |
|---|---|---|
| An uncontrolled study in which 20 patients with UC who were in remission but intolerant or allergic to 5-ASA[a] were administered the probiotic during a 12-mo period. Selected bacteria were enumerated in fecal samples. | Streptococci, lactobacilli, and bifidobacteria increased in numbers in the feces from day 20 of treatment relative to baseline values, and average fecal pH was lowered approximately 0.8 unit ($P < 0.05$). Populations of other bacterial groups that were measured were not altered. Fifteen of the 20 patients (75%) remained in remission. | 121 |
| Forty patients in remission (antibiotic therapy for 1 mo) from chronic relapsing pouchitis were included in a double-blind, randomized, placebo-controlled study. VSL#3 or placebo was administered to the subjects for 9 mo. | Three of 20 patients in the probiotic group relapsed (15%), whereas all in the placebo group relapsed (100%; $P < 0.001$). | 30 |
| Twenty patients with pouches were administered VSL#3 for 1 yr immediately following ileostomy closure. A comparable group received a placebo. | Two of the patients administered the probiotic had an episode of acute pouchitis (10%), compared to eight of the placebo group (40%; $P < 0.05$). Overall, the IBDQ[b] score improved in the probiotic group but not in the placebo group. | 31 |
| Remission was induced in patients with recurrent pouchitis by 1 mo of treatment with antibiotics. Then 20 patients received VSL#3 for 1 yr and 16 received a placebo. | Remission was maintained in 17 patients administered the probiotic (85%) and in 1 patient in the placebo group (6%; $P < 0.0001$). The IBDQ score remained high in the probiotic group but deteriorated in the placebo group. | 69 |

[a] 5-ASA, 5-aminosalicylic acid.
[b] IBDQ, Inflammatory Bowel Disease Questionnaire.

sociated with millions of bacterial cells stimulates the development of the immune system (32). We do not yet have a clear picture, however, about how these molecular communications between bacteria and host affect health in the longer term. To understand health, we must learn about the mechanisms that operate in the healthy bowel and by which a stable and potentially beneficial ecosystem is sustained and maintained. We need to delineate the blueprints that underpin the healthy bowel. Then, bowel health could be guarded by reason, perhaps by dietary interventions that would produce predictable outcomes on the basis of knowledge of molecular networks in the bowel.

Fortunately, the technological approaches to achieve these goals are at the fingertips of microbiologists. Metagenomics, for example, is a facet of microbial ecology in which a microbial community is studied in terms of its collective genomes, rather than focusing on the diversity of species and their individual genomes (33). The metagenomic approach traditionally entails the cloning of large fragments of community genomic DNA that have been extracted directly from the ecosystem of choice. This abrogates the problem of noncultivability of the majority of the community inhabitants. The cloned DNA fragments are large enough to encode operons and therefore might result in the expression, by a surrogate bacterial host, of several enzymes that could catalyze a relatively complex metabolic process, including the synthesis of secondary metabolites. Metagenomic libraries derived from microbial community genomes can be screened for heterologous phenotypic traits that include enzymes and other proteins that are essential to the functioning of the ecosystem (29, 33, 125). Hence, they provide a means of accessing and assessing details of community biochemistry through its underpinning genetics.

Acquisition of a comprehensive knowledge of food webs in the bowel will, however, probably require somewhat more complex and integrative approaches. Stable isotope (does not radioactively decay) probing is a technique that can be used to identify bacteria that use a particular substance as a growth substrate ("who is doing what"). The potential substrate is highly enriched in a stable isotope such as $^{13}C$, and the detection of the active bacteria relies on the selective recovery and analysis of isotope-enriched cellular components (usually DNA and RNA). Nucleic acids are informative biomarkers, and $^{13}C$-labeled molecules can be separated from unlabeled molecules by density gradient centrifugation. The purified $^{13}C$-DNA contains the genomes of the bacteria that have utilized the substrate directly or indirectly (cross-feeding from primary degraders). PCR can be used to amplify 16S rRNA genes from the $^{13}C$-labeled genomes of the bacterial consortium which, when sequenced, reveal the taxonomy of its members (21).

Developments in the use of fluorescent oligonucleotide probes in combination with fluorescence-activated cell sorting mean that cells of the consortium membership can be separated from the rest of the community (46). DNA extracted from the sorted cells would permit random shotgun sequencing to proceed and the reconstruction of near-complete genomes, as has been reported for an acidophilic biofilm community dominated by a small number of microbial species. As reported by Tyson and colleagues (119), analysis of the gene complement of each of the community members revealed the pathways for carbon and nitrogen fixation and energy production and provided knowledge of survival strategies used by the bacteria in their hazardous environment. Genomic sequences would allow the derivation of DNA microarrays by which global gene expression of the consortium could be measured in order to obtain data by which bacterial responses to fluctuations in specific substrate concentrations in the bowel could be predicted (93). Although some technical improvements may still be necessary, stable isotope probing has the potential to be used to monitor the flow of $^{13}C$ through the bowel community and to thus examine and delineate bacterial food chains that may ultimately impinge on mammalian host welfare.

Metabolomics is the nontargeted, holistic, quantitative analysis of changes in the complete set of metabolites in the cell (the metabolome) in response to environmental or cellular changes (120). Metabolites are low-molecular-mass organic compounds (<1,000 Da) that participate in general metabolic reactions or are required for maintenance, growth, and normal functioning of a cell. Changes in cellular physiology are amplified through transcription of genes and translation to proteins, but due to regulatory mechanisms and/or substrate availability, a 10-fold increase in concentration of a transcript or enzyme is not necessarily reflected in a 10-fold increase in a particular cell activity. Alterations in transcriptome or proteome do, however, have large effects on the concentrations of intermediary metabolites in the cell because they reflect the activities of metabolic pathways. Of particular importance is the ability of metabolomics to penetrate the mechanisms of intracellular signaling in which both concentrations of metabolites and their associated dynamics are important. Whereas knowledge of the intracellular metabolites (metabolic fingerprint, the endometabolome) is essential in this work, changes in the physiology of bowel bacterial consortia could also, and perhaps more easily, be revealed by investigation of the exometabolome (metabolic footprint), represented by the extracellular milieu which contains metabolites secreted or consumed by bacteria in the bowel (49).

The metabolic footprint of the bowel bacteria is reflected in the metabonome (metabolome) of the animal host because bacterial metabolites are absorbed from the gut lumen into the lymph and blood circulations. Hence, the body fluids (blood, lymph, bile, sweat, and urine) of the host contain numerous bacterial products that may provide biomarkers of food-microbe-host interrelationships and possible indicators of health or disease (75). For example, as reported by Wang and colleagues (127), infection of mice with the nematode *Schistosoma mansoni* could be diagnosed on the basis of alterations to the urine metabonome in which shifts in bacterial metabolite concentrations were included. These shifts doubtless relate to the displacement of baseline murine-bacterial relationships by the introduction of parasites into the bowel ecosystem. The host metabonome is the sum of the interacting metabolomes of the whole organism and thus represents the end product of genetic, environmental, and host-bacterial relationships. The study of the metabonome might therefore contribute to a full systems biology approach to understanding and maintaining bowel health (48).

Drawing the blueprint of the healthy bowel will require a systems biology approach encompassing a diversity of scientists from different disciplines. The primary aim of the research will be to understand how all of the heterogeneous parts (dietary components, bacterial consortia, and animal physiology and development) are integrated, with a supplementary aim of identifying biomarkers of health or disease. Delineation of these biomarkers will enable microbiologists to better investigate ways by which the bowel community could be manipulated to promote health. The probiotics of today may disappear in the face of clinically validated interventions with specific bioactive molecules of microbial origin.

## REFERENCES

1. **Achkar, J. P., M. Al-Haddad, B., Lashner, F. H. Remzi, A. Brzezinski, B. Shen, F. Khandwala, and V. Fazio.** 2005. Differentiating risk factors for acute and chronic pouchitis. *Clin. Gastroenterol. Hepatol.* **3:**60–66.
2. **Alander, M., R. Satokari, R. Korpela, M. Saxelin, T. Vilpponen-Salmela, T. Mattila-Sandholm, and A. von Wright.** 1999. Persistence of colonization of human colonic mucosa by a probiotic strain, *Lactobacillus rhamnosus* GG, after oral consumption. *Appl. Environ. Microbiol.* **65:**351–354.
3. **Altermann, E., W. M. Russell, M. A. Azcarate-Peril, R. Barrangou, B. L. Buck, O. McAuliffe, N. Souther, A. Dobson, T. Duong, M. Callanan, S. Lick, A. Hamrick, R. Cano, and T. R. Klaenhammer.** 2005. Complete genome sequence of the probiotic lactic acid bacterium *Lactobacillus acidophilus* NCFM. *Proc. Natl. Acad. Sci. USA* **102:**3906–3912.
4. **Ambroze, W. L., Jr., R. R. Dozois, J. H. Pemberton, R. W. Beart, Jr., and D. M. Il-**

**strup.** 1992. Familial adenomatous polyposis: results following ileal pouch-anal anastomosis and ileorectostomy. *Dis. Colon Rectum* **35:**12–15.

5. **Ashelford, K. E., N. A. Chuzhanova, J. C. Fry, A. J. Jones, and A. J. Weightman.** 2005. At least 1 in 20 16S rRNA sequence records currently held in public repositories is estimated to contain substantial anomalies. *Appl. Environ. Microbiol.* **71:**7724–7736.
6. **Ashelford, K. E., N. A. Chuzhanova, J. C. Fry, A. J. Jones, and A. J. Weightman.** 2006. New screening software shows that most recent large 16S rRNA gene clone libraries contain chimeras. *Appl. Environ. Microbiol.* **72:**5734–5741.
7. **Backhed, F., H. Ding, T. Wang, L. V. Hooper, G. Y. Koh, A. Nagy, C. F. Semenkovich, and J. I. Gordon.** 2004. The gut microbiota as an environmental factor that regulates fat storage. *Proc. Natl. Acad. Sci. USA* **101:** 15718–15723.
8. **Ben-Amor, K., H. Heilig, H. Smidt, E. E. Vaughan, T. Abee, and W. M. de Vos.** 2005. Genetic diversity of viable, injured, and dead fecal bacteria assessed by fluorescence-activated cell sorting and 16S rRNA gene analysis. *Appl. Environ. Microbiol.* **71:**4679–4689.
9. **Bibiloni, R., M. Mangold, K. L. Madsen, R. N. Fedorak, and G. W. Tannock.** 2006. The bacteriology of biopsies differs between newly diagnosed, untreated, Crohn's disease and ulcerative colitis patients. *J. Med. Microbiol.* **55:** 1141–1149.
10. **Bik, E. M., P. B. Eckburg, S. R. Gill, K. E. Nelson, E. A. Purdom, F. Francois, G. Perez-Perez, M. J. Blaser, and D. A. Relman.** 2006. Molecular analysis of the bacterial microbiota in the human stomach. *Proc. Natl. Acad. Sci. USA* **103:**732–737.
11. **Bouma, G., and W. Strober.** 2003. The immunological and genetic basis of inflammatory bowel disease. *Nat. Rev. Immunol.* **3:**521–533.
12. **Brandtzaeg, P.** 2002. Current understanding of gastrointestinal immuno-regulation and its relation to food allergy. *Ann. N. Y. Acad. Sci.* **964:** 13–45.
13. **Brouwer, M. L., S. A. A. Wolt-Plompen, A. E. J. Dubois, S. van der Heide, D. F. Jansen, M. A. Hoijer, H. F. Kauffman, and E. J. Duiverman.** 2006. No effects of probiotics on atopic dermatitis in infancy: a randomized placebo-controlled trial. *Clin. Exp. Allergy* **36:** 899–906.
14. **Campieri, C., M. Campieri, B. Bertuzzi, E. Swennen, D. Matteuzzi, S. Stefoni, F. Pirovano, C. Centi, S. Ulisse, G. Famularo, and C. De Simone.** 2001. Reduction of oxaluria after an oral course of lactic acid bacteria at high concentration. *Kidney Int.* **60:**1097–1105.
15. **Conte, M. P., S. Schippa, I. Zamboni, M. Penta, F. Chiarini, L. Seganti, J. Osborn, P. Falconieri, O. Borrelli, and S. Cucchiara.** 2006. Gut-associated bacterial microbiota in paediatric patients with inflammatory bowel disease. *Gut* **55:**1760–1767.
16. **Cooperstock, M. S., and A. J. Zedd.** 1983. Intestinal flora of infants, p. 79–99. *In* D. J. Hentges (ed.), *Human Intestinal Microflora in Health and Disease.* Academic Press, New York, NY.
17. **Cummings, J. H., and G. T. Macfarlane.** 1991. The control and consequences of bacterial fermentation in the human colon. *J. Appl. Bacteriol.* **70:**443–459.
18. **De Roos, N. M., and M. B. Katan.** 2000. Effects of probiotic bacteria, lipid metabolism, and carcinogenesis: a review of papers published between 1988 and 1998. *Am. J. Clin. Nutr.* **71:** 405–411.
19. **De Vrese, M., A. Stegelmann, B. Richter, S. Fenselau, C. Laue, and J. Schrezenmeir.** 2001. Probiotics—compensation for lactase insufficiency. *Am. J. Clin. Nutr.* **73:**421S–429S.
20. **Duchmann, R., I. Kaiser, E. Hermann, W. Mayet, K. Ewe, and K.-H. Meyer zum Buschenfelde.** 1995. Tolerance exists towards resident intestinal flora but is broken in active inflammatory bowel disease (IBD). *Clin. Exp. Immunol.* **102:**448–455.
21. **Dumont, M. G., and J. C. Murrell.** 2005. Stable isotope probing—linking microbial identity to function. *Nat. Rev. Microbiol.* **3:**499–504.
22. **Dunn, S. R., M. L. Simenhoff, K. E. Ahmed, W. J. Gaughan, B. O. Eltayeb, M. E. D. Fitzpatrick, S. M. Emery, J. W. Ayresc, and K. E. Holtc.** 1998. Effect of oral administration of freeze-dried *Lactobacillus acidophilus* on small bowel bacterial overgrowth in patients with end stage kidney disease: reducing uremic toxins and improving nutrition. *Int. Dairy J.* **8:**545–553.
23. **Dunne, C., L. Murphy, S. Flynn, L. O'Mahoney, S. O'Halloran, M. Feeney, D. Morrissey, G. Thornton, G. Fitzgerald, C. Daly, B. Kiely, E. M. Quigley, G. C. O'Sullivan, F. Shanahan, and J. K. Collins.** 1999. Probiotics: from myth to reality. Demonstration of functionality in animal models of disease and in human clinical trials. *Antonie van Leeuwenhoek* **76:**279–292.
24. **Eckburg, P. B., E. M. Bik, C. N. Bernstein, E. Purdom, L. Dethlefsen, M. Sargent, S. R. Gill, K. E. Nelson, and D. A. Relman.** 2005. Diversity of the human intestinal microbial flora. *Science* **308:**1635–1638.
25. **Finegold, S. M., R. Attebury, and V. L. Sutter.** 1974. Effect of diet on human fecal flora:

comparison of Japanese and American diets. *Am. J. Clin. Nutr.* **27:**1456–1469.

26. **Flint, H. J., S. H. Duncan, K. P. Scott, and P. Louis.** 2007. Interactions and competition within the microbial community of the human colon: links between diet and health. *Environ. Microbiol.* **9:**1101–1111.
27. **Franks, A. H., H. J. M. Harmsen, G. C. Raangs, G. J. Jansen, F. Schut, and G. W. Welling.** 1998. Variations of bacterial populations in human feces measured by fluorescent in situ hybridization with group-specific 16S rRNA-targeted oligonucleotide probes. *Appl. Environ. Microbiol.* **64:**3336–3345.
28. **Fruin, A. B., O. El-Zammer, A. F. Stucchi, M. O'Brien, and J. M. Becker.** 2003. Colonic metaplasia in the ileal pouch is associated with inflammation and is not the result of long-term adaptation. *J. Gastrointest. Surg.* **7:**246–253.
29. **Gill, S. R., M. Pop, R. T. DeBoy, P. B. Eckburg, P. J. Turnbaugh, B. S. Samuel, J. I. Gordon, D. A. Relman, C. M. Fraser-Liggett, and K. E. Nelson.** 2006. Metagenomic analysis of the human distal gut microbiome. *Science* **312:**1355–1359.
30. **Gionchetti, P., F. Rizzello, U. Helwig, A. Venturi, K. Manon Lammers, P. Brigidi, B. Vitali, G. Poggioli, M. Miglioli, and M. Campieri.** 2003. Prophylaxis of pouchitis onset with probiotic therapy: a double-blind, placebo-controlled trial. *Gastroenterology* **124:**1202–1209.
31. **Gionchetti, P., F. Rizzello, A. Venturi, P. Brigidi, D. Matteuzzi, G. Bazzocchi, G. Poggiolo, M. Migioli, and M. Campieri.** 2000. Oral bacteriotherapy as maintenance treatment in patients with chronic pouchitis: a double-blind, placebo-controlled trial. *Gastroenterology* **119:**305–309.
32. **Gordon, H. A., and L. Pesti.** 1971. The gnotobiotic animal as a tool in the study of host microbial relationships. *Bacteriol. Rev.* **35:**390–429.
33. **Handelsman, J.** 2004. Metagenomics: application of genomics to uncultured microorganisms. *Microbiol. Mol. Biol. Rev.* **68:**669–685.
34. **Hardin, G.** 1960. The competitive exclusion principle. *Science* **131:**1292–1297.
35. **Harmsen, H. J. M., A. C. M. Wildeboer, G. C. Raangs, A. A. Wagendorp, N. Klijn, J. G. Bindels, and G. W. Welling.** 2000. Analysis of intestinal flora development in breast-fed and formula-fed infants by using molecular identification and detection methods. *J. Pediatr. Gastroenterol. Nutr.* **30:**61–67.
36. **Hata, Y., M. Yamamoto, M. Ohni, K. Nakajima, Y. Nakamura, and T. Takano.** 1996. A placebo-controlled study of the effect of sour milk on blood pressure in hypertensive subjects. *Am. J. Clin. Nutr.* **64:**767–771.
37. **Hayashi, H., R. Takahashi, T. Nishi, M. Sakamoto, and Y. Benno.** 2005. Molecular analysis of jejunal, ileal, caecal and recto-sigmoidal human colonic microbiota using 16S rRNA gene libraries and terminal restriction fragment length polymorphism. *J. Med. Microbiol.* **54:**1093–1101.
38. **Hillman, J. D., J. Mo, E. McDonnell, D. Cvitkovitch, and C. H. Hillman.** 2007. Modification of an effector strain for replacement therapy of dental caries to enable clinical safety trials. *J. Appl. Microbiol.* **102:**1209–1219.
39. **Hooper, L. V., M. H. Wong, A. Thelin, L. Hansson, P. G. Falk, and J. I. Gordon.** 2001. Molecular analysis of commensal host-microbial relationships in the intestine. *Science* **291:**881–884.
40. **Hooper, L. V., J. Xu, P. G. Falk, T. Midtvedt, and J. I. Gordon.** 1999. A molecular sensor that allows a gut commensal to control its nutrient foundation in a competitive ecosystem. *Proc. Natl. Acad. Sci. USA* **96:**9833–9838.
41. **Hopkin, J. M.** 1997. Mechanisms of enhanced prevalence of asthma and atopy in developed countries. *Curr. Opin. Immunol.* **9:**788–792.
42. **Isolauri, E., S. Rautava, M. Kalliomaki, P. Kirjavainen, and S. Salminen.** 2002. Role of probiotics in food hypersensitivity. *Curr. Opin. Allergy Clin. Immunol.* **2:**263–271.
43. **Isolauri, E., Y. Sutas, P. Kankaanpaa, H. Arvilommi, and S. Salminen.** 2001. Probiotics: effects on immunity. *Am. J. Clin. Nutr.* **73:**444S–450S.
44. **Kalliomaki, M., S. Salminen, H. Arvilommi, P. Kero, P. Koskinen, and E. Isolauri.** 2001. Probiotics in primary prevention of atopic disease: a randomized placebo-controlled trial. *Lancet* **357:**1076–1079.
45. **Kalliomaki, M., S. Salminen, T. Poussa, and E. Isolauri.** 2007. Probiotics during the first 7 years of life: a cumulative risk reduction of eczema in a randomized, placebo-controlled trial. *J. Allergy Clin. Immunol.* **119:**1019–1021.
46. **Kalyuzhnaya, M. G., R. Zabinsky, S. Bowerman, D. R. Baker, M. E. Lidstrom, and L. Chistoserdova.** 2006. Fluoresence in situ hybridization-flow cytometry-cell sorting-based method for separation and enrichment of type I and type II methanotroph populations. *Appl. Environ. Microbiol.* **72:**4293–4301.
47. **Katz, J. A.** 2006. Antibiotics, probiotics, prebiotics, fish oil and micronutrients (past). *Inflamm. Bowel Dis.* **12:**S12.
48. **Kell, D. B.** 2006. Metabolomics, modeling and machine learning in systems biology—towards an understanding of the languages of cells. *FEBS J.* **273:**873–894.

49. **Kell, D. B., M. Brown, H. M. Davey, W. B. Dunn, I. Spasic, and S. G. Oliver.** 2005. Metabolic footprinting and systems biology: the medium is the message. *Nat. Rev. Microbiol.* **3:**557–565.
50. **Kleeson, B., A. J. Kroesen, H. J. Buhr, and M. Blaut.** 2002. Mucosal and invading bacteria in patients with inflammatory bowel disease compared with controls. *Scand. J. Gastroenterol.* **37:** 1034–1041.
51. **Kruis, W., E. Schutz, P. Fric, B. Fixa, G. Judmaier, and M. Stolte.** 1997. Double-blind comparison of an oral Escherichia coli preparation and mesalazine in maintaining remission of ulcerative colitis. *Aliment. Pharmacol. Ther.* **11:** 853–858.
52. **Lay, C., L. Rigottier-Gois, K. Holmstrom, M. Rajilic, E. E. Vaughan, W. M. de Vos, M. D. Collins, R. Thiel, P. Namsolleck, M. Blaut, and J. Dore.** 2005. Colonic microbiota signatures across five northern European countries. *Appl. Environ. Microbiol.* **71:**4153–4155.
53. **Lee, A.** 1999. *Helicobacter pylori:* opportunistic member of the normal microflora or agent of communicable disease? p. 128–163. *In* G. W. Tannock (ed.), *Medical Importance of the Normal Microflora.* Kluwer Academic Publishers, Dordrecht, The Netherlands.
54. **Lepage, P., P. Seksik, M. Sutren, M. F. de la Cochetiere, R. Jian, P. Marteau, and J. Dore.** 2005. Biodiversity of the mucosa-associated microbiota is stable along the distal digestive tract in healthy individuals and patients with IBD. *Inflamm. Bowel Dis.* **11:**473–480.
55. **Lesniewska, V., I. Rowland, P. D. Cani, A. M. Neyrinck, N. M. Delzenne, and P. J. Naughton.** 2006. Effect on components of the intestinal microflora and plasma neuropeptide levels of feeding *Lactobacillus delbrueckii, Bifidobacterium lactis,* and inulin to adult and elderly rats. *Appl. Environ. Microbiol.* **72:**6533–6538.
56. **Macdonald, T. T., and G. Monteleone.** 2005. Immunity, inflammation, and allergy in the gut. *Science* **307:**1920–1925.
57. **Macfarlane, S., E. Furrie, J. H. Cummings, and G. T. Macfarlane.** 2004. Chemotaxonomic analysis of bacterial populations colonizing the rectal mucosa in patients with ulcerative colitis. *Clin. Infect. Dis.* **38:**1690–1699.
58. **Macfarlane, S., and G. T. Macfarlane.** 2006. Composition and metabolic activities of bacteria biofilms colonizing food residues in the human gut. *Appl. Environ. Microbiol.* **72:**6204–6211.
59. **Madden, J. A. J., and J. O. Hunter.** 2002. A review of the role of the gut microflora in irritable bowel syndrome and the effects of probiotics. *Br. J. Nutr.* **88:**S67–S72.
60. **Madden, M. V., A. S. McIntyre, and R. J. Nicholls.** 1994. Double-blind crossover trial of metronidazole versus placebo in chronic unremitting pouchitis. *Dig. Dis. Sci.* **39:**1193–1196.
61. **Manichanh, C., L. Rigottier-Gois, E. Bonnaud, K. Gloux, E. Pelletier, L. Frangeul, R. Nalin, C. Jarrin, P. Chardon, P. Marteau, J. Roca, and J. Dore.** 2005. Reduced diversity of faecal microbiota in Crohn's disease revealed by a metagenomic approach. *Gut* **55:**205–211.
62. **Marteau, P., P. Seksik, and R. Jian.** 2002. Probiotics and intestinal disease: a clinical perspective. *Br. J. Nutr.* **88:**S51–S57.
63. **Matricardi, P. M.** 2002. Probiotics against allergy: data, doubts, and perspectives. *Allergy* **57:** 185–187.
64. **McBee, R. H.** 1977. Fermentation in the hindgut, p. 185–222. *In* R. T. J. Clarke and T. Bauchop (ed.), *Microbial Ecology of the Gut.* Academic Press, London, United Kingdom.
65. **McCarthy, J., L. O'Mahony, L. O'Callaghan, B. Sheil, E. E. Vaughan, N. Fitzsimons, J. Fitzgibbon, G. C. O'Sullivan, B. Kiely, J. K. Collins, and F. Shanahan.** 2003. Double blind, placebo controlled trial of two probiotic strains in interleukin 10 knockout mice and mechanistic link with cytokine balance. *Gut* **52:**975–980.
66. **Metchnikoff, E.** 1907. *The Prolongation of Life. Optimistic Studies.* William Heinemann, London, United Kingdom.
67. **Metchnikoff, E.** 1908. *The Nature of Man. Studies in Optimistic Philosophy.* William Heinemann, London, United Kingdom.
68. **Miller, T. L., and M. J. Wolin.** 1983. Stability of *Methanobacter smithii* populations in the microbial flora excreted from the human large bowel. *Appl. Environ. Microbiol.* **45:**317–318.
69. **Mimura, T., F. Rizzello, U. Helwig, G. Poggioli, S. Schreiber, I. C. Talbot, R. J. Nicholls, P. Giochetti, M. Campieri, and M. A. Kamm.** 2004. Once daily high dose probiotic therapy (VSL#3) for maintaining remission in recurrent or refractory pouchitis. *Gut* **53:**108–114.
70. **Moore, W. E. C., and L. V. Holdeman.** 1974. Special problems associated with the isolation and identification of intestinal bacteria in fecal flora studies. *Am. J. Clin. Nutr.* **27:**1450–1455.
71. **Murch, S. H.** 2005. Probiotics as mainstream allergy therapy? *Arch. Dis. Child.* **90:**881–882.
72. **Murray, C. S., and A. Woodcock.** 2002. Gut microflora and atopic disease, p. 239–261. *In* G. W. Tannock (ed.), *Probiotics and Prebiotics: Where Are We Going?* Caister Academic Press, Wymondham, United Kingdom.

73. **Mylonaki, M., N. B. Rayment, D. S. Rampton, B. N. Hudspith, and J. Brostoff.** 2005. Molecular characterization of rectal mucosa-associated bacterial flora in inflammatory bowel disease. *Inflamm. Bowel Dis.* **11:**481–487.

74. **Nanji, A. A., U. Khettry, and S. M. H. Sadrazadeh.** 1994. Lactobacillus feeding reduces endotoxaemia and severity of experimental alcoholic liver (disease). *Proc. Soc. Exp. Biol. Med.* **205:**243–247.

75. **Nicholson, J. K., E. Holmes, J. C. Lindon, and I. D.Wilson.** 2004. The challenges of modeling mammalian biocomplexity. *Nat. Biotechnol.* **22:**1268–1274.

76. **Norin, K. E., B. E. Gustafsson, B. S. Lindblad, and T. Midtvedt.** 1985. The establishment of some microflora associated biochemical characteristics in feces from children during the first years of life. *Acta Paediatr. Scand.* **74:**207–212.

77. **Oozeer, R., J. P. Furet, N. Goupli-Feuillerat, J. Anba, J. Mengaud, and G. Corthier.** 2005. Differential activities of four *Lactobacillus casei* promoters during bacterial transit through the gastrointestinal tracts of human-microbiota-associated mice. *Appl. Environ. Microbiol.* **71:**1356–1363.

78. **Ott, S. J., M. Musfeldt, D. F. Wenderoth, J. Hampe, O. Brant, U. R. Folsch, K. N. Timmis, and S. Schreiber.** 2004. Reduction in diversity of the colonic mucosa associated bacterial microflora in patients with active inflammatory bowel disease. *Gut* **53:**685–693.

79. **Ouwehand, A. C., E. Isolauri, F. He, H. Hashimoto, Y. Benno, and S. Salminen.** 2001. Differences in *Bifidobacterium* flora composition in allergic and healthy infants. *J. Allergy Clin. Immunol.* **108:**144–145.

80. **Parks, T. P., Q. Xu, L. A. Lagenaur, and P. P. Lee.** 2005. Bacterial therapeutics for the treatment and prevention of urogenital infections, p. 171–194. *In* G. W. Tannock (ed.), *Probiotics and Prebiotics: Scientific Aspects.* Caister Academic Press, Wymondham, United Kingdom.

81. **Podolsky, D. K.** 2002. Inflammatory bowel disease. *N. Engl. J. Med.* **347:**417–428.

82. **Prescott, S. L., C. Macaubus, B. J. Holt, T. B. Smallacombe, R. Loh, P. D. Sly, and P. G. Holt.** 1998. Transplacental priming of the human immune system to environmental allergens: universal skewing of initial T cell responses toward the Th2 cytokine profile. *J. Immunol.* **160:**4730–4737.

83. **Prindiville, T., M. Cantrell, and K. H. Wilson.** 2004. Ribosomal DNA sequence analysis of mucosa-associated bacteria in Crohn's disease. *Inflamm. Bowel Dis.* **10:**824–833.

84. **Reid, G.** 2002. Probiotics for urogenital health. *Nutr. Clin. Care* **5:**3–8.

85. **Reid, G., D. Beuerman, C. Heineman, and A. W. Bruce.** 2001. Probiotic *Lactobacillus* dose required to restore and maintain a normal vaginal flora. *FEMS Immunol. Med. Microbiol.* **32:**37–41.

86. **Reid, G., M. E. Sanders, H. R. Gaskins, G. R. Gibson, A. Mercenier, R. A. Rastall, M. B. Roberforid, I. Rowland, C. Cherbut, and T. R. Klaenhammer.** 2003. New scientific paradigms for probiotics and prebiotics. *J. Clin. Gastroenterol.* **37:**105–118.

87. **Rinne, M., M. Kalliomaki, S. Salminen, and E. Isolauri.** 2006. Probiotic intervention in the first months of life: short-term effects on gastrointestinal symptoms and long-term effects on gut microbiota. *J. Pediatr. Gastroenterol.* **43:**200–205.

88. **Roberton, A. M., and A. P. Corfield.** 1999. Mucin degradation and its significance in inflammatory conditions of the gastrointestinal tract, p. 222–261. *In* G. W. Tannock (ed.), *Medical Importance of the Normal Microflora.* Kluwer Academic Publishers, Dordrecht, The Netherlands.

89. **Rochet, V., L. Rigottier-Gois, M. Sutren, M.-N. Krementscki, C. Andrieux, J. P. Furet, P. Tailliez, F. Levenez, A. Mogenet, J.-L. Bresson, S. Méance, C. Cayuela, A. Leplingard, and J. Doré.** 2006. Effects of orally administered *Lactobacillus casei* DN-114001 on the composition or activities of the dominant faecal microbiota in healthy humans. *Br. J. Nutr.* **95:** 421–429.

90. **Roediger, W. E.** 1980. Role of anaerobic bacteria in the metabolic welfare of the colonic mucosa in man. *Gut* **21:**793–798.

91. **Rousseaux, C., X. Thuru, A. Gelot, N. Barnich, C. Neut, L. Dubuquoy, C. Dubuquoy, E. Merour, K. Geboes, M. Chamaillard, A. Ouwehand, G. Leyer, D. Carcano, J. F. Colombel, D. Ardid, and P. Desreumaux.** 2007. *Lactobacillus acidophilus* modulates intestinal pain and induces opioid and cannabinoid receptors. *Nat. Med.* **13:**35–37.

92. **Sakamoto, I., M. Igarashi, K. Kimura, A. Takagai, T. Miwa, and Y. Koga.** 2001. Suppressive effect of *Lactobacillus gasseri* OLL 2716 (LG21) on *Helicobacter pylori* infection in humans. *J. Antimicrob. Chemother.* **47:**709–710.

93. **Samuel, B. S., and J. I. Gordon.** 2006. A humanized gnotobiotic mouse model of host-archaeal-bacterial mutualism. *Proc. Natl. Acad. Sci. USA* **103:**10011–10016.

94. **Sandborn, W. J.** 1994. Pouchitis following ileal pouch-anal anastomosis: definition, pathogenesis, and treatment. *Gastroenterology* **107:**1856–1860.

95. **Sanders, M. E.** 2000. Considerations for use of probiotic bacteria to modulate human health. *J. Nutr.* **130:**384S–390S.

96. **Santosa, S., E. Farnworth, and P. J. H. Jones.** 2006. Probiotics and their potential health claims. *Nutr. Rev.* **64:**265–274.

97. **Sartor, R. B.** 2004. Microbial influences in inflammatory bowel disease: role in pathogenesis and clinical implications, p. 138–162. *In* R. B. Sartor and W. J. Sandborn (ed.), *Kirstner's Inflammatory Bowel Diseases.* Elsevier Publishers, London, United Kingdom.

98. **Satokari, R. M., E. E. Vaughan, A. D. L. Akkermans, M. Saarela, and W. M. de Vos.** 2001. Bifidobacterial diversity in human feces detected by genus-specific PCR and denaturing gradient gel electrophoresis. *Appl. Environ. Microbiol.* **67:**504–513.

99. **Savage, D. C.** 1977. Microbial ecology of the gastrointestinal tract. *Annu. Rev. Microbiol.* **31:** 107–133.

100. **Savage, D. C., R. Dubos, and R. W. Schaedler.** 1968. The gastrointestinal epithelium and its autochthonous bacterial flora. *J. Exp. Med.* **127:**67–76.

101. **Schell, M. A., M. Karamirantzou, B. Snel, D. Vilanova, B. Berger, G. Pessi, M. C. Zwahlen, F. Desiere, P. Bork, M. Delby, and R. D. Pridmore.** 2002. The genome sequence of *Bifidobacterium longum* reflects its adaptation to the human gastrointestinal tract. *Proc. Natl. Acad. Sci. USA* **99:**14422–14427.

102. **Sepp, E., K. Julge, M. Vasar, P. Naaber, B. Bjorksten, and M. Mikelsaar.** 1997. Intestinal microflora of Estonian and Swedish infants. *Acta Paediatr.* **86:**856–961.

103. **Shanahan, F.** 2002. Probiotics and inflammatory bowel disease: from fads and fancy to facts and future. *Br. J. Nutr.* **88:**S5–S9.

104. **Shanahan, F., F. F. Guarner, A. Von Wright, T. Vilpponen-Salmela, D. O'Donoghue, and B. Kiely.** 2006. A one year, randomized, double-blind, placebo controlled trial of a Lactobacillus or a Bifidobacterium probiotic for maintenance of steroid-induced remission of ulcerative colitis. *Gastroenterology* **130:**A44.

105. **Simon, G. L., and S. L. Gorbach.** 1984. Intestinal flora in health and disease. *Gastroenterology* **86:**174–193.

106. **Spanhaak, S., R. Havenaar, and G. Schaafsma.** 1998. The effect of consumption of milk fermented by *Lactobacillus casei* Shirota on the intestinal microflora and immune parameters in humans. *Eur. J. Clin. Nutr.* **52:**899–907.

107. **Stappenbeck, T. S., L. V. Hooper, and J. I. Gordon.** 2002. Developmental regulation of intestinal angiogenesis by indigenous microbes via Paneth cells. *Proc. Natl. Acad. Sci. USA* **99:** 15451–15455.

108. **Stebbings, S., K. Munro, M. A. Simon, G. Tannock, J. Highton, H. Harmsen, G. Welling, P. Seksik, J. Dore, G. Grame, and A. Tilsala-Timisjarvi.** 2002. Comparison of the faecal microflora of patients with ankylosing spondylitis and controls using molecular methods of analysis. *Rheumatology* **41:**1395–1401.

109. **Strachan, D. P.** 1989. Hay fever, hygiene, and household size. *Br. Med. J.* **299:**1259–1260.

110. **Suau, A., R. Bonnet, M. Sutren, J.-J. Godon, G. R. Gibson, M. D. Collins, and J. Dore.** 1999. Direct analysis of genes encoding 16S rRNA from complex communities reveals many novel molecular species within the human gut. *Appl. Environ. Microbiol.* **65:**4799–4807.

111. **Tannock, G. W.** 2003. Probiotics: time for a dose of realism. *Curr. Issues Intest. Microbiol.* **4:** 33–42.

112. **Tannock, G. W.** 2003. The intestinal microflora, p. 1–23. *In* R. Fuller and G. Perdigon (ed.), *Gut Flora. Nutrition, Immunity and Health.* Blackwell Press, Oxford, United Kingdom.

113. **Tannock, G. W.** 2004. A special fondness for lactobacilli. *Appl. Environ. Microbiol.* **70:**3189–3194.

114. **Tannock, G. W., K. Munro, H. J. M. Harmsen, G. W. Welling, J. Smart, and P. K. Gopal.** 2000. Analysis of the fecal microflora of human subjects consuming a probiotic product containing *Lactobacillus rhamnosus* DR20. *Appl. Environ. Microbiol.* **66:**2578–2588.

115. **Thompson-Fawcett, M. W.** 2003. Pouchitis and pouch dysfunction, p. 567–585. *In* J. Satsangi and L. R. Sutherland (ed.), *Inflammatory Bowel Diseases.* Elsevier, New York, NY.

116. **Thompson-Fawcett, M. W., V. Marcus, M. Redston, Z. Cohen, and R. S. McLeod.** 2001. Risk of dysplasia in long-term ileal pouches and pouches with chronic pouchitis. *Gastroenterology* **121:**275–281.

117. **Thompson-Fawcett, M. W., V. A. Marcus, M. Redston, Z. Cohen, and R. S. McLeod.** 2001. Adenomatous polyps develop commonly in the ileal pouch of patients with familial adenomatous polyposis. *Dis. Colon Rectum* **44:**347–353.

118. **Tiveljung, A., J. D. Soderholm, G. Olaison, J. Jonasson, and H.-J. Monstein.** 1999. Presence of eubacteria in biopsies from Crohn's disease inflammatory lesions as determined by 16S rRNA gene-based PCR. *J. Med. Microbiol.* **48:**263–268.

119. **Tyson, G. W., J. Chapman, P. Hugenholtz, E. E. Allen, R. J. Ram, P. M. Richardson, V. V. Solovyev, E. M. Rubin, D. S. Rokhsar, and J. F. Banfield.** 2004. Community structure and metabolism through re-

construction of microbial genomes from the environment. *Nature* **428:**37–43.

120. **van der Werf, M. J., R. H. Jellema, and T. Hankemeier.** 2005. Microbial metabolomics: replacing trial-and-error by the unbiased selection and ranking of targets. *J. Ind. Microbiol. Biotechnol.* **32:**234–252.

121. **Venturi, A., P. Gionchetti, F. Rizzello, R. Johansson, E. Zucconi, P. Brigidi, D. Matteuzzi, and M. Campieri.** 1999. Impact on the composition of the faecal flora by a new probiotic preparation: preliminary data on maintenance treatment of patients with ulcerative colitis. *Aliment. Pharmacol. Ther.* **13:**1103–1108.

122. **Viljanen, M., E. Pohjavuori, T. Haahtela, R. Korpela, M. Kuitunen, A. Sarnesto, O. Vaarala, and E. Savilahti.** 2005. Induction of inflammation as a possible mechanism of the probiotic effect in atopic eczema-dermatitis syndrome. *J. Allergy Clin. Immunol.* **115:**1254–1259.

123. **Viljanen, M., E. Savilahti, T. Haahtela, K. Juntunen-Backman, R. Korpela, T. Poussa, T. Tuure, and M. Kuitunen.** 2005. Probiotics in the treatment of atopic eczema/dermatitis syndrome in infants: a double-blind placebo-controlled trial. *Allergy* **60:**494–500.

124. **Walter, J., C. Hertel, G. W. Tannock, C. M. Lis, K. Munro, and W. P. Hammes.** 2001. Detection of *Lactobacillus, Pediococcus, Leuconostoc,* and *Weissella* species in human feces by using group-specific PCR primers and denaturing gradient gel electrophoresis. *Appl. Environ. Microbiol.* **67:**2578–2585.

125. **Walter, J., M. Mangold, and G. W. Tannock.** 2005. Construction, analysis and β-glucanase screen of a bacterial artificial chromosome library from the large bowel microbiota of mice. *Appl. Environ. Microbiol.* **71:**2347–2354.

126. **Wang, M., S. Ahrne, B. Jeppsson, and G. Molin.** 2005. Comparison of bacterial diversity along the human intestinal tract by direct cloning and sequencing of 16S rRNA genes. *FEMS Microbiol. Ecol.* **54:**219–231.

127. **Wang, Y., E. Holmes, J. K. Nicholson, O. Cloarec, J. Chollet, M. Tanner, B. H. Singer, and J. Utzinger.** 2004. Metabonomic investigations of mice infected with *Schistosoma mansoni:* an approach for biomarker identification. *Proc. Natl. Acad. Sci. USA* **101:**12676–12681.

128. **Wollowski, I., G. Rechkemmer, and B. L. Pool-Zobel.** 2001. Protective role of probiotics and prebiotics in colon cancer. *Am. J. Clin. Nutr.* **73:**451S–455S.

129. **Xu, J., M. K. Bjursell, J. Himrod, S. Deng, L. K. Carmichael, H. C. Chiang, L. V. Hooper, and J. I. Gordon.** 2003. A genomic view of the human-*Bacteroides thetaiotaomicron* symbiosis. *Science* **299:**2074–2076.

130. **Yabuhara, A., C. Macaubas, S. L. Prescott, T. J. Venaille, B. J. Holt, W. Habre, P. D. Sly, and P. G. Holt.** 1997. Th2-polarized immunological memory to inhalant allergens in atopics is established during infancy and early childhood. *Clin. Exp. Allergy* **27:**1237–1239.

131. **Yamano, T., H. Iino, M. Takada, S. Blum, F. Rochat, and Y. Fukushima.** 2006. Improvement of the human intestinal flora by ingestion of the probiotic strain *Lactobacillus johnsonii* La1. *Br. J. Nutr.* **95:**303–312.

132. **Young, S. L., M. A. Simon, M. A. Baird, G. W. Tannock, R. Bibiloni, K. Spencely, J. M. Lane, P. Fitzharris, J. Crane, I. Town, E. Addo-Yobo, C. S. Murray, and A. Woodcock.** 2004. Bifidobacterial species differentially affect expression of cell surface markers and cytokines of dendritic cells harvested from cord blood. *Clin. Diagn. Lab. Immunol.* **11:**686–690.

133. **Zoetendal, E. G., A. D. Akkermans, and W. M. de Vos.** 1998. Temperature gradient gel electrophoresis analysis of 16S rRNA from human fecal samples reveals stable and host specific communities of active bacteria. *Appl. Environ. Microbiol.* **64:**3854–3859.

# INTERACTIONS BETWEEN ENVIRONMENTAL MICROBIAL ECOSYSTEMS AND HUMANS: THE CASE OF THE WATER ENVIRONMENT AND ANTIBIOTIC RESISTANCE

*Chuanwu Xi, Kathleen Bush, Karen L. Lachmayr, Yongli Zhang, and Timothy E. Ford*

# 5

Since microorganisms first emerged on Earth approximately 3.5 billion years ago, they have evolved to become very diverse and inhabit most environments on this planet. They can be found in even the most inhospitable environments demonstrating highly specialized metabolic adaptations. For example, some, such as the iron-oxidizing *Ferroplasma,* can survive at the extremely acidic pH of 0 (53). Others have adapted to inhabit anoxic mud, hot springs, salt flats, roots of plants, guts of animals, and ocean depths up to 10,000 m (22). It was estimated that the number of prokaryotes on Earth is $4 \times 10^{30}$ to $6 \times 10^{30}$ and comprises millions of species (78). Microscopy and cultivation-based methods provide only a small glimpse of the overall diversity in the microbial world; the recent development of molecular methods (72), for example, 16S rRNA gene sequencing and metagenomics, allows for a far more detailed assessment of microbial diversity in a vast array of environments. The list of discovered species, including pathogens, will continue to increase as new detection methods are developed and applied to explore the microbial diversity in the environment.

The human body not only provides a unique environment and diverse range of habitats for microorganisms but also is continuously exposed to and inoculated with billions of microorganisms from the outside environment. One of the major sources of this exposure is food intake, and humans consume thousands of times more microorganisms from food than from the drinking water; however, waterborne microorganisms are also significant sources of inoculation and present some unique concerns with respect to the environment's contribution to human microbiota and health. The human body is exposed to waterborne microorganisms mainly through ingestion of drinking water but also through other activities of daily life. These routes include inhalation, food intake, and direct contact with microorganisms during showering, swimming, and other recreational activities (Fig. 1).

Completing the cycle, humans alter environmental microbial ecosystems by inoculation and the addition of selective pressures. The most direct inoculation of microbial eco-

*Chuanwu Xi, Kathleen Bush, and Yongli Zhang,* Department of Environmental Health Sciences, School of Public Health, University of Michigan, 109 Observatory St., 6626 SPHI, Ann Arbor, MI 48109. *Karen L. Lachmayr.* Department of Organismic and Evolutionary Biology, Harvard University, 16 Divinity Ave., Room 4083, Cambridge, MA 02138. *Timothy E. Ford,* University of New England, Pickus Hall, Room 105, Biddeford, ME 04005.

*Food-Borne Microbes: Shaping the Host Ecosystem,* Edited by L.-A. Jaykus, H. H. Wang, and L. S. Schlesinger, © 2009 ASM Press, Washington, DC

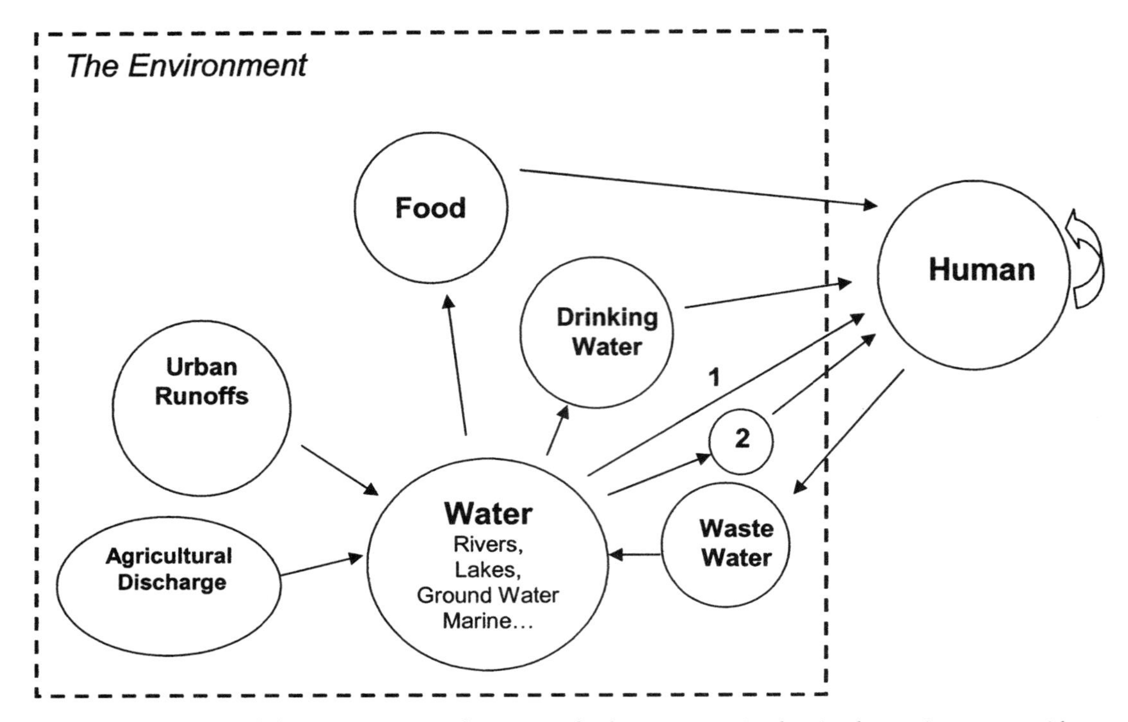


**FIGURE 1** A model for major routes of exposure for humans to microbes in the environment, with an emphasis on the water cycle. The direction of arrows indicates the movement of microbes in the system. 1, recreational water direct exposure; 2, other indirect exposures.

systems by humans occurs through wastewater, which introduces high concentrations of bacteria and genetic material into the environment. The other significant route of inoculation occurs through agricultural discharge, particularly animal waste. Wastewater, agricultural discharge, and urban runoff, including runoff from landfills, also alter microbial ecosystems through the leaching of compounds that exert selective pressures, for example, metals, antibiotics, and organics. Other selective pressures on microbial communities are introduced through anthropogenically induced temperature fluctuations and the physical remodeling of the environment, including deforestation, erosion, dredging, and flooding. Ultimately, it is diverse environmental microbial communities, augmented by human influences, which inoculate the host, predominantly via food and water consumption and to a lesser extent by other routes of exposure.

Not surprisingly, of primary concern is the interaction of waterborne pathogens and human populations. The persistence and proliferation of human pathogens in water are well-documented causes of disease, and a number of comprehensive journal and book reviews exist on waterborne diseases and their associated microbial agents (10, 37, 69, 71). In addition, numerous food-borne diseases can also be attributed to the contamination of water used for irrigation and food processing. However, few studies exist on the potential beneficial effects of waterborne microbes on human health, e.g., the contribution of waterborne microbes to the establishment of diverse human microbiota. Better characterization and comparison of microbial communities in the diverse water systems with which humans interact, namely, drinking water, surface water (wastewater), and marine water, will contribute to a better understanding of both adverse and potential beneficial effects of natural microbiota on human health.

In this chapter, we briefly review current knowledge of microbial diversity in different aquatic environments, with a focus on general bacterial populations. This broader understanding of the composition of microbial communities is primarily a result of 16S rRNA gene sequence analysis. We use the case of antibiotic resistance to demonstrate how host ecosystems are influenced by external microbial communities and, in turn, how hosts influence microbial ecology. Lastly, we discuss knowledge gaps and suggest future research directions to address the link between microbial populations in the aquatic environment and the human microbiota.

## MICROBIAL DIVERSITY IN AQUATIC ENVIRONMENTS

Water shows extreme variability with respect to chemistry, light, temperature, nutrients, and biological composition, all of which promote a diverse community of microorganisms. These changes between light and dark and between oxidized and reduced conditions make water a dynamic environment. Bacteria, archaea, fungi, viruses and phages, and protists (protozoans and algae) are the major types of microorganisms commonly found in water. Previous culture-based methods provided glimpses of the diversity and abundance of bacterial populations in different aquatic environments (57). However, molecular data (including 16S rRNA gene sequence analysis) have provided us with additional interesting findings and a better understanding of the extraordinary complexity and vastness of microbial communities (24, 80).

The cellular density in different aquatic environments has been reported to be in the range of $10^4$ to $10^7$ cells/ml (see review in reference 78). In freshwaters and saline lakes, the average density of prokaryotes is about $10^6$ cells/ml, with an estimated total number of $2.3 \times 10^{26}$ cells in these lakes. In marine waters, the average cellular density is about $5 \times 10^5$ cells/ml for the continental shelf and the upper 200 m of the open ocean and $5 \times 10^4$ cells/ml for deep (>200 m) ocean. The total estimated number of cells in upper oceanic waters is about $3.6 \times 10^{28}$.

Exemplifying the more intricate and complex picture provided by 16S data, Hahn recently reviewed the microbial diversity of freshwater habitats (23) and found that most bacterial sequences retrieved from these waters were novel and not represented by phylogenetic groups previously obtained using culture-based methods. Data collected using modern tools also suggest that the bacterial diversity in these aquatic environments is rather ubiquitous compared to that in soil environments (19). Zwart et al. found nearly identical 16S rRNA sequences recovered from lakes in North America and Europe (80). Using habitat-specific cluster analysis of available 16S rRNA gene sequences from plankton of lakes and rivers, Zwart et al. found that the majority of bacterial sequences were most closely related to other freshwater clones or isolates (79). The authors have listed 34 phylogenetic clusters of closely related sequences that are either restricted to freshwater or dominated by freshwater sequences, which they called "typical freshwater bacteria," suggesting the existence of clades of globally distributed freshwater bacteria. Additionally, the abundance of 16S rRNA gene sequences shared by clones or isolates from other environments, such as soil and marine water, indicates that those bacteria are able to survive in a wide range of habitats (79).

Groundwater is generally believed to have less microbial contamination than other freshwater sources and in many places is the only source for drinking water. However, very few studies have investigated the general bacterial diversity of groundwater. One study obtained 247 16S rRNA gene sequences from groundwater and clustered these sequences into 37 operational taxonomic units (44). The majority of these were affiliated with the *Betaproteobacteria* and candidate divisions OD1 and OP11.

Although water is not consumed directly from the marine environment, human intake concerns often revolve around the microbial contamination of seafood. Recreational use of marine water is another potential route of exposure to ocean-dwelling microorganisms. In the marine environment, certain areas of the ocean seem to be dominated by particular groups of microorganisms. In the upper surface of the open ocean, bacteria are dominant and comprise about 80% of total prokaryotic cells; up to 50% of the total bacterial community consists of members of the SAR11 clade (30, 46). The archaeal population increases as depth increases, and members of the archaeal phylum *Crenarchaeota* were shown to comprise 39% of the total microbial population in the deep ocean. The metagenomics study of the marine planktonic microbiota which was done as part of the *Sorcerer II* global ocean sampling expedition revealed an overall consistent taxonomic distribution of ribotypes compared to previous PCR-based 16S rRNA gene sequencing studies of marine environments (59) and a smaller amount of novel sequences (16% of ribotypes and 3.4% of sequences) at the family level. The most dominant phyla or classes included *Alphaproteobacteria* (32%), *unclassified Proteobacteria* (15.5%), *Gammaproteobacteria* (13.2%), *Bacteroidetes* (13%), *Cyanobacteria* (7.9%), *Firmicutes* (7.5%), *Actinobacteria* (1.7%), and another five (less than 1%) (59).

## MICROBIAL DIVERSITY IN DRINKING WATER AND WATER DISTRIBUTION SYSTEMS

There are many reviews and reports on the microbiological safety of drinking water (20, 21, 69), and the reliable and abundant supply of safe and clean drinking water is a major, and relatively recent, public health milestone. Removal of microorganisms from source water normally is achieved through various treatment steps, including coagulation, filtration, and disinfection. Control of microbial regrowth in drinking water distribution systems is often achieved through the addition of disinfectants with residual activity and the reduction of biodegradable organic matter. Normally, operators use indicator species like total coliforms or heterotropic plate counts (HPC) to monitor drinking water quality.

Although drinking water is of high quality in developed countries, a number of outbreaks of waterborne diseases attributed to drinking water are still reported every year (41, 64, 67). One report estimates that 26.0 million infections/year and 13.0 million illnesses/year occur as a result of municipal surface water systems in the United States alone (56), which suggests that contaminated potable drinking water is still a major public health concern. A better understanding of the microbial diversity of drinking water is necessary to design innovative and effective control strategies that will ensure safe and high-quality drinking water. The conventional approaches to drinking water treatment are often ineffective at controlling microbial regrowth (36). Berry et al. reviewed the complexities associated with controlling microbial regrowth in drinking water (5) and suggested that biofilms grown on pipe surfaces contributed to several issues of microbial quality of drinking water. HPC in drinking water vary from less than 1 to $10^4$ CFU per ml. Actual counts depend on the quality of the source water, the types and efficacy of treatment, the type and concentration of disinfection residuals, the age and the condition of the storage and distribution system, the concentration of dissolved organics in the treated drinking water, the ambient temperature of the raw and finished water, the travel time between the water treatment plant and sampling locations, and, of course, the HPC detection method and time and temperature of incubation (2, 15). Based on the species composition of isolated colonies, a wide range of bacterial genera comprise HPC, including bacterial pathogens (2, 15).

Most research using molecular techniques to study the microbiology of drinking water focuses on pathogens and indicators. One of few reported studies using 16S rRNA gene sequencing analysis found that gram-positive bacteria and *Alpha-, Beta-,* and *Gammaproteobacteria* were the major groups representing the heterotrophic bacteria in drinking water (73). However, the limited number of sequences retrieved in this study (approximately 100) limited generalizability regarding the broader picture of microbial diversity in drinking water. Another study characterized the bacterial communities in biofilms in drinking water distribution systems using 16S rRNA gene sequence analysis and found that the majority of the microbes were closely related to the *Proteobacteria* and only a small fraction were highly homologous to rRNA sequences from *Actinobacteria,* low-G+C gram-positive organisms, and the *Cytophaga-Flavobacterium-Bacteroides* group (62).

## MICROBIAL DIVERSITY IN WASTEWATER

Municipal and industrial sewage contains a variety of chemical and microbial contaminants, and wastewater treatment is one of the most important engineered systems to ameliorate potential anthropogenic damage to the environment. The cell density in aerobic tanks is very high ($10^6$ to $10^7$ CFU/ml). In conventional municipal wastewater treatment, the total concentration of microbes discharged into receiving waters is reduced by settlement and/or chlorine (or UV) disinfection, while organic matter and nutrients are removed by the activated sludge process. Ultimately, this method of wastewater treatment can reduce the cellular density by 4 to 6 logs. However, a significant concentration of microbes is still released via effluent into the environment. In some municipal treatment plants, the final disinfection process is not required. There are also regions in the United States with combined sewage systems, where a large portion of the sewage is released into receiving waters without any treatment, especially under extreme weather conditions, such as heavy rain- or snowfall, when the systems are more likely to overflow.

The complex milieu of wastewater, characterized by high bacterial loading and other selective pressures, leads to interesting microbial interactions and community structures that are constantly in flux. Scientists have found that a particularly large portion of sewage isolates harbor plasmids (61). In addition, studies have confirmed that horizontal gene transfer in both raw and treated sewage does, in fact, take place (42, 60, 70). There is also some evidence to suggest that the stress of the changing wastewater environment causes some bacteria to enter a viable but nonculturable state (8, 47). Most research on the microbiology of wastewater treatment has focused on pathogens and specific groups of organisms that impact treatment efficiency. Some 16S rRNA gene sequencing analysis of general microbial diversity in activated sludge and biofilm systems has been reported since 1995 (see reviews in references 76 and 77). From the total 750 16S rRNA gene sequences analyzed, the most frequently retrieved sequences were affiliated with the phyla *Beta-, Alpha-,* and *Gammaproteobacteria; Bacteroidetes; Actinobacteria; Planctomycetes; Firmicutes;* and *Chloroflexi* and *Thermomicrobia.*

One source of selective pressure in wastewater is the presence of various household and industrial chemicals that flow down the drain. For example, pharmaceuticals, household cleaners, personal care products, and food additives, such as caffeine, are likely to persist in significant concentrations in wastewater. Even the world's most modern sewage treatment plants are not designed to remove these so-called "micropollutants." However, absorption to activated sludge, particularly by less polar compounds, could serve as one mechanism of removal (25). This raises questions with regard to sludge management, since processed sludge is often applied to crops as fertilizer. Clearly much more research is needed to understand how humans may be shaping microbial communities by the release of various chemicals into wastewater and ultimately the environment.

## ANTIBIOTIC RESISTANCE IN THE WATER ENVIRONMENT

The occurrence and spread of multidrug-resistant bacteria currently constitute one of the world's most pressing public health problems. In the past 60 years, since the onset of therapeutic antibiotic use, antibiotic resistance among pathogens in a clinical setting has been well documented. However, antibiotic resistance genes can also occur in nonpathogenic bacteria, which can then be transferred via lateral gene transfer (39, 40). Therefore, knowledge of the environmental distribution of such genes in all bacteria is also important. The results of recent studies suggest that the aquatic environment acts as a reservoir for antibiotic resistance genes, which is intensified by anthropogenic effects.

Today's society uses enormous quantities of antibiotics for a variety of purposes. The total annual use of antibiotics in the United States alone is estimated to be 38,720,000 lbs. The breakdown of this estimate for specific applications includes outpatient medicine (18%), inpatient medicine (2.3%), nontherapeutic animal use (71%), therapeutic animal use (7.8%), aquaculture (0.82%), and on crops (0.15%) (35). Clearly, the majority of antibiotics used in the United States are fed to animals for purposes of growth promotion and disease prevention, i.e., nontherapeutic animal use. After antibiotics serve their intended purpose and leave the treated person, animal, or plant, they generally enter the environment still in their active form. Ultimately, antibiotics are likely to end up in aquatic ecosystems, via agricultural discharge/surface runoff, the leaching of landfills, or water used for aquaculture or from sewage.

The presence of antibiotics in aquatic environments has, in fact, been detected (33). Despite the fact that most antibiotics are naturally occurring compounds, measured quantities can be assumed to be of anthropogenic origin, since in situ production is minimal. The concentration of several antibiotics, for example, erythromycin, roxithromycin, and sulfamethoxazole, in wastewater effluents and surface waters can reach several micrograms per liter (25, 32). Generally, the concentration of antibiotics in surface waters is 1 order of magnitude less than the median concentrations reported in sewage treatment effluent (25).

The presence of antibiotics is, of course, a powerful selective pressure for the propagation of resistance. The emergence of antibiotic resistance has time and again been linked to the use of antibiotics. For example, even though β-lactamases are estimated to have existed for the past two billion years, their evolution and spread have been highly correlated to the anthropogenic development and prolificacy of β-lactam antibiotics in the past half-century (6). In one British hospital, the incidence of β-lactamase-producing *Staphylococcus aureus* rose from less than 8% to almost 60% in the 4- to 5-year period immediately following World War II, when administering penicillin was becoming routine practice (3). Similarly, in the aquatic environment, released antibiotics may exert a selective pressure on microorganisms carrying antibiotic resistance genes, thus promoting the growth of antibiotic-resistant bacteria (ARB) and the further emergence and spread of antibiotic resistance genes. Evidence linking antibiotic use to the emergence of resistance in aquatic systems is now documented. For example, an association has been observed between the frequent use of sulfamethoxazole in clinical medicine and the high prevalence of resistance to this drug among sewage microbes (43). In addition to antibiotics, heavy metal exposure is also likely to select for antibiotic resistance in the environment, as metal resistance and antibiotic resistance are known to be highly correlated (4, 14, 27, 45, 55).

Most previous studies analyzing antibiotic resistance in aquatic environments relied entirely on phenotypic analysis. Two major drawbacks of assessing resistance by phenotype are that it relies on culturing and that it does not provide information on the mechanism of resistance. For example, characterizing phenotype would not distinguish between β-lactam resistance mediated by β-lactamases and that mediated by penicillin binding proteins, and these two mechanisms have substantially different clinical treatment implications (75).

However, recently there have been a handful of studies on this subject which also utilize molecular data, and the distribution of antibiotic resistance genes in the water environment is beginning to be elucidated. The vast majority of antibiotic resistance genotyping studies have focused on tetracycline resistance, most in animal waste lagoons (9, 17, 26, 38, 66), with a couple of studies from alternative aquatic environments such as rivers and drinking water (18, 52). Additionally, a recent review summarized the potential ecological and human health impacts of antibiotics and ARB found in wastewater treatment plants (32).

One study, using *Acinetobacter* spp. as indicator species, utilized a combination of cultivation and molecular techniques to investigate how the wastewater treatment process (WWTP) may enhance antibiotic resistance (Y. Zhang and C. Xi, unpublished data). Using selective agar, 366 *Acinetobacter* isolates were collected from different stages of a single wastewater treatment plant as well as from upstream and downstream sites in the receiving river. Resistance to amoxicillin-clavulanic acid (AMC), chloramphenicol (CHL), ciprofloxacin (CIP), colistin, gentamicin, rifampin (RIF), sulfisoxazole, tetracycline, and trimethoprim (TMP) was tested. The prevalence of antibiotic resistance to AMC, CHL, and RIF and of multidrug resistance (three drugs and more) was highest in the final effluent and consistently increased from raw influent to final effluent in WWTP (AMC, 8.7 to 37.9%; CHL, 25.2 to 69.0%; RIF, 63.1 to 84.5%; multidrug, 33.0 to 72.4%) and from upstream to downstream in the river (AMC, 9.5 to 25.8%; CHL, 27.0 to 48.4%; RIF, 65.1 to 85.5%; tetracycline, 3.2 to 14.5%; multidrug, 28.6 to 56.5%). The percentage of resistance of HPC populations to CIP, CHL, RIF, and sulfisoxazole doubled or tripled in the final effluent compared to the raw influent.

The results of another study also indicate that the WWTP may select for ARB. Again sampling along the wastewater stream, this study quantified genes that encode resistance to tetracycline *(tet)* and β-lactam *(bla)* antibiotics isolated from both culturable and nonculturable microorganisms. Additionally, DNA and 16S rRNA gene concentrations were quantified, so proportions of antibiotic resistance genes could be normalized to proxies of bacterial count and biomass, respectively. While sewage treatment reduced the overall concentrations of *bla* and *tet* genes, the ratio of these antibiotic resistance genes to 16S rRNA genes increased with treatment, suggesting that bacteria harboring *bla* and *tet* genes are more likely to survive the treatment process. Furthermore, the results of this study suggest that antibiotic resistance genes are being introduced into the environment through sewage effluent in significantly higher concentrations than would occur naturally, thus creating reservoirs of increased resistance potential (35).

There is also increasing evidence for the presence of ARB in drinking water (28, 50, 54, 65). For example, the frequency of ARB among HPC populations is very high (16, 51), and bacteria isolated from drinking water using HPC were resistant to a wide range of antibiotics. In particular, resistances to ampicillin, penicillin, and oxacillin were quite high, at 43, 55, and 77.7% of strains, respectively (51). The proportion of the HPC population showing resistance increases if there is chemical treatment before sand filtration (16). One study surveying the antibiotic resistance profile of HPC populations in a drinking water distribution system reported widespread resistance for bacteria found in finished drinking water and water distribution systems. Furthermore, there was a significant increase in AMC, CHL, and RIF resistance following water purification (Zhang and Xi, unpublished). Characterization of the molecular mechanism(s) responsible for the increased antibiotic resistance of drinking water HPC populations is under way in the same laboratory, including investigation regarding the risks associated with such resistant bacteria in drinking water.

## FUTURE DEVELOPMENTS AND RESEARCH NEEDS

Microorganisms are ubiquitous in the environment. The human body is exposed to and continuously inoculated with billions of microorganisms and provides a number of unique ecosystems and diverse habitats for microbial colonization. The human body also releases significant amounts of microbes back into the environment. Only a subset of the bacterial phyla, namely, *Proteobacteria, Bacteroides, Firmicutes, Actinobacteria,* and *Verrucomicrobia,* are found to dominate the human gut, demonstrating the uniqueness of that ecosystem (12, 13, 49). There is much that remains unknown about the roles of colonization, infection, persistence, and transmission of individual pathogens as they relate to human health.

Attention should be focused on understanding individual pathogens in the context of the microbial community at large and developing an ecological perspective regarding the mechanisms by which microorganisms in the natural environment shape human microbial populations both in healthy persons and in those with abnormal disease states. It is believed that at birth, infants receive early inoculation mainly from maternal vaginal, fecal, and skin microbiota (vertical transmission of microbes), the organisms of which are well adapted to the internal environment and generally dominate the adult microbiota (49). Few studies document the role of early inoculation of infants with natural microbiota derived from water and food, and the subsequent colonization, competition, and integration of these microorganisms in the growing child. In order to better understand these relationships, we need to compare the microbial populations in humans inhabiting different environments and define both differences and similarities in these populations.

Several methods have been used to compare bacterial populations derived from different aquatic environments (26, 74, 79). Comparison at different taxonomic levels, such as phylum, class, and species, should continue to be performed. However, to ensure sound comparison, DNA sequences covering a thorough, if not complete, array of microbial diversity must also be analyzed. The development of typical 16S rRNA gene clone-and-sequencing approaches has revolutionized research into microbial ecology and revealed many novel bacterial species in varied environments (48, 63); however, these techniques suffer from methodological biases, such as those associated with DNA extraction method, PCR amplification conditions, and cloning procedures (29, 68). Furthermore, the low number of clones analyzed in some studies limits the ability to provide sufficient enough data to accurately characterize the actual community composition and diversity of complex ecosystems. The development of high-density universal 16S rRNA microarrays (11) and high-throughput pyrosequencing (58) allows a much broader coverage of microbial diversity and should provide sufficient sequence data for comparative analyses. The techniques will also provide more accurate quantitative data for the relative abundance of individual taxa.

Studies on the composition of microbial communities often give only a snapshot of information. But dynamic analysis of microbial populations (31, 49) should provide more insight into the development and restoration of microbiota, especially when perturbation of the human microbiota occurs as the result of a change in health status and/or the use of antibiotics and other drugs. These types of studies should lead to a better understanding of how microbes from natural environments will adapt to the human ecosystem and become an integral part of the human microbiota.

We also need additional information on the survival, persistence, and transmission of human microbiota. Bacteriophages likely exert a strong influence on the diversity and population structure of bacterial communities in the human gut (7); characterization of bacteriophage populations in both the environment (1) and the human body, as well as their role in microbial ecology, should provide further insights into population dynamics, including the potential for horizontal gene transfer.

Antibiotic resistance and the risks of gene exchange accelerated through water treatment are growing concerns (32). The recent findings of increased abundance of ARB and antibiotic resistance genes in surface water and drinking water strengthen these concerns. It seems likely that ARB will have a selective advantage in colonizing the human body, especially during or following antibiotic treatment. Mechanisms of invasion and colonization by ARB, and how ARB influence the overall diversity and dynamics of the human microbiota, warrant further study. The spread of antibiotic resistance in hospitals and within communities, combined with evidence linking early antibiotic use and human immunosuppressive/abnormal diseases, e.g., asthma (34), warrants further study of the link between antibiotics, ARB present in the environment, and antibiotic resistance in humans.

Understanding the population structure and dynamics of the human microbiota has very important implications for human health and the development of probiotics for the treatment and prevention of many human diseases. The interaction between environmental microbes and the human microbiota will likely reshape our thinking on the relevance of environmental microbiology to public health.

## REFERENCES

1. **Albinana-Gimenez, N., P. Clemente-Casares, S. Bofill-Mas, A. Hundesa, F. Ribas, and R. Girones.** 2006. Distribution of human polyomaviruses, adenoviruses, and hepatitis E virus in the environment and in a drinking-water treatment plant. *Environ. Sci. Technol.* **40:**7416–7422.
2. **Allen, M. J., S. C. Edberg, and D. J. Reasoner.** 2004. Heterotrophic plate count bacteria—what is their significance in drinking water? *Int. J. Food Microbiol.* **92:**265–274.
3. **Barber, M., and J. E. M. Whitehead.** 1949. Bacteriophage types in penicillin-resistant staphylococcal infections. *Br. Med. J.* **2:**565–569.
4. **Belliveau, B. H., M. E. Starodub, and J. T. Trevors.** 1991. Occurrence of antibiotic and metal resistance and plasmids in Bacillus strains isolated from marine sediment. *Can. J. Microbiol.* **37:**513–520.
5. **Berry, D., C. Xi, and L. Raskin.** 2006. Microbial ecology of drinking water distribution systems. *Curr. Opin. Biotechnol.* **17:**297–302.
6. **Bomo, A. M., M. V. Storey, and N. J. Ashbolt.** 2004. Detection, integration and persistence of aeromonads in water distribution pipe biofilms. *J. Water Health* **2:**83–96.
7. **Breitbart, M., I. Hewson, B. Felts, J. M. Mahaffy, J. Nulton, P. Salamon, and F. Rohwer.** 2003. Metagenomic analyses of an uncultured viral community from human feces. *J. Bacteriol.* **185:**6220–6223.
8. **Chandrasekaran, S., B. Venkatesh, and D. Lalithakumari.** 1998. Transfer and expression of a multiple antibiotic resistance plasmid in marine bacteria. *Curr. Microbiol.* **37:**347–351.
9. **Chee-Sanford, J. C., R. I. Aminov, I. J. Krapac, N. Garrigues-Jeanjean, and R. I. Mackie.** 2001. Occurrence and diversity of tetracycline resistance genes in lagoons and groundwater underlying two swine production facilities. *Appl. Environ. Microbiol.* **67:**1494–1502.
10. **Cloete, E., J. Rose, L. H. Nel, and T. Ford.** 2004. *Microbial Waterborne Pathogens.* International Water Association, London, United Kingdom.
11. **DeSantis, T. Z., E. L. Brodie, J. P. Moberg, I. X. Zubieta, Y. M. Piceno, and G. L. Andersen.** 2007. High-density universal 16S rRNA microarray analysis reveals broader diversity than typical clone library when sampling the environment. *Microb. Ecol.* **53:**371–383.
12. **Dethlefsen, L., P. B. Eckburg, E. M. Bik, and D. A. Relman.** 2006. Assembly of the human intestinal microbiota. *Trends Ecol. Evol.* **21:**517–523.
13. **Dethlefsen, L., M. McFall-Ngai, and D. A. Relman.** 2007. An ecological and evolutionary perspective on human-microbe mutualism and disease. *Nature* **449:**811–818.
14. **de Vicente, A., M. Aviles, J. C. Codina, J. J. Borrego, and P. Romero.** 1990. Resistance to antibiotics and heavy metals of *Pseudomonas aeruginosa* isolated from natural waters. *J. Appl. Bacteriol.* **68:**625–632.
15. **Edberg, S. C., and M. J. Allen.** 2004. Virulence and risk from drinking water of heterotrophic plate count bacteria in human population groups. *Int. J. Food Microbiol.* **92:**255–263.
16. **El-Zanfaly, H. T., D. J. Reasoner, and E. E. Geldreich.** 1998. Bacteriological changes asso-

ciated with granular activated carbon in a pilot treatment plant. *Water Air Soil Pollut.* **107:**73–80.

17. **Engemann, C. A., L. Adams, C. W. Knapp, and D. W. Graham.** 2006. Disappearance of oxytetracycline resistance genes in aquatic systems. *FEMS Microbiol. Lett* **263:**176–182.
18. **Ferguson, M. A., and J. G. Hering.** 2006. $TiO_2$-photocatalyzed As(III) oxidation in a fixed-bed, flow-through reactor. *Environ. Sci. Technol.* **40:**4261–4267.
19. **Finlay, B. J., and K. J. Clarke.** 1999. Ubiquitous dispersal of microbial species. *Nature* **400:** 828.
20. **Ford, T., and R. Colwell.** 1995. *A Global Decline in Microbiological Safety of Water: a Call for Action.* American Academy of Microbiology, Washington, DC.
21. **Ford, T. E.** 1999. Microbiological safety of drinking water: United States and global perspectives. *Environ. Health Perspect.* **107**(Suppl. 1):191–206.
22. **Freeman, S.** 2002. Bacteria and Archaea, p. 484–497. *In* S. L. Snavely (ed.), *Biological Sciences.* Prentice Hall, Upper Saddle River, NJ.
23. **Hahn, M. W.** 2006. The microbial diversity of inland waters. *Curr. Opin. Biotechnol.* **17:**256–261.
24. **Hiorns, W. D., B. A. Methe, S. A. Nierzwicki-Bauer, and J. P. Zehr.** 1997. Bacterial diversity in Adirondack mountain lakes as revealed by 16S rRNA gene sequences. *Appl. Environ. Microbiol.* **63:**2957–2960.
25. **Hirsch, R., T. Ternes, K. Haberer, and K. L. Kratz.** 1999. Occurrence of antibiotics in the aquatic environment. *Sci. Total Environ.* **225:** 109–118.
26. **Hongoh, Y., P. Deevong, T. Inoue, S. Moriya, S. Trakulnaleamsai, M. Ohkuma, C. Vongkaluang, N. Noparatnaraporn, and T. Kudo.** 2005. Intra- and interspecific comparisons of bacterial diversity and community structure support coevolution of gut microbiota and termite host. *Appl. Environ. Microbiol.* **71:**6590–6599.
27. **Iwane, T., T. Urase, and K. Yamamoto.** 2001. Possible impact of treated wastewater discharge on incidence of antibiotic resistant bacteria in river water. *Water Sci. Technol.* **43:**91–99.
28. **Jones, O. A., J. N. Lester, and N. Voulvoulis.** 2005. Pharmaceuticals: a threat to drinking water? *Trends Biotechnol.* **23:**163–167.
29. **Juretschko, S., G. Timmermann, M. Schmid, K.-H. Schleifer, A. Pommerening-Röser, H.-P. Koops, and M. Wagner.** 1998. Combined molecular and conventional analyses of nitrifying bacterium diversity in activated sludge: *Nitrosococcus mobilis* and *Nitrospira*-like bacteria as dominant populations. *Appl. Environ. Microbiol.* **64:**3042–3051.
30. **Karner, M. B., E. F. DeLong, and D. M. Karl.** 2001. Archaeal dominance in the mesopelagic zone of the Pacific Ocean. *Nature* **409:** 507–510.
31. **Kent, A. D., A. C. Yannarell, J. A. Rusak, E. W. Triplett, and K. D. McMahon.** 2007. Synchrony in aquatic microbial community dynamics. *ISME J.* **1:**38–47.
32. **Kim, S., and D. S. Aga.** 2007. Potential ecological and human health impacts of antibiotics and antibiotic-resistant bacteria from wastewater treatment plants. *J. Toxicol. Environ. Health Part B* **10:**559–573.
33. **Kolpin, D. W., E. T. Furlong, M. T. Meyer, E. M. Thurman, S. D. Zaugg, L. B. Barber, and H. T. Buxton.** 2002. Pharmaceuticals, hormones, and other organic wastewater contaminants in US streams, 1999–2000: a national reconnaissance. *Environ. Sci. Technol.* **36:**1202–1211.
34. **Kozyrskyj, A. L., P. Ernst, and A. B. Becker.** 2007. Increased risk of childhood asthma from antibiotic use in early life. *Chest* **131:** 1753–1759.
35. **Lachmayr, K.** 2007. *Anthropogenically Induced Reservoirs of Antibiotic Resistance: the Case of Massachusetts Bay.* Ph.D. dissertation. Harvard University, Boston, MA.
36. **LeChevallier, M. W., C. D. Cawthon, and R. G. Lee.** 1988. Factors promoting survival of bacteria in chlorinated water supplies. *Appl. Environ. Microbiol.* **54:**649–654.
37. **Leclerc, H., L. Schwartzbrod, and E. Dei-Cas.** 2002. Microbial agents associated with waterborne diseases. *Crit. Rev. Microbiol.* **28:**371–409.
38. **Lehman, E. M.** 2007. Seasonal occurrence and toxicity of Microcystis in impoundments of the Huron River, Michigan, USA. *Water Res.* **41:** 795–802.
39. **Levy, S. B.** 2002. Factors impacting on the problem of antibiotic resistance. *J. Antimicrob. Chemother.* **49:**25–30.
40. **Levy, S. B., G. B. FitzGerald, and A. B. Macone.** 1976. Changes in intestinal flora of farm personnel after introduction of a tetracycline-supplemented feed on a farm. *N. Engl. J. Med.* **295:**583–588.
41. **Liang, J. L., E. J. Dziuban, G. F. Craun, V. Hill, M. R. Moore, R. J. Gelting, R. L. Calderon, M. J. Beach, and S. L. Roy.** 2006. Surveillance for waterborne disease and outbreaks associated with drinking water and water not intended for drinking—United States, 2003–2004. *MMWR Surveill. Summ.* **55:**31–65.

42. **Linton, A. H.** 1988. Plasmids in the environment. *Schriftenr. Ver Wasser Boden Lufthyg.* **78:** 197–224.

43. **Martinez, M., M. A. Mondaca, and R. Zemelman.** 1994. Antibiotic-resistant gram-negative bacilli in the sewage of the City of Concepcion, Chile. *Rev. Latinoam. Microbiol.* **36:**39–46.

44. **Miyoshi, T., T. Iwatsuki, and T. Naganuma.** 2005. Phylogenetic characterization of 16S rRNA gene clones from deep-groundwater microorganisms that pass through 0.2-micrometer-pore-size filters. *Appl. Environ. Microbiol.* **71:** 1084–1088.

45. **Morozzi, G., G. Cenci, G. Caldini, G. Losito, and A. Morosi.** 1986. Relationship between environment spread and presence in hosts of *Escherichia coli* strains resistant to antibiotics and metals. *Zentralbl. Bakteriol. Mikrobiol. Hyg. B* **182:** 393–400.

46. **Morris, R. M., M. S. Rappe, S. A. Connon, K. L. Vergin, W. A. Siebold, C. A. Carlson, and S. J. Giovannoni.** 2002. SAR11 clade dominates ocean surface bacterioplankton communities. *Nature* **420:**806–810.

47. **Munro, P. M., G. N. Flatau, R. L. Clement, and M. J. Gauthier.** 1995. Influence of the RpoS (KatF) sigma factor on maintenance of viability and culturability of *Escherichia coli* and *Salmonella typhimurium* in seawater. *Appl. Environ. Microbiol.* **61:**1853–1858.

48. **Pace, N. R., D. A. Stahl, D. J. Lane, and G. J. Olsen.** 1986. The analysis of natural microbial populations by ribosomal RNA sequences. *Adv. Microb. Ecol.* **9:**1.

49. **Palmer, C., E. M. Bik, D. B. Digiulio, D. A. Relman, and P. O. Brown.** 2007. Development of the human infant intestinal microbiota. *PLoS Biol.* **5:**e177.

50. **Pathak, S. P., and K. Gopal.** 2008. Prevalence of bacterial contamination with antibiotic-resistant and enterotoxigenic fecal coliforms in treated drinking water. *J. Toxicol. Environ. Health Part A* **71:**427–433.

51. **Pavlov, D., C. M. de Wet, W. O. Grabow, and M. M. Ehlers.** 2004. Potentially pathogenic features of heterotrophic plate count bacteria isolated from treated and untreated drinking water. *Int. J. Food Microbiol.* **92:**275–287.

52. **Pei, R., S. C. Kim, K. H. Carlson, and A. Pruden.** 2006. Effect of river landscape on the sediment concentrations of antibiotics and corresponding antibiotic resistance genes (ARG). *Water Res.* **40:**2427–2435.

53. **Prescott, L. M., J. P. Harley, and D. A. Klein.** 2002. *Microbiology,* 5th ed. McGraw-Hill Higher Education, New York, NY.

54. **Pruden, A., R. Pei, H. Storteboom, and K. H. Carlson.** 2006. Antibiotic resistance genes as emerging contaminants: studies in northern Colorado. *Environ. Sci. Technol.* **40:**7445–7450.

55. **Rasmussen, L. D., and S. J. Sorensen.** 1998. The effect of longterm exposure to mercury on the bacterial community in marine sediment. *Curr. Microbiol.* **36:**291–297.

56. **Reynolds, K. A., K. D. Mena, and C. P. Gerba.** 2008. Risk of waterborne illness via drinking water in the United States. *Rev. Environ. Contam. Toxicol.* **192:**117–158.

57. **Rheinheimer, G.** 1980. *Aquatic Microbiology,* 2nd ed. Wiley, New York, NY.

58. **Roesch, L. F., R. R. Fulthorpe, A. Riva, G. Casella, A. K. Hadwin, A. D. Kent, S. H. Daroub, F. A. Camargo, W. G. Farmerie, and E. W. Triplett.** 2007. Pyrosequencing enumerates and contrasts soil microbial diversity. *ISME J.* **1:**283–290.

59. **Rusch, D. B., A. L. Halpern, G. Sutton, K. B. Heidelberg, S. Williamson, S. Yooseph, D. Wu, J. A. Eisen, J. M. Hoffman, K. Remington, K. Beeson, B. Tran, H. Smith, H. Baden-Tillson, C. Stewart, J. Thorpe, J. Freeman, C. Andrews-Pfannkoch, J. E. Venter, K. Li, S. Kravitz, J. F. Heidelberg, T. Utterback, Y. H. Rogers, L. I. Falcon, V. Souza, G. Bonilla-Rosso, L. E. Eguiarte, D. M. Karl, S. Sathyendranath, T. Platt, E. Bermingham, V. Gallardo, G. Tamayo-Castillo, M. R. Ferrari, R. L. Strausberg, K. Nealson, R. Friedman, M. Frazier, and J. C. Venter.** 2007. The Sorcerer II Global Ocean Sampling expedition: northwest Atlantic through eastern tropical Pacific. *PLoS Biol.* **5:**e77.

60. **Sagik, B., C. Sorber, and B. Morse.** 1981. The survival of EK1 and EK2 systems in sewage treatment plant models, p. 449–460. *In* S. B. Levy, R. C. Clowes, and E. L. Koenig (ed.), *Molecular Biology, Pathogenicity and Ecology of Bacterial Plasmids.* Plenum, New York, NY.

61. **Saye, D. J., and R. V. Miller.** 1989. The aquatic environment: consideration of horizontal gene transmission in a diversified habitat, p. 223–259. *In* S. B. Levy and R. V. Miller (ed.), *Gene Transfer in the Environment.* McGraw Hill, New York, NY.

62. **Schmeisser, C., C. Stockigt, C. Raasch, J. Wingender, K. N. Timmis, D. F. Wenderoth, H. C. Flemming, H. Liesegang, R. A. Schmitz, K. E. Jaeger, and W. R. Streit.** 2003. Metagenome survey of biofilms in drinking-water networks. *Appl. Environ. Microbiol.* **69:**7298–7309.

63. **Schmidt, T. M., E. F. DeLong, and N. R. Pace.** 1991. Analysis of a marine picoplankton

community by 16S rRNA gene cloning and sequencing. *J. Bacteriol.* **173:**4371–4378.

**64. Schuster, C. J., A. G. Ellis, W. J. Robertson, D. F. Charron, J. J. Aramini, B. J. Marshall, and D. T. Medeiros.** 2005. Infectious disease outbreaks related to drinking water in Canada, 1974–2001. *Can. J. Public Health* **96:**254–258.

**65. Schwartz, T., W. Kohnen, B. Jansen, and U. Obst.** 2003. Detection of antibiotic-resistant bacteria and their resistance genes in waste-water, surface water, and drinking water biofilms. *FEMS Microbiol. Ecol.* **43:**325–335.

**66. Smith, M. S., R. K. Yang, C. W. Knapp, Y. Niu, N. Peak, M. M. Hanfelt, J. C. Galland, and D. W. Graham.** 2004. Quantification of tetracycline resistance genes in feedlot lagoons by real-time PCR. *Appl. Environ. Microbiol.* **70:** 7372–7377.

**67. Soller, J. A.** 2006. Use of microbial risk assessment to inform the national estimate of acute gastrointestinal illness attributable to microbes in drinking water. *J. Water Health* **4**(Suppl. 2)**:**165–186.

**68. Suzuki, M. T., and S. J. Giovannoni.** 1996. Bias caused by template annealing in the amplification of mixtures of 16S rRNA genes by PCR. *Appl. Environ. Microbiol.* **62:**625–630.

**69. Szewzyk, U., R. Szewzyk, W. Manz, and K. H. Schleifer.** 2000. Microbiological safety of drinking water. *Annu. Rev. Microbiol.* **54:**81–127.

**70. Tamanai-Shacoori, Z., M. Arturo, M. Pommepuy, C. Mamez, and M. Cormier.** 1995. Conjugal transfer of natural plasmids between *Escherichia coli* strains in sterile environmental water. *Curr. Microbiol.* **30:**155–160.

**71. Theron, J., and T. E. Cloete.** 2002. Emerging waterborne infections: contributing factors, agents, and detection tools. *Crit. Rev. Microbiol.* **28:**1–26.

**72. Theron, J., and T. E. Cloete.** 2000. Molecular techniques for determining microbial diversity and community structure in natural environments. *Crit. Rev. Microbiol.* **26:**37–57.

**73. Tokajian, S. T., F. A. Hashwa, I. C. Hancock, and P. A. Zalloua.** 2005. Phylogenetic assessment of heterotrophic bacteria from a water distribution system using 16S rDNA sequencing. *Can. J. Microbiol.* **51:**325–335.

**74. Tringe, S. G., C. von Mering, A. Kobayashi, A. A. Salamov, K. Chen, H. W. Chang, M. Podar, J. M. Short, E. J. Mathur, J. C. Detter, P. Bork, P. Hugenholtz, and E. M. Rubin.** 2005. Comparative metagenomics of microbial communities. *Science* **308:**554–557.

**75. U.S. Congress, Office of Technology Assessment.** 1995. *Impacts of Antibiotic-Resistant Bacteria.* OTA-H-629. U.S. Government Printing Office, Washington, DC.

**76. Wagner, M., and A. Loy.** 2002. Bacterial community composition and function in sewage treatment systems. *Curr. Opin. Biotechnol.* **13:** 218–227.

**77. Wagner, M., A. Loy, R. Nogueira, U. Purkhold, N. Lee, and H. Daims.** 2002. Microbial community composition and function in wastewater treatment plants. *Antonie van Leeuwenhoek* **81:**665–680.

**78. Whitman, W. B., D. C. Coleman, and W. J. Wiebe.** 1998. Prokaryotes: the unseen majority. *Proc. Natl. Acad. Sci. USA* **95:**6578–6583.

**79. Zwart, G., B. C. Crump, M. P. Kamst-van Agterveld, F. Hagen, and S.-K. Han.** 2002. Typical fresh water bacteria: an analysis of available 16S rRNA gene sequences from plankton of lakes and rivers. *Aquat. Microb. Ecol.* **28:**141–155.

**80. Zwart, G., W. D. Hiorns, B. A. Methe, M. P. van Agterveld, R. Huismans, S. C. Nold, J. P. Zehr, and H. J. Laanbroek.** 1998. Nearly identical 16S rRNA sequences recovered from lakes in North America and Europe indicate the existence of clades of globally distributed freshwater bacteria. *Syst. Appl. Microbiol.* **21:**546–556.

# INTERACTIONS AND MODIFICATIONS WITHIN MICROBIAL ECOSYSTEMS

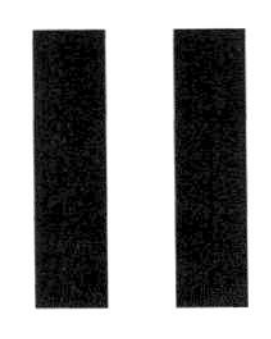

# BIOFILMS IN THE FOOD ENVIRONMENT

*Joseph F. Frank*

# 6

Growth on surfaces offers numerous advantages to microorganisms, making biofilms the predominant growth form of microorganisms in natural environments. Advantages of biofilm growth include access to nutrients under the low-nutrient conditions that predominate in nature, protection from environmental or host stress, and ease of genetic exchange (48). Biofilm can be considered the default form of growth for many microorganisms, with the planktonic growth typical of laboratory broth culture of secondary importance. Microbes in nature that are not attached to surfaces are at the mercy of liquid flows and airflows that, although they assist with disbursement, provide an unstable and unpredictable growth environment. Biofilms form on abiotic and biotic surfaces that provide sufficient water at a growth-permissive temperature. Biofilms in nature usually exhibit a high level of organization that allows diffusion of nutrients and oxygen to underlying cell layers and diffusion of metabolic waste products away from cells. The microbial diversity of biofilms in nature reflects the growth environment, whereas the microbial diversity of biofilms in food processing systems, although strongly influenced by the growth environment, can be limited by the lack of diversity of microorganisms that enter the system. Microbial access to food processing systems can be limited by physical enclosure (i.e., sealed processing systems) and hygienic practices (34).

Biofilms form on plant and meat surfaces in an effort to utilize nutrients in the underlying tissue. A slippery surface is often the initial sensory manifestation of biofilm formation on food tissues. Living plants and animals defend themselves against invasive biofilms by using protective surfaces and immune system responses; however, there are numerous examples of animal-biofilm and plant-biofilm relationships that benefit the host. These include biofilms that digest cellulose in the rumen (103), bioluminescent biofilms of some fish species (41), and nitrogen-fixing biofilms that stimulate plant growth (47).

Biofilms form on food processing surfaces in areas that retain moisture and are not sufficiently cleaned at regular intervals. Most such sites are environmental rather than direct food contact surfaces, as food contact surfaces usually receive a high level of hygienic attention. Raw foods entering a processing facility contain a variety of microorganisms, many of

*Joseph F. Frank,* Department of Food Science and Technology, University of Georgia, Athens, GA 30602.

*Food-Borne Microbes: Shaping the Host Ecosystem,* Edited by L.-A. Jaykus, H. H. Wang, and L. S. Schlesinger, © 2009 ASM Press, Washington, DC

which are capable of attaching to food contact surfaces. However, the combination of food residues and attached microorganisms which often appears on food processing surfaces will usually be removed through cleaning before the microbes form a biofilm. Processing line dead ends (areas containing static nutrient and/or moisture accumulations) are found only in poorly designed or poorly maintained systems, as they support microbial growth and are difficult to clean. In these environments, the biofilm phenotype will be manifest in cells at the soil-surface interface but may not be apparent for cells growing on gross amounts of food residue. In practice, such distinctions are not critical, as the procedures for detecting and cleaning growth niches are dependent not on microbial phenotype but, rather, on identification of the growth niche and the ability to apply cleaning agents at sufficient concentration, time, and temperature to effectively clean the site (36). For processing plant growth niches where biofilms form, such as gasketed joints, vacuum hoses, equipment exteriors, and floor drains, the characteristics and microbial content of the biofilms are likely to be highly diverse, depending on the microbial and nutrient contents of the food being processed, previous treatment of the food, temperature of the growth site, and chemical applications (cleaning). This makes generalization about food processing biofilms difficult and provides a challenge for the development of laboratory simulations.

## BIOFILM DEVELOPMENT

Biofilms are structured microbial communities of cells differentiated to play specific roles in the maintenance of the community and its structure. Even a single-species biofilm will include cells of various phenotypes. Davey and O'Toole (22) characterize biofilm development as a complex process that requires collective bacterial behavior which may involve coordination and communication among multiple bacterial species. The biofilm development process has been characterized for various single-species systems as a series of phenotypic transformations. Sauer et al. (90) grouped these transformations into five stages, including reversible attachment, irreversible attachment, maturation 1, maturation 2, and dispersal. They observed that more than 50% of the proteome is impacted by this development process and that there are specific physiologies distinctive to each stage. In this chapter the discussion of biofilm development is organized into three sections: initiation (reversible and irreversible attachment), structure development (maturation), and dispersal. Although there is a general process for how biofilms develop, mechanisms underlying the process differ among microorganisms.

### Initiation of Biofilm Formation

Attachment of bacteria involves various cell envelope structures, often including flagella, pili, capsules, outer membrane polymers, and S-layer proteins (35). Initiation of attachment by *Staphylococcus aureus* is associated with the production of a surface protein, Bap (biofilm-associated protein), whose synthesis is positively controlled by the SarA protein, which also activates virulence genes (109). Reversible attachment for *Escherichia coli* and *Pseudomonas aeruginosa* is promoted by motility, presumably to help overcome repulsive forces at the surface (84, 90). Type I pili are involved in the initial attachment process for *E. coli* and *Salmonella enterica* serovar Typhimurium (61, 84). Once attached, *P. aeruginosa* employs twitching motility utilizing type IV pili to move across the surface to form aggregations that appear as microcolonies (82). A similar process occurs with *E. coli* but appears to involve flagellum-associated movement (84). One implication of this process is that if an immature biofilm is observed at a single time point, the observed microcolonies may have been formed by either surface growth, aggregation of initially dispersed attached cells, recruitment of cells from the surrounding liquid, or a combination of these mechanisms.

Irreversible attachment is characterized by the loss of flagella by cells and the synthesis of cell envelope molecules that promote bonding to the surface. *E. coli* synthesizes curli at this stage, followed by the synthesis of colonic acid, which eventually becomes an integral part of the biofilm structure (85). *P. aeruginosa* initiates synthesis of alginate at this stage (90). The transition from a reversibly attached cell to an irreversibly attached cell ready to initiate biofilm formation is under complex regulatory control. Pruss et al. (85) make the following observations in their review of this process. *E. coli* employs 16 gene regulators in the transition to biofilm. Regulators C-reactive protein and H-NS affect hundreds of genes, whereas LrhA and HdtR regulate only a few genes. Global regulation of two-component signal transduction pathways is also involved. The cell is able to sense being at a surface via histidine kinase associated with the cytoplasmic membrane. Response to this sensor molecule is regulated by aspartyl kinase. Twenty-nine histidine kinases and 32 response regulators have been identified for *E. coli* K-12 to be involved in the transition to biofilm. The two-component global regulators also regulate outer membrane porin proteins, quorum sensing, flagellum synthesis, and initial sensing of stress to the cell envelope. The most important regulators of attachment include FlhDC, which regulates synthesis of flagella and curli, and EnvZ/OmpR, which regulates outer membrane porins, carbon source usage, and permease proteins. EnvA/OmpR-deficient mutants exhibit decreased resistance to several antibiotics, indicating the importance of porin and permease regulation to survival. A third major regulator, RcsCDB, regulates colonic acid synthesis along with up to 5% of the cell's genome. This global regulator controls numerous processes associated with stress adaptation and biofilm growth.

Although attachment at the early stage of biofilm formation is described as irreversible, in the sense that the cells do not readily detach into surrounding liquid, the biofilm itself can move along a surface when subjected to the shear stress of fluid flow (86).

## Development of Biofilm Structure

After cells have attached, the biofilm forms by a combination of growth at the surface and recruitment of planktonic cells. Flagella may assist with recruitment to the biofilm, possibly by helping overcome repulsive forces between the cell and surface (84). If flagellum-defective mutants of *Listeria monocytogenes* are forcibly attached to a surface via centrifugation, typical biofilm will still not develop, presumably because the lack of functioning flagella impairs recruitment of planktonic cells to the surface (65). *Staphylococcus epidermidis* produces adhesins that promote intracellular interactions necessary for biofilm structure to develop (74), whereas curli and pili provide this function in *E. coli* (52). Attached cells can also recruit genetically unrelated species to the biofilm via carbohydrate receptors that specifically interact with adhesive proteins in a process referred to as coaggregation (88). Genetically distinct cells coaggregated in suspension can also be recruited into the biofilm.

Biofilms exhibit an ordered structure that facilitates the mass transfer processes required to maintain cell viability. This structure promotes access to nutrients and oxygen and diffusion of waste products away from the cells. Typical structures include columns of cells separated by channels through which water flows, delivering nutrients and removing waste products. The columns may expand at the top to give a mushroom-like appearance or form filaments. Biofilms can also be structured as a honeycomb-like network of connected cells with the interstitial spaces acting as water channels. These water channels can be observed even within microcolonies (Fig. 1). Most of the information obtained on structural development of biofilms is from single-species studies that may not adequately represent biofilms in natural or food processing environments. In addition, the structural development of food system biofilms that incorporate food residues has not been determined.

Lasa (59) summarized some common features in the development of biofilm structure. The structural components include exopolysaccharides such as cellulose, β-1,6-linked *N*-acetylglucosamine, colonic acid, and alginate. Modifications of β-1,6-linked *N*-acetylglucosamine act as intercellular adhesins within biofilms of various species. Since bac-

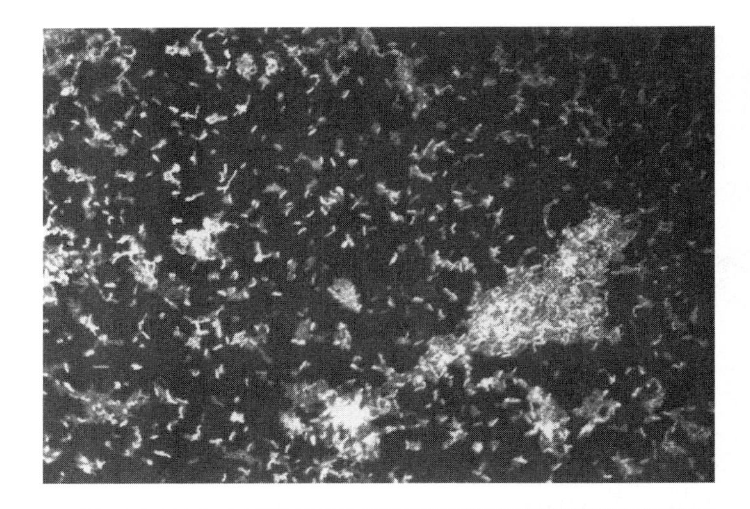

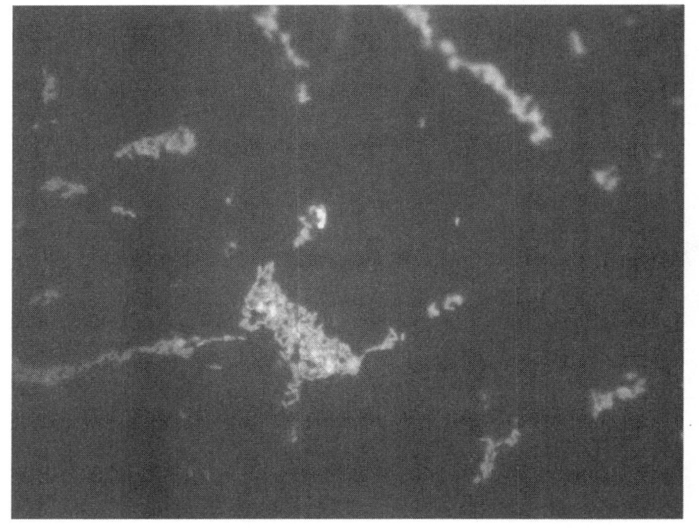

**FIGURE 1** Monospecies biofilms produced on polyvinyl chloride by unidentified isolates from a chicken house water system (stained with acridine orange). Incubation was in R2A broth for 7 days at 12°C. (Top) Dispersal of single cells and small microcolonies between larger porous microcolonies; (bottom) open space between porous microcolonies. (Micrographs by Nathanon Trachoo and J. F. Frank.)

terial cells may produce many types of exopolysaccharides, their synthesis must be regulated to produce an organized structure. One common molecular signal for this regulation across different species is cyclic di-GMP (c-di-GMP). Levels of cytoplasmic c-di-GMP are controlled by proteins having a structural domain in common with Bap proteins. Bap proteins of diverse bacterial species share common structural features. These include being present on the cell surface and having a high molecular weight, a core domain of tandem repeats, and a role in virulence (60). Bap proteins are required for biofilm formation because of their involvement in primary attachment and intercellular aggregation (60).

Quorum sensing is the means by which cell processes are regulated in a population-dependent manner. Quorum sensing is mediated by small molecule autoinducers. Autoinducers for gram-negative bacteria are primarily in the acyl homoserine lactone group produced by the LuxI/LuxR system, and those for gram-positive bacteria are primarily modified oligopeptides (autoinducing peptides) mediated by two-component sensing systems (15). These systems facilitate intraspecies communication. Autoinducer 2 is a furanosyl borate diester believed to mediate interspecies communication (29). Quorum sensing (inter- and intraspecies communication) is a common, but perhaps not universal, element in the development of biofilm structure; lack of quorum sensing may cause disorganized surface growth (23, 43, 78, 93, 96, 112). The development of the capped column (mushroom appearance) architecture observed in many biofilms is linked to quorum sensing in *P. aeruginosa* (66). Genes for rhamnolipid synthesis are expressed in the stalk (column) but not in the cap. Growing cells with twitching motility migrate up the stalk to form the cap, whereas cells in the stalk neither divide nor migrate. Rhamnolipid in the stalk may act as a surfactant to provide a low-friction path to the cap.

In certain cases, inhibition of quorum sensing can inhibit biofilm-associated infectious processes (6), an indication that maintenance of a biofilm-associated infection may rely on an organized biofilm ultrastructure.

### Cell Dispersion From the Biofilm

Food system biofilms would be of less concern if there were no dispersal stage. Dispersal of cells from the biofilm can be either an autoinduced or an externally induced process. One function of cleaning the plant environment and equipment is to disrupt and disperse biofilm cells so that they can be removed from the system or killed by chemical sanitizers. Plant activities that promote dispersal without a kill step are generally to be avoided. External forces that disperse biofilms include water sprays; fluid shear forces; direct contact with cleaning implements, vehicles, and other items; and human contact. Biofilms autoinduce dispersal after the structure has matured or in response to environmental signals. The dispersal stage is characterized by large-scale down-regulation of genes, with cells expressing proteins in a pattern similar to the planktonic form. Cells of *P. aeruginosa* disperse by becoming motile and swimming away, leaving behind an empty shell-like structure (90). Autoinduced dispersal can be in the form of either single cells continually detaching over a period (erosion) or sudden loss of large cell aggregates (sloughing) (104). Although sloughing of cell aggregates may be infrequent, in one multispecies model system these aggregates were sufficiently large (up to 1,000 $\mu m^2$) to account for 60% of the detached biofilm area (104). The signal for dispersal of *P. aeruginosa* is the elevation of nitric oxide levels associated with the anaerobic respiration (7). Biofilm *P. aeruginosa* cells lacking a gene locus associated with dispersal (*bdlAI,* biofilm dispersion locus) exhibit increased intracellular c-di-GMP and increased adherence (79). Thormann et al. (106), working with *Shewanella oneidensis,* provide convincing evidence that c-di-GMP is a key intracellular regulator of biofilm formation and detachment. Since c-di-GMP exerts control over exopolysaccharide production that is critical to biofilm structure and stability, Thormann et al. (106) speculate that detachment is a response to the cessation of exopolysaccharide biosynthesis resulting from reduced c-di-GMP intracellular levels.

## THE BIOFILM MATRIX

The composition of the biofilm matrix is complex and variable, making generalizations difficult. Even single-species biofilms may have different matrix components depending on the growth environment. A common feature of biofilm structure is that living cells are a minor component, whether determined by volume or mass. For example, Jahn and Nielsen (46) estimated that 70 to 98% of the total organic carbon in natural wastewater biofilms was extracellular. The matrix of food system biofilms most likely has increased complexity because of the potential for incorporation of food components, including fats, proteins, and minerals. Although little is understood about the matrix of food system biofilms, such an understanding could lead to an improvement in cleaning processes, since the removal of the matrix is an important aspect of surface hygiene. Biofilm structural components not removed by cleaning could act as attachment sites for planktonic microorganisms, thus allowing for more rapid reestablishment of the biofilm (35).

Hunter and Beveridge (44) observed the structure of a monoculture biofilm of *P. aeruginosa* using freeze-substitution transmission electron microscopy, a technique that maintains the structural integrity of the sample while providing high-resolution images. These images showed biofilms containing tightly and loosely packed cells, lysed cells, and cell-free matrix with suspended particulate matter. The exopolymeric substance also exhibited various structural organizations. Further examination of biofilms revealed that a substantial volume of the matrix was filled by membrane vesicles derived from gram-negative bacteria (91). These vesicles were observed in laboratory and natural (multispecies) biofilms. Membrane vesicles in biofilms could interact with matrix polymers and may contain enzymes and other active agents derived from the periplasm. The authors speculate that the periplasmic component of membrane vesicles could degrade or sequester antibiotics. One might also expect these components to degrade or sequester chemical sanitizing agents.

Another important aspect of the biofilm matrix is that it contains a large amount of exogenous DNA that contributes to its structural stability (114). Evidence indicates that DNA is released and integrated into the matrix in association with quorum sensing signals for both gram-positive and gram-negative cells (100). In addition to contributing to the structural stability of the matrix, exogenous biofilm DNA provides a mechanism for horizontal gene transfer (105).

Proteins are also a key component of the biofilm matrix (99). Wild strains of *Bacillus subtilis* form a biofilm (pellicle) at the air-liquid interface in which nonmotile cells are bundled together in a parallel arrangement. The cells are held together by an extracellular matrix composed primarily of polysaccharide and protein. The study of Branda et al. (10) provides a good illustration of how biofilm matrix is constructed. *B. subtilis* cells deficient in either polysaccharide or exocellular protein production produced fragile pellicles, whereas doubly deficient mutants did not produce pellicles. However, when single complementary mutants were combined, a normal pellicle was formed, indicating that polymers synthesized by different cells could combine to produce a normal structure. The potential complexity of biofilm matrices becomes apparent when the variety of cell types that could contribute to structural components in natural systems is considered. In addition, the matrix composition is dynamic, as individual organisms can produce several different exopolysaccharides and the relative proportion of these polymers can change over time (70).

## THE BIOFILM ECOSYSTEM

### Microbial Interactions

Biofilms formed in food processing environments will often contain multiple species, depending on the growth conditions and level of sanitation associated with the growth niche; i.e., a lack of restrictive growth conditions and infrequent or ineffective sanitation will likely give rise to multispecies biofilms. The structure of the biofilm puts different microbial types into close proximity, allowing for metabolic and genetic interactions. Biofilm microcolonies may contain multiple microbial species, or different species may occupy different growth niches within the biofilm structure. For example, *Klebsiella pneumoniae* and *P. aeruginosa* coexist in biofilms in a stable relationship even though in broth culture *K. pneumoniae* has twice the growth rate of *P. aeruginosa* (101). This stable coexistence is related to the spatial segregation of the two species in the biofilm structure. Metabolic interactions may be mediated by quorum sensing and genetic interaction by exogenous DNA, as noted above. Microbial consortia can form a biofilm as a result of cooperative growth arising from nutrient exchange or metabolic product removal. Examples of these interactions have been presented by Davey and O'Toole (22). Biofilm consortia are instrumental in the degradation of complex organic matter whether by aerobic or anaerobic digestion. In the case of methanogenic degradation, individual members of the biofilm are dependent on each other's metabolic processes for their own energy production, the net result being an efficient conversion of organic matter to methane. This type of cooperation and mutual dependence is called syntrophism.

## Implications for Food Processing Environments

Little is known about the predominant microflora in food processing environments, as the microflora is highly diverse, and there is no "typical" biofilm growth niche that can be characterized as representative of this environment. Nutrient and growth conditions substantially differ throughout a processing facility, and different facilities utilize different raw materials and employ different process operations and sanitation procedures, all of which impact biofilm microflora. Research on food processing biofilms has centered on the ability of pathogenic and spoilage microorganisms to grow or survive in these environments, with emphasis on the influence of sanitation procedures. Little information is available on interactions of commensal biofilm microflora with pathogenic and spoilage microorganisms. This is not surprising, as the primary strategy for controlling spoilage and pathogenic microorganisms in food processing environments is to reduce levels of all surface-associated microorganisms by cleaning and chemical sanitation. This strategy is generally successful when the environment and equipment are designed and maintained to allow effective cleaning. Evidence indicates that if biofilms are allowed to develop, multispecies biofilms develop tolerance to antimicrobial agents, including chemical sanitizers, through synergistic relationships (14) and that the commensal biofilm microflora can protect pathogens from inactivation. Such protection has been observed for *Campylobacter jejuni* in model water system biofilms (107) and for *Pseudomonas fluorescens* cocultured with *Bacillus cereus* (72).

The predominant heterotrophic bacteria isolated by Jeong and Frank (49, 50) from the wet surfaces in meat and dairy plants consisted of *Staphylococcus, Flavobacterium, Pseudomonas, Bacillus, Micrococcus,* and *Streptococcus* spp. as well as coryneform bacteria. *Listeria monocytogenes* was able to coexist in biofilms produced by these isolates for 8 to 10 days at 21°C and 25 days at 10°C, though often at a low level. Carpentier and Chassaing (16) sampled the predominant microflora from cleaned processing plant surfaces and isolated *Bacillus, Sphingomonas, Pseudomonas, Staphylococcus, Sphingobacterium, Comamonas, Stenotrophomonas, Ralstonia,* and *Chryseomonas* spp. These isolates had significantly different effects on *L. monocytogenes* in the biofilm. Of 29 isolates tested, 3 reduced levels of the pathogen by 3 log units compared to the monoculture listeria biofilms; 11 strains had no effect on *L. monocytogenes* levels, and 4 isolates stimulated listeria growth. Leriche and Carpentier (68) selected a strain of *Staphylococcus sciuri* from this group of isolates for its ability to reduce levels of *L. monocytogenes* in biofilms. Their results indicated that *S. sciuri* produced exocellular materials with antiadhesive properties toward *L. monocytogenes*. These studies provide evidence that the predominant biofilm microflora in a processing facility can have a significant impact on the amount of *L. monocytogenes* organisms that inhabit the facility. Bremer et al. (12) observed that growth of *L. monocytogenes* in biofilm was enhanced by the presence of a *Flavobacterium* sp. and also that survival of the pathogen was enhanced, even under conditions in which the *Flavobacterium* sp. was inactivated. If the *Listeria* cells were being protected by exocellular polymeric substances produced by the *Flavobacterium,* then live *Flavobacterium* would not be required to maintain the protective effect.

It is an objective of the food sanitarian to make the food processing environment as inhospitable as possible for microbial growth and to limit the transmission of microorganisms throughout the facility. One means that has been employed to accomplish the latter goal is the use of disinfectant footbaths. Langsrud et al. (57) isolated biofilm-producing *Serratia marcescens* from footbaths containing quaternary ammonium disinfectant. Though resistant to benzalkonium chloride, these isolates were sensitive to peroxyacetic acid/hydrogen peroxide disinfectant. These observations illustrate the adaptive ability of biofilm microflora. Steps to counteract such adaptation could include greater frequency of cleaning and rotation of chemical sanitizers.

Efforts have been made to select microflora that produce biofilms that reduce the presence of *L. monocytogenes*. This has been accomplished primarily through the use of bacteriocin-producing bacteria. Lee and Frank (63) and Leriche et al. (69) observed that nisin-producing *Lactococcus lactis* reduced levels of *L. monocytogenes* in biofilms. The most effective reduction was observed when initial levels of the pathogen were low (69). Ammor et al. (3) isolated various biofilm bacteria that produced bacteriocin-like substances and found that growth of spoilage and pathogenic microorganisms was inhibited in these biofilms. Biofilms produced by *Lactobacillus sakei* were most inhibitory to undesirable microflora. Zhao et al. (115) applied two cultures with antilisteria properties *(L. lactis* and *Enterococcus durans)* to poultry plant floor drains, a common site for biofilm growth. The treatment successfully reduced levels of *Listeria* spp. in the drains.

*Campylobacter jejuni* is a food-borne and waterborne pathogen that in laboratory culture is sensitive to oxidation and other environmental stresses. However, *C. jejuni* survives environmental stresses when it is incorporated into existing biofilms (108). Microbial interactions within a biofilm can lead to protection against food industry cleaning chemicals. This protective effect is associated with oxygen consumption within the biofilm (67), which is also a plausible mechanism for the survival of *Campylobacter* in biofilms.

The ability of the biofilm matrix to entrap and hold microorganisms has potential public health consequences. Biofilms act as a reservoir for protozoa that harbor pathogens (13). Searcy et al. (92) found that oocysts of *Cryptosporidium parvum,* a pathogen responsible for waterborne disease outbreaks, were captured at the surface of *P. aeruginosa* biofilms in a flowthrough system. Although the oocysts did not penetrate into the biofilm matrix, they were not readily released from the biofilm, even with increased fluid flow.

Microbial interactions in biofilms also have implications for food spoilage microorganisms. Many food spoilage microorganisms, including *Pseudomonas, Shewanella,* lactic acid bacteria, and yeasts, are excellent biofilm producers. Protocooperative growth in milk between *Lactococcus lactis* subsp. *cremoris* and *P. fluorescens* was observed at 7°C in biofilms at low inoculum levels (so insufficient acid was produced to inhibit *P. fluorescens*) (53). *P. fluorescens* stimulated growth of *L. lactis* in biofilms at high inoculum levels. This is an example of how interactions of microorganisms in biofilms often differ from interactions observed in broth culture.

## STRAIN VARIATION IN BIOFILM FORMATION

Environmental adaptation of bacteria often results in the emergence of variant subpopulations (phase variants). Phase variants may occur at high frequency, and often the variant strain readily reverts to the normal phenotype. These phenotypic variants differ from the precursor strain in colony morphology, adherence properties, and biofilm-forming ability. Phase variants arise spontaneously within the biofilm and, in the case of biofilm-based infections of *P. aeruginosa,* upon exposure to an antibiotic (42). A hyperpiliated small-colony variant of *P. aeruginosa* was isolated from biofilms by Deziel et al. (24). This variant also exhibited autoaggregative behavior and enhanced biofilm formation. Small-colony variants of *P. aeruginosa* exhibit diverse phenotypes that may include changes in growth rate and increased antibiotic resistance. The enhanced biofilm variant is associated with increased intracellular c-di-GMP, an observation consistent with the previous discussion of biofilm regulation (42).

A red dry rough phase variant of *Escherichia coli* O157:H7 formed large aggregates in broth and dense biofilms on glass and polystyrene (110). This phenotype is associated with production of curli. Curli enhance adhesion and also contribute to biofilm structure (89, 110). The red dry rough phase variant exhibited significantly decreased biofilm formation under low-nutrient conditions, in contrast to the wild type, which responded to low nutrient levels with greater biofilm production. This observation indicates complexity in the mechanism of biofilm regulation in *E. coli* in response to environmental signals.

Phase variants of *L. monocytogenes* differ in their ability to attach and produce biofilms. Dickson and Siragusa (25) observed that rough strains of *L. monocytogenes* did not attach as well to beef tissue as the wild type. The impaired attachment may be associated with filament formation, as it was not associated with surface chemistry characteristics. Monk et al. (77) isolated two types of rough colony variants of *L. monocytogenes* from biofilms, distinguished by short-chain and long-chain cell morphologies. Both types of rough variants exhibited enhanced biofilm formation, with the variants exhibiting increased cell chain length (filamentous growth) when grown as biofilms. The authors note that increased chain length could result in significant underestimation of cell numbers when using conventional swab and CFU determinations. The formation of long chains by rough variants of *L. monocytogenes* is associated with mutations in genes coding for autolysins MurA and extracellular p60 (73).

Folsom and Frank (32) isolated phase variants of *L. monocytogenes* that were tolerant to hypochlorous acid and produced increased amounts of biofilm. Hypochlorous acid tolerance was not observed when the variants were grown in suspension. A smooth colony spontaneous mutant of a rough variant was found to be most tolerant to chlorine treatment and to produce the most biofilm when compared to other colony variants (rough and smooth) and the wild type. Proteomic analysis of the smooth colony mutant indicated downregulation of a peroxide resistance-associated protein, indicating that the mechanisms for hypochlorous acid tolerance differ from those for peroxide tolerance. Increased expression of sugar binding membrane proteins was observed, a possible mechanism for increasing sugar availability for exocellular polysaccharide production by the biofilm.

*L. monocytogenes* exhibits substantial strain variation with regard to biofilm forming ability. Micrographs presented in Fig. 2 illustrate that *L. monocytogenes* strains produce at least five distinct types of biofilms that can be described as (i) disorganized single cells and small aggregates, (ii) dispersed small aggregates and microcolonies, (iii) dispersed large microcolonies, (iv) connected microcolonies, and (v) discrete microcolonies (31). Specific serotypes and genetic subtypes of *L. monocytogenes* also exhibit diversity in biofilm-forming ability. However, evidence that biofilm forming ability contributes to survival of this pathogen in food processing facilities, or to the ability of the pathogen to cause food-borne disease, is not consistent (8, 26, 40). *L. monocytogenes* strains respond to nutrient conditions differently with regard to biofilm formation. Djordjevic et al. (26) found that all 31 strains of *L. monocytogenes* studied produced more biofilm in minimal defined medium (modified Welshimer's broth) than in nutrient-rich medium (Trypticase soy broth [TSB] with yeast extract). Moltz and Martin (76) observed that 6 of 8 strains gave similar results, with 2 strains producing more biofilm in TSB (no yeast extract added). Folsom et al. (33) determined biofilm formation by 30 strains of *L. monocytogenes* in TSB (no yeast extract added) and TSB diluted 1:10. They found that 14 strains

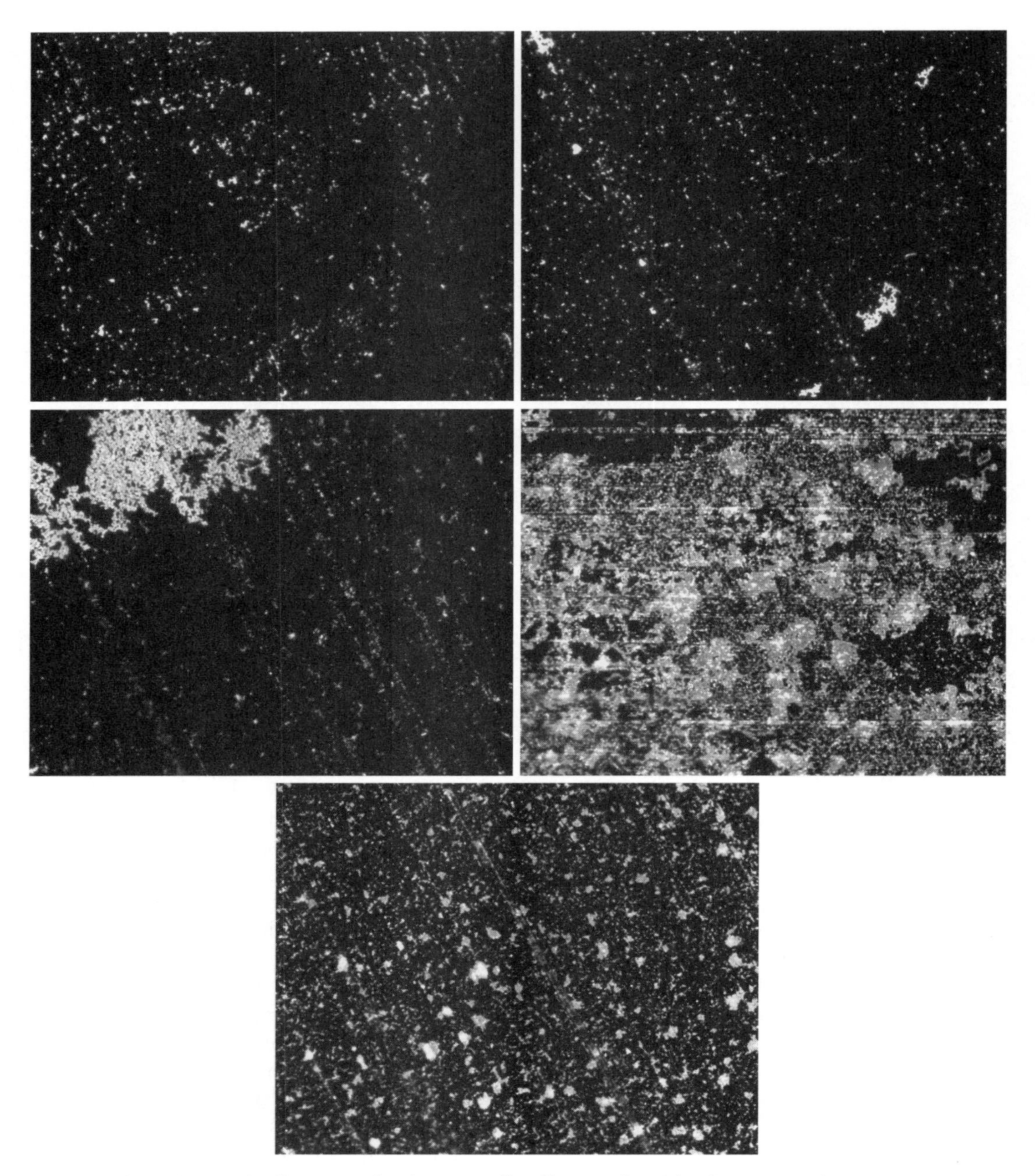

**FIGURE 2** Micrographs illustrating the diversity of biofilms produced by five strains of *Listeria monocytogenes* grown on stainless steel at 32°C for 24 h in TSB. Reprinted from the *Journal of Food Protection* (31) with permission of the publisher.

produced more biofilm in TSB than in diluted TSB, 11 strains produced the same amount of biofilm in both media, and 5 produced more biofilm in diluted TSB than in normal TSB. The low-nutrient medium decreased biofilm formation in serotype 4b strains but not 1/2a strains, which may contribute to the greater isolation of 1/2a than 4b strains from food processing environments.

## WATER SYSTEM BIOFILMS

Most food processing facilities utilize large volumes of water. Regardless of the source, potable water entering a plant is likely to be delivered by a system that supports biofilm growth. Unless the water is treated immediately before use, it will contain microflora originating from distribution system biofilms. Biofilms in water systems are nutrient limited and structured so as to adsorb nutrients from the water flow. Operating conditions that result in turbulent flow, as opposed to laminar flow, across surfaces increase biofilm growth, presumably due to increased exposure of the attached cells to nutrients (97). The microflora of water distribution system biofilms is diverse, and once formed, the biofilms can harbor microorganisms of public health significance, including indicator microorganisms and primary and secondary pathogens. Langmark et al. (56) observed that the type of water disinfection process—UV light or chlorination—did not influence biofilm development within a water system; in both cases, a biofilm of $10^4$ to $10^5$ cells/cm$^2$ formed. The predominant microflora of water system biofilms can be characterized as having low physiological activity and as being difficult to culture using conventional plating methods. However, characterizations using molecular techniques may not differentiate active from inactive or dead biotypes. Keinanen-Toivola et al. (51) used sequence analysis of 16S rRNA clones to identify *Nevskia ramosa* and related organisms as the dominant metabolically active bacteria in biofilms of a municipal water system. A metabolically active *Mycobacterium* clone was also recovered from this system, along with numerous clones of unknown identity. Nonpathogenic mycobacteria commonly inhabit water distribution systems as a result of their inherent chlorine resistance, biofilm formation ability, and association with protozoa (111).

Martiny et al. (75) and Lee et al. (62) investigated microbial succession in water system biofilms. Lee et al. (62) observed that colonization by aerobic *Sphingomonas,* anaerobic *Rhodobacter,* and nonculturable bacteria occurred within the first day of surface exposure. Microbial diversity increased for 70 days, with *Enterobacteriaceae,* possibly *E. coli* and *Shigella,* detected in some samples. Martiny et al. (75) observed initial surface colonization of a water distribution system with a diverse microflora including *Pseudomonas, Sphingomonas, Aquabacterium,* and *Nitrospira* among the culturable bacteria. Initial colonization was followed by growth of *Nitrospira,* which dominated the biofilm after 256 days, with a concurrent decrease in microbial diversity. Between days 256 and 700 a mature diverse biofilm developed, which included heterotrophic and autotrophic organisms.

Water system biofilms are potential harborages for pathogenic microorganisms, including *Helicobacter pylori* (113), *Campylobacter jejuni* (64), and *Legionella pneumophila* (83), as well as conventional indicator microorganisms. The growth of indicator microorganisms in a water system biofilm does not necessarily indicate a public health risk, just as the presence of pathogens in these biofilms would not necessarily be associated with a positive indicator test.

Microorganisms in water system biofilms most likely contribute to the microflora in our food supply. Most of these microorganisms are adapted to low-nutrient environments and cannot multiply in foods. Others, such as *Pseudomonas* and related species, are important spoilage organisms in chilled foods. Pathogenic microorganisms, including bacteria, viruses, and parasites, rarely occur in treated municipal water systems but occasionally cause outbreaks or sporadic illness. Municipal water used in food processing that makes direct contact with food should be treated to remove or inactivate pathogenic microorganisms. Treatment of water at the point of entrance into the processing facility does not preclude the development of biofilms in the in-plant distribution system.

## BIOFILMS ON EDIBLE PLANTS

Multistate disease outbreaks associated with fresh produce consumption along with increased consumption of fresh and minimally processed produce have resulted in increased interest in biofilms on edible plant tissues. Biofilms containing mainly commensal microorganisms can form on roots, leaves, and the internal vascular tissues of edible plants. Biofilms that form in the rhizosphere often consist of plant growth-promoting bacteria that might include nitrogen-fixing bacteria, ammonia-oxidizing bacteria, saprophytes such as *Pseudomonas* spp., and other growth-promoting bacteria such as *Enterobacter agglomerans* (80). Because of numerous disease outbreaks associated with the consumption of leafy greens, there is great interest in biofilms that form on leaf surfaces. Microorganisms on leaf surfaces exist either as dispersed cells or in biofilm communities. Just as a biofilm immersed in liquid develops a microbial community that, although derived from the liquid phase, differs in microbial composition, so does the leaf biofilm differ in microbial composition from attached but dispersed cells on that leaf (9). Edible leaves are aerial habitats with intermittent excess moisture (from rain or dew), periods of low humidity, and exposure to UV radiation and temperature extremes. For example, there may be lengthy periods of dryness or wetness depending on weather conditions and type of irrigation. Leaves are covered with a waxy cuticle which discourages microbial attachment and limits nutrient availability by preventing leaching from the underlying plant tissue. Biofilms that form on leafy greens before harvest are therefore limited by the nature of the habitat. Morris and Monier (80) cite the preference of *P. aeruginosa* for biofilm formation on trichomes as evidence that nutrient availability is limited on much of the leaf surface. Biofilms on leaves usually form as isolated microcolonies with large areas of the leaf free of growth, another indication of heterogeneity of nutrient availability. The protective effects of biofilms are apparent on leaf surfaces, as cells in small aggregates tend to die in times of dehydration stress, whereas cells in large microcolonies tend to survive these conditions (80). A greater amount of exocellular polymeric substances in the large microcolonies is the likely explanation for this phenomenon.

Biofilms on the leaves of edible plants are common. Morris et al. (81) observed biofilms on leaves of spinach, Chinese cabbage, celery, leeks, basil, parsley, and broad-leaved endive. Other types of fresh produce, including lettuce, tomatoes, carrots, and mushrooms, also allow biofilms to develop (87). The conditions in which sprouts are produced are conducive to biofilm proliferation on all parts of the sprout tissue (30). The microflora of fresh produce biofilms is diverse, with *Pseudomonas* and related bacteria (including *Pantoea agglomerans*), filamentous fungi, and gram-positive bacteria commonly isolated (80, 81). The influence of biofilm microflora on survival or growth of human pathogens on fresh produce is largely unknown. Seo and Frank (94) observed that attachment of *E. coli* O157:H7 to lettuce leaf was not influenced by the presence of a *P. fluorescens* biofilm. Cooley et al. (21) found that *Wautersia paucula* enhanced survival of *E. coli* O157:H7 on lettuce roots and leaves, whereas *Enterobacter asburiae* reduced the ability of the pathogen to survive on the lettuce plant 20- to 30-fold.

*Salmonella enterica* serovar Thompson demonstrated good ability to survive on cilantro plants, and under conditions of sufficient temperature and moisture it was able to grow (11). The pathogen became incorporated into *P. agglomerans* biofilms on leaf veins, a location where surface water would accumulate. This report provides evidence that lengthy survival of *Salmonella* during periods of dehydration may also contribute to its ability to survive on plants in the field. The ability of *Salmonella* spp. to survive on fresh produce pre- and postharvest is not necessarily associated with biofilm formation. Lapidot et al. (58) compared survival of wild-type *Salmonella* serovar Typhimurium and a biofilm formation-deficient isogenic mutant on parsley. The strains had similar survival properties when the parsley was subjected to rinsing with water or chlorine solution about 1 h after inoculation; however, the wild type was able to survive chlorination only marginally better than the mutant after the parsley was stored for 1 week at 4 or 25°C.

Microorganisms can grow as biofilms within plant tissues, a phenomenon usually associated with plant disease processes. The possibility that internal biofilms in plant tissues interact with human pathogens internalized during postharvest cooling is worthy of future research.

## BIOFILMS IN FOOD PROCESSING

Though much has been written about biofilms in food processing (2, 18), there are few direct observations of the frequency, microbial composition, and structure of these biofilms. A major impediment to proposing general principles of biofilm structure and composition as related to food processing is the diversity of the potential growth niches with regard to nutrient and water availability, temperature, and chemical stresses. The frequency of cleaning food contact surfaces ranges from daily to weekly. The frequency of cleaning environmental surfaces may range from daily to rarely. Potential sites for biofilm formation range from condensate, in which the main nutrient supply is absorbed from the air and there is no fluid flow, to stress cracks in metal equipment that are saturated with nutrients from product flow. Biofilms may form in equipment used to freeze foods as well as on pasteurization heat exchangers. Probably the most common sites for biofilm formation in a food processing plant are floor drains, a unique environment because of the cyclic exposure to nutrients, water, and cleaning and sanitizing chemicals. There is a tendency in the food industry to refer to any microbial contamination on a surface as a biofilm. However, in modern food processing facilities, most surfaces are cleaned and sanitized sufficiently often so that contaminating microfloras do not have ample opportunity (adequate time, temperature, and nutrient and water availability) to establish a biofilm before being removed or killed by the next cleaning cycle. The amount of CFU per square centimeter on a surface provides an indication of biofilm presence, as high counts ($>10^6$ CFU/cm$^2$) are often the result of surface growth. However, the CFU level that might be considered indicative of a biofilm must be arbitrary because of lack of knowledge of initial contamination levels. (If initial contamination is 100 CFU/cm$^2$, growth up to $10^5$ CFU/cm$^2$ could be a biofilm, whereas a surface could also have a contamination level of $10^5$ CFU/cm$^2$ without growth and therefore no biofilm.) In some cases, poor equipment design allows food to accumulate in "dead ends" that become microbial growth niches. A dead end can be defined as a macro- or microscopic site that entraps nutrients due to lack of fluid flow or product movement. Cleaning and sanitizing solutions are ineffective at dead ends because entrapped nutrients are not removed. Examples include hollow tubing used in conveyor and other equipment framing, hollow conveyor rollers, poorly welded seams, stress cracks in metal, joints with worn gaskets, and pumps with leaky seals. Microbial growth that occurs on food residues entrapped in a dead end is often assumed to have characteristics of a biofilm.

The complexity of the food processing environment is reflected in the complexity of its predominant microflora (16). Most of the information on microbial diversity in food processing environments has been obtained from sampling surfaces that have not been verified to meet the definition of a biofilm, in that neither growth on the surface nor the presence of exocellular matrix is observed (39, 49, 50, 95).

### Biofilms in Fluid Handling Systems

Fluid handling systems in the food industry are generally designed for clean-in-place sanitation. Clean-in-place systems are highly effective at removing organic matter and killing residual microorganisms as long as the system is designed to maintain turbulent fluid flow over all food contact surfaces. Places where flow may be insufficient to obtain adequate cleaning may include crevices at gasketed joints and the gasket material itself (5). Rubber is especially prone to serving as a biofilm harborage because small cracks can appear when it is under pressure. Systems can be built without the need for gaskets by welding lines together. Though this may seem to be a wise strategy for eliminating growth niches, if the internal seam of the welded joint is not smooth, the resulting biofilm at this harborage point will go undetected until product quality is compromised, and at that time the biofilm will be difficult to detect and remove. Microorganisms will attach and grow in heat exchangers used for processing milk if the equipment is operated for sufficiently long times between cleanings. For example, Langeveld et al. (55) found evidence for biofilm formation in the tubular portion of a heat exchanger after 20 h of operation. Biofilms may also form in the postpasteurization portion of milk handling systems (95). Microorganisms in heat exchange systems do not necessarily influence product shelf life, as they usually do not grow well at refrigeration temperatures.

Biofilms in food processing are more likely to contribute to the food product microflora when portions of the production/processing system are open to the environment. Such systems allow access to a diversity of environmental microfloras and therefore a high likelihood that available growth niches will be colonized. In addition, many food processing systems utilize complex equipment combined with processes that are nonlethal to a portion of the microflora. Cheese manufacture provides an example of this, as the cheese manufacturing plant provides numerous opportunities for biofilm formation. Since milk and whey are the predominant nutrient sources in this environment, growth of lactic acid bacteria in these biofilms is favored. Evidence indicates that non-starter culture lactic acid bacteria in biofilms survive the cleaning/sanitizing process and then contaminate the cheese during manufacture (1, 98). *Lactobacillus curvatus* can be isolated from these biofilms and grows in the cheese during ripening, causing the formation of white calcium lactate crystals (white haze defect).

Contamination of ready-to-eat foods with *L. monocytogenes* has resulted in an emphasis on biofilm control in the processing/packaging environment. However, direct evidence that growth of *L. monocytogenes* in biofilms is responsible for product contamination is lacking. The image in Fig. 3 is of a biofilm produced in the drip pan of a cooling unit in a ready-to-eat meat processing facility. In just 1 day, a complex microbial biofilm community is beginning to form on this surface. The flow of air through a cooling unit provides opportunity for aerosol transmission. The isolation of *Listeria* spp. from processing plant surfaces is widely reported, but these reports seldom provide information on total microbial populations on the surface or other evidence for the presence of biofilms (54). Frank et al. (37) found a positive correlation between the presence of *Listeria* spp. and high microbial populations on processing plant surfaces. Relevant research and practical information on the control of *L. monocytogenes* in food processing environments has been summarized by Chmielewski and Frank (18, 19) and Kornacki and Gurtler (54).

The behavior of biofilm microorganisms in food products is of interest because of the possibility that these microorganisms directly contaminate the product. Dykes (28) compared the behavior of biofilm and planktonic *L. monocytogenes* inoculated onto bologna (a product that allows growth) and summer sausage (a product that causes inactivation). There was no difference in growth between biofilm and planktonic cells after inoculation onto the bologna. Biofilm cells that were 48 h old survived marginally better on the summer sausage than planktonic cells or biofilm cells that were 96 h old. Jacobsen and Koch (45) determined the ability of an inoculum of *L. monocytogenes,* prepared by growing as a biofilm and drying,

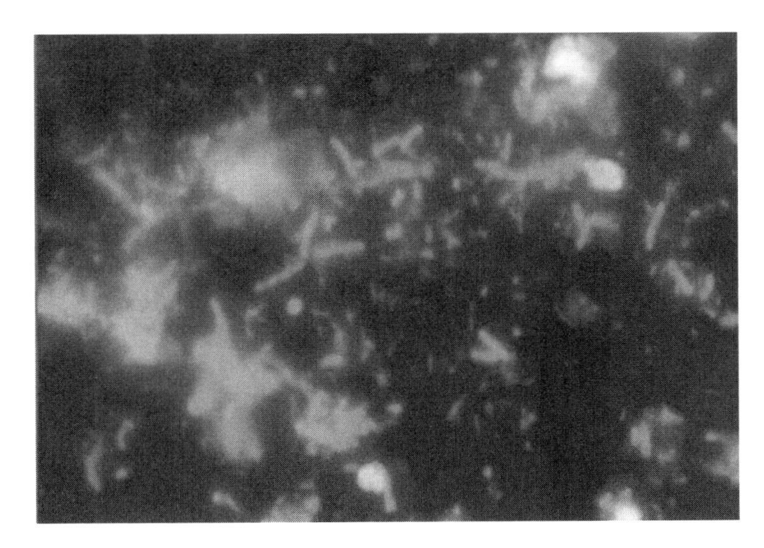

**FIGURE 3** Biofilm formed in 1 day on a stainless steel coupon attached to the inner surface of the drip pan of a chiller unit. The unit was located on the ceiling of a room in a ready-to-eat meat processing plant. The coupon was stained with acridine orange and observed using the oil immersion objective (100×) and a 10× ocular lens on a Carl Zeiss Standard Microscope equipped for epifluorescence. Image courtesy of Amy C. Wong, Food Research Institute, University of Wisconsin, Madison.

to grow in meat products. The dehydration-stressed biofilm cells exhibited a longer lag phase than inocula prepared at 5°C but behaved in a manner similar to that of an inoculum of starved cells. Ashton et al. (4) determined the behavior of *E. coli* O157:H7 prepared as wet and dry biofilms when inoculated onto beef that was subsequently chilled under aerobic and vacuum packaging conditions. The biofilm-prepared inocula behaved similarly to planktonic inocula when product was vacuum packaged but decreased to a greater extent than the planktonic inocula under aerobic conditions.

The diversity of potential biofilm growth niches in food processing environments presents a challenge in developing relevant model systems for research. Model systems for studying food system biofilms may be static (test tubes or microtiter plates) or static with periodic refreshment of nutrients, or they may utilize a continuous flow of nutrients. They may employ low or high nutrient levels and allow growth for hours or weeks. The biofilm substratum may be stainless steel, Buna-N rubber, polycarbonate, polyvinyl chloride, or another polymer. Growth temperatures can range from cold to warm. Many combinations of these conditions occur in a food processing environment. One aspect often missing from model system biofilms is the incorporation of soluble and insoluble food residues into the biofilm. At many biofilm growth sites there is potential for incorporation of lipids and protein of food product origin into the biofilm structure. These components could have a significant role in protecting biofilm cells from removal and inactivation. Since most research on biofilms of relevance to the food processing environment employ model systems that are limited in their representation of the actual diversity of the food processing growth niches, these research findings should be interpreted with caution.

## ANTIMICROBIAL RESISTANCE

Cells in biofilms are more difficult to inactivate by application of antimicrobial chemicals and physical stresses than their planktonic counterparts. This resistance is not associated with mutation or plasmid transfer and is probably based on multiple mechanisms (2). The biofilm matrix can protect cells by inactivating or binding to antimicrobials (102). If biofilm cells are released from the surface and disrupted (as should occur in a cleaning process), the cells immediately lose their protection against antimicrobials such as quaternary ammonia and acid anionic sanitizers (38, 102). Other antimicrobial agents rapidly penetrate the biofilm yet still fail to kill cells at the same rate as the planktonic counterparts (20, 116). A portion of biofilm cells are slow growing or dormant and therefore not susceptible to antibiotics. Lewis (71) presents evidence for a small population of physiologically distinct "persister" cells in biofilms that tolerate exposure to antibiotics by shutting down the targeted cellular functions. In addition, biofilm environments subject a portion of the cells to stresses such as reduced oxygen, starvation, and toxic metabolites that activate the general stress response (27). Chambless et al. (17) developed three-dimensional models to describe the inactivation of biofilm cells assuming four different hypothetical protective mechanisms. The mechanisms that were modeled included slow penetration of the antimicrobial, induction of stress response by the antimicrobial, induction of stress response by nutrient limitation or other alterations to the microenvironment, and the presence of a small population of persister cells. The model simulations indicate a distinctive inactivation curve for each protective mechanism, with each mechanism providing a different spatial distribution of survivors within the treated biofilm. Each protective mechanism produced a plausible result.

## CONCLUSIONS

There is substantial evidence that microorganisms derived from biofilms are ingested with our food and water on a daily basis and may therefore influence the human ecosystem. The major contributor of biofilm microorganisms in our diet is most likely fresh produce, since biofilms form on these foods before harvest, postharvest growth is likely, and the products are consumed without heat treatment. Drinking water is also a likely contributor of biofilm-derived microorganisms to our diet, as water distribution systems readily support biofilm growth. Processed foods that support microbial growth may occasionally be recontaminated with microorganisms of biofilm origin, but current microbial control efforts in food processing facilities (which have led to reductions in food-borne listeriosis over the past decade) minimize the amount of biofilm microflora in these foods. Microbial growth niches in food processing environments are highly diverse in terms of nutrient availability, temperature, and periodic chemical and physical stresses; these conditions, in turn, lead to diversity of microbial content in the biofilms. Most biofilm research with implications for food processing has used model systems that do not reflect the diversity of the biofilm-forming environments present in these facilities.

## REFERENCES


1. **Agarwal, S., K. Sharma, B. G. Swanson, G. U. Yuksei, and S. Clark**. 2006. Nonstarter lactic acid bacteria biofilms and calcium lactate crystals in cheddar cheese. *J. Dairy Sci.* **89:**1452–1466.
2. **Agle, M.** 2007. Biofilms in the food industry, p. 3–18. *In* H. P. Blaschek, H. H. Wang, and M. E. Agle (ed.), *Biofilms in the Food Environment.* IFT Press, Blackwell Publishing, Ames, IA.
3. **Ammor, S., G. Tauveron, E. Dufour, and I. Chevallier.** 2006. Antibacterial activity of lactic acid bacteria against spoilage and pathogenic bacteria isolated from the same meat small-scale facility. 2. Behaviour of pathogenic and spoilage bacteria in dual species biofilms including a bacteriocin-like-producing lactic acid bacteria. *Food Control* **17:**462–468.
4. **Ashton, L. V., I. Geornaras, J. D. Stopporth, P. N. Skandamis, K. E. Belk, J. A. Scanga, G. C. Smith, and J. N. Sofos.** 2006. Fate of inoculated *Escherichia coli* O157:H7, cultured under different conditions, on fresh and decontaminated beef transitioned from vacuum to aerobic packaging. *J. Food Prot.* **69:**1273–1279.
5. **Austin, J. W., and G. Bergeron.** 1995. Development of bacterial biofilms in dairy processing lines. *J. Dairy Res.* **62:**509–519.
6. **Balaban, N., P. Stoodley, C. A. Fux, S. Wilson, J. W. Costerton, and G. Dell'Acqua.** 2005. Prevention of staphylococcal biofilm-associated infections by the quorum sensing inhibitor RIP. *Clin. Orthop. Relat. Res.* **437:**48–54.
7. **Barraud, N., D. J. Hassett, S.-H. Hwang, S. A. Rice, S. Kjeleberg, and J. S. Webb.** 2006. Involvement of nitric oxide in biofilm dispersal of *Pseudomonas aeruginosa*. *J. Bacteriol.* **188:** 7344–7353.

8. **Borucki, M. K., J. D. Peppin, D. White, F. Loge, and D. R. Call.** 2003. Variation in biofilm formation among strains of *Listeria monocytogenes*. *Appl. Environ. Microbiol.* **69:**7336–7342.
9. **Boureau, T., M.-A. Jacques, R. Beruyer, Y. Dessaux, H. Dominguez, and C. E. Morris.** 2003. Comparison of the phenotypes and genotypes of biofilm and solitary epiphytic bacterial populations on broad-leaved endive. *Microbiol. Ecol.* **47:**87–95.
10. **Branda, S. S., F. Chu, D. B. Kearns, R. Losick, and R. Kolter.** 2006. A major protein component of the *Bacillus subtilis* biofilm matrix. *Mol. Microbiol.* **59:**1229–1238.
11. **Brandl, M. T., and R. E. Mandrell.** 2002. Fitness of *Salmonella enterica* serovar Thompson in the cilantro phyllosphere. *Appl. Environ. Microbiol.* **68:**3614–3621.
12. **Bremer, P. J., I. Monk, and C. M. Osborne.** 2001. Survival of *Listeria monocytogenes* attached to stainless steel surfaces in the presence or absence of *Flavobacterium* spp. *J. Food Prot.* **64:**1369–1376.
13. **Brown, M. R. W., and J. Barker.** 1999. Unexplored reservoirs of pathogenic bacteria: protozoa and biofilms. *Trends Microbiol.* **7:**46–50.
14. **Burmolle, M., J. S. Webb, D. Rao, L. H. Hansen, S. J. Sorensen, and S. Kjelleberg.** 2006. Enhanced biofilm formation and increased resistance to antimicrobial agents and bacterial invasion are caused by synergistic interactions in multispecies biofilms. *Appl. Environ. Microbiol.* **72:** 3916–3923.
15. **Camilli, A., and B. L. Bassler.** 2006. Bacterial small-molecule signaling pathways. *Science* **311:** 1113–1116.
16. **Carpentier, B., and D. Chassaing.** 2004. Interactions in biofilms between *Listeria monocytogenes* and resident microorganisms from food industry premises. *Int. J. Food Microbiol.* **97:**111–122.
17. **Chambless, J. D., S. M. Hunt, and P. S. Stewart.** 2006. A three-dimensional computer model of four hypothetical mechanisms protecting biofilms from antimicrobials. *Appl. Environ. Microbiol.* **72:**2005–2013.
18. **Chmielewski, R. A. N., and J. F. Frank.** 2003. Biofilm formation and control in food processing facilities. *Comp. Rev. Food Sci. Safety* **2:**22–32.
19. **Chmielewski, R. A. N., and J. F. Frank.** 2007. Inactivation of *Listeria monocytogenes* biofilms using chemical sanitizers and heat, p. 73–104. *In* H. P. Blaschek, H. H. Wang, and M. E. Agle (ed.), *Biofilms in the Food Environment*. IFT Press, Blackwell Publishing, Ames, IA.
20. **Cochran, W. L., G. A. McFeters, and P. S. Stewart.** 2000. Reduced susceptibility of thin *Pseudomonas aeruginosa* biofilms to hydrogen peroxide and monochloramine. *J. Appl. Microbiol.* **88:**22–30.
21. **Cooley, M. B., D. Chao, and R. E. Mandrell.** 2006. *Escherichia coli* O157:H7 survival and growth on lettuce is altered by the presence of epiphytic bacteria. *J. Food Prot.* **69:**2329–2335.
22. **Davey, M. E., and G. A. O'Toole.** 2000. Microbial biofilms: from ecology to molecular genetics. *Microbiol. Mol. Biol. Rev.* **64:**847–867.
23. **Davies, D. G., M. R. Parsek, J. P. Pearson, B. H. Iglewski, J. W. Costerton, and E. P. Greenberg.** 1998. The involvement of cell-to-cell signals in the development of bacterial biofilm. *Science* **280:**295–298.
24. **Deziel, E., Y. Comeau, and R. Villemur.** 2001. Initiation of biofilm formation by *Pseudomonas aeruginosa* 57RP correlates with emergence of hyperpiliated and highly adherent phenotypic variants deficient in swimming, swarming, and twitching motilities. *J. Bacteriol.* **183:**1195–1204.
25. **Dickson, J. S., and G. R. Siragusa.** 1994. Cell surface charge and initial attachment characteristics of rough strains of *Listeria monocytogenes*. *Lett. Appl. Microbiol.* **19:**192–196.
26. **Djordjevic, D., M. Wiedmann, and L. A. McLandsborough.** 2002. Microtiter plate assay for assessment of *Listeria monocytogenes* biofilm formation. *Appl. Environ. Microbiol.* **68:**2950–2958.
27. **Drenkard, E.** 2003. Antimicrobial resistance of *Pseudomonas aeruginosa* biofilms. *Microb. Infect.* **5:** 1213–1219.
28. **Dykes, G. A.** 2003. Behaviour of *Listeria monocytogenes* on two processed meat products as influenced by temperature or attached growth during preincubation. *Food Microbiol.* **20:**91–96.
29. **Federle, M. J., and B. L. Bassler.** 2003. Interspecies communication in bacteria. *J. Clin. Investig.* **112:**1291–1299.
30. **Fett, W. W.** 2000. Naturally occurring biofilms on alfalfa and other types of sprouts. *J. Food Prot.* **63:**625–632.
31. **Folsom, J. P., and J. F. Frank.** 2006. Chlorine resistance of *Listeria monocytogenes* biofilms and relationship to subtype, cell density, and planktonic cell chlorine resistance. *J. Food Prot.* **69:**1292–1296.
32. **Folsom, J. P., and J. F. Frank.** 2007. Proteomic analysis of a hypochlorous acid-tolerant *Listeria monocytogenes* cultural variant exhibiting enhanced biofilm production. *J. Food Prot.* **70:** 1129–1136.
33. **Folsom, J. P., G. R. Siragusa, and J. F. Frank.** 2006. Formation of biofilm at different nutrient levels by various genotypes of *Listeria monocytogenes*. *J. Food Prot.* **69:**826–834.

34. **Frank, J. F.** 2000. Control of biofilms in the food and beverage industry, p. 205–224. *In* J. Walkeer, S. Surman, and J. Jass (ed.), *Industrial Biofouling*. John Wiley & Sons, Ltd., West Sussex, United Kingdom.
35. **Frank, J. F.** 2001. Microbial attachment to food and food contact surfaces. *Adv. Food Nutr. Res.* **43:**319–370.
36. **Frank, J. F., J. Ehlers, and L. Wicker.** 2003. Removal of *Listeria monocytogenes* and poultry soil-containing biofilms using chemical cleaning and sanitizing agents under static conditions. *Food Prot. Trends* **8:**654–663.
37. **Frank, J. F., R. A. N. Gillett, and G. O. Ware.** 1990. Association of *Listeria* spp. contamination in the dairy processing plant environment with the presence of staphylococci. *J. Food Prot.* **53:**928–932.
38. **Frank, J. F., and R. Koffi.** 1990. Surface-adherent growth of *Listeria monocytogenes* is associated with increased resistance to surfactant sanitizers and heat. *J. Food Prot.* **53:**550–554.
39. **Gunduz, G. T., and G. Tuncel.** 2006. Biofilm formation in an ice cream plant. *Antonie van Leeuwenhoek* **89:**329–336.
40. **Hanna, S. E., and H. H. Wang.** 2007. Biofilm development by *Listeria monocytogenes,* p. 47–71. *In* H. P. Blaschek, H. H. Wang, and M. E. Agle (ed.), *Biofilms in the Food Environment*. IFT Press, Blackwell Publishing, Ames, IA.
41. **Hastings, J. W., and G. Mitchell.** 1971. Endosymbiotic bioluminescent bacteria from the light organ of pony fish. *Biol. Bull.* **141:**261–268.
42. **Haussler, S.** 2004. Biofilm formation by the small colony variant phenotype of *Pseudomonas aeruginosa*. *Environ. Microbiol.* **6:**546–551.
43. **Heydorn, A., B. Ersboll, J. Kato, M. Hentzer, M. R. Parsek, T. Tolker-Nielsen, M. Givskov, and S. Molin.** 2002. Statistical analysis of *Pseudomonas aeruginosa* biofilm development: impact of mutations in genes involved in twitching motility, cell-to-cell signaling, and stationary-phase sigma factor expression. *Appl. Environ. Microbiol.* **68:**2008–2017.
44. **Hunter, R. C., and T. J. Beveridge**. 2005. High-resolution visualization of *Pseudomonas aeruginosa* PAO1 biofilms by freeze-substitution transmission electron microscopy. *J. Bacteriol.* **187:**7619–7630.
45. **Jacobsen, T., and A. G. Koch.** 2006. Influence of histories of the inoculum on lag phase and growth of *Listeria monocytogenes* in meat models. *J. Food Prot.* **69:**532–541.
46. **Jahn, A., and P. H. Nielsen.** 1998. Cell biomass and exopolymer composition in sewer biofilms. *Water Sci. Technol.* **37:**17–24.
47. **Jayasinghearachchi, H. S., and G. Seneviratne.** 2004. A bradyrhizobial-*Penicillium* spp. biofilm with nitrogenase activity improves $N_2$ fixing symbiosis of soybean. *Biol. Fertil. Soils* **40:**432–434.
48. **Jefferson, K. K.** 2004. What drives bacteria to produce a biofilm? *FEMS Microbiol. Lett.* **236:** 163–173.
49. **Jeong, D. K., and J. F. Frank.** 1994. Growth of *Listeria monocytogenes* at 21 EC in biofilms with microorganisms isolated from meat and dairy processing environments. *Lebensm.-Wiss. Technol.* **27:**415–423.
50. **Jeong, D. K., and J. F. Frank.** 1994. Growth of *Listeria monocytogenes* at 10 EC in biofilms with microorganisms isolated from meat and dairy processing environments. *J. Food Prot.* **57:**576–586.
51. **Keinanen-Toivola, M., R. P. Revetta, and J. W. Santo Domingo.** 2006. Identification of active bacterial communities in a model drinking water biofilm system using 16S rRNA-based clone libraries. *FEMS Microbiol. Lett.* **257:**182–188.
52. **Kikuchi, T., Y. Mizunoe, A. Takade, S. Naito, and S. Yoshida.** 2005. Curli fibers are required for development of biofilm architecture in *Escherichia coli* K-12 and enhance bacterial adherence to human uroepithelial cells. *Microbiol. Immunol.* **49:**875–884.
53. **Kives, J., D. Guadarrama, B. Orgaz, A. Rivera-Sen, J. Vazquez, and C. SanJose.** 2005. Interactions in biofilms of *Lactococcus lactis* ssp. *cremoris* and *Pseudomonas fluorescens* cultured in cold UHT milk. *J. Dairy Sci.* **88:**4165–4171.
54. **Kornacki, J. L., and J. Gurtler.** 2007. Incidence and control of *Listeria* in food processing facilities, p. 681–766. *In* E. T. Ryser and E. H. Marth (ed.), Listeria, *Listeriosis, and Food Safety*. CRC Press, Boca Raton, FL.
55. **Langeveld, L. P., R. M. van Montfort-Quasig, A. H. Weerkamp, R. Waalesijn, and J. S. Wever.** 1995. Adherence, growth and release of bacteria in a tube heat-exchanger for milk. *Neth. Milk Dairy J.* **49:**207–220.
56. **Langmark, J., M. V. Storey, N. J. Ashboit, and T. A. Stanstrom.** 2005. Biofilms in an urban water distribution system: measurement of biofilm biomass, pathogens and pathogen persistence within the greater Stockholm area, Sweden. *Water Sci. Technol.* **52:**181–189.
57. **Langsrud, S., T. Moreto, and G. Sundheim.** 2003. Characterization of *Serratia marcescens* survival in disinfecting footbaths. *J. Appl. Microbiol.* **95:**186–195.
58. **Lapidot, A., U. Romling, and S. Yaron.** 2006. Biofilm formation and the survival of *Salmonella* Typhimurium on parsley. *J. Food Microbiol.* **109:**229–233.

59. **Lasa, I.** 2006. Towards the identification of the common features of bacterial biofilm development. *Int. Microbiol.* **9:**21–28.
60. **Lasa, I., and J. R. Penades.** 2006. Bap: a family of surface proteins involved in biofilm formation. *Res. Microbiol.* **157:**99–107.
61. **Ledebor, N. A., J. G. Frye, M. McClelland, and B. D. Jones.** 2006. *Salmonella enterica* serovar Typhimurium requires the Lpf, Pef, and Tafi fimbriae for biofilm formation on HEp-2 tissue culture cells and chicken intestinal epithelium. *Infect. Immun.* **74:**3156–3169.
62. **Lee, D.-G., J.-H. Lee, and S.-J. Kim.** 2005. Diversity and dynamics of bacterial species in a biofilm at the end of the Seoul water distribution system. *World J. Microbiol. Biotechnol.* **21:**155–162.
63. **Lee, S. H., and J. F. Frank.** 1992. Competitive growth and attachment of *Listeria monocytogenes* and *Lactococcus lactis* ssp. *lactis* ATCC 11454. *J. Microbiol. Biotechnol.* **2:**73–77.
64. **Lehtola, M. J., T. Pitkanen, L. Miebach, and I. T. Miettinen.** 2006. Survival of *Campylobacter jejuni* in potable water biofilms: a comparative study with different detection methods. *Water Sci. Technol.* **54:**57–61.
65. **Lemon, K. P., D. E. Higgins, and R. Kolter.** 2007. Flagellar motility is critical for *Listeria monocytogenes* biofilm formation. *J. Bacteriol.* **189:** 4418–4424.
66. **Lequette, Y., and E. P. Greenberg.** 2005. Timing and localization of rhamnolipid synthesis gene expression in *Pseudomonas aeruginosa* biofilms. *J. Bacteriol.* **187:**37–44.
67. **Leriche, V., R. Briandet, and B. Carpentier.** 2003. Ecology of mixed biofilms subjected daily to a chlorinated alkaline solution: spatial distribution of bacterial species suggests a protective effect of one species to another. *Environ. Microbiol.* **5:**64–71.
68. **Leriche, V., and B. Carpentier.** 2000. Limitation of adhesion and growth of *Listeria monocytogenes* on stainless steel surfaces by *Staphylococcus sciuri* biofilms. *J. Appl. Microbiol.* **88:**594–605.
69. **Leriche, V., D. Chassaing, and B. Carpentier.** 1999. Behaviour of *L. monocytogenes* in an artificially made biofilm of a nisin-producing strain of *Lactococcus lactis*. *Int. J. Food Microbiol.* **51:** 169–182.
70. **Leriche, V., P. Sibille, and B. Carpentier.** 2000. Use of an enzyme-linked lectinsorbent assay to monitor the shift in polysaccharide composition in bacterial biofilms. *Appl. Environ. Microbiol.* **66:**1851–1856.
71. **Lewis, K.** 2005. Persister cells and the riddle of biofilm survival. *Biochemistry* (Moscow) **70:**267–274.
72. **Lindsay, D., V. S. Brozel, J. F. Mostert, and A. von Holy.** 2002. Differential efficacy of a chlorine dioxide-containing sanitizer against single species and binary biofilms of a dairy-associated *Bacillus cereus* and a *Pseudomonas fluorescens* isolate. *J. Appl. Microbiol.* **92:**352–361.
73. **Machata, S., T. Hain, M. Rohde, and T. Chakraborty.** 2005. Simultaneous deficiency of both MurA and p60 proteins generates a rough phenotype in *Listeria monocytogenes*. *J. Bacteriol.* **187:**8385–8394.
74. **Mack, D., P. Becker, I. N. Chatterjee, S. Dobinsky, J. K.-M. Knobloch, G. Peters, R. Holger, and M. Herrmann.** 2004. Mechanisms of biofilm formation in *Staphylococcus epidermidis* and *Staphylococcus aureus*: functional molecules, regulatory circuits, and adaptive responses. *Int. J. Med. Microbiol.* **294:**203–212.
75. **Martiny, A. C., T. M. Jorgensen, H.-J. Albrechsten, E. Arvin, and S. Molin.** 2003. Long-term succession of structure and diversity of a biofilm formed in a model drinking water distribution system. *Appl. Environ. Microbiol.* **69:** 6899–6907.
76. **Moltz, A. G., and S. E. Martin.** 2005. Formation of biofilms by *Listeria monocytogenes* under various growth conditions. *J. Food Prot.* **68:**92–97.
77. **Monk, I. R., G. M. Cook, B. C. Monk, and P. J. Bremer.** 2004. Morphotypic conversion of *Listeria monocytogenes* biofilm formation: biological significance of rough colony isolates. *Appl. Environ. Microbiol.* **70:**6686–6694.
78. **Moons, P., R. Van Houdt, A. Aertsen, K. Vanoirbeek, Y. Engelborghs, and C. W. Michiels.** 2006. Role of quorum sensing and antimicrobial component production by *Serratia plymuthica* in formation of biofilms, including mixed biofilms with *Escherichia coli*. *Appl. Environ. Microbiol.* **72:**7294–7300.
79. **Morgan, R., S. Kohn, S.-H. Hwang, D. J. Hassett, and K. Sauer.** 2006. BdlA, a chemotaxis regulator essential for biofilm dispersion in *Pseudomonas aeruginosa*. *J. Bacteriol.* **188:**7335–7343.
80. **Morris, C. E., and J.-M. Monier.** 2003. The ecological significance of biofilm formation by plant-associated bacteria. *Annu. Rev. Phytopathol.* **41:**429–453.
81. **Morris, C. E., J.-M. Monier, and M.-A. Jacques.** 1997. Methods for observing microbial biofilms directly on leaf surfaces and recovering them for isolation of culturable microorganisms. *Appl. Environ. Microbiol.* **63:**1570–1576.
82. **O'Toole, G. A., and R. Kolter.** 1998. Flagellar and twitching motility are necessary for *Pseudomonas aeruginosa* biofilm development. *Mol. Microbiol.* **30:**295–304.
83. **Piao, Z., C. C. Sze, O. Barysheva, K. Iida, and S. Yoshida.** 2006. Temperature-regulated

formation of mycelial mat-like biofilms by *Legionella pneumophila*. *Appl. Environ. Microbiol.* **72:** 1613–1622.

84. **Pratt, L. A., and R. Kolter.** 1998. Genetic analysis of *Escherichia coli* biofilm formation: roles of flagella, motility, chemotaxis and type I pili. *Mol. Microbiol.* **30:**285–293.
85. **Pruss, B. M., C. Besemann, A. Denton, and A. J. Wolfe.** 2006. A complex transcription network controls the early stages of biofilm development by *Escherichia coli*. *J. Bacteriol.* **188:**3731–3739.
86. **Purevdorj, B., J. W. Costerton, and P. Stoodley.** 2002. Influence of hydrodynamics and cell signaling on the structure and behavior of *Pseudomonas aeruginosa* biofilms. *Appl. Environ. Microbiol.* **68:**4457–4464.
87. **Rayner, J., R. Veeh, and J. Flood.** 2004. Prevalence of microbial biofilms on selected fresh produce and household surfaces. *Int. J. Food Microbiol.* **95:**29–39.
88. **Rickard, A. H., P. Gilbert, N. J. High, P. E. Kolenbrander, and P. S. Handley.** 2003. Bacterial coaggregation: an integral process in the development of multi-species biofilms. *Trends Microbiol.* **11:**94–99.
89. **Ryu, J. H., H. Kim, J. F. Frank, and L. R. Beuchat.** 2004. Attachment and biofilm formation on stainless steel by *Escherichia coli* O157:H7 as affected by curli production. *Lett. Appl. Microbiol.* **39:**359–362.
90. **Sauer, K., A. K. Camper, G. D. Ehrlich, J. W. Costerton, and D. G. Davies.** 2002. *Pseudomonas aeruginosa* displays multiple phenotypes during development as a biofilm. *J. Bacteriol.* **184:**1140–1154.
91. **Schooling, S. R., and T. J. Beveridge.** 2006. Membrane vesicles: an overlooked component of matrices of biofilms. *J. Bacteriol.* **188:**5945–5957.
92. **Searcy, K. E., A. I. Packman, E. R. Atwill, and T. Harter.** 2006. Capture and retention of *Cryptosporidium parvum* oocysts by *Pseudomonas aeruginosa* biofilms. *Appl. Environ. Microbiol.* **72:** 6242–6247.
93. **Sela, S., S. Frank, E. Belausoz, and R. Pinto.** 2006. A mutation in the *lux*S gene influences *Listeria monocytogenes* biofilm formation. *Appl. Environ. Microbiol.* **72:**5653–5658.
94. **Seo, K. H., and J. F. Frank.** 1999. Attachment of *Escherichia coli* O157:H7 to lettuce leaf surface and bacterial viability in response to chlorine treatment as demonstrated using confocal scanning laser microscopy. *J. Food Prot.* **62:**3–9.
95. **Sharma, M., and S. K. Anand.** 2002. Characterization of constitutive microflora of biofilms in dairy processing lines. *Food Microbiol.* **19:**627–636.
96. **Shirtiff, M. E., J. T. Mader, and A. K. Camper.** 2002. Molecular interactions in biofilms. *Chem. Biol.* **9:**859–871.
97. **Simoes, L. C., N. Azedo, A. Pacheco, C. W. Keevil, and M. J. Vieira.** 2006. Drinking water biofilm assessment of total and culturable bacteria under different operating conditions. *Biofouling* **22:**91–99.
98. **Somers, E. B., M. E. Johnson, and A. C. L. Wong.** 2001. Biofilm formation and contamination of cheese by nonstarter lactic acid bacteria in the dairy environment. *J. Dairy Sci.* **84:** 1926–1936.
99. **Southerland, I. W.** 2001. The biofilm matrix—an immobilized but dynamic microbial environment. *Trends Microbiol.* **9:**222–227.
100. **Spoering, A. L., and M. S. Gilmore.** 2006. Quorum sensing and DNA release in bacterial biofilms. *Curr. Opin. Microbiol.* **9:**133–137.
101. **Stewart, P. S., A. K. Camper, S. D. Handran, C.-T. Huang, and M. Warnecke.** 1997. Spatial distribution and coexistence of *Klebsiella pneumoniae* and *Pseudomonas aeruginosa* in biofilms. *Microb. Ecol.* **33:**2–10.
102. **Stewart, P. S., J. Rayner, F. Roe, and W. M. Rees.** 2001. Biofilm penetration and disinfection efficacy of alkaline hypochlorite and chlorosulfamates. *J. Appl. Microbiol.* **91:**525–532.
103. **Stoodley, P., K. Sauer, D. G. Davies, and J. W. Costerton.** 2002. Biofilms as complex differentiated communities. *Annu. Rev. Microbiol.* **56:**187–209.
104. **Stoodley, P., S. Wilson, L. Hall-Stoodley, J. D. Boyle, H. M. Lappin-Scott, and J. W. Costerton.** 2001. Growth and detachment of cell clusters from mature mixed-species biofilms. *Appl. Environ. Microbiol.* **67:**5608–5613.
105. **Thomas, C. M., and K. M. Nielsen.** 2005. Mechanisms of, and barriers to, horizontal gene transmission between bacteria. *Nat. Rev. Microbiol.* **3:**711–721.
106. **Thormann, K. M., S. Duttler, R. M. Saville, M. Hyodo, S. Shukla, Y. Hayakawa, and A. M. Spormann.** 2006. Control of formation and cellular detachment from *Shewanella oneidensis* MR-1 biofilms by cyclic di-GMP. *J. Bacteriol.* **188:**2681–2691.
107. **Trachoo, N., and J. F. Frank.** 2002. Effectiveness of chemical sanitizers against *Campylobacter jejuni*-containing biofilms. *J. Food Prot.* **65:** 1117–1121.
108. **Trachoo, N., J. F. Frank, and N. J. Stern.** 2002. Survival of *Campylobacter jejuni* in biofilms isolated from chicken houses. *J. Food Prot.* **65:** 1110–1116.
109. **Trotonda, M. P., A. C. Manna, A. L. Cheung, I. Lasa, and J. R. Penades.** 2005.

SarA positively controls Bap-dependent biofilm formation in *Staphylococcus aureus*. *J. Bacteriol.* **187:**5790–5798.

110. **Ulrich, G. A., P. H. Cooke, and E. B. Solomon.** 2006. Analyses of the red-dry-rough phenotype of an *Escherichia coli* O157:H7 strain and its role in biofilm formation and resistance to antibacterial agents. *Appl. Environ. Microbiol.* **72:**2564–2572.

111. **Vaerewijck, M. J. M., G. Huys, J. C. Palomino, J. Swings, and F. Portaels.** 2005. Mycobacteria in drinking water distribution systems: ecology and significance for human health. *FEMS Microbiol. Rev.* **29:**911–934.

112. **Van Houdt, R., A. Aertsen, A. Jansen, A. L. Quintana, and C. W. Michiels.** 2004. Biofilm formation and cell-to-cell signaling in Gram-negative bacteria isolated from a food processing environment. *J. Appl. Microbiol.* **96:** 177–184.

113. **Watson, C. L., R. J. Owen, B. Said, S. Lai, J. V. Lee, S. Surman-Lee, and G. Nichols.** 2004. Detection of *Helicobacter pylori* by PCR but not culture in water and biofilm samples from drinking water distribution systems in England. *J. Appl. Microbiol.* **97:**690–698.

114. **Whitchurch, C. B., T. Tolker-Nielsen, P. C. Ragas, and J. S. Mattick.** 2002. Extracellular DNA required for bacterial biofilm formation. *Science* **295:**1487.

115. **Zhao, T., T. C. Podtburg, P. Zhao, B. E. Schmidt, D. A. Baker, B. Cords, and M. P. Doyle.** 2006. Control of *Listeria* spp. by competitive-exclusion bacteria in floor drains of a poultry processing plant. *Appl. Environ. Microbiol.* **72:**3314–3320.

116. **Zheng, Z. L., and P. S. Stewart.** 2002. Penetration of rifampin through *Staphylococcus epidermidis* biofilms. *Antimicrob. Agents Chemother.* **46:**900–903.

# QUORUM SENSING AND SIGNAL TRANSDUCTION IN BIOFILMS: THE IMPACTS OF BACTERIAL SOCIAL BEHAVIOR ON BIOFILM ECOLOGY

*Yung-Hua Li*

# 7

Bacteria in nature can exist in two physical environments: a planktonic environment, in which bacteria are suspended in fluid phase as free-living cells, and a sessile biofilm environment, in which bacteria attach to a surface and form an integrated community (6, 8). Although they frequently shift from one environment to another, bacteria in most natural environments are predominantly associated with surfaces and live in a multispecies biofilm community embedded in a self-derived extracellular polymeric matrix (6–8, 54). It was believed for many years that bacteria, unlike eukaryotic organisms, behaved as self-sufficient individuals and maintained a unicellular lifestyle (8, 16). During infections, bacterial mass was considered nothing more than the sum of these individuals. Our perception of bacteria as keeping a unicellular lifestyle was deeply rooted in the pure-culture paradigm of Robert Koch's era, when Koch established his "golden criteria" to define a bacterial pathogen by using pure-culture approaches (6, 16). Indeed, Koch's concept has led to the great success in the identification of bacterial pathogens and development of antibiotic treatments in acute bacterial infections (20). However, pure-culture planktonic growth of bacteria rarely exists in natural environments. In fact, bacteria in nature almost always live in a complex and dynamic surface-associated community or biofilm (6, 8, 54). It has not been until the last three decades that our view of a self-sufficient unicellular lifestyle of bacteria has changed. The advances from at least two major research areas, biofilms and bacterial quorum sensing, have led us to begin to appreciate, in much more detail for the first time, the concept that bacteria can organize into groups, form well-organized communities, and communicate with each other for coordinated activities or social life that was once believed to be restricted to multicellular organisms (16).

## BACTERIAL SOCIAL BEHAVIORS: THE CONNECTION BETWEEN QUORUM SENSING AND BIOFILMS

Most surfaces on this planet teem with microbial biofilms, which account for over 99% of microbial life (8). If taking an intact biofilm for view under a microscope, one immediately finds that bacteria in biofilms do not stick together randomly but, rather, form well-

*Yung-Hua Li*, Department of Microbiology and Immunology and Department of Applied Oral Sciences, Dalhousie University, 5981 University Ave., Rm. 5215, Halifax, NS B3H 3J5, Canada.

*Food-Borne Microbes: Shaping the Host Ecosystem*, Edited by L.-A. Jaykus, H. H. Wang, and L. S. Schlesinger, © 2009 ASM Press, Washington, DC

organized communities with numerous specialized configurations. One may also find that bacterial cells in biofilms physically interact with each other and maintain intimate relationships (24, 25). Even without physical contact, bacteria living at the same habitat likely secrete extracellular components or small molecules to interact with each other. Work over the last three decades indicates that bacteria have evolved sophisticated mechanisms to form biofilm communities, to communicate with each other, and to exchange their genetic information (8, 24). Especially, many bacteria have been found to regulate diverse physiological processes through a mechanism called quorum sensing, in which bacterial cells produce, detect, and respond to small diffusible signal molecules to communicate with each other for coordinated activities in a cell density-dependent manner (2, 11). It has long been known that in infectious diseases, the invading bacteria need to reach a critical cell density before they express virulence and overwhelm the host defense mechanisms to initiate an infection. Since quorum sensing systems are widespread in prokaryotic organisms, it is not surprising that bacterial cell-cell communication mechanisms have important implications in microbial infections. The ability of bacteria to communicate with each other through quorum sensing and behave as a group like a multicellular organism has provided significant benefits to bacteria in host colonization, formation of biofilms, defense against competitors, and adaptation to changing environments (8–13). Microbiologists have discovered an unexpectedly high degree of coordinated multicellular behaviors that have led to the perception of biofilms as "cities" of microorganisms (54). A growing body of excellent reviews has plotted the highlights of quorum sensing and its roles in biofilm development over the last decades (2, 23, 31, 47). A connection between quorum sensing and microbial biofilms has brought together investigators who have a common interest in how bacteria function as a group for social activities (41). Quorum sensing and biofilm biology have now become very active areas in microbiology. The integration of scientists who are interested in bacterial social behavior into the biofilm research field represents a powerful force in the development of new strategies to prevent and treat biofilm-associated infections (41).

## PROBLEMS FROM BACTERIAL SOCIAL ACTIVITIES

Since the discovery of penicillin in 1928, antibiotics have proven tremendously successful in controlling acute bacterial infections (5, 6, 20). Microbiologists have learned to predict antibiotic effects in vivo by evaluating the MIC or the minimal bactericidal concentration in vitro. MIC and minimal bactericidal concentration assess the effect of antibiotics against planktonic organisms in the exponential growth phase and therefore correctly predict antibiotic efficacy against rapidly dividing bacteria in acute infections. However, clinicians who deal with chronic biofilm-associated infections, such as medical device or implant infections, frequently fail to cure their patients by using the same treatments (5, 6). There is increasing evidence that biofilm infections often resist to the highest deliverable levels of antibiotics. The infections may persist for months or years, resulting in long-term suffering and tissue damage (5). There are many examples of biofilm infections threatening human health, including infections of bone, airway/lung tissue, cardiac tissues, the middle ear, the gastrointestinal tract, the eyes, the urogenital tract, prosthetic devices, indwelling catheters, and implants and dental diseases (8). As reported by the Centers for Disease Control and Prevention (CDC), biofilms have been involved in more than 65% of hospital infections (8). Because of inherent resistance to antibiotics, biofilm infections can be life threatening to immunocompromised patients (5, 6, 8).

Studies of microbial biofilms revealed that bacteria present in biofilms have characteristics different from those of their free-living counterparts, including a significantly increased resistance to antibiotics and the host immune response (5, 6, 8). Living in biofilms allows bacteria to have several advantages to interact with each other and function as a group for coordinated activities. Bacteria with altered physiological activities (biofilm phenotypes) are known to result largely from bacterial social behaviors controlled by quorum sensing or other mechanisms when they are living in biofilms (16). More importantly, these changed phenotypes are usually associated with the virulence and pathogenicity of bacteria (10, 40). In modern clinical microbiology, the establishment of bacterial biofilms has been considered an important pathogenic trait in chronic infections (3, 5–8, 10). The armament of therapeutic agents available to treat bacterial infections today does not take into account the unique biology of bacterial social behaviors in biofilms. This becomes a problem, because biofilms resulting in persistent infections cannot be resolved with standard antibiotic treatment. Since we have not considered the problem of bacterial group behaviors until recently, effective therapeutic strategies to prevent or treat biofilm infections are currently not available. Therefore, understanding bacterial social behaviors and their molecular mechanisms in the development of biofilms will greatly facilitate the development of novel strategies in the prevention and treatment of biofilm infections.

## QUORUM SENSING AS A CENTRAL MECHANISM FOR SOCIAL ACTIVITIES

Bacteria in a community may convey their presence to one another by producing, detecting, and responding to the accumulation of chemical signaling molecules called autoinducers (2, 37). This process of intercellular communication, called quorum sensing, was first described for the bioluminescent marine bacterium *Vibrio fischeri* (2, 12, 37, 46). *V. fischeri* lives in symbiotic associations with a number of marine animal hosts. In these partnerships, the host uses light produced by *V. fischeri* for specific purposes such as attracting prey, avoiding predators, or finding a mate (12, 37). In exchange for the light it provides, *V. fischeri* obtains a nutrient-rich environment, in which it resides (12, 37, 46). A luciferase enzyme complex is found to be responsible for light production in *V. fischeri*. Bioluminescence occurs only when *V. fischeri* is at high cell density, which is controlled by quorum sensing. Specifically, the production and accumulation of, and the response to, a minimum threshold concentration of an autoinducer regulate density-dependent light production in *V. fischeri* and enables *V. fischeri* to emit light (37, 46). Remarkably, such a quorum sensing-mediated social activity for bioluminescence emission by marine bacteria has been found on a global scale (38). Over the centuries, mariners have reported witnessing mystery nocturnal displays in which the surface of the sea produces an intensive, uniform, and sustained glow, called "milky sea," which extends horizontally over 100 km of sea surface. Miller and colleagues have recently used a satellite sensor system and detected such a massive bioluminescence emission of milky sea, which extended over an area as large as the state of Connecticut in the northwestern Indian Ocean (38). The milky sea is an excellent manifestation of quorum sensing-mediated bioluminescence bloom produced by massive numbers of a marine bacterium, *Vibrio harveyi,* living in association with microalga colonies on the surface of the sea (38). Recent studies have well documented that such a global scale of bacterial social activities for bioluminescence glowing is tightly regulated by multiple quorum sensing pathways that form a complex regulatory network (12, 37, 46).

It is now known that many bacteria regulate diverse physiological processes through one or more quorum sensing mechanisms, involving symbiosis, spore formation, genetic competence, antibiotic production, virulence, and biofilm formation (7, 9–14, 21). Detection of autoinducers allows bacteria to distinguish between low and high cell densities and allows a bacterial population to coordinately control the gene expression of the entire population. In many bacteria, quorum sensing represents a central mechanism to regulate social behaviors and allows bacteria to behave as multicellular organisms and to reap benefits that would be unattainable to them as individuals (12, 21, 37, 46).

## COMMON THEMES IN BACTERIAL QUORUM SENSING

Quorum sensing relies upon the interaction of a small diffusible signal molecule with a sensor or transcriptional activator to initiate gene expression for coordinated activities. Studies of bacterial quorum sensing have led to the discovery that bacteria produce, detect, and respond to a vast array of chemical molecules as signals or "languages" for cell-cell communication (2, 13, 46). The processes controlled by quorum sensing are diverse and reflect the specific needs of particular communities. In all cases, a signal molecule activates a complex adaptation response involving many genes and their regulatory network. Quorum sensing systems in bacteria can generally be divided into three classes: (i) LuxI/LuxR-type quorum sensing in gram-negative bacteria, which use *N*-acyl-homoserine lactones (AHL) as signal molecules (Fig. 1); (ii) oligopeptide–two-component-type quorum sensing in gram-positive bacteria, which use small peptides as signal molecules (Fig. 2); and (iii) *luxS*-encoded autoinducer 2 (AI-2) quorum sensing in both gram-negative and gram-positive bacteria (Fig. 3). Each type of signal molecule is detected and responded to by a precise sensing apparatus and regulatory network (40, 46).

In gram-negative bacteria, signal molecules are AHL whose synthesis is dependent on a LuxI-like protein (42). AHL freely diffuse across the cell membrane and increase in concentration in the environment in proportion to cell density. A cognate LuxR-like protein is responsible for recognition of the AHL, and when bound to the AHL, LuxR-like protein binds to specific promoter DNA elements and activates transcription of target genes (Fig. 1). The biochemical mechanism of action of the LuxI/LuxR pairs is conserved. The LuxI-like enzymes produce a specific AHL by coupling the acyl side chain of a specific acyl-acyl carrier protein from the fatty acid biosynthetic machinery to the homocysteine moiety of *S*-adenosylmethionine. This intermediate lactonizes to form AHL, releasing methylthioadenosine. There are over 100 species of gram-negative bacteria identified as using LuxI/LuxR-type quorum sensing to control a wide range of cellular processes. Each species produces a unique AHL or a unique combination of AHL, and as a result, only the members of the same species recognize and respond to their own signal molecule (21, 46). Many other examples of gram-negative circuits exist that utilize a basic LuxI/LuxR quorum sensing mechanism onto which additional regu-

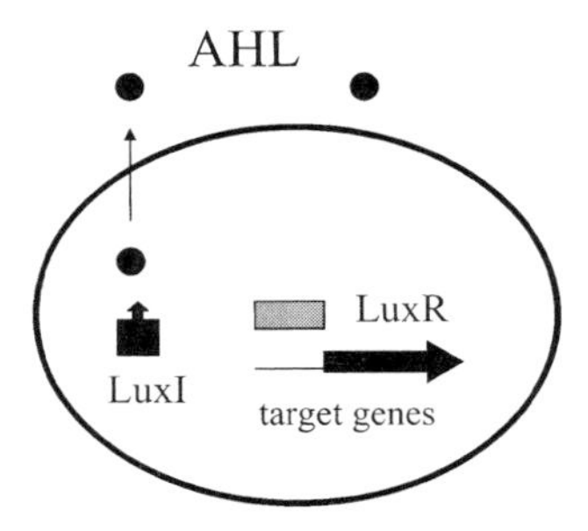


Transcription is not activated at low cell density.

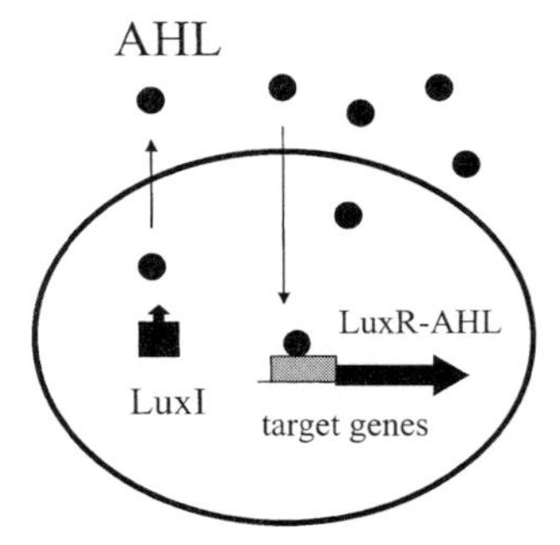


Transcription is activated at high cell density.

**FIGURE 1** LuxI/LuxR-type quorum sensing in gram-negative bacteria. The LuxI-like protein is an autoinducer synthase that catalyzes the formation of a specific AHL. The AHL freely diffuses through the cell membrane at high cell density. LuxR is a transcriptional regulator protein that binds to the diffusing AHL and, in turn, activates the transcription of its target genes.

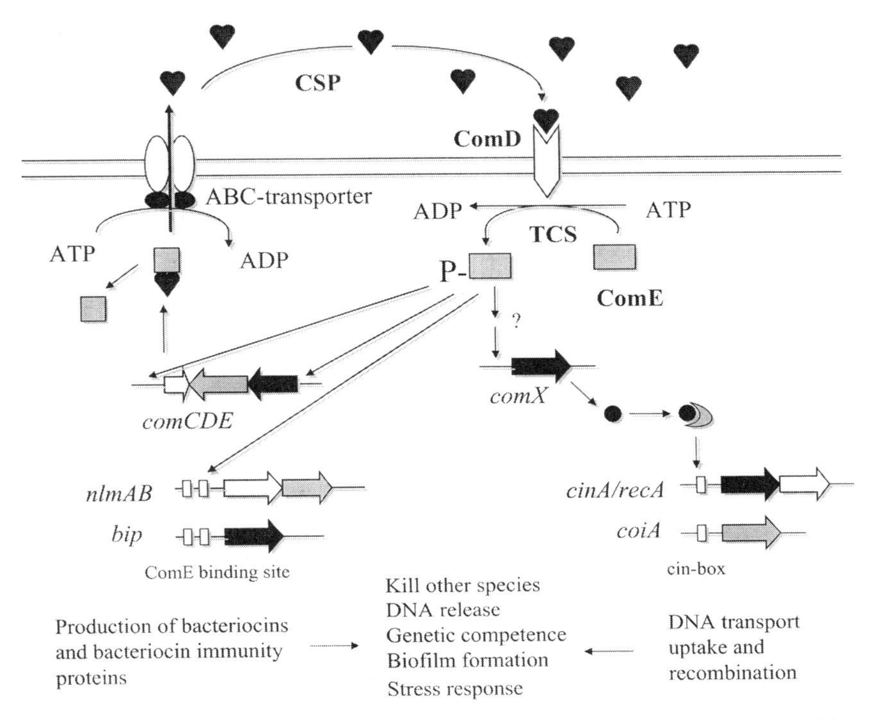


**FIGURE 2** Oligopeptide–two-component-type quorum sensing in gram-positive bacteria. Here is a hypothetical model of a quorum sensing system and its controlled phenotypes in *Streptococcus mutans*. ComD is the histidine kinase protein to sense CSP. ComE is the cognate response regulator to control the transcription of its target genes with the promoter containing a conserved 9-bp repeat element of accgttnag-12 bp-accgttnag (ComE binding site). ComX is an alternative sigma factor that is presumably regulated by ComE and directs RNA polymerase to drive transcription of the late competence genes, such as *cinA, recA,* and *coiA,* which contain the consensus sequence of tacgaata (cin-box). *nlmAB* and *bip* are the genes encoding CSP-dependent bacteriocin and bacteriocin immunity protein.

latory factors have been layered. These "designer" regulatory components enable a wide assortment of behaviors to be controlled by a quorum sensing mechanism in a particular species of bacteria (13, 14, 42, 52).

In contrast to those in gram-negative bacteria, quorum sensing systems in gram-positive bacteria generally consist of three components (Fig. 2): a signaling peptide known as autoinducing peptide (AIP) and a two-component signal transduction system that specifically detects and responds to an AIP (7, 11, 40, 51). In further contrast to AHL signals, cell membrane is not permeable by AIP, but, rather, a dedicated oligopeptide transporter, largely an ABC transporter, is required to export AIP into the extracellular environment (7, 11, 40, 47). Gram-positive bacteria normally produce a signal peptide precursor which is cleaved from the double-glycine consensus sequence, and the active AIP is then exported through a peptide-specific ABC transporter into their environments. Most signaling peptides in gram-positive bacteria typically consist of 5 to 25 amino acids, and some contain unusual side chains (7, 9, 11). Detection of signaling peptides in gram-positive bacteria is almost exclusively mediated

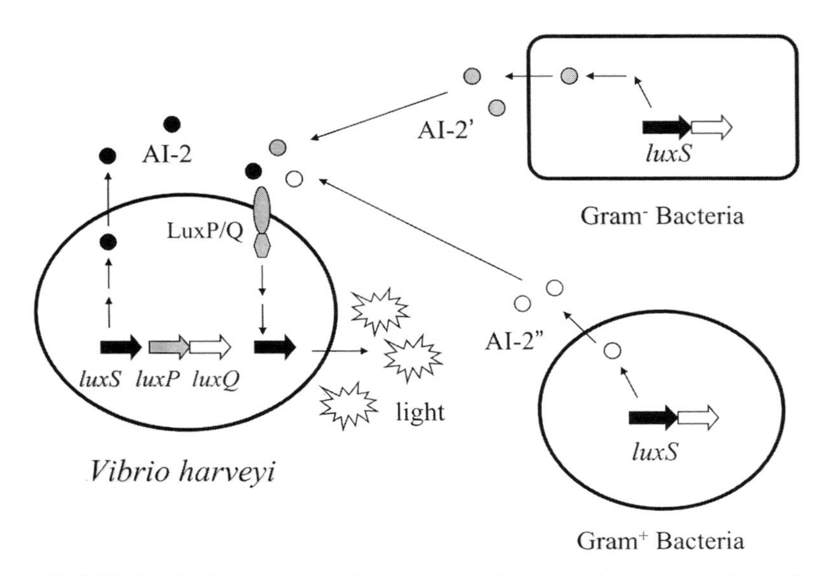


**FIGURE 3** *luxS*-encoded AI-2 quorum sensing in both gram-negative and -positive bacteria. AI-2 is synthesized by the enzyme LuxS and chemically is a furanone. Many gram-negative and -positive bacteria have been found to harbor LuxS homologues in their genomes, but their cognate receptors are unknown with the exception of that in *Vibrio harveyi,* which uses the LuxPQ two-component-like system to sense and respond to AI-2 for bioluminescence emission. AI-2′ and AI-2″ stand for AI-2 homologues from different species, and they can be sensed by LuxPQ of *V. harveyi.*

by a two-component signal transduction system which consists of a membrane-associated histidine kinase protein sensing the AIP and a cytoplasmic response regulator protein enabling the cell to respond to the peptide via regulation of gene expression (11, 40, 47).

In addition to these two quorum sensing mechanisms, another type of quorum sensing, mediated by AI-2, has been described for both gram-negative and gram-positive organisms (12, 25, 46, 47). Different from the above-described two systems, which are specifically for intraspecies signaling, AI-2 allows for interspecies communication, so it is called a "universal language" used for cross-species communication (Fig. 3). AI-2, which was first characterized in the marine bacterium *V. harveyi,* is a furanosyl borate that regulates cell density-dependent bioluminescence (4). The synthesis of AI-2 depends on a *luxS*-encoded synthase, which is a metabolic enzyme involved primarily in the conversion of ribosylhomocysteine into homocysteine and 4,5-dihydroxy-2,3-pentanedione (DPD), the precursor of AI-2 (4). A *luxS* mutation interrupting this metabolic pathway changes the whole metabolism of the bacteria. Homologues of LuxS have been found in many species of bacteria, suggesting the prevalence of AI-2 quorum sensing among prokaryotes (12–14, 46). With such a wide distribution, it is not surprising that the *luxS*-encoded quorum sensing mechanism has important roles in microbial ecology.

## QUORUM SENSING IN REGULATION OF BIOFILM DEVELOPMENT

In 1998, Davies and colleagues first described the role of *las* quorum sensing in biofilm formation of *Pseudomonas aeruginosa* (9). These researchers found that a *lasI* mutant defective in the synthesis of 3-oxododecanoyl-HSL formed biofilms that were flat, densely packed, and homogeneous relative to the highly structured, heterogeneous biofilms of the wild-type parent, PAO1. In contrast to those of the wild-type strain, the biofilms formed by the mutant were also dispersed by the addition of the detergent sodium dodecyl sulfate. This finding suggests that quorum sensing plays an important role in the development of bacterial biofilms (9). More importantly, this study suggests an inextricable connection between two bacterial social behaviors, quorum sensing and biofilm formation. Since that study, many studies on quorum sensing and biofilm formation of various species of bacteria have been published (7, 21, 31, 45, 56). Many groups have examined whether quorum sensing controls biofilm formation. However, investigations of the roles quorum sensing plays in biofilm systems for different organisms and how quorum sensing works mechanistically in a biofilm community remain in their infancy. In many cases, quorum sensing has been shown to influence biofilm development. In some cases, quorum sensing does not appear to be involved in biofilm formation. Many questions remain to be answered.

## AHL-Based Quorum Sensing in Biofilm Formation

Quorum sensing systems that are most extensively studied in gram-negative bacteria are those in *Pseudomonas aeruginosa,* which is a model organism for studying bacterial social behaviors (21, 23). *P. aeruginosa* is ubiquitous in a wide variety of environments and is considered one of the most common opportunistic pathogens in human infections causing fatal systemic disease under certain conditions (5, 48). Lung infections with biofilms of this pathogen are particularly common in patients with cystic fibrosis, and virtually all people with this respiratory disease develop a *P. aeruginosa* infection (10). It is the cause of death in more than 90% of these patients (5). In this organism, quorum sensing is highly complex and consists of two interlinked AHL-dependent regulatory circuits, which are modulated by numerous regulators acting at both the transcriptional and posttranscriptional levels (21–23). This genetic complexity may be one of the key elements responsible for the tremendous environmental versatility of *P. aeruginosa.*

Two intimately linked quorum sensing systems operate in *P. aeruginosa* (14, 21). The *las* system consists of the transcriptional activator LasR and the synthase LasI, which directs the synthesis of *N*-3-oxo-dodecanoyl-homoserine lactone (3-oxo-C12-HSL). Another is the *rhl* system, which consists of the transcriptional regulator RhlR and the synthase RhlI, which directs the synthesis of *N*-butanoyl-homoserine lactone (C4-HSL). These two systems work in concert to up-regulate a number of pseudomonal factors that enable this pathogen to survive in highly diverse environments (13, 21, 42, 49). The hierarchy of the *las/rhl* system is aided by the inhibitory action of the unbound *las* autoinducer 3-oxo-C12-HSL on the binding of C4-HSL to the RhlR transcriptional activator. The up-regulation of *lasR* transcription and the elevated concentrations of the LasR-3-oxo-C12-HSL complex allow the *rhl* system to be subsequently activated (21). *las*IR was the first system that was found to regulate the architecture and maturation of *P. aeruginosa*'s biofilms characterized by void spaces and towers of cells encased in an extracellular matrix (9). The development and maintenance of this fully mature biofilm phenotype were under the direction of this quorum sensing system. The architecture of this biofilm also affected the distribution and antimicrobial tolerance of bacteria in the biofilm. The *lasI* mutant that was deficient in the production of the autoinducer 3-oxo-C12-HSL formed biofilm cell clusters that were 20% of the thickness of the wild-type biofilm and were sensitive to detergent removal. When 3-oxo-C12-HSL was added to the system, the *lasI* mutant was once again able to form structured biofilms (9). However, it quickly became apparent that the influence of quorum sensing on the biofilm structure of *P. aeruginosa* was dependent upon experimental conditions in several subsequent studies (21, 23, 52).

## Signal Peptide-Mediated Quorum Sensing in Biofilm Formation

Gram-positive bacteria, such as *Staphylococcus aureus* and various streptococci, use a signal peptide-mediated three-component system for quorum sensing (7, 11, 29, 35, 40). Among these organisms, *Streptococcus mutans* is one of the best-studied bacteria that use quorum sensing in regulating biofilm formation (7, 27–31). *S. mutans* is a bacterium that depends on a "biofilm lifestyle" for survival and persistence in its natural ecosystem, dental plaque (1, 7, 34). Although it is a resident organism in dental biofilms, *S. mutans* under appropriate conditions can rapidly produce acids from dietary fermentable carbohydrates and initiate demineralization of the tooth surface. *S. mutans* is therefore considered an important etiologic agent of dental caries (1, 27–31). The ability of *S. mutans* to initiate dental caries depends on several virulence traits, including (i) initiation of biofilm formation (7), (ii) production of bacteriocins killing other species in dental biofilms (27, 28, 53), (iii) high efficiency in using carbohydrates and producing acids (1, 30, 33), and (iv) tolerance to acids (1, 30). These virulence-associated traits provide *S. mutans* with overwhelming ecological advantages for competing and dominating in dental biofilms that cause dental caries (34). *S. mutans* also causes corrosion of dental materials, leading to secondary caries around restorations. Moreover, it can be a cause of subacute infective endocarditis (1, 39, 41).

*S. mutans* has a well-conserved quorum sensing system that consists of at least six gene products encoded by *comCDE*, *comAB*, and *comX* (Fig. 2). The *comC* encodes a signal peptide precursor which is cleaved and exported to release a 21-amino-acid quorum sensing signal peptide or competence-stimulating peptide (CSP) through a peptide-specific ABC transporter encoded by *csl*AB *(comAB)* (1, 39). The *comDE* genes encode a two-component transduction system that specifically senses and responds to CSP. Another gene in the quorum sensing cascade is *comX*, which encodes an alternative sigma factor that directs transcription of numerous genes required for uptake and integration of foreign DNA (7). When it reaches a critical concentration, CSP interacts with the ComD histidine kinase receptor of neighboring cells, resulting in autophosphorylation of ComD at the expense of ATP. ComD activates its cognate response regulator, ComE, through phosphorylation and, in turn, activates its target genes, presumably *comCDE*, *comAB*, and *comX*, to trigger the signaling cascade for genetic competence as well as formation of a positive-feedback loop for quorum sensing (7, 29), although multilevel regulation may be involved in competence development (1, 28, 52). A recent study of the structure-activity relationship of the quorum sensing signaling peptide (CSP) from *S. mutans* has shown that the CSP of *S. mutans* displays at least two functional domains (50). The C-terminal structural motif appears to be crucial for activation of the quorum sensing signal transduction pathway, while the core, amphipathic α-helical structure with the hydrophobic domain is required for receptor binding (51). This work has provided new insights into the molecular detail of signal molecules from streptococci, which may facilitate the development of quorum sensing inhibitors that can interfere with *Streptococcus*-associated infections.

Perhaps the most fascinating finding in *S. mutans* is that this signaling system appears to be the key connection to genetic competence, bacteriocin production, biofilm formation, and stress response, which are key virulence factors in the pathogenesis of this organism (27–31). Studies showed that the biofilms consisting of competence-defective mutants had reduced biomass; conversely, increasing competence by the addition of synthetic CSP increased the formation of biofilms (31). Interestingly, the increased biomass could be abolished in the presence of DNase I, an endonuclease that cleaves double-stranded DNA, suggesting a role for the quorum sensing system in regulating DNA release and biofilm formation (43, 44). Recent studies of *S. pneumoniae* have shown that CSP-mediated competence induces a programmed cell lysis of a subpopulation of *S. pneumoniae* along with the release of DNA from lysed cells (17). This phenotype, called allolysis or fratricide, is suggested to be an important mechanism to ensure that competent cells obtain available DNA during genetic transformation (17). This work has provided an explanation for extracellular DNA release from *S. pneumoniae*. However, whether *S. mutans* uses the same mechanism to trigger cellular lysis and DNA release from a subpopulation during genetic competence remains to be studied. *S. mutans* has long been known to produce an array of bacteriocins and bacteriocin immunity proteins, including *nlmAB,* encoding mutacin IV; *smB,* encoding a class I bacteriocin; and *bip,* encoding a bacteriocin immunity protein (1, 27, 28, 53). Recent studies have shown that the quorum sensing system that regulates genetic competence in *S. mutans* also controls the production of bacteriocins and bacteriocin immunity proteins (1, 28, 53). This unique regulatory mechanism may provide an ample explanation for the connection between quorum sensing and genetic competence, DNA release, and biofilm formation (Fig. 2). It is reasonable to assume that the ComCDE quorum sensing mechanism may provide *S. mutans* with an ecological advantage for competition in multispecies dental biofilms. Interestingly, the role of quorum sensing and DNA release in biofilm formation has been described for both gram-positive and gram-negative bacteria (44, 50). Further study is necessary to determine whether this is a general trend in biofilm formation.

Quorum sensing via the accessory gene regulator *(agr)* of *Staphylococcus aureus* is another well-studied example of quorum sensing that plays a global regulatory role in biofilm formation and virulence factors (33, 40). The *agr* locus consists of two divergent transcription units using promoters P2 and P3. The P2 operon, consisting of *agrA, -B, -C,* and *-D,* encodes four proteins that constitute the *agr*-sensing mechanism. The autoinducer (AIP) from *S. aureus* is an octapeptide with a unique thioester ring structure which is generated from its precursor, AgrD, and secreted out of the cell through the action of the AgrB membrane protein. As its concentration increases in the extracellular environment, AIP binds to the histidine kinase receptor, AgrC, activating its autophosphorylation. Phosphorylated AgrC, in turn, activates the response regulator AgrA, driving transcription at the P2 promoter. AgrAP also increases transcription at the P3 promoter, resulting in elevated intracellular levels of RNAIII (40, 56). Interestingly, AIP from one strain not only is capable of activating the *agr* regulon in itself but also can inhibit *agr* activation of other *S. aureus* strains (39). Sequence variation in *agrB, agrD,* and *agrC* has led to the identification of at least four *S. aureus agr* specificity groups in which AIP produced by one group inhibits *agr* expression in others (3, 33, 39). There is mounting evidence that the *agr* phenotype and expression patterns may influence several aspects of biofilm behavior, including attachment of cells to surfaces, biofilm dispersal, and even the chronic nature of biofilm-associated infections (40, 56). Many infections by staphylococci are caused not by the free-living organism but, rather, by biofilms, including endocarditis, osteomyelitis, and implanted-device-related infections (3, 40, 56). Many of the products involved in biofilm development, including surface-associated adhesins, hemolysin, toxins, and autolysins, are regulated by the *agr* system. The expression of most of these staphylococcal products is under partial or complete control of the staphylococcal accessory regulator *agr* quorum sensing system. However, the studies addressing the quorum response and staphylococcal biofilms appear to be somewhat conflicting, and the role of *agr* in biofilm formation varies with the conditions. Therefore, it remains to be shown whether quorum sensing control in staphylococci represents a promising target for the development of novel antibacterial agents.

### AI-2-Mediated Quorum Sensing in Biofilm Formation

AI-2 is proposed to mediate cell-cell communication across species (2, 12, 46). The discovery of the interspecies signal molecule AI-2, which is produced and detected by a large number of diverse bacteria, implies that bacteria have a means to assess the cell density of other species in microbial communities (Fig. 3). By facilitating interspecies communication, AI-2 may enhance social cooperation among bacterial species in mixed communities. AI-2 has also proved important in the development of structured biofilms, especially multispecies biofilms in natural ecosystems (25, 36, 57). For example, AI-2 levels in a mixed culture of *Actinomyces naeslundii* T14V and *Streptococcus oralis* 34 are critical to the dual-species phenotype of mutualistic interdigitated biofilm growth of these two organisms when saliva is used as the sole nutrient source (45). A dual-species biofilm containing an *S. oralis* 34 *luxS* mutant and *A. naeslundii* T14V does not show mutualistic interdigitated growth, but this defect can be restored by adding synthetic DPD, a product of the LuxS enzyme, into the culture, and such a complementation is concentration dependent (45). *S. mutans* has also been found to harbor a highly conserved homologue of the *luxS* gene encoding an AI-2 signaling system (36, 57). Inactivation of the *luxS* gene affects *S. mutans* biofilms, suggesting that the AI-2 quorum sensing system also plays a role in the biofilm formation of *S. mutans* (36, 57). However, the mechanism behind this connection between AI-2 activity and biofilm formation is unknown. Many species of bacteria in natural biofilms like dental plaque have been found to have a *luxS*-encoded quorum sensing system. These systems may play important roles in interspecies cooperation.

## HOW MIGHT QUORUM SENSING SIGNAL MOLECULES FUNCTION IN BIOFILMS?

Almost all of the quorum sensing mechanisms described so far have been studied in the context of planktonic cultures. This is understandable because it simplifies the signaling process. In liquid cultures, all bacteria are presumed to be physiologically similar and produce signal molecules at the same rate. However, quorum sensing and signal transduction in biofilms might be much more complicated because of a range of physical, chemical, and nutritional factors that may influence signal production, stability, distribution, and efficiency to interact with their cognate receptors in a biofilm. How quorum sensing signal molecules function in a biofilm and how frequently quorum sensing is activated in a biofilm are largely unknown. Biofilms normally consist of bacterial cells and an extracellular matrix, including a mixture of secreted proteins, polysaccharides, nucleic acids, and dead cells (5–8). AHL molecules are known to diffuse freely across the cell membrane, so they are assumed to have little problem in reaching their target receptors via free diffusion through the biofilm matrix. These molecules vary in their hydrophobicity, depending on the length of the acyl side chain. Long-chain AHL appear to partition the hydrophobic environment (52). For species possessing multiple AHL signals, such as *P. aeruginosa,* this feature might be central to understanding signaling patterns in biofilms. However, signaling peptides produced from many gram-positive bacteria are likely influenced by physical, chemical, and biological factors within a biofilm because of the feature that small peptides likely interact with other charged molecules. Currently, little is known about whether signal peptides can be affected by diffusion limitation or nonspecific binding to polysaccharides, proteins, DNA, and even cell walls of other organisms within the biofilm. In addition, the cost for a gram-positive bacterium to produce an active signal peptide is very high. Keller and Surette have recently estimated that the production of a signal peptide in *S. aureus* costs 184 ATP, but the cost for an AHL in *P. aeruginosa* is only 8 ATP (22). Clearly, the production of a signal peptide is much more expensive in gram-positive bacteria. It is therefore reasonable to assume that the nutrient or energy source is a significant factor to influence signal peptide-mediated quorum sensing in gram-positive biofilms.

Theoretically, signal molecules that function to estimate population density could be affected by the concentration of a signal molecule, diffusion limitation, accessibility to the receptor, degradation, and production of the same autoinducer, such as AI-2, by third parties, whether intentionally or by chance. Some workers have used mathematic models to estimate the potential influence and possible mechanisms behind quorum sensing (18, 22). Quorum sensing could be considered diffusion sensing, since quorum sensing activation depends on the diffusion of a signal molecule to interact with the cognate receptor. This implies that quorum sensing is an autonomous activity of single cells to detect mass transfer limitation. However, the quorum sensing and diffusion sensing concepts may encompass an evolutionary conflict. Quorum sensing postulates that bacteria sense their density to allow them to engage in social behavior; accordingly, quorum sensing assumes that sensing evolved because of the group benefits (18). In contrast, diffusion sensing assumes that sensing evolved because of a direct fitness benefit for the individual cells, so it does not invoke group benefits for the evolution of autoinducer sensing. By unifying these conflicting concepts, Hense et al. (18) have recently proposed a new hypothesis of efficiency sensing concept, in which some of the problems associated with signaling in complex environments, as well as the problem of maintaining honesty in signaling, could be avoided when the signaling cells grow in microcolonies or in biofilms. Using a mathematical model, these authors have proposed that the spatial distribution of cells can be more important than their density and that spatial distribution and density are independent measures. As a consequence, efficiency sensing is a functional hypothesis that acknowledges the fact that autoinducers can measure a combination of cell density, diffusion limitation, and spatial distribution of autoinducer. Efficiency sensing is also a unifying evolutionary hypothesis, as it argues that quorum sensing is favored by both individual and group benefits. This new theory has described a typical mode of biofilm growth and formation of clonal clusters but avoids the problems of complexity and cheating that autoinducer-sensing bacteria encounter in situ (18). However, this model remains to be experimentally tested.

## SOCIAL INTERACTIONS OF MICROBES IN BIOFILMS

Biofilms are spatially structured communities of microbes whose function depends on a complex web of symbiotic interactions (8). High cell density and close proximity of diverse species of microorganisms are typical of life in natural biofilms, where organisms are involved in complex social interactions that occur both within and between species and can be either competitive or cooperative (8, 24–27). Competition for nutrients and other growth parameters is certainly an important driving force for the development of biofilm structure. Increased cell density enables chemical signals to cross the border of strains or species for quorum sensing-based group behaviors in biofilms, likely adding another level of complexity to biofilms (25). Furthermore, the expression of different adhesins, their cognate receptors, and exopolymeric components by individual cell types within a biofilm community can contribute to overall biofilm development (7, 8, 54). Importantly, many bacteria are capable of using quorum sensing signal-mediated mechanisms for horizontal gene transfer through either conjugation or transformation (15, 29, 32, 43). These processes not only act as driving forces for microbial evolution but also favor biofilm formation (15, 29, 32). Under such complex conditions, bacteria could benefit from division of labor, collective actions, and other forms of altruistic behavior or cooperation with their neighbors (26). Increasing evidence shows that quorum sensing-mediated social activities favor microbial interactions and are believed to be major mechanisms to maintain microbial homeostasis

(26, 34). For example, dental plaque is a well-recognized biofilm community characterized by its vast biodiversity (>700 species) and high cell density ($10^{11}$ cells/g, wet weight) (24, 25). The high cell density and species diversity within dental biofilms coupled with environmental fluctuations should create an environment that is conducive to inevitable intra- and interspecies interactions. Indeed, cooperative interactions among oral bacteria have been well studied, including bacterial coaggregation that facilitates a cell's attachment to the tooth surface, nutritional synergy and complementation to enable cell growth in saliva, and formation of food chains through metabolic cooperation between two or more species (24, 25, 27–31). These cooperative interactions probably play very important roles in the development of dental biofilms. However, competition or antagonistic interactions among different species may be equally important for the maintenance of a balanced relationship between microbes in dental biofilms and between dental biofilms and host defense mechanisms in the oral cavity (34). Many bacteria in dental biofilms produce peptide bacteriocins, which may play important roles in interspecies competition, biodiversity, and ecological fitness of microbes in dental biofilms (27, 28, 53). Many studies have shown that production of bacteriocins by naturally transformable streptococci, such as *S. mutans, S. gordonii, S. sanguinis,* and *S. mitis,* is tightly controlled by the quorum sensing system that regulates genetic competence in these species (27–31, 44, 53). Interestingly, all of these species are considered primary colonizers in dental biofilms, although bacteriocins produced by one species kill other species (27, 28, 53). These sophisticated interspecies interactions may play an essential part in balancing competition and coexistence of these organisms within microbial communities like dental biofilms. These interactions may be necessary for survival and perpetuation of diverse species in the same ecosystem.

## IMPACTS OF MICROBIAL SOCIAL ACTIVITIES ON ECOLOGY AND EVOLUTION

Quorum sensing relies upon the interaction of a small diffusible signal molecule with a sensor or transcriptional activator to initiate gene expression for coordinated activities (11–14, 55). During quorum sensing, microbes cooperate to obtain group-specific benefits. For example, the bacterium *Myxococcus xanthus* requires cell-cell signaling and social cooperation to form a fruiting body containing hardy spores in response to starvation (21–23). During such social cooperation, a part of the population survives starvation by forming the fruiting bodies, but most cells in the population, which provide the cooperation, are sacrificed (26). This phenomenon is called altruism, in which cooperation benefits the group but has a cost for the cooperating individual (16, 26). The idea of microbial social behaviors that provide altruism has gained widespread acceptance in recent years. From an evolutionary point of view, however, such a social behavior may create conflicts and potential risks to the population. Evolution is based on a fierce competition between individuals and rewards only selfish individuals or "cheaters" that do not cooperate but obtain the benefit from cooperators (22, 26). Quorum sensing in many bacteria appears to provide such altruism to the population of the same species, including spore formation, genetic competence, and other phenotypes (21–26, 46). In addition, the production of signal molecules and the response to these molecules are expensive processes which represent a metabolic burden for the signal-producing bacteria (22). The question then arises as to why bacteria produce costly molecules and respond to such signals if it increases the fitness of other individuals at their own cost. This creates a conflict of interest between the fitness of the individual and the fitness of the group. Natural selection implies competition and opposes cooperation unless a specific mechanism is at work. Nevertheless, we observe cooperation on many levels of biological organization (11–17, 22–28). Natural biofilms in many environments are characterized by high density and diversity of microbial species. The biofilm communities allow close cell-cell contact within the same species or between different species, resulting in inevitable intra- and interspecies interactions, including quorum sensing-mediated activities. These cooperative interactions play very important roles in maintaining homeostasis of microbes in a biofilm community (22–28). Recent research has heightened interest in how and when microbes cooperate to obtain group-specific fitness advantages. Through different mechanisms, cooperative behaviors can be stabilized, emphasizing the aspects most relevant to organisms. The diversity and interactions that can arise in biofilm communities represent unique opportunities for testing ecological and evolutionary theories on a "real-time" timescale.

## QUORUM SENSING INTERFERENCE

Quorum sensing systems of bacteria rely on signal molecules and their cognate receptor proteins for gene regulation and coordinated activities (55). Any compound that prevents accumulation of or recognition between signal molecules and receptor proteins might block bacterial quorum sensing and its gene expression. The discovery of bacterial quorum sensing mechanisms has led to identification of some compounds or enzymes that quench quorum sensing; such quenching is called signaling interference (58). Evidence is accumulating that such signal interference can be developed as promising approaches to control biofilm formation and microbial infections. Interestingly, anti-quorum sensing compounds exist in nature. Both plants and algae produce compounds that mimic quorum sensing signals of bacteria, so they interfere with bacterial quorum sensing and colonization (19, 58). For example, the red seaweed *Delisea pulchra* (Greville), which grows under the sea around Australia, produces a range of biologically active furanones (19). These natural compounds are found to be powerful signal antagonists for prevention against bacterial colonization by interfering with AHL signaling systems (19, 58). This biological understanding has led to the application of furanones as inhibitors of bacterial fouling. Furanones inhibit bacterial colonization and biofilm formation through interference with AHL quorum sensing pathways in gram-negative bacteria (19, 49, 58). They also interfere with AI-2 signaling systems in both gram-negative and gram-positive bacteria (47, 58). Additionally, furanones inhibit the expression of bacterial exoenzymes that actively degrade components of the immune system, thereby enhancing the immune response. Surface-attached furanones immobilized on catheters also inhibit bacterial attachment and retain activity for extended periods. Furanones are strong deterrents of the settlement and growth of macrofouling organisms and as such have potential application as a marine antifouling technology. Laboratory antifouling assays have been used to identify effective and safe furanone analogues, while field trials of furanones incorporated into coatings and polymers demonstrate efficacies similar to those of commercial biocides (19). There is growing evidence that bacterial quorum sensing systems are involved in cross-kingdom signaling with eukaryotic organisms (51). Likewise, eukaryotes are capable of actively responding to the presence of these signal molecules and produce compounds interfering with bacterial quorum sensing by acting as agonists or antagonists (58).

Humankind fights bacterial infections by using antibiotics or antimicrobial agents. The success of this treatment is largely based on antibiotics or antimicrobial agents that aim to inhibit bacterial growth. The major concern of this approach is the frequent development of antibiotic resistance. Furthermore, a big obstacle in fighting bacterial infections is that over 99% of bacteria found in nature and in the human body are associated with surfaces and form biofilms, which enable bacteria to resist inhibition or removal by the highest deliverable levels of antibiotic agents. As we began to gain control over epidemic infectious diseases, biofilm infections came to the fore. A global concern has emerged that we are entering a postantibiotic era with a reduced capability to combat persistent biofilm infections (5, 6, 20). Because biofilms are refractory to antibiotics, biofilm infections can be life threatening to immunocompromised patients. Given the many bacteria that employ quorum sensing mechanisms in controlling virulence, pathogenicity, and biofilm formation, quorum sensing constitutes a new target for the development of antibacterial agents with potential application in many fields. Currently, at least four strategies aiming at interference with quorum sensing have been proposed, including (i) inhibition of signal generation, (ii) interference with signal dissemination, (iii) blocking signal receptors, and (iv) inhibition of the signaling response system (19, 32, 49, 58). The key of these strategies is to interfere with bacterial quorum sensing and its controlled pathogenic activities. Knowing the molecular details of communication systems and their control of virulence and pathogenicity opens a new avenue for controlling microbial infections. The development of signal analogues that specifically block or override the bacterial command line will enable us to control the unwanted activities without affecting bacterial growth. A major difference between these compounds and antibiotics is that they do not directly inhibit bacterial growth, so there is no strong selection pressure to create resistant microbes. Compounds that can inhibit signals of quorum sensing systems can be developed into potent antagonists against infectious bacteria. Such novel drugs that specifically target quorum sensing systems are capable of attenuating bacterial infections in a manner that is less likely to result in the development of resistant mutants. Several studies have recently described the application of AHL analogues or signal peptide analogues to achieve inhibition of quorum sensing circuits in some bacteria (19, 32, 34, 49, 58). These studies have generated substantial knowledge about the structure-function relationships of quorum sensing signals, which is of great value in the search for potent quorum sensing inhibitors.

## CONCLUDING REMARKS

In the past decade, significant advances have been made regarding bacterial quorum sensing and group behaviors. Quorum sensing is emerging as an integral component of bacterial global gene regulatory networks responsible for bacterial adaptation in biofilms. However, research on how bacterial quorum sensing works mechanistically in a biofilm remains in its infancy. A clear challenge is to determine what factors of a biofilm influence the onset of quorum sensing and subsequent gene expression. The regulation of the target genes of these systems and their roles during biofilm development and maturation must be determined. Another key challenge is to determine the functional consequences of quorum sensing in multispecies biofilms. Does quorum sensing influence the pathogenic potential of biofilms, or perhaps alter the antimicrobial resistance of the biofilm? Does quorum sensing occur frequently in natural ecosystems, or do signal-consuming organisms limit the extent of signaling? How does bacterial social activity influence ecology and evolution of bacteria in biofilms? Future research will clearly address these questions in the emerging field of bacterial social behaviors. The answer to these questions will undoubtedly provide new insights and surprises.

## ACKNOWLEDGMENTS

This work was supported in part by grant MOP-74487 from the Canadian Institute for Health Research (CIHR), by grant RGPIN 311682-07 from the Natural Sciences and Engineering Research of Canada (NSERC), by New Opportunity grant 10313 from the Canadian Foundation for Innovation (CFI), and by research funds from the Faculty of Dentistry of Dalhousie University.

## REFERENCES


1. **Ajdic, D., W. M. McShan, R. E. McLaughlin, G. Savic, J. Chang, M. B. Carson, C. Primeaux, R. Tian, S. Kenton, H. Jia, S. Lin, Y. Qian, S. Li, H. Zhu, F. Najar, H. Lai, J. White, B. A. Roe, and J. J. Ferretti.** 2002. Genome sequence of *Streptococcus mutans* UA159, a cariogenic dental pathogen. *Proc. Natl. Acad. Sci. USA* **99:**14434–14439.
2. **Bassler, B. L.** 2002. Small talk. Cell-to-cell communication in bacteria. *Cell* **109:**421–424.
3. **Chan, W. C., B. J. Coyle, and P. Williams.** 2004. Virulence regulation and quorum sensing in staphylococcal infections: competitive AgrC antagonists as quorum sensing inhibitors. *J. Med. Chem.* **47:**4633–4641.
4. **Chen, X., S. Schauder, N. Potier, A. Van Dorssealaer, I. Pelczer, B. L. Bassler, and F. M. Hughson.** 2002. Structural identification of a bacterial quorum-sensing signal containing boron. *Nature* **415:**545–549.
5. **Costerton, J. W., P. S. Stewart, and E. P. Greenberg.** 1999. Bacterial biofilms: a common cause of persistent infections. *Science* **284:**1318–1322.
6. **Costerton, W., R. Veeh, M. Shirtliff, M. Pasmore, C. Post, and G. Ehrlich.** 2003. The application of biofilm science to the study and control of chronic bacterial infections. *J. Clin. Investig.* **112:**1466–1477.
7. **Cvitkovitch, D. G., Y. H. Li, and R. P. Ellen.** 2003. Quorum sensing and biofilm formation in streptococcal infections. *J. Clin. Investig.* **112:**1626–1632.
8. **Davey, M. E., and G. A. O'Toole.** 2000. Microbial biofilms: from ecology to molecular genetics. *Microbiol. Mol. Biol. Rev.* **64:**847–867.
9. **Davies, D. G., M. R. Parsek, J. P. Pearson, B. H. Iglewski, J. W. Costerton, and E. P. Greenberg.** 1998. The involvement of cell-to-cell signals in the development of a bacterial biofilm. *Science* **280:**295–298.
10. **de Kievit, T. R., and B. H. Iglewski.** 2000. Bacterial quorum sensing in pathogenic relationships. *Infect. Immun.* **68:**4839–4849.
11. **Dunny, G. M., and B. A. Leonard.** 1997. Cell-cell communication in Gram-positive bacteria. *Annu. Rev. Microbiol.* **51:**527–564.
12. **Federle, M. J., and B. L. Bassler.** 2003. Interspecies communication in bacteria. *J. Clin. Investig.* **112:**1291–1299.
13. **Fuqua, C., and E. P. Greenberg.** 2002. Listening in on bacteria: acyl-homoserine lactone signalling. *Nat. Rev. Mol. Cell Biol.* **3:**685–695.
14. **Fuqua, C., M. R. Parsek, and E. P. Greenberg.** 2001. Regulation of gene expression by cell-to-cell communication: acyl-homoserine lactone quorum sensing. *Annu. Rev. Genet.* **35:**439–468.
15. **Ghigo, J. M.** 2001. Natural conjugative plasmids induce bacterial biofilm development. *Nature* **412:**442–445.
16. **Greenberg, E. P.** 2003. Bacterial communication and group behavior. *J. Clin. Investig.* **112:** 1288–1290.
17. **Guiral, S., T. J. Mitchell, B. Martin, and J. P. Claverys.** 2005. Competence-programmed predation of noncompetent cells in the human pathogen *Streptococcus pneumoniae*: genetic requirements. *Proc. Natl. Acad. Sci. USA* **102:**8710–8715.
18. **Hense, B. A., C. Kuttler, J. Müller, M. Rothballer, A. Hartmann, and J. U. Kreft.** 2007. Does efficiency sensing unify diffusion and quorum sensing? *Nat. Rev. Microbiol.* **5:**230–239.
19. **Hentzer, M., and M. Givskov.** 2003. Pharmacological inhibition of quorum sensing for the treatment of chronic bacterial infections. *J. Clin. Investig.* **112:**1300–1307.
20. **Hogan, D., and R. Kolter.** 2002. Why are bacteria refractory to antimicrobials? *Curr. Opin. Microbiol.* **5:**472–477.
21. **Juhas, M., L. Eberl, and B. Tümmler.** 2005. Quorum sensing: the power of cooperation in the world of *Pseudomonas*. *Environ. Microbiol.* **7:**459–471.
22. **Keller, L., and M. G. Surette.** 2006. Communication in bacteria: an ecological and evolutionary perspective. *Nat. Rev. Microbiol.* **4:**249–258.
23. **Kirists, M. J., and M. R. Parsek.** 2006. Does *Pseudomonas aeruginosa* use intercellular signaling to build biofilm communities? *Cell. Microbiol.* **8:** 1841–1849.
24. **Kolenbrander, P. E.** 2000. Oral microbial communities: biofilms, interactions, and genetic systems. *Annu. Rev. Microbiol.* **54:**413–437.
25. **Kolenbrander, P. E., R. N. Andersen, D. S. Blehert, P. G. Egland, J. S. Foster, and R. J. Parmer, Jr.** 2002. Communication among oral bacteria. *Microbiol. Mol. Biol. Rev.* **66:**486–505.
26. **Kreft, J. U.** 2004. Biofilms promote altruism. *Microbiology* **150:**2751–2760.
27. **Kreth, J., J. Merritt, W. Shi, and F. Qi.** 2005. Competition and coexistence between *Streptococcus mutans* and *Streptococcus sanguinis* in the dental biofilm. *J. Bacteriol.* **187:**7193–7203.
28. **Kreth, J., J. Merritt, W. Y. Shi, and F. X. Qi.** 2005. Coordinated bacteriocin production and competence development: a possible mechanism for taking up DNA from neighbouring species. *Mol. Microbiol.* **57:**392–404.

29. **Li, Y.-H., P. C. Y. Lau, J. H. Lee, R. P. Ellen, and D. G. Cvitkovitch.** 2001. Natural genetic transformation of *Streptococcus mutans* growing in biofilms. *J. Bacteriol.* **183:**897–908.
30. **Li, Y.-H., M. N. Hanna, G. Svensäter, R. P. Ellen, and D. G. Cvitkovitch.** 2001. Cell density modulates acid adaptation in *Streptococcus mutans*: implications for survival in biofilms. *J. Bacteriol.* **183:**6875–6884.
31. **Li, Y.-H., N. Tang, M. B. Aspiras, P. C. Y. Lau, J. H. Lee, R. P. Ellen, and D. G. Cvitkovitch.** 2002. A quorum-sensing signaling system essential for genetic competence in *Streptococcus mutans* is involved in biofilm formation. *J. Bacteriol.* **184:**2699–2708.
32. **Luo, H., K. Wan, and H. H. Wang.** 2005. High-frequency conjugation system facilitates biofilm formation and pAMβ1 transmission by *Lactococcus lactis*. *Appl. Environ. Microbiol.* **71:** 2970–2978.
33. **Lyon, G. J., P. Mayville, T. W. Muir, and R. P. Novick.** 2000. Rational design of a global inhibitor of the virulence response in *Staphylococcus aureus,* based in part on localization of the site of inhibition to the receptor-histidine kinase, AgrC. *Proc. Natl. Acad. Sci. USA* **97:**13330–13335.
34. **Marsh, P. D.** 1994. Microbial ecology of dental plaque and its significance in health and disease. *Adv. Dent. Res.* **8:**263–271.
35. **Mayville, P., G. Ji, R. Beavis, H. Yang, M. Goger, R. P. Novick, and T. W. Muir.** 1999. Structure-activity analysis of synthetic autoinducing thiolactone peptides from *Staphylococcus aureus* responsible for virulence. *Proc. Natl. Acad. Sci. USA* **96:**1218–1223.
36. **Merritt, J., F. Qi, S. D. Goodman, M. H. Anderson, and W. Shi.** 2003. Mutation of *luxS* affects biofilm formation in *Streptococcus mutans*. *Infect. Immun.* **71:**1972–1979.
37. **Miller, M. B., and B. L. Bassler.** 2001. Quorum sensing in bacteria. *Annu. Rev. Microbiol.* **55:** 165–199.
38. **Miller, S. D., S. D. Haddock, C. D. Elvidge, and T. F. Lee.** 2005. Detection of a bioluminescent milky sea from space. *Proc. Natl. Acad. Sci. USA* **102:**14181–14184.
39. **Mitchell, T. J.** 2003. The pathogenesis of streptococcal infections: from tooth decay to meningitis. *Nat. Rev. Microbiol.* **1:**219–230.
40. **Novick, R. P.** 2003. Autoinduction and signal transduction in the regulation of staphylococcal virulence. *Mol. Microbiol.* **48:**1429–1449.
41. **Parsek, M. R., and E. P. Greenberg.** 2005. Sociomicrobiology: the connections between quorum sensing and biofilms. *Trends Microbiol.* **13:**27–33.
42. **Parsek, M. R., D. L. Val, B. L. Hanzelka, J. E. Cronan, Jr., and E. P. Greenberg.** 1999. Acyl homoserine-lactone quorum-sensing signal generation. *Proc. Natl. Acad. Sci. USA* **96:**4360–4365.
43. **Petersen, F. C., and A. A. Scheie.** 2004. Biofilm mode of growth of *Streptococcus intermedius* favored by a competence-stimulating signaling peptide. *J. Bacteriol.* **186:**6327–6331.
44. **Petersen, F. C., L. Tao, and A. A. Scheie.** 2005. DNA binding-uptake system: a link between cell-to-cell communication and biofilm formation. *J. Bacteriol.* **187:**4392–4400.
45. **Rickard, A. H., R. J. Palmer, Jr., D. S. Blehert, S. R. Campagna, M. F. Semmelhack, P. G. Egland, B. L. Bassler, and P. E. Kolenbrander.** 2006. Autoinducer 2: a concentration-dependent signal for mutualistic bacterial biofilm growth. *Mol. Microbiol.* **60:** 1446–1456.
46. **Schauder, S., and B. L. Bassler.** 2001. The languages of bacteria. *Genes Dev.* **15:**1468–1480.
47. **Shirtliff, M. E., J. T. Mader, and A. K. Camper.** 2002. Molecular interactions in biofilms. *Chem. Biol.* **9:**859–871.
48. **Singh, P. K., A. L. Schaefer, M. R. Parsek, T. O. Moninger, M. J. Welsh, and E. P. Greenberg.** 2000. Quorum-sensing signals indicate that cystic fibrosis lungs are infected with bacterial biofilms. *Nature* **407:**762–764.
49. **Smith, R. S., and B. H. Iglewski.** 2003. *Pseudomonas aeruginosa* quorum sensing as a potential antimicrobial target. *J. Clin. Investig.* **112:**1460–1465.
50. **Sponering, A. L., and M. S. Gilmore.** 2006. Quorum sensing and DNA release in bacterial biofilms. *Curr. Opin. Microbiol.* **9:**133–137.
51. **Syvitski, R. T., X.-L. Tian, K. Sampara, A. Salman, S. F. Lee, D. L. Jakeman, and Y.-H. Li.** 2007. Structure-activity analysis of quorum-sensing signaling peptides from *Streptococcus mutans*. *J. Bacteriol.* **189:**1441–1450.
52. **Taga, M. E., and B. L. Bassler.** 2003. Chemical communication among bacteria. *Proc. Natl. Acad. Sci. USA* **100:**14549–14554.
53. **van der Ploeg, J. R.** 2005. Regulation of bacteriocin production in *Streptococcus mutans* by the quorum-sensing system required for development of genetic competence. *J. Bacteriol.* **187:**3980–3989.
54. **Watnick, P., and R. Kolter.** 2000. Biofilm, city of microbes. *J. Bacteriol.* **182:**2675–2679.
55. **Withers, H., S. Swift, and P. Williams.** 2001. Quorum sensing as an integral component of gene regulatory networks in Gram-negative bacteria. *Curr. Opin. Microbiol.* **4:**186–193.

**56. Yarwood, J. M., D. J. Bartels, E. M. Volper, and E. P. Greenberg.** 2004. Quorum sensing in *Staphylococcus aureus* biofilms. *J. Bacteriol.* **186:** 1838–1850.

**57. Yoshida, A., T. Ansai, T. Takehara, and H. K. Kuramitsu.** 2005. LuxS-based signaling affects *Streptococcus mutans* biofilm formation. *Appl. Environ. Microbiol.* **71:**2372–2380.

**58. Zhang, L.-H., and Y.-H. Dong.** 2004. Quorum sensing and signal interference: diverse implications. *Mol. Microbiol.* **53:**1563–1571.

# MOLECULAR MECHANISMS OF MICROBIAL SURVIVAL IN FOODS

*Francisco Diez-Gonzalez and Julie Kuruc*

# 8

Bacteria are the most ubiquitous group of living organisms on Earth, and they have been isolated from almost every imaginable ecological niche on our planet. Accordingly, bacteria have an extremely diverse array of metabolic capabilities (86). An enormous assortment of heterotrophic bacteria is typically associated with more complex life forms such as plants and animals, with which they have commensal, mutual, symbiotic, or parasitic relationships (87). As a result of these close associations between bacteria and higher life forms, most human foodstuffs that originate from animals and plants are inhabited by a wide variety of microorganisms.

The bacteria present in foods not only originate from the living plants and animals but also occur as the result of postharvest and postslaughter contamination, often arising from many different sources (64). The presence of bacteria in foods may be considered undesirable due to potential spoilage of the product, or, in the case of food-borne pathogens, their presence can be a public health concern. Throughout history, humans have adapted certain beneficial processes that preserve food by delaying spoilage and preventing food-borne illnesses. As a result, a variety of technologies have been developed to modify or treat foods for the control of microorganisms (53). Early technologies such as fermentation, cooking, drying, and salting provided preservation of foods for long periods after they had been harvested. In the last 100 years, other effective preservation techniques have been adopted to supplement the still widely used traditional methods.

The principles of food preservation are in part based on changes to the intrinsic properties of foods, such as water content, pH, oxidation-reduction (O/R) potential, solute concentration, and the presence of antimicrobial substances, all of which stress microbial cells (64). The term "stress" can be defined in the broader sense as any extracellular influence that threatens the ability of microorganisms to perform their living functions. The current food chain also relies heavily on the use of heating and cooling technologies to control microorganisms. In general, preservation technologies are intended to cause severe damage to bacterial cells which may lead to extensive reduction of their metabolic capabilities and eventually to cell death (91). Thermal pro-

*Francisco Diez-Gonzalez,* Department of Food Science and Nutrition, University of Minnesota, St. Paul, MN 55108. *Julie Kuruc,* PepsiCo Chicago, 617 W. Main St., Barrington, IL 60010.

*Food-Borne Microbes: Shaping the Host Ecosystem,* Edited by L.-A. Jaykus, H. H. Wang, and L. S. Schlesinger, © 2009 ASM Press, Washington, DC

cessing, irradiation, high-pressure processing, and some antimicrobial compounds are among the food preservation technologies that are intended to cause rapid bacterial death. Technologies that are based on the reduction of temperature (refrigeration and freezing), water activity ($a_w$) (drying, addition of solutes, and concentration), pH (fermentation and acidification), and control of O/R potential do not normally cause a rapid reduction of viability, but they markedly inhibit the ability of bacteria to grow (14).

Over the course of millions of years, natural exposure of heterotrophic bacteria to the elements has allowed them to evolve and tolerate different forms of stress, most of which are the same stresses as implemented by current food preservation techniques (27). Frequent changes in temperature, humidity, oxygen content, and hydrogen ion concentrations caused by climate or exposure to other organisms have constantly challenged the survival of bacteria. As simple unicellular prokaryotes, their survival in the face of natural stresses depends on their ability to adapt, which is ultimately directed by mutation or the acquisition of genes from other bacteria or even other organisms (87, 97).

Depending on the type of stress and its magnitude, the complex network of molecular stress responses allows microorganisms to recover from the initial insult and survive (111). In many cases, preexposure to sublethal doses of a specific stress allows bacterial cells to survive harsher conditions upon exposure to the same stress. As an example, the ability of *Escherichia coli* O157:H7 to survive an acid shock at a pH as low as 2 is enhanced by previous exposure to pH 5 (26). As microorganisms acquire increased stress tolerance due to preexposure to a sublethal stress, they also often develop enhanced resistance to other types of stress, a phenomenon referred as "cross-protection" or "cross-resistance" (111).

The use of multiple genetic systems which control an array of genes (global regulators and regulons) has proven to be a very successful strategy for evolution and survival of bacteria (72). A number of these systems are present globally in both gram-positive and gram-negative microorganisms. Species-specific stress response genes have also been found. This chapter describes the most important global regulator systems in representative food-borne pathogens and reviews the individual molecular mechanisms of survival against specific food-related stresses. Our discussion includes representative gram-positive (*Listeria monocytogenes* and *Staphylococcus aureus*) and gram-negative (*Escherichia coli, Salmonella,* and *Campylobacter*) food-borne pathogens which have been thoroughly studied.

## GLOBAL STRESS REGULATORS

Upon exposure to stresses imposed in food preservation, the bacterial cell may undergo a number of harmful changes. For example, traditional thermal processing methods such as cooking and pasteurization cause extensive structural and functional changes to many cellular components (64). At high temperatures, the two primary chemical alterations responsible for cell death include denatured proteins and DNA breakage. In the case of ionizing radiation, cells are killed directly by energy adsorbed by critical macromolecules, in particular, DNA; indirect lethal effects occur as the result of radiolysis reaction products which release free radicals that, in turn, interfere with biological function (32).

DNA damage triggers massive coordinated cellular and molecular responses referred to as the "global stress response" (39). The SOS response system is probably the best-characterized global response to DNA damage, and this is manifested by *Enterobacteriaceae* species (38). This system is induced to minimize the lethal and mutagenic effects of DNA breakage. As many as 30 different genes form the SOS regulon, which is normally repressed in the absence of extensive DNA damage (39, 40). DNA breakage results in a high concentration of single-stranded DNA molecules which triggers a series of events that results in induction of the SOS regulon (140).

Upon formation of nucleoprotein complexes between single-stranded DNA and RecA, a bacterial recombinase protein, self-proteolysis of the SOS repressor, LexA, occurs. Inactivation of LexA leads to the induction of a variety of genes encoding enzymes involved in DNA repair, recombination, and replication (38, 140). Among the SOS-induced gene products, several error-prone DNA polymerases are synthesized as a result of LexA cleavage (39). In *E. coli,* the two most important DNA polymerases are Pol IV and Pol V, which are encoded by the *dinB* and *umuDC* genes. These polymerases have a high processing rate but also have a high degree of inaccuracy, resulting in a relatively high mutation rate. Other genes that are induced by the SOS global regulator response are *sulA,* which is responsible for inhibiting cell division, as well as *recN* and *ruvAB,* both of which act to repair breaks in double-stranded DNA (140).

Most of the current knowledge on bacterial stress responses has been acquired using *Escherichia coli* and *Salmonella* as model organisms. In addition to the SOS global response regulator, at least four other stress response systems have been identified in these organisms: the general stress response, the stringent response, the heat shock response, and the polyphosphate-mediated nutrient limitation response (38, 39). Other regulatory systems involved in stress responses include PhoPQ, Fur, and OmpR/EnvZ (113). In addition to these protein-mediated regulatory systems, small RNA molecules have been recently identified to be involved in a complex network of bacterial stress responses (45). These systems are discussed below.

## General Stress Regulator

During exponential growth, bacterial cells require very complex and coordinated synthesis of macromolecules (140). Rapid DNA replication is catalyzed by DNA Pol III, and transcription is mediated by RNA polymerases. Bacteria typically have a single type of RNA polymerase, and in the case of *E. coli,* this is a large holoenzyme that is formed by five protein subunits: two $\alpha$ subunits, one $\beta$, one $\beta'$, and a $\sigma$ chain (86). Tightly bound $\alpha_2\beta\beta'$ subunits form the RNA polymerase catalytic core, and the $\sigma$ factor is loosely associated. The role of the $\sigma$ subunit is to recognize promoter sequences in order to initiate transcription of specific genes. Bacterial cells can synthesize as many as 14 $\sigma$ factor variants that are involved in the activation of different genes. Investigators have identified and characterized seven $\sigma$ factors in *E. coli.*

In *Escherichia coli,* $\sigma^{70}$, which is encoded by *rpoD,* is the main $\sigma$ factor responsible for housekeeping gene synthesis and is present during exponential cellular growth (127). When exponentially growing cultures of *E. coli* reach stationary phase, they induce the major component of the "general stress response," the $\sigma^S$ factor (59). This particular $\sigma$ factor regulates cell survival during short periods of growth and under stressful conditions (107). The $\sigma^S$ factor is encoded by *rpoS,* which has also been referred to in the literature as $\sigma^{38}$ and KatF (38). The transcription of *rpoS* is initiated during late exponential phase but does not reach maximum levels until the start of the stationary phase (60). Binding of $\sigma^S$ subunits to the RNA polymerase triggers the synthesis of a variety of regulatory and metabolic proteins (59).

RpoS activates the transcription of as many as 87 genes or regulons which are involved in a variety of cellular functions, accounting for approximately 10% of the *E. coli* genome (135). At this time, at least 15 separate genetic systems having a major role in stationary-phase metabolism, and more than 15 different stress response genes or operons, all of which are regulated by $\sigma^S$, have been characterized (28). RpoS also controls genes that are associated with maintaining cell wall structure and stability as well as bacterial shape. In *Salmonella* and *Yersinia,* several genes encoding virulence factors have also been reported to be under the influence of $\sigma^S$ (28). The specific functions of approximately 40% of the genes controlled by RpoS still remain unknown (135).

The network of genes regulated by $\sigma^S$ is found throughout the *E. coli* genome, but there are several clusters with a relatively high density of RpoS-regulated operons (135). At chromosomal locations 18.6 and 79.3 minutes, two regions of 89 and 91 kb contain 27 and 29 $\sigma^S$-controlled genes, respectively. Weber et al. (135) have classified all the RpoS-regulated genes into six major categories based on their functions: metabolism, regulation, transport, adaptation to stress, protein processing, and unknown. Further genomic studies will likely uncover the function of some of those unknown-function genes as well as a number of additional $\sigma^S$ genes.

Among the RpoS-dependent genes associated with stress responses, some are related to the SOS system, while others are linked to specific stresses such as acid, oxidative, and osmolarity stresses (28). Layton and Foster (77) recently reported that the concentration of the error-prone DNA polymerase Pol IV was 50 times lower in *rpoS* mutant cells than in wild-type cells, and this difference was independent of the SOS system. Because the Pol IV protein is unstable, continuous $\sigma^S$-regulated transcription of *dinB* is required in order to provide sufficient levels of Pol IV during the stationary phase, which is necessary for survival under long-term starvation conditions (38). The synthesis of two catalases, HPI and HPII, which are involved in the hydrolysis of hydrogen peroxide and are encoded by *katG* and *katE,* respectively, was found to be tightly regulated by RpoS, also referred to as KatF (115). In addition, it has been found that the presence of RpoS is required in order to activate the expression of *otsA* and *otsB,* which are responsible for the production of the osmoprotectant trehalose in several bacteria (124). The glutamate decarboxylase operon, which is involved in acid resistance of several gram-negative microorganisms, includes the *gadA, gadB,* and *gadC* genes, which are indirectly regulated by $\sigma^S$ by means of *gadX* and *gadW* (135). Some of these specific stress response systems are discussed later in this chapter.

The intracellular concentration of $\sigma^S$ is regulated at every step before its synthesis as well as during its degradation (Fig. 1) (59). A decline in cellular growth rate has been identified as the primary signal that triggers *rpoS* transcription. However, stimulation of transcription only occurs if the growth reduction is gradual, as demonstrated by experiments involving controlled growth shift and sudden growth termination (28, 59). At the molecular level there are several promoters, regulatory systems, and small molecules that direct *rpoS* transcription. As many as three different promoters have been found upstream of the *rpoS* gene—i.e., *nldD*p1, *nlpD*p2, and *rpoS*p—with the last appearing to stimulate the highest level

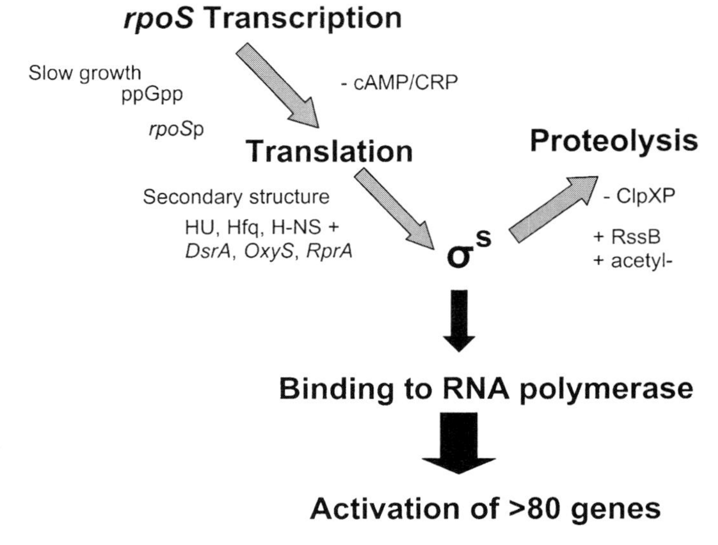


**FIGURE 1** Diagram of $\sigma^S$ regulation and some of the main factors involved in its activation (+) or inhibition (−). cAmP, cyclic AMP; CRP, catabolite repression protein.

of mRNA synthesis (59). The catabolite repression complex formed by catabolite repression protein and cyclic AMP has an important role in repressing the transcription of *rpoS* during exponential phase by binding at a few regions located up- and downstream of *rpoSp* (59, 75, 135). The "alarmon" molecule guanosine tetraphosphate (ppGpp) also enhances *rpoS* mRNA synthesis by affecting transcriptional elongation or transcript stability (59).

A variety of external signals can influence the translation of *rpoS* mRNA. A drop in temperature and exposure to high osmolarity, low pH, or high cell densities stimulate the rate of translation, while carbon starvation during stationary phase inhibits further synthesis of $\sigma^S$ after early stationary phase (28, 59). At the molecular level, *rpoS* mRNA translation appears to be negatively regulated by the formation of secondary structures that prevent binding to ribosomes (28, 59). Exposure to stress factors favors the synthesis of $\sigma^S$ and appears to have a major role in modifying the aforementioned secondary structure. In addition to the formation of a secondary structure, the translation of *rpoS* mRNA is regulated by several rather complex factors, including a ribosome-associated protein, HfQ; a nucleoid protein, HU; the histone-like protein, H-NS; and small regulatory RNAs, *dsrA, oxyS,* and *rprA* (45, 59).

Posttranslational regulatory mechanisms are also quite important in mediating the activation of $\sigma^S$-dependent stress responses. The rate of degradation of $\sigma^S$ protein can be affected by the growth stage of the cell, with its half-life ranging from an estimated 1 to 20 min (74). One of the main roles of the ATP-dependent ClpXP protease is the degradation of $\sigma^S$ (59). However, ClpXP requires the presence of a recognition factor, RssB, to bind to RpoS. RssB mutants have an abnormally high concentration of $\sigma^S$, and some studies suggest that phosphorylation is needed to activate RssB. Since the best substrate for RssB phosphorylation is acetyl phosphate, this compound is therefore an indirect regulator of $\sigma^S$ proteolysis (59).

In gram-positive microorganisms such as *Bacillus* and *Listeria,* $\sigma^B$ or SigB, a general stress regulator equivalent to $\sigma^S$ of gram-negative bacteria, has been identified and reported to control as many as 150 genes (55, 57). The SigB general stress system is activated by a wide variety of growth-limiting (starvation for glucose, phosphate, and oxygen) and environmental factors, including high and low temperatures, acid pH, high osmolarity, and exposure to high salt, ethanol, and nitric oxide, among others (55). Antibiotics such as vancomycin and bacitracin are also known to induce SigB stress response genes. In general, all the SigB-inducing stress factors can be divided into two major categories: energy limitations and environmental stresses (14). The gram-positive organisms reported to have SigB variants include *Bacillus, Staphylococcus, Listeria, Mycobacterium, Ralstonia,* and *Streptomyces* (55). Interestingly, some of the strict anaerobic and facultatively anaerobic gram-positive bacteria, including *Clostridium, Lactococcus,* and *Streptococcus,* do not appear to have SigB regulation (55), so it has been speculated that one of the main roles of alternative stress sigma factors is protection against oxidative stress.

Typically $\sigma^B$ contributes to cell function throughout all phases of cell growth but plays a more critical role at the end of exponential growth, once the cell becomes stressed (6, 9, 21). The *sigB* gene is located in an operon that includes seven *rsb* genes. These genes act to regulate $\sigma^B$ activity once the appropriate stress signals are detected (27). The $\sigma^B$-dependent proteins contribute to acid resistance in *Listeria monocytogenes* but are not essential for its pathogenicity, whereas in the case of *Salmonella enterica* serovar Typhimurium, this general stress factor is required for pathogenicity (27).

The role of sigma factors in the survival of bacteria in the food production and processing chain has been studied using mutant strains. Because of its public health relevance, *Escherichia coli* O157:H7 has been used to study the function of RpoS (4, 134). Price et al. (102) reported that colonization of the gastrointestinal tract of cattle was markedly reduced with *rpoS* mutants compared to wild-type strains. Cheville and coworkers (16) published one of the first reports that indicated the importance of *rpoS* in the survival of serotype O157:H7 against a variety of stressful conditions present in the food environment. A mutation in *rpoS* caused *E. coli* O157:H7 to become less heat tolerant than a wild-type strain (142).

The direct role of RpoS in survival of other bacterial pathogens has also been reported. With *Vibrio parahaemolyticus,* an *rpoS* mutant had significantly diminished survival capability in fish as compared to the parent strain (130). The heat tolerance of *Vibrio vulnificus,* however, appeared to be independent of *rpoS,* but this sigma factor was still needed for complete cross-protection to oxidative stresses (112). *Salmonella* has been associated with outbreaks linked to the consumption of alfalfa sprouts, and it appears that RpoS plays an important role in allowing this organism to attach to plant tissues (6). Because of the complex network of RpoS-regulated genes, the interpretation of some of these studies should be reevaluated using novel genomic and proteomic approaches.

## Other Global Stress Responses

The synthesis of mRNA and tRNA in *Escherichia coli* and other bacteria is regulated by the unusual guanosine nucleotide alarmones ppGpp and pppGpp, which act as global regulators (12). When bacterial cells are deprived of amino acid sources, the synthesis of all types of RNA is markedly repressed. This effect is referred to as the stringent response (38, 140). This response was first observed in the laboratory when cells were shifted from rich to minimal media and is one of the most widely characterized global stress responses. The stringent response is particularly important during periods of starvation and stress, and it is often induced at the same time as RpoS because of nutrient limitation during stationary phase (12). The stringent response has been observed in at least 12 different species of bacteria, including both gram-positive and gram-negative organisms.

The complex series of events that characterize the stringent response begins when "uncharged" tRNA builds up inside of the cell as a result of a lack of amino acids. This limitation causes ribosome "stalling" and the excess of uncharged tRNA activates RelA, an enzyme that synthesizes (p)ppGpp after binding to ribosomes containing uncharged tRNA (12, 140). The newly synthesized (p)ppGpp regulates transcription by directly binding to the RNA polymerase at the active site and preventing further enzyme interaction with promoter regions (12). Several indirect effects of (p)ppGpp have also been suggested, such as inhibition of RNA synthesis via reducing the availability of free RNA polymerase.

While the stringent response reduces RNA synthesis, the lack of specific amino acids causes a derepression of biosynthetic genes and (p)ppGpp selectively allows the synthesis of their corresponding mRNA (38). As a global stress response, the stringent response not only reduces RNA synthesis but also inhibits translation and stimulates protein degradation. Starvation is not the only condition that induces the stringent response, as stress and impaired growth phenotypes both trigger its effects (38). As mentioned above, (p)ppGpp is also involved in the RpoS general stress response.

The late Arthur Kornberg, who won the Nobel Prize in 1959 for his work on characterizing DNA synthesis, spent the last years of his life studying polyphosphate (polyP) molecules because he believed that this group of compounds plays a very important role in most living organisms (13). The presence of polyP in bacterial cells was detected some time ago in the form of "volutin granules," but its function has yet to be elucidated (13). The synthesis of polyP is catalyzed by phosphate kinase enzymes, and more than 100 different bacterial species have two homologous enzymes, PPK1 and PPK2 (13). The realization that polyP had a major role in the stress response stems from the observation that *ppk1* mutants of gram-positive and gram-negative bacteria had diminished responses to a variety of stresses (70). In pathogenic bacteria such as

*Salmonella, Shigella,* and *Vibrio cholerae,* a *ppk1* mutation often results in a significant reduction of pathogenicity (13, 70).

Among the stress responses affected by polyP deficiency are increased sensitivity to acids, oxidation, osmotic shock, and heat (103, 105). PPK1-deficient *Salmonella* mutants grew significantly slower in the presence of organic acids and were more easily inactivated by heat than were wild-type strains (103). One *Shigella flexneri* and two *Salmonella enterica* strains had diminished survival at 55°C due to *ppk1* mutations (67). In the case of *Vibrio cholerae,* mutations in *ppk* caused a reduced tolerance to low pH and oxidative stress but did not affect toxin production, motility, or intestinal colonization (63).

As the complex network of stress response mechanisms continues to be discovered, we are realizing that the individual global stress response systems are tightly interconnected. The regulatory connections that take place at the molecular level can occur at any stage of metabolism, from DNA replication to protein degradation. The different global and general stress responses exert control over a variety of bacterial components that allow for the cell to survive harsh conditions. Not only are DNA-protein and protein-protein interactions quite prevalent but also regulation of transcription is controlled by small RNA segments (45). In the remainder of this chapter, the molecular components that are involved in responding to different stress exposures are discussed. However, it should be noted that even if the specific stress response mechanisms currently appear to be independent of each other, it is quite likely that additional investigation will find them to be interconnected via at least one of the global systems.

## ACID STRESS

Drastic changes in the pH of the cytoplasm can harm microorganisms by disrupting the plasma membrane or inhibiting the activity of enzymes and membrane transport proteins (35). The bacterial response to acid stress involves a complex system made up of multiple levels of resistance and a large number of genes. The efficacy of the response depends on several factors, including the organism and strain, the type of acid stress, growth phase, and other environmental factors (35, 99). Two levels of resistance which have been characterized include the acid tolerance response (ATR) and acid resistance (AR).

The ATR is the ability of an organism to survive mildly acidic conditions. If the external pH decreases to 4.5 or lower, chaperones (such as acid shock proteins) are produced in order to prevent protein denaturation and help refold any proteins that have already been denatured. Acid adaptation occurs when cells are gradually grown to stationary phase under mildly acidic conditions (114). This phenomenon has been shown to increase the chance of survival in acidic environments as much as 1,000-fold compared to the chance of survival of nonadapted cells (26, 83). In addition, it has been found that cells that have been acid adapted also demonstrate increased resistance to other stresses (oxidative, alcohol, heat, and osmotic) as well as increased tolerance to more severe acidic environments (34). The ATR is also an important signal for pathogens, since the response prompts cells to begin expressing virulence factors (48, 114).

When bacteria are subjected to more severe acidic conditions for an extended period, cells require protection beyond that provided by the ATR. Thus, some bacteria also have AR mechanisms. AR systems enable survival at or below pH 2.5 (15, 26, 82). In these systems an exogenous substrate, such as an amino acid, must be supplemented in order for the AR system to be induced and/or to perform its function (82). Studies have shown that once an AR system has been induced, the organism can remain acid resistant and active for prolonged periods, even at low temperatures such as 4°C (71, 82).

The ability to tolerate acid is an essential characteristic of food-borne pathogens. ATR and AR systems allow an organism to survive in foods that may be naturally acidic, fermented, or subjected to preservation methods that include direct or gradual acidification (84). After the food is ingested, the organism must withstand exposure to acid stress encountered in the gastrointestinal tract of the host if it is to eventually adhere to intestinal wall and produce toxin. AR is very important to pathogenicity, as it offers the pathogen an advantage by increasing the likelihood of survival in the gastrointestinal tract, thereby increasing the opportunity for colonization and subsequent illness (8, 34, 84).

The specific components involved in the response to acidic conditions for *Escherichia coli, Listeria monocytogenes,* and *Salmonella* spp. are discussed below. It should be noted that even though one can attempt to characterize the response mechanism of a particular genus, the system is extremely complex. Variations in the acid stress response can change depending on even seemingly minor differences such as the bacterial strain and the type of acid in the environment (114).

One commonality among the aforementioned pathogens is that they all possess general stress responses that are constitutively expressed once cells reach stationary-phase growth. This resistance system is induced independent of pH (57, 59). These AR systems are controlled in part by stationary-phase alternative sigma factors which interact with a core polymerase in order to direct transcription of new regulons and change gene expression to respond to the new environment (141). Even though RpoS- and SigB-based resistances are separate responses and involve different genes and regulators, it has been found that they share a subset of genes that encode molecules such as catalases and DNA binding proteins (135).

### *Escherichia coli*

The minimum pH at which *E. coli* can grow is 4.4. The point at which *E. coli* fully activates the ATR is between pH 4.5 and 5.5 (71). However, once the surrounding environment drops to pH 5.5 to 6, *E. coli* synthesizes an assortment of new proteins as part of the ATR (83). Several systems contribute to survival of *E. coli* in different acidic environments. In addition to the $\sigma^S$-dependent global stress response, there are four distinct mechanisms that have been associated with protecting *E. coli* cells against an acid shock: three amino acid decarboxylase-dependent systems based on arginine, glutamate, or lysine and a glucose-repressed oxidative response that protects cells when amino acids are not present in the medium (82, 83). Regardless of which system is activated, survivability of cells depends directly on the growth stage reached prior to acid exposure (15, 26, 76).

While the oxidative system requires RpoS to impart AR, RpoS is believed to be only partially involved in the arginine-, glutamate-, and lysine-dependent decarboxylase systems (41, 83). These systems have the ability to protect cells from a pH as low as 2.0, but only if the respective amino acid is present in the medium. In the presence of exogenous glutamate, lysine, or arginine, the mechanism consumes an intracellular proton, thereby converting the amino acid into $CO_2$ and $\gamma$-aminobutyric acid, cadaverine, or agmatine, respectively, depending on the system being utilized. These compounds are then transported out of the cell by an antiporter, while new amino acid substrates are brought into the cell to continue the cycle (35, 93). This process continues until the net proton concentration in the cytoplasm has been lowered to provide a more favorable intercellular pH (22, 88, 108).

The arginine-dependent system requires arginine decarboxylase, encoded by *adi,* as well as an arginine-agmatine antiporter *(adiC)* (44). Two regulatory genes, *cysB* and *adiY,* activate the expression of *adiA* and *adiC* (122). In the case of AdiY, this activation occurs only when AdiY is overexpressed. CysB activates the transcription of *adiA/C* only under anaerobiosis and acidic conditions in rich medium (109). In *Escherichia coli* O157:H7, the expression of arginine-dependent AR requires previous growth of stationary cells on rich medium and under aerobic conditions and at low pH (26).

A similar structure is found in the lysine-dependent system. The *cadBA* operon encodes a lysine decarboxylase *(cadA)* and a lysine-cadaverine antiporter *(cadB)* (93). Unlike CysB of the arginine-dependent system, CadA is induced under both aerobic and anaerobic conditions. CadC is required for *cadBA* transcription; however, *cadC* is located upstream of *cadBA,* and they are transcribed separately from each other (93). Additionally, CadC and the permease (LysP) are involved in sensing the signals produced when extracellular lysine is present in the medium (125). Cadaverine, one of the end products of lysine decarboxylation, inhibits *cadBA* expression and therefore has the ability to repress this response once excess cadaverine has accumulated in the system (88).

Glutamate decarboxylase is required in the glutamate-dependent (GAD) system and appears to act independently of $\sigma^S$ (26, 61). This system involves several proteins, including a γ-aminobutyric acid antiporter (GadC) located at the cell membrane and glutamate decarboxylase enzymes (GadA and GadB) present in the cytoplasm (82, 120). It has been observed that only one of the two glutamate decarboxylase genes, *gadA* or *gadB,* is necessary for survival of *E. coli* down to pH 2.5, but both have to be active for survival at pH 2 (3, 120). Two DNA binding proteins, GadX (AraC-like regulator) and GadY, are integral to this system (126). GadX constitutively activates *gadA* and *gadBC,* while GadY represses *gadX* expression and interacts closely with both the *gadA* and *gadX* promoters (85, 95). Even though the GAD response is independent of $\sigma^S$, it has been observed that upon entry into the stationary phase, *gadX* expression becomes repressed if $\sigma^S$ is repressed, suggesting multiple mechanisms that allow for various levels of acid response in the cell (85). The GAD enzymes and their activity have also been recognized in bacteria other than *E. coli,* including several nonenteric organisms such as *Providencia alcalifaciens, Bacteroides fragilis, Clostridium perfringens, Lactococcus* spp., *Lactobacillus lactis,* and *Listeria* spp. (98, 109).

### *Listeria monocytogenes*

The most favorable range for ATR induction in *Listeria monocytogenes* is pH 5 to 6 (71). Under mildly acidic conditions (pH 5.5), over 37 proteins are induced, and once the pH is decreased to 3.5, over 47 proteins are activated. Of the 47 proteins, 23 were found to be induced under both acid stress levels. It can be surmised that these proteins impart survival at lethal acid doses after the cell has been exposed to mild acid stress (34). Differences in protein expression can also be seen depending on the type of acid stress. It has been recognized that $\sigma^B$ may have the ability to mediate different mechanisms which, in turn, impart protection based on whether the acid is organic or inorganic (34).

There are at least four mechanisms that are a part of the acid stress response in *L. monocytogenes* (8). The first system is activated only in exponentially growing cells, is induced by acidic pH, and relies, at least in part, on $\sigma^B$ (34). Once in the stationary phase, enhanced AR is provided to cells by the $\sigma^B$-dependent general stress response system. In this system, acid stress creates an environmental signal that triggers an Rsb cascade (RsbUVWX) inside the cell. In the Rsb regulon, RsbW is the primary regulator of *sigB* and the $\sigma^B$ response (14). It has been determined that a second, $\sigma^B$-independent mechanism also contributes to AR during the stationary phase, but the specifics of its regulation have yet to be elucidated (8, 34). Additionally, a GAD-based AR mechanism composed of three glutamate decarboxylase genes *(gadD1, gadD2,* and *gadD3)* and two antiporters *(gadT1* and *gadT2),* which is similar to that of *E. coli,* has been found to be active in *Listeria* spp. (19, 20).

### *Salmonella enterica* Serovar Typhimurium

Normal growth of *Salmonella* occurs at pH 5 to 9, although the organism can grow at pH values as low as 4 (52, 82). The range at which the ATR is induced in *Salmonella* serovar Typhimurium cells is pH 4 to 5 (71). When *Salmonella* cells are directly subjected to an acid shock at pH 4.5 or below, over 52 proteins are synthesized. All but four of those known proteins are different from those expressed during the acid adaptation response (36).

*Salmonella* has at least three systems to survive lethal acid exposure: acid tolerance due to the general $\sigma^S$ stress response and two pH-induced growth phase-dependent ATR systems (78, 79, 82). The pH-induced stationary-phase-dependent and pH-induced log-phase-dependent ATR systems are considered to be genetically distinct from each other, as it has been found that genes involved in one response have little to no effect on the other system (5). The log-phase-dependent ATR is induced by log-phase cells subjected to a low-pH environment and thus is well suited for a quick response to acid stress. This response system involves the induction of more than 50 proteins. The system is RpoS dependent and requires the regulatory genes *rpoS* and *fur* (52). After *fur* is induced by the presence of acid, the Fur regulator positively controls several RpoS-independent acid shock proteins in the system (37, 52). The *fur* response in acid stress is a mechanism separate from that for iron regulation.

The stationary-phase-dependent ATR is acid inducible, similar to the log-phase-dependent ATR; however, activation requires exposure to low pH over a long period (5, 79, 82). Therefore, this system is apt in situations where cells are slowly transitioning to more severe acid conditions (79). The stationary-phase ATR provides protection down to pH 4 and involves synthesis of fewer proteins than does the log-phase ATR (17). This response is independent of RpoS and *fur* but is *ompR* dependent. When stationary-phase *Salmonella* cells are exposed to stresses such as acid, *ompR,* a response regulator, is increased, which can trigger the expression of genes needed for this response via the OmpR/EnvZ regulatory system (5). OmpR also induces the expression of *ompC* and *ompF,* both of which are genes that encode outer membrane proteins. OmpR is also an important virulence factor, as it controls the expression of the acid-induced virulence operon *ssrAB* (5). During acid stress, both the general stress response and the acid-inducible stationary-phase response contribute to the overall AR of stationary-phase cells (52).

It should be noted that *Salmonella* serovar Typhimurium cells possess an arginine decarboxylase and antiporter, similar to that of *E. coli,* but upon acid challenge, the organism does not survive (82). Recent work has indicated that the arginine-dependent AR system in *Salmonella* requires anaerobic conditions to be fully functional (66). More studies are necessary to determine what constraints limit the use of this system for protection in acidic environments.

In minimally processed foods, bacteria are more likely to be stressed than killed. Organic acids, such as lactic and acetic acids, are commonly added to foods as preservatives (5, 114). Since pH is often used as a hurdle, the ability of food-borne pathogens to adapt, survive, and possibly increase their virulence at suboptimal pH has direct health and safety implications (84). Hence, further studies on the mechanisms of the acid stress response in bacterial pathogens under specific processing conditions and in different food matrices are necessary (114). At the molecular level, many of the nonprimary acid shock proteins involved in the intricate mechanisms of AR, as well as those involved in environmental sensing, have yet to be determined (5). These include those of the OmpR/EnvZ and Fur systems as well as those yet to be identified in *L. monocytogenes* and other important pathogens.

## OSMOTIC STRESS

The bacterial cell membrane is semipermeable, allowing for the passive diffusion of water into and out of the cell while restricting the movement of most solutes (73, 121). In nature, bacteria are constantly subjected to dynamic changes in the amount of water and nutrients available in their environment. An extreme change in the osmolality of the bacterial environment, known as osmotic stress, causes a drastic flux of water and solutes across the membrane (73). The exposure to severe osmotic stress inhibits DNA replication and cell division, leading to cell death (131). Over time, bacteria have evolved a number of mechanisms, some of which have yet to be discovered, which allow them to adapt and ultimately survive. Specifically, the change in osmolality is sensed by the bacterium, triggering a response that includes the activation of the enzymes and transport proteins. This response quickly counteracts the imbalance and minimizes damage to the cell.

Gram-negative and gram-positive bacteria use global strategies in their response to osmotic stress as well as some unique species-specific responses (11). The sensitivity to osmotic stress is a function of both osmotic pressure and temperature (121). This section reviews the current knowledge on the molecular response to osmotic stress as manifested by a few food-borne pathogens.

### Signals and Primary Responses to Osmotic Stress

Cells react to osmotic stress by regulating changes in cellular volume, based on water content and intracellular solute concentration (73). Osmoregulation can occur only once a bacterium senses osmotic stress. There are many stimuli that have been purported to act as signals of osmotic stress, including a change in external osmotic pressure, internal osmolality, turgor pressure, membrane strain, concentration of a specific signal molecule, or $K^+$ insufficiency (73, 121). In the case of *Escherichia coli,* the mechanism that detects changes in solute concentration involves transmembrane chemoreceptors. The receptors control the activity of a cytoplasmic kinase, CheA, which phosphorylates CheY, a response regulator. Phosphorylated CheY then binds to flagellar motors of the cell. This affects the motile behavior of the bacterium, allowing it to move away from the areas of high solute concentration (129).

Mechanosensitive channels also mediate rapid solute efflux under hypo-osmotic-stress conditions (119). In *E. coli, mscL* and *mscS* encode mechanosensitive protein channels which are also thought to play a role in sensing and adapting to changing osmotic environments. It has been observed that upon disruption of these proteins, the rate of cell growth is severely disrupted (10). The regulators of MscL and MscS have not been determined (23). Investigation of other possible signaling systems characteristic of *E. coli* and other bacteria remains an area of active research.

One of the primary responses of bacteria to osmotic stress is the activation of transport systems necessary for rapid uptake or excretion of compatible solutes, which are also known as osmoprotectants. The actual increase in the intracellular concentration of solutes reduces the amount of water lost due to osmosis, counterbalancing osmotic pressure (11, 68). The solute molecules themselves, such as glycine betaine, proline, choline, glutamate, and carnitine, stabilize the cell by reducing unfolding and denaturation of proteins, thus maintaining cellular homeostasis (11, 68, 73, 131). In most bacteria, glycine betaine is the most effective osmoprotectant, as it increases the volume of available water in the cytoplasm (68, 131). The necessary accumulation of compatible solutes requires the activation of transport systems.

### Osmotic Responses of *Escherichia coli* and *Salmonella enterica* Serovar Typhimurium

In *E. coli* and *Salmonella* serovar Typhimurium, a two-step process is involved in the response to osmotic stress. The first step involves the activation of transport systems, which allows the rapid accumulation of amino acids and their derivatives and potassium ions ($K^+$) (50). Second, the cell increases turgor pressure by uptake of compatible solutes from the growth medium via transport systems into the cytoplasm (121).

In serovar Typhimurium and *E. coli,* $K^+$ transport is regulated by Kdp, Trk, and Kup (73). Kdp responds to signals of cellular turgor pressure and the availability of $K^+$ but is only a temporary response to osmotic stress (21, 119). In addition to $K^+$ transport, *E. coli* has two known glycine betaine transporters which also have dual specificity for proline ProU and ProP (73). The *proU* gene is strongly inducible by osmotic stress as a result of changes in intracellular concentration of $K^+$, and it is thought to primarily function during osmoadaptation (21, 50, 68). As long as cells are exposed to media of high osmolarity, the *proU* operon is induced. It is thought that *kdp* and *proU* might be regulated by a common signal (50). The *proP* gene is weakly inducible in the case of osmotic stress (68).

In *E. coli,* the synthesis of only a few proteins increases as a response to hyperosmotic shock (50). The alternative sigma factor RpoS regulates stationary-phase-induced genes as well as genes involved in the osmotic stress response (7, 23). It is thought that genes controlled by *rpoS* can be induced by osmotic shock since osmotic stress causes cells to stop growing, which mimics entry into the stationary phase (18). Two RpoS-dependent genes, *otsA* and *otsB,* are involved in trehalose synthesis (21, 23). ProP is regulated by RpoS as well as by catabolite repression protein and Fis. ProU has multiple regulators, including TopA, GyrAB, IHF, HU, and H-NS (18, 23, 119). The *rpoS* gene controls the repression of inhibition of H-NS, allowing its induction under hyperosmotic conditions. However, it has also been found that H-NS has a second mechanism with which to control genes that is independent of $\sigma^S$ (7).

OmpR/EnvZ is a two-component regulatory system in *E. coli,* as well as other gram-negative bacteria, which is involved in the response to changes in external osmolarity. OmpR and EnvZ are positive regulators for the synthesis of the major outer membrane porin proteins, OmpC and OmpF (21, 118). OmpR is a regulatory protein with DNA binding properties, while EnvZ is a sensor protein which controls the phosphorylation of OmpR. In this system, OmpR becomes phosphorylated by EnvZ under osmotic stress (118). Phosphorylated OmpR activates OmpF when osmolarity levels are low and acts as a repressor under conditions of high osmolarity. In contrast, OmpC expression increases with increasing osmolarity and is repressed at low osmolarity (21, 118). It has been suggested that in the case of *Salmonella,* the production of *ompC* and *ompF* gene products might be primarily associated with stationary phase instead of specifically induced by osmotic stress (11). In addition to regulation of porins, phosphorylated OmpR negatively regulates *flhDC,* the master operon responsible for expression of flagellar synthesis under stress conditions (118).

Other genes that may become involved in the osmotic stress response include *putP, putA,* and genes in the *bet* regulon. When proline is the only available compatible solute, PutP and PutA of *E. coli* and *Salmonella* allow for the use of proline as the sole carbon and nitrogen source. The *putP* gene product mediates $Na^+$-proline symport, while *putA* catalyzes proline oxidation and acts as a repressor (50). A high-affinity proline transporter is encoded by *putP* (80). The *bet* regulon (BetTIBA) of *E. coli* is regulated by BetI and ArcA and encodes proteins involved in the transport of choline and its oxidation to glycine betaine (21). ArcA inhibits betaine synthesis under anaerobic conditions.

## Osmotic Stress Responses of *Listeria monocytogenes* and *Staphylococcus aureus*

Twelve proteins that are induced by osmotic stress have been identified in *L. monocytogenes* (43, 119). These include *betL,* genes in the *gbu* operon, *opuC, proBA, relA, clpC, clpP,* and *sigB.* It has been determined that the alternative sigma factor $\sigma^B$ is a major contributor to the osmotic stress response, as a $\sigma^B$-dependent promoter has been found to be located upstream of *betL* and the *opuC* operon. Entry into stationary phase induces $\sigma^B$, enabling transport and ultimately the use of osmoprotectants (43). The *sigB* gene product induces the osmoregulatory *opuC* (ABC carnitine transporter) operon as well as other genes encoding peptidases and transport systems that protect against osmotic stress (51).

BetL is an $Na^+$-dependent membrane protein involved in temporary betaine uptake immediately following osmotic shock. Long-term protection and adaptation to osmotic shock are dependent on GbuABC, another betaine porter which is ATP dependent.

OpuC is the main carnitine uptake system and has the ability to transport a very small amount of betaine. OpuC also contributes to the pathogenicity of *L. monocytogenes* (119). The *proBA* gene encodes a low-affinity proline transporter. The *clpC* and *clpP* genes encode general stress proteins, chaperones, and proteases. *relA* is a gene that encodes (p)ppGpp (guanosine 5′-triphosphate−3′-diphosphate and guanosine 5′-diphosphate−3′-diphosphate) synthetase, involved in the growth of this bacterium under high osmotic pressure (31, 43). Additionally, as a result of hyperosmotic stress, FlaR is involved in modulating the helicity of DNA, similar to the action of H-NS of *E. coli* (119).

Recently it has been observed that *ctc* expression is strongly induced by osmotic stress, suggesting that the Ctc protein might be regulated by $\sigma^B$. Ctc becomes involved in the osmotic stress response when the osmoprotectants glycine betaine and carnitine are not present in the medium. Although its actual function is unknown, it is thought to be involved in growth (43). Other proteins with possible roles in the osmotic response of *Listeria* include OpuB, which is involved in carnitine uptake, and OppA, another protein that promotes the intracellular accumulation of peptides and also plays a role in *Listeria* virulence (119).

*Staphylococcus aureus* uses mechanisms similar to those of the gram-negative bacteria to survive osmotic stress, but *S. aureus* is very adaptive. *S. aureus* is the only food-borne pathogen capable of growing at an $a_w$ below 0.90 and surviving at an $a_w$ as low as 0.86 (131, 132). This ability may be due in part to an increase in expression of PdhA and PdhB, protein molecules which are subunits of pyruvate dehydrogenase, a primary metabolism enzyme (132). Two glycine betaine transporters of *S. aureus* have been identified and are distinguished by their affinities for glycine betaine. The low-affinity transporter is induced by osmotic pressure and thus probably plays a role in osmoadaptation. The second system, the high-affinity transporter, is induced simply by the presence of glycine betaine. This is in contrast to the glycine betaine system of *L. monocytogenes,* which is only weakly induced by osmotic stress (68).

### Future Considerations

Food is a rich source of compatible solutes for bacteria, and therefore, knowledge of the solute-specific characteristics of a food is essential when determining food preservation approaches which limit available water (121). Understanding the intricacies of the osmotic stress response can aid in the design of effective control measures which act to prevent the growth and survival of pathogens in foodstuffs with low $a_w$ (119).

It is clear that the osmotic stress response of bacteria is a diverse system involving many external signals and coinciding mechanisms which, in large part, have yet to be elucidated (119). At the molecular level, the mechanisms of compatible solute function and their transport in various bacterial species, as well as the possible roles of osmoprotectants in bacterial pathogenicity, are not fully understood (43, 119, 131). Additionally, there is not much information regarding possible osmoregulation of porins in nonenteric gram-negative species (21). On a more holistic scale, it is unknown whether osmoregulated operons, such as the OmpR/EnvZ and ProU operons of *E. coli,* possess common regulatory elements (21). The subject of osmotic stress response and regulation will remain an area of interesting, and necessary, development for years to come.

## OXIDATIVE STRESS

Oxygen is one of the elements most toxic to all living forms, but in the course of evolution early microorganisms evolved exquisite mechanisms to neutralize its effects (90). According to different theories of evolution, when primitive forms of life evolved the ability to harvest the sun's energy by photosynthesis, a critical step was the ability to dispose of oxygen as the main by-product (86). The remaining life forms that did not produce oxygen also had to evolve mechanisms by which they were capable of detoxifying molecular oxygen. As a result, modern bacteria have a variety of molecular responses that allow them to survive when exposed to high concentrations of oxygen.

The oxygen forms most deleterious to living organisms are referred to as reactive oxygen species (ROS), and they include the superoxide anion and hydrogen peroxide (33). Evolution has selected for organisms that have enzymes capable of detoxificating ROS, such as superoxide dismutase, catalase, and hydroperoxide reductase (123). Bacteria have also evolved different mechanisms to sense the presence of ROS, and many of these are based on the ability of cysteine residues to react with superoxide dismutase. In less than 30 min after exposure to hydrogen peroxide, at least 30 proteins are induced in *Escherichia coli* and *Salmonella* (143). In gram-negative bacteria such as *Escherichia coli,* sensing and detoxification of hydrogen peroxide are controlled by the OxyR regulon (123, 143).

The sensing function of the OxyR regulon is actually performed by OxyR itself. All the evidence indicates that the activation of OxyR is triggered by oxidation of the cysteine residue Cys-199 in one of the its four subunits (90). Oxidized Cys-199 then reacts with a neighboring Cys-208 of another OxyR subunit. The formation of a disulfide bond leads to promoter binding and activation of specific stress response genes. It has been estimated that a concentration of only 100 nM hydrogen peroxide is required for OxyR activation. Once the concentration of hydrogen peroxide decreases, glutathione reductase (GorA) and glutaredoxin 1 (GrxA) catalyze the reduction and inactivation of OxyR (143).

The family of OxyR-dependent stress response genes includes those encoding proteins involved in hydrogen peroxide detoxification and reducing agent synthesis and regulatory proteins (143). The catalase (KatG) and alkyl hydroxyperoxide reductases (AhpCF) are responsible for metabolizing hydrogen peroxide and neutralizing the free-radical by-products (123). In addition to GorA and GrxA, OxyR also induces the production of a thioredoxin 2 *(trxC)* involved in intracellular redox balance (123). Dps is a nonspecific DNA binding protein with a ferritin-like domain that is also regulated by OxyR, induced when cells are exposed to high temperatures and controlled by RpoS (58, 135). The small regulatory RNA OxyS is also part of the OxyR family, which negatively regulates RpoS translation (45).

One of the most important gene networks regulated by OxyR is involved in iron uptake and metabolism. Specifically, OxyR directly activates a ferric uptake repressor, Fur, which is part of an important superfamily of proteins involved in metal ion uptake (90). The Fur regulon includes the iron sensor (Fur), zinc sensor (Zur), and a hydrogen peroxide sensor (PerR). OxyR also activates genes involved in heme biosynthesis *(hemH),* Fe-S synthesis and repair *(suf),* and a disulfide chaperone isomerase *(dsbG)* (143). Recent evidence in several genera, including *Salmonella, Vibrio, Yersinia,* and *Pseudomonas,* indicates that Fur is linked to the induction of virulence genes (29).

In recent years, a family of organic hydroperoxide resistance (Ohr) proteins that are OxyR independent has been identified in several gram-positive and gram-negative bacteria (90). The Ohr regulon includes two enzymes involved in organic hydroperoxide detoxification, OhrA and OhrB, which are separately regulated. While the synthesis of OhrB is only induced by the presence of organic hydroperoxides, OhrA is activated by heat shock and once cells are in stationary phase (90). All of the Ohr protective proteins are repressed by OhrR, which is deactivated by the reaction of a cysteine residue with peroxides in a manner similar to OxyR activation.

## HEAT SHOCK STRESS

Heat shock proteins (HSPs) are generally expressed at all temperatures, since they actively participate in protein synthesis, transport, folding, and degradation (100). Upon an abrupt increase in environmental temperature, bacteria respond by temporarily increasing the expression of HSPs. After 5 to 10 min at the higher temperature, normal cellular growth is discontinued by decreasing the expression of the general HSPs (25, 117, 140). The cell then enters into a heat-induced stressed state by increasing the expression of particular genes within heat shock regulons specific to the organism (117, 140).

The stress of heat shock can cause deleterious physiological changes in the bacterial cell, immediately compromising survival. These changes include the improper folding or unfolding of proteins, aggregation, and, depending on the level of heat and the organism under stress, alterations in metabolic pathways, protein phosphorylation, and membrane integrity (116, 133, 140). On the other hand, it has also been shown that exposing bacterial cells to an intermediary heat shock can enhance the organism's survival when the organism is subsequently exposed to a more drastic form of heat stress or other type of stress. This response has been called thermotolerance (116). Below, the most recent developments on the microbial strategies that allow gram-positive (*Bacillus* and *Staphylococcus*) and gram-negative (*Escherichia*, *Campylobacter*, and *Salmonella*) microorganisms to survive mild heat stress are summarized.

### Gram-Positive Bacteria

At least four classes of heat-inducible genes have been identified for *Bacillus* spp. (25, 56, 58, 92) (Table 1). Each class is differentiated from the other by the mechanism that regulates the synthesis of the HSPs (140).

#### CLASS I

Class I is made up of the HrcA regulon, which includes the *hrcA* and *groEL* operons. HrcA serves as a repressor of transcription and thus negatively regulates the heat shock genes by binding to a conserved *cis*-active sequence known as a CIRCE (controlling inverted repeat of chaperone expression) element (25, 58, 92). Transcription of the *groEL* operon is dependent on $\sigma^A$ (92).

#### CLASS II

It is estimated that over 200 general stress genes are included in the SigB regulon. These genes are controlled by the $\sigma^B$ transcription factor and are expressed within 3 min following heat shock. Deactivation of these genes begins after approximately 10 min, while the other classes are up-regulated (58). This class not only is involved in response to sublethal heat but also is activated by other types of stress, including exposure to high pressure; low levels of oxygen, phosphate, or sugar; and the presence of salts or alcohol (25, 56).

#### CLASS III

Class III gene expression is highly conserved in low-G+C gram-positive bacteria (25). In this class, CtsR (class three stress gene repressor) is transcribed from both $\sigma^A$ and $\sigma^B$ promoters and acts as a negative regulator of the system (58, 92). ClpC, ClpP, and ClpX (ClpX is in class IV) are stationary-phase proteins that are required for motility, sporulation, and growth at high temperatures (56). It has been found that the CtsR regulon is also present in other bacteria, such as *Streptococcus*, *Clostridium*, and *Enterococcus* spp., but the functional role of the regulon in these organisms has not yet been determined (25).

#### CLASS IV

Other stress response genes whose specific regulation mechanisms are unknown at present are placed into class IV, also referred to as class U (unidentified) (25, 58). The commonality between these genes is that they do not have the CIRCE or CtsR operator sequence and induction is independent of $\sigma^B$ (25, 56). In addition, many of the genes in this class code for transport functions (58).

In contrast to *Bacillus subtilis*, the heat stress response of *Staphylococcus aureus* is dually regulated, involving characteristics of both class I and class III. In this system, genes are regulated

**TABLE 1** Heat shock genes of gram-positive bacteria as organized by class[a]

| Class I | Class II | | Class III | Class I/III | Class IV (U)[b] |
|---|---|---|---|---|---|
| *hrcA* | *bmrU* | *trxA* | *ctsR* | *dnaK (S. aureus)* | *ahpF* |
| *grpE* | *bmr* | *yacH* | *yacH* | *clpP (Streptococcus salivarius)* | *clpX* |
| *dnaK* | *bmrR* | *yacI* | *yacI* | | *ftsH* |
| *dnaJ* | *bofC* | *yacL* | *clpC* | | *htpG* |
| *yqeT* | *csbA* | *ycdF* | *sms* | | *htrA* |
| *yqeU* | *csbB* | *ycdG* | *yacK* | | *lonA* |
| *yqeV* | *yfhO* | *ydaP* | *clpP* | | *yvtA* |
| *groEL* | *csbC (yxcC)* | *ydaD* | *clpE* | | *yvtB* |
| *groES* | *csbD (ywmG)* | *ydaE* | *clpC* | | *ywcG* |
| | *csbX* | *ydaF* | | | *ywcH* |
| | *ctc* | *ydaG* | | | |
| | *dps* | *yfkM* | | | |
| | *gsiB* | *yflT* | | | |
| | *gspA* | *yhdF* | | | |
| | *gtaB* | *yhdG* | | | |
| | *katB* | *yhdN* | | | |
| | *katX* | *yjbC* | | | |
| | *opuE* | *yjbD* | | | |
| | *rsbR* | *ykzA* | | | |
| | *rsbS* | *yocK* | | | |
| | *rsbT* | *ysdB* | | | |
| | *rsbU* | *ytkL* | | | |
| | *rsbV* | *ytxG* | | | |
| | *rsbW* | *ytxH* | | | |
| | *sigB* | *ytxJ* | | | |
| | *rsbX* | *yvyD* | | | |
| | *yxkO* | | | | |

[a] Adapted from reference 58 with permission from the authors and publisher.
[b] Unidentified.

by HrcA as well as CtsR, and as in class I, regulation involves a CIRCE element (25).

## Gram-Negative Bacteria

*Escherichia coli,* the most widely studied gram-negative organism, is currently known to have a single response to heat stress involving two regulons and approximately 20 genes (25, 140). Many of these genes are highly conserved among other gram-negative bacteria and thus expressed during heat stress. *Campylobacter jejuni* synthesizes 24 proteins, and based on microarray analysis, a change in expression of more than 336 genes occurs as a result of exposure to elevated temperature (133). In *E. coli,* the location and accumulation of misfolded proteins dictate which heat response is activated. Unfolded proteins in the envelope trigger the $\sigma^E$ regulon, while those in the cytoplasm activate $\sigma^{32}$ transcription (24, 25, 69, 89).

### $\sigma^E$ REGULON

The genes in the $\sigma^E$ regulon are induced only if the organism is subjected to temperatures above 40°C (24, 58). The *rpoE* gene product regulates $\sigma^E$, and induction of $\sigma^E$ leads to the synthesis of at least 10 proteins and the transcription of more than 43 genes (24, 58). The *rpoE* operon contains *rseA* and *rseB,* which encode proteins that integrate into the cytoplasmic membrane and periplasm, respectively.

The *htrA* and *fkpA* genes encode important enzymes that are involved in the folding and degradation of proteins (24). Two newly discovered outer membrane proteins are expressed in the cell as a result of heat shock, both of which act to repair DNA or membrane lipids. A decrease in the expression of these outer membrane proteins actually decreases expression of $\sigma^E$ (140). The $\sigma^E$ factor also transcribes *rpoH,* which encodes transcription factor $\sigma^{32}$ of the second heat stress regulon (24).

### $\sigma^{32}$ REGULON

Most genes in the heat stress response are positively regulated by $\sigma^{32}$(81). This gene is transcribed from four known promoters, i.e., P1, P3, P4, and P5. However, at temperatures over 50°C, P3 is the only active promoter (81, 104). The $\sigma^{32}$ polypeptide is negatively regulated by a subset of HSPs, including those encoded by *dnaK, dnaJ,* and *grpE* (81). These genes, along with *groEL,* encode "chaperones," which are molecules that degrade damaged proteins and prevent aggregation or misfolding of newly made proteins (17, 47, 81, 117). The Lon and Clp proteins (ClpB, ClpP, and ClpX) also have important roles in proteolysis (17, 25, 100). Moreover, studies have demonstrated that the Clp proteins also contribute to virulence in pathogens (100). In this regulon, the heat shock genes *htrC* and *htpY* are located upstream of the DnaK operon. Interestingly, overexpression of *htpY* induces only the *htrA* promoter in the $\sigma^E$ regulon but leads to increased expression of both $\sigma^{32}$- and $\sigma^E$-dependent heat shock regulons, while *htrC* negatively regulates the response (140).

Although the regulatory mechanisms of the heat stress response have been studied in representative gram-positive *(Bacillus)* and gram-negative *(Escherichia)* bacteria, further study is needed. For example, the means by which the bacterial cell initially senses a change in environmental temperature are unknown. Likewise, we know little about the mechanism(s) of signal transport or induction of specific regulatory responses (17, 24). Further, the major regulators of the heat stress response in important food-borne pathogens such as *Campylobacter* have yet to be elucidated (133).

## LOW-TEMPERATURE STRESS

Low-temperature storage of foods is an extremely successful preservation technology. Refrigerated foods are typically maintained at 7°C or lower, while frozen foods are maintained at less than 0°C. The survival of bacteria during low-temperature storage is possible because of specific stress response mechanisms. As the previous section discussed the molecular responses occurring as a consequence of exposure to higher temperatures, this section summarizes what is known about the mechanisms enabling bacterial survival at low temperatures.

A temperature shift from a mesophilic (30 to 40°C) range to below 10°C causes a series of changes to the bacterial cell that involve metabolic and structural alterations as well as modifications in membrane fluidity (62). For most bacteria, the sudden downshift of temperature triggers the production of specific cold shock proteins while simultaneously inhibiting the action of most other proteins. In *Escherichia coli,* as many as 26 different proteins are synthesized after exposure to reduced temperatures (30, 62). Within the first hour, a relatively large and well-conserved family of eight homologous cold shock proteins (CspA to CspI) are expressed. In addition to these Csps, the synthesis of several nucleic acid-associated factors is also stimulated. These proteins include DNA gyrase (GyrA), RNA helicase (DeaD), a transcription factor (NusA), and a translation factor (InfB) (62).

Csps are among the earliest-identified stress response proteins and are probably one of the best-understood molecular responses. The structure and function of most of these proteins have been fully elucidated and detailed by three-dimensional imaging. Csps are relatively small, approximately 7.4 kDa, and are highly conserved among bacterial species (62).

All Csps have the ability to bind to either single-stranded DNA or RNA, but not to double-stranded DNA. All Csps are induced by low-temperature shifts, but CspC and CspE are also synthesized after heat shock. The production of CspD is enhanced by starvation and stationary phase (30, 62). In *Listeria monocytogenes,* the levels of Csp1 and Csp2 increase after exposure to high hydrostatic pressure (HHP) (138).

The main function of the Csps appears to be the regulation of transcription during periods of low temperature (30). The evidence that Csps have a role in transcription stems from the observation that the chromatin-associated protein H-NS is also associated with the cold shock response. The expression of H-NS is increased by CspA, a transcription activator which binds to the promoter region of the *h-ns* gene. Most Csps also play an important role as RNA chaperones by assisting in the formation of secondary and tertiary structures.

Csp regulation at the translational stage is critical for the survival of bacterial cells exposed to low temperature (30). The first sign of Csp regulation is the stabilization of mRNA, which includes Csp's own mRNA (62). Because low temperature increases the stability of DNA hybrids, the role of Csp in translation is to facilitate DNA strand separation, promote the continuous RNA synthesis of household genes, and minimize misfolding of mRNA. Other cold-temperature-induced proteins such as DeaD are helicases involved in unwinding double-stranded nucleic acids (62).

## HIGH-PRESSURE PROCESSING SURVIVAL

HHP is an emerging food preservation method that, in some instances, is a very effective alternative to traditional technologies. HHP does not require the use of chemicals, and unlike with heat treatment, changes in the appearance, flavor, texture, and nutrition of the food are frequently negligible, even in the case of minimally processed foods (1, 9). From a food safety standpoint, HHP has been used successfully for pathogen inactivation in a variety of foods (2, 9, 42, 128).

The effect of high pressure on microbial cells varies in accordance with the degree of pressure, treatment temperature, treatment time, suspension medium, and the target organism (101). The application of mild pressure injures bacteria by increasing the permeability of the cytoplasmic membrane, which substantially inhibits metabolic activity. This level of pressure treatment is ideal for applications in which a delay in the growth of spoilage organisms is the priority (9). A higher-pressure treatment causes cell death by permanently damaging ribosomes, nucleic acids, and the cell wall, causing cellular leakage and inactivating integral membrane transport systems (106).

Microbial growth is usually inhibited after exposure to pressures ranging from 20 to 130 MPa (54). Depending on the species, vegetative cell death usually occurs between 130 and 800 MPa, but the inactivation of bacterial endospores often requires pressures above 1,200 MPa (46, 106). In general, stationary-phase cells are more resistant to HHP than are cells growing exponentially (2, 65). Several reports have indicated that gram-negative bacteria are more sensitive to HHP than gram-positive cells, suggesting that the variability observed among strains might be largely due to susceptibility of the outer membrane to pressure damage (9). Overall, the effects of pressure on bacterial physiology and the response elicited by organisms after pressure treatment are not well understood and deserve investigation. In this section, some concepts about the effect of HHP on the bacterial cell as well as the particular stress responses of *Escherichia coli* and *Listeria monocytogenes* to HHP are discussed.

### Physiological Effects on Cells

Protein synthesis by the bacterial cell is extremely susceptible to elevated pressure (94). This is because high pressure causes conformational changes to proteins and ribosomes, perhaps even denaturation, which leads directly to the inactivation of enzymes and growth inhibition (46, 49). In *E. coli,* a pressure as low as 150 to 250 MPa is enough to induce ribosomal dissociation (94). Although sublethal pressure has been shown to completely inhibit protein synthesis in some bacteria, others have the ability to resume synthesis after the pressure is relieved. However, in all cases, if the pressure is extremely high, irreversible cell death will ensue (94).

Another deleterious effect of high pressure is increased permeability of the bacterial outer membrane. This, in turn, affects protein synthesis and overall cellular stability. For example, it is thought that high levels of exogenous $Mg^{2+}$ help to stabilize the ribosomal structure (49, 94), but membrane permeabilization results in $Mg^{2+}$ leakage from the cell, causing disruptions in ribosomal functionality (54, 94). This effect might be more notable for exponential-phase cells than for stationary-phase cells, since the latter may be able to reseal their membranes after exposure to sublethal pressure, resulting in less severe membrane disruption (94, 96). Clearly, the severity of membrane damage depends directly on the amount of pressure applied (96).

**Adaptive Response**

As a result of exposure to low degrees of pressure, the bacterial cell produces new proteins, many of which are also synthesized in response to very high or very low temperature stress (1, 94, 136). Therefore, it is not surprising that most pressure-resistant strains are also more resistant to other environmental stresses (2). Conversely, when cells are exposed to a stress other than pressure, such as heat shock or mild acidity, a heightened resistance to subsequent pressure stress may occur as a consequence of activating the general stress response, yet another example of cross-protection (96, 139).

The variation in pressure resistance among *Escherichia coli* strains also depends on RpoS activity (110). RpoS, which regulates the stationary-phase-dependent response in gram-negative bacteria, alters RNA polymerase, which, in turn, activates over 80 genes involved in the general stress response. In the case of *E. coli,* a total of 55 proteins have been identified as part of the pressure stress response (110). Fifteen of these proteins have already been characterized and identified as key components in the heat shock (ClpB, ClpP, DnaK, F10.1, GroEL, GroES, GrpE, G21.0, Lon, and RpoH) and cold shock (F10.6, G41.2, H-NS, and RecA) responses. The remaining proteins appear to be induced by pressure alone (137).

*E. coli* also induces a pressure-mediated SOS response that is separate from the RpoS-mediated response, as evidenced by its dependence on RecA and LexA activity (1). This particular SOS response is slightly different from the response elicited after UV exposure and as a consequence of DNA damage. The pressure-induced SOS response results in the activation of the RecA protein, which then cleaves LexA (repressor). Together these proteins activate the transcription of more than 40 genes, including *uvrA, recA,* and *sulA* (1).

Another high-pressure effect is a decrease in the amount of several outer membrane proteins (OmpC, OmpF, and OmpX) based on a mechanism that is independent of EnvZ or OmpR (1, 136). It has been suggested that the loss of outer membrane proteins might be associated with the SOS response, although this has yet to be proven (1). Similar to that of *E. coli,* a general stress response is also activated in *L. monocytogenes* after HHP processing. In this case the alternative sigma factor $\sigma^B$, which is encoded by *sigB,* controls the transcription of genes located in the general stress regulon (139). Specific to *L. monocytogenes,* it was found that high levels of cold shock proteins are produced by the organism after HHP treatment even if the organism did not undergo cold shock (138). It is possible that the partial inactivation of ribosomes might trigger induction of Csps, which is a common result of both cold shock and exposure to sublethal pressure. There are four known Csps that have been identified in *L. monocytogenes,* i.e., Csp1,

Csp2, Csp3, and Csp4. However, only Csp1 and Csp2 levels were found to be elevated after exposure to high-pressure stress alone (138).

A thorough understanding of natural variability and strain-to-strain differences in pressure resistance is necessary to establish standard HHP processes (106), as is a greater understanding of adaptive responses to HHP. It is quite possible that the combination of pressure and other processes, including temperature, pH, or antibacterial compounds, may actually work synergistically to promote pathogen destruction (2, 9, 139). For example, pressure resistance of *E. coli, L. monocytogenes, Salmonella* spp., and *S. aureus* at 25°C was reduced significantly after treatment at 50°C (2, 54). At the molecular level, studies that shed light on the mechanisms of the bacterial response to high-pressure stress are needed (54, 128).

## SUMMARY

Advancement of our knowledge about the molecular basis for bacterial survival under stressful conditions is critical to the assurance of safe and palatable foods, whether using traditional or novel preservation technologies. Organisms such as *Escherichia coli, Salmonella,* and *Bacillus subtilis* have served as excellent models, but there are many other important organisms, and it is likely that more species-specific regulators have yet to be discovered. The advent of genomics-, proteomics-, and metabolomics-based techniques has accelerated our knowledge of the components involved in novel stress responses. In addition, the availability of an increasing number of fully sequenced bacterial genomes should facilitate further advances in the field of bacterial stress responses.

## REFERENCES

1. **Aertsen, A., R. Van Houdt, K. Vanoirbeck, and C. W. Michiels.** 2004. An SOS response induced by high pressure in *Escherichia coli. J. Bacteriol.* **186:**6133–6141.
2. **Alpas, H., N. Kalchayanand, F. Bozoglu, A. Sikes, C. P. Dunne, and B. Ray.** 1999. Variation in resistance to hydrostatic pressure among strains of food-borne pathogens. *Appl. Environ. Microbiol.* **65:**4248–4251.
3. **Arnold, C. N., J. McElhanon, A. Lee, R. Leonhart, and D. A. Siegele.** 2001. Global analysis of *Escherichia coli* gene expression during the acetate-induced acid tolerance response. *J. Bacteriol.* **183:**2178–2186.
4. **Arnold, K. W., and C. W. Kaspar.** 1995. Starvation- and stationary-phase-induced acid tolerance in *Escherichia coli* O157:H7. *Appl. Environ. Microbiol.* **61:**2037–2039.
5. **Bang, I. S., J. P. Audia, Y. K. Park, and J. W. Foster.** 2002. Autoinduction of the ompR response regulator by acid shock and control of the *Salmonella enterica* acid tolerance response. *Mol. Microbiol.* **44:**1235–1250.
6. **Barak, J. D., L. Gorski, P. Naraghi-Arani, and A. O. Charkowski.** 2005. *Salmonella enterica* virulence genes are required for bacterial attachment to plant tissue. *Appl. Environ. Microbiol.* **71:**5685–5691.
7. **Barth, M., C. Marschall, A. Muffler, D. Fischer, and R. Hengge-Aronis.** 1995. Role for the histone-like protein H-NS in growth phase-dependent and osmotic regulation of $\sigma^S$ and many $\sigma^S$-dependent genes in *Escherichia coli. J. Bacteriol.* **177:**3455–3464.
8. **Becker, L. A., M. S. Çetin, R. W. Hutkins, and A. K. Benson.** 1998. Identification of the gene encoding the alternative sigma factor $\sigma^B$ from *Listeria monocytogenes* and its role in osmotolerance. *J. Bacteriol.* **180:**4547–4554.
9. **Benito, A., G. Ventoura, M. Casadei, T. Robinson, and B. Mackey.** 1999. Variation in resistance of natural isolates of *Escherichia coli* O157 to high hydrostatic pressure, mild heat, and other stresses. *Appl. Environ. Microbiol.* **65:**1564–1569.
10. **Blount, P., M. J. Schroeder, and C. Kung.** 1997. Mutations in a bacterial mechanosensitive channel change the cellular response to osmotic stress. *J. Biol. Chem.* **272:**32150–32157.
11. **Botsford, J. L., M. Alvarez, R. Hernandez, and R. Nichols.** 1994. Accumulation of glutamate by *Salmonella typhimurium* in response to osmotic stress. *Appl. Environ. Microbiol.* **60:**2568–2574.
12. **Braeken, K., M. Moris, R. Daniels, J. Vanderleyden, and J. Michiels.** 2006. New horizons for (p)ppGpp in bacterial and plant physiology. *Trends Microbiol.* **14:**45–54.
13. **Brown, M. R., and A. Kornberg.** 2004. Inorganic polyphosphate in the origin and survival of species. *Proc. Natl. Acad. Sci. USA* **101:**16085–16087.
14. **Brul, S., and J. Wells.** 2005. Understanding pathogen survival and resistance in the food

chain, p. 391–420. *In* M. Griffiths (ed.), *Understanding Pathogen Behaviour.* Woodhead Publishing Ltd., Cambridge, England.

15. **Castanie-Cornet, M.-P., T. A. Penfound, D. Smith, J. F. Elliott, and J. W. Foster.** 1999. Control of acid resistance in *Escherichia coli. J. Bacteriol.* **181:**3525–3535.
16. **Cheville, A. M., K. W. Arnold, C. Buchrieser, C.-M. Cheng, and C. W. Kaspar.** 1996. *rpoS* regulation of acid, heat, and salt tolerance in *Escherichia coli* O157:H7. *Appl. Environ. Microbiol.* **62:**1822–1824.
17. **Chuang, S., and F. R. Blattner.** 1993. Characterization of twenty-six new heat shock genes of *Escherichia coli. J. Bacteriol.* **175:**5242–5252.
18. **Chuang, S., D. L. Daniels, and F. R. Blattner.** 1993. Global regulation of gene expression in *Escherichia coli. J. Bacteriol.* **175:**2026–2036.
19. **Cotter, P. D., K. O'Reilly, and C. Hill.** 2001. Role of the glutamate decarboxylase acid resistance system in the survival of *Listeria monocytogenes* LO28 in low pH foods. *J. Food Prot.* **64:** 1362–1368.
20. **Cotter, P. D., S. Ryan, C. G. Gahan, and C. Hill.** 2005. Presence of GadD1 glutamate decarboxylase in selected *Listeria monocytogenes* strains is associated with an ability to grow at low pH. *Appl. Environ. Microbiol.* **71:**2832–2839.
21. **Csonka, L. N.** 1989. Physiological and genetic responses of bacteria to osmotic stress. *Microbiol. Rev.* **53:**121–147.
22. **Cui, S., J. Meng, and A. A. Bhagwat.** 2001. Availability of glutamate and arginine during acid challenge determines cell density-dependent survival phenotype of *Escherichia coli* strains. *Appl. Environ. Microbiol.* **67:**4914–4918.
23. **Culham, D. E., A. Lu, M. Jishage, K. A. Krogfelt, A. Ishihama, and J. M. Wood.** 2001. The osmotic stress response and virulence in pyelonephritis isolates of *Escherichia coli:* contributions of RpoS, ProP, ProU and other systems. *Microbiology* **147:**1657–1670.
24. **Dartigalongue, C., D. Missiakas, and S. Raina.** 2001. Characterization of the *Escherichia coli* $\sigma^E$ regulon. *J. Biol. Chem.* **276:**20866–20875.
25. **Derré, I., G. Rapoport, and T. Msadek.** 1999. CtsR, a novel regulator of stress and heat shock response, controls *clp* and molecular chaperone gene expression in Gram-positive bacteria. *Mol. Microbiol.* **31:**117–131.
26. **Diez-Gonzalez, F., and Y. Karaibrahimoglu.** 2004. Comparison of the glutamate-, arginine- and lysine-dependent acid resistance systems in *Escherichia coli* O157:H7. *J. Appl. Microbiol.* **96:**1237–1244.
27. **Dobrindt, U., and J. Hacker.** 2001. Whole genome plasticity in pathogenic bacteria. *Curr. Opin. Microbiol.* **4:**550–557.
28. **Dodd, C.** 2005. Factors affecting stress response, p. 115–127. *In* M. Griffiths (ed.), *Understanding Pathogen Behaviour.* Woodhead Publishing Ltd., Cambridge, England.
29. **Ellermeier, J. R., and J. M. Slauch.** 2008. Fur regulates expression of the *Salmonella* pathogenicity island 1 type III secretion system through HilD. *J. Bacteriol.* **190:**476–486.
30. **Ermolenko, D. N., and G. I. Makhatadze.** 2002. Bacterial cold-shock proteins. *Cell. Mol. Life Sci.* **59:**1902–1913.
31. **Fang, W., H. Siegumfeldt, B. B. Budde, and M. Jakobsen.** 2004. Osmotic stress leads to decreased intracellular pH of *Listeria monocytogenes* as determined by fluorescence ratio-imaging microscopy. *Appl. Environ. Microbiol.* **70:**3176–3179.
32. **Farkas, J.** 2001. Physical methods of food preservation, p. 567–591. *In* M. P. Doyle, L. R. Beuchat, and T. J. Montville (ed.), *Food Microbiology: Fundamentals and Frontiers,* 2nd ed. ASM Press, Washington, DC.
33. **Fedoroff, N.** 2006. Redox regulatory mechanisms in cellular stress responses. *Ann. Bot.* **98:** 289–300.
34. **Ferreira, A., D. Sue, C. P. O'Byrne, and K. J. Boor.** 2003. Role of *Listeria monocytogenes* $\sigma^B$ in survival of lethal acidic conditions and in the acquired acid tolerance response. *Appl. Environ. Microbiol.* **69:**2692–2698.
35. **Foster, J. W.** 2000. Microbial responses to acid stress, p. 99–115. *In* G. Storz and R. Hengge-Aronis (ed.), *Bacterial Stress Responses.* ASM Press, Washington, DC.
36. **Foster, J. W.** 1991. *Salmonella* acid shock proteins are required for the adaptive acid tolerance response. *J. Bacteriol.* **173:**6896–6902.
37. **Foster, J. W., and H. K. Hall.** 1991. Inducible pH homeostasis and the acid tolerance response of *Salmonella typhimurium. J. Bacteriol.* **173:**5129–5135.
38. **Foster, P. L.** 2007. Stress-induced mutagenesis in bacteria. *Crit. Rev. Biochem. Mol. Biol.* **42:**372–397.
39. **Foster, P. L.** 2005. Stress responses and genetic variation in bacteria. *Mutat. Res.* **569:**3–11.
40. **Friedberg, E. C., G. C. Walker, W. Siede, R. D. Wood, R. A. Schultz, and T. Ellenberger.** 2006. *DNA Repair and Mutagenesis,* 2nd ed. ASM Press, Washington, DC.
41. **Gale, E. F.** 1946. The bacterial amino acid decarboxylases. *Adv. Enzymol.* **VI:**1–31.
42. **Garcia-Graells, C., K. J. A. Hauben, and C. W. Michiels.** 1998. High-pressure inactivation and sublethal injury of pressure-resistant *Escherichia coli* mutants in fruit juices. *Appl. Environ. Microbiol.* **64:**1566–1568.

43. **Gardan, R., O. Duché, S. Leroy-Sétrin, the European *Listeria* Genome Consortium, and J. Labadie.** 2003. Role of *ctc* from *Listeria monocytogenes* in osmotolerance. *Appl. Environ. Microbiol.* **69:**154–161.

44. **Gong, S., H. Richard, and J. W. Foster.** 2003. YjdE (AdiC) is the arginine:agmatine antiporter essential for arginine-dependent acid resistance in *Escherichia coli. J. Bacteriol.* **185:**4402–4409.

45. **Gottesman, S.** 2004. The small RNA regulators of *Escherichia coli:* roles and mechanisms. *Annu. Rev. Microbiol.* **58:**303–328.

46. **Gould, G.** 2005. Pathogen resistance and adaptation to emerging technologies, p. 442–459. *In* M. Griffith (ed.), *Understanding Pathogen Behaviour.* Woodhead Publishing Ltd., Cambridge, England.

47. **Gragerov, A., E. Nudler, N. Komissarova, G. A. Gaitanaris, M. E. Gottesman, and V. Nikiforov.** 1992. Cooperation of GroEL/GroES and DnaK/DnaJ heat shock proteins in preventing protein misfolding in *Escherichia coli. Proc. Natl. Acad. Sci. USA* **89:**10341–10344.

48. **Greenacre, E. J., T. F. Brocklehurst, C. R. Waspe, D. R. Wilson, and P. D. G. Wilson.** 2003. *Salmonella enterica* serovar Typhimurium and *Listeria monocytogenes* acid tolerance response induced by organic acids at 20°C: optimization and modeling. *Appl. Environ. Microbiol.* **69:**3945–3951.

49. **Gross, M., K. Lehle, R. Jaenicke, and K. H. Nierhaus.** 1993. Pressure-induced dissociation of ribosomes and elongation cycle intermediates. Stabilizing conditions and identification of the most sensitive functional state. *Eur. J. Biochem.* **218:**463–468.

50. **Grothe, S., R. L. Krogsrud, D. J. McClellan, J. L. Milner, and J. M. Wood.** 1986. Proline transport and osmotic stress response in *Escherichia coli* K-12. *J. Bacteriol.* **166:**253–259.

51. **Hain, T., H. Hossain, S. S. Chatterjee, S. Machata, U. Volk, S. Wagner, B. Brors, S. Haas, C. T. Kuenne, A. Billion, S. Otten, J. Pane-Farre, S. Engelmann, and T. Chakraborty.** 2008. Temporal transcriptomic analysis of the *Listeria monocytogenes* EGD-e $\sigma^B$ regulon. *BMC Microbiol.* **8:**20.

52. **Hall, H. K., and J. W. Foster.** 1996. The role of Fur in the acid tolerance response of *Salmonella typhimurium* is physiologically and genetically separable from its role in iron acquisition. *J. Bacteriol.* **178:**5683–5691.

53. **Hartman, P. A.** 2001. The evolution of food microbiology, p. 3–12. *In* M. P. Doyle, L. R. Beuchat, and T. J. Montville (ed.), *Food Microbiology: Fundamentals and Frontiers,* 2nd ed. ASM Press, Washington, DC.

54. **Hauben, K. J. A., D. H. Bartlett, C. C. F. Soontjens, K. Cornelis, E. Y. Wuytack, and C. W. Michiels.** 1997. *Escherichia coli* mutants resistant to inactivation by high hydrostatic pressure. *Appl. Environ. Microbiol.* **63:**945–950.

55. **Hecker, M., J. Pané-Farré, and U. Völker.** 2007. SigB-dependent general stress response in *Bacillus subtilis* and related gram-positive bacteria. *Annu. Rev. Genet.* **61:**215–236.

56. **Hecker, M., W. Schumann, and U. Völker.** 1996. Heat-shock and general stress response in *Bacillus subtilis. Mol. Microbiol.* **19:**417–428.

57. **Hecker, M., and U. Völker.** 2001. General stress response of *Bacillus subtilis* and other bacteria. *Adv. Microb. Physiol.* **44:**35–91.

58. **Helmann, J. D., M. F. W. Wu, P. A. Kobel, F. Gamo, M. Wilson, M. M. Morshedi, M. Navre, and C. Paddon.** 2001. Global transcriptional response of *Bacillus subtilis* to heat shock. *J. Bacteriol.* **183:**7318–7328.

59. **Hengge-Aronis, R.** 2002. Signal transduction and regulatory mechanisms involved in control of the $\sigma^S$ (RpoS) subunit of RNA polymerase. *Microbiol. Mol. Biol. Rev.* **66:**373–395.

60. **Hengge-Aronis, R.** 1993. Survival of hunger and stress: the role of *rpoS* in early stationary phase gene regulation in *E. coli. Cell* **72:**165–168.

61. **Hersh, B. M., F. T. Farooq, D. N. Barstad, D. L. Blankenhorn, and J. L. Slonczewski.** 1996. A glutamate-dependent acid resistance gene in *Escherichia coli. J. Bacteriol.* **178:**3978–3981.

62. **Horn, G., R. Hofweber, W. Kremer, and H. R. Kalbitzer.** 2007. Structure and function of bacterial cold shock proteins. *Cell. Mol. Life Sci.* **64:**1457–1470.

63. **Jahid, I. K., A. J. Silva, and J. A. Benitez.** 2006. Polyphosphate stores enhance the ability of *Vibrio cholerae* to overcome environmental stresses in a low-phosphate environment. *Appl. Environ. Microbiol.* **72:**7043–7049.

64. **Jay, J. M., M. J. Loessner, and D. A. Golden.** 2005. *Modern Food Microbiology,* 7th ed. Springer, New York, NY.

65. **Karatzas, K. A. G., and M. H. J. Bennik.** 2002. Characterization of a *Listeria monocytogenes* Scott A isolate with high tolerance towards high hydrostatic pressure. *Appl. Environ. Microbiol.* **68:**3183–3189.

66. **Kieboom, J., and T. Abee.** 2006. Arginine-dependent acid resistance in *Salmonella enterica* serovar Typhimurium. *J. Bacteriol.* **188:**5650–5653.

67. **Kim, K. S., N. N. Rao, C. D. Fraley, and A. Kornberg.** 2002. Inorganic polyphosphate is essential for long-term survival and virulence factors in *Shigella* and *Salmonella* spp. *Proc. Natl. Acad. Sci. USA* **99:**7675–7680.

68. **Ko, R., L. Tombras Smith, and G. M. Smith.** 1994. Glycine betaine confers enhanced osmotolerance and cryotolerance on *Listeria monocytogenes*. *J. Bacteriol.* **176:**426–431.
69. **Konkel, M. E., B. J. Kim, J. D. Klena, C. R. Young, and R. Ziprin.** 1998. Characterization of the thermal stress response of *Campylobacter jejuni*. *Infect. Immun.* **66:**3666–3672.
70. **Kornberg, A.** 2008. Abundant microbial inorganic polyphosphate, poly P kinase are underappreciated. *Microbe* **3:**119–123.
71. **Koutsoumanis, K. P., and J. N. Sofos.** 2004. Comparative acid stress response of *Listeria monocytogenes, Escherichia coli* O157:H7 and *Salmonella* Typhimurium after habituation at different pH conditions. *Lett. Appl. Microbiol.* **38:**321–326.
72. **Kultz, D.** 2003. Evolution of the cellular stress proteome: from monophyletic origin to ubiquitous function. *J. Exp. Biol.* **206:**3119–3124.
73. **Kültz, D., and L. Csonka.** 1999. What sets the TonE during osmotic stress? *Proc. Natl. Acad. Sci. USA* **96:**1814–1816.
74. **Lange, R., and R. Hengge-Aronis.** 1994. The cellular concentration of the $\sigma^S$ subunit of RNA polymerase in *Escherichia coli* is controlled at the levels of transcription, translation, and protein stability. *Genes Dev.* **8:**1600–1607.
75. **Lange, R., and R. Hengge-Aronis.** 1991. Identification of a central regulator of stationary-phase gene expression in *Escherichia coli*. *Mol. Microbiol.* **5:**49–59.
76. **Large, T. M., S. T. Walk, and T. S. Whittam.** 2005. Variation in acid resistance among Shiga toxin-producing clones of pathogenic *Escherichia coli*. *Appl. Environ. Microbiol.* **71:**2493–2500.
77. **Layton, J. C., and P. L. Foster.** 2003. Error-prone DNA polymerase IV is controlled by the stress-response sigma factor, RpoS, in *Escherichia coli*. *Mol. Microbiol.* **50:**549–561.
78. **Lee, I. S., J. Lin, H. K. Hall, B. Bearson, and J. W. Foster.** 1995. The stationary-phase sigma factor $\sigma^S$ (RpoS) is required for a sustained acid tolerance response in virulent *Salmonella typhimurium*. *Mol. Microbiol.* **17:**155–167.
79. **Lee, I. S., J. L. Slonczewski, and J. W. Foster.** 1994. A low-pH-inducible, stationary-phase acid tolerance response in *Salmonella typhimurium*. *J. Bacteriol.* **176:**1422–1426.
80. **Liao, M. K., and S. Maloy.** 2001. Substrate recognition by proline permease in *Salmonella*. *Amino Acids* **21:**161–174.
81. **Liberek, K., T. P. Galitski, M. Zylicz, and C. Georgopoulos.** 1992. The DnaK chaperone modulates the heat shock response of *Escherichia coli* by binding to the $\sigma^{32}$ transcription factor. *Biochemistry* **89:**3516–3520.
82. **Lin, J., I. S. Lee, J. Frey, J. L. Slonczewski, and J. W. Foster.** 1995. Comparative analysis of extreme acid survival in *Salmonella typhimurium, Shigella flexneri,* and *Escherichia coli*. *J. Bacteriol.* **177:**4097–4104.
83. **Lin, J., M. P. Smith, K. C. Chapin, H. S. Baik, G. N. Bennett, and J. W. Foster.** 1996. Mechanisms of acid resistance in enterohemorrhagic *Escherichia coli*. *Appl. Environ. Microbiol.* **62:** 3094–3100.
84. **Lou, Y., and A. E. Yousef.** 1997. Adaptation to sublethal environmental stresses protects *Listeria monocytogenes* against lethal preservation factors. *Appl. Environ. Microbiol.* **63:**1252–1255.
85. **Ma, Z., R. Hope, D. L. Tucker, T. Conway, and J. W. Foster.** 2002. Collaborative regulation of *Escherichia coli* glutamate-dependent acid resistance by two AraC-like regulators, GadX and GadW (YhiW). *J. Bacteriol.* **184:**7001–7012.
86. **Madigan, M. T., J. M. Martinko, P. V. Dunlap, and D. P. Clark.** 2008. *Brock Biology of Microorganisms,* 12th ed., p. 34–115. Pearson Education, Inc., Upper Saddle River, NJ.
87. **McArthur, J. V.** 2006. *Microbial Ecology: an Evolutionary Approach*. Elsevier, Boston, MA.
88. **Meng, S. Y., and G. N. Bennett.** 1992. Nucleotide sequence of the *Escherichia coli cad* operon: a system for neutralization of low extracellular pH. *J. Bacteriol.* **174:**2659–2669.
89. **Missiakas, D., and S. Raina.** 1997. Protein misfolding in the cell envelope of *Escherichia coli:* new signaling pathways. *Trends Biochem. Sci.* **22:** 59–63.
90. **Mongkolsuk, S., and J. D. Helmann.** 2002. Regulation of inducible peroxide stress responses. *Mol. Microbiol.* **45:**9–15.
91. **Montville, T. J., and K. R. Matthews.** 2005. *Food Microbiology—an Introduction*. ASM Press, Washington, DC.
92. **Narberhaus, F.** 1999. Negative regulation of bacterial heat shock genes. *Mol. Microbiol.* **31:**1–8.
93. **Neely, M. N., C. L. Dell, and E. R. Olson.** 1994. Roles of LysP and CadC in mediating the lysine requirement for acid induction of the *Escherichia coli cad* operon. *J. Bacteriol.* **176:**3278–3285.
94. **Niven, G. W., C. A. Miles, and B. M. Mackey.** 1999. The effect of hydrostatic pressure on ribosome conformation in *Escherichia coli*: an in vivo study using differential scanning calorimetry. *Microbiology* **145:**419–425.
95. **Opdyke, J. A., J. G. Kang, and G. Storz.** 2004. GadY, a small-RNA regulator of acid response genes in *Escherichia coli*. *J. Bacteriol.* **186:** 6698–6705.
96. **Pagán, R., and B. Mackey.** 2000. Relationship between membrane damage and cell death in

pressure-treated *Escherichia coli* cells: difference between exponential- and stationary-phase cells and variation among strains. *Appl. Environ. Microbiol.* **66:**2829–2834.

97. **Pal, C., B. Papp, and M. J. Lercher.** 2005. Adaptive evolution of bacterial metabolic networks by horizontal gene transfer. *Nat. Genet.* **37:**1372–1375.

98. **Park, G., and F. Diez-Gonzalez.** 2004. A novel glutamate-dependent acid resistance among strains belonging to the *Proteeae* tribe of *Enterobacteriaceae*. *FEMS Microbiol. Lett.* **237:** 303–309.

99. **Park, Y.-K., B. Bearson, S. H. Bang, I. S. Bang, and J. W. Foster.** 1996. Internal pH crisis, lysine decarboxylase and the acid tolerance response of *Salmonella typhimurium*. *Mol. Microbiol.* **20:**605–611.

100. **Porankiewicz, J., J. Wang, and A. K. Clarke.** 1999. New insights into the ATP-dependent Clp protease: *Escherichia coli* and beyond. *Mol. Microbiol.* **32:**449–458.

101. **Préstamo, G., P. D. Sanz, M. Fonberg-Broczek, and G. Arroyo.** 1999. High pressure response of fruit jams contaminated with *Listeria monocytogenes*. *Lett. Appl. Microbiol.* **28:** 313–316.

102. **Price, S. B., C.-M. Cheng, C. W. Kaspar, J. C. Wright, F. J. DeGraves, T. A. Penfound, M.-P. Castanie-Cornet, and J. W. Foster.** 2000. Role of *rpoS* in acid resistance and fecal shedding of *Escherichia coli* O157:H7. *Appl. Environ. Microbiol.* **66:**632–637.

103. **Price-Carter, M., T. G. Fazzio, E. I. Vallbona, and J. R. Roth.** 2005. Polyphosphate kinase protects *Salmonella enterica* from weak organic acid stress. *J. Bacteriol.* **187:**3088–3099.

104. **Raina, S., D. Missiakas, and C. Georgopoulos.** 1995. The *rpoE* gene encoding the $\sigma^E$ ($\sigma^{24}$) heat shock sigma factor of *Escherichia coli*. *EMBO J.* **14:**1043–1055.

105. **Rao, N. N., and A. Kornberg.** 1999. Inorganic polyphosphate regulates responses of *Escherichia coli* to nutritional stringencies, environmental stresses and survival in the stationary phase. *Prog. Mol. Subcell. Biol.* **23:**183–195.

106. **Rastogi, N. K., K. S. Raghavarao, V. M. Balasubramaniam, K. Niranjan, and D. Knorr.** 2007. Opportunities and challenges in high pressure processing of foods. *Crit. Rev. Food Sci. Nutr.* **47:**69–112.

107. **Rees, C. E. D., C. E. R. Dodd, P. T. Gibson, I. R. Booth, and G. S. A. B. Stewart.** 1995. The significance of bacteria in stationary phase to food microbiology. *Int. J. Food Microbiol.* **28:**263–275.

108. **Richard, H., and J. W. Foster.** 2004. *Escherichia coli* glutamate- and arginine-dependent acid resistance systems increase internal pH and reverse transmembrane potential. *J. Bacteriol.* **186:** 6032–6041.

109. **Richard, H. T., and J. W. Foster.** 2003. Acid resistance in *Escherichia coli*. *Adv. Appl. Microbiol.* **52:**167–186.

110. **Robey, M., A. Benito, R. H. Hutson, C. Pascual, S. F. Park, and B. M. Mackey.** 2001. Variation in resistance to high hydrostatic pressure and *rpoS* heterogeneity in natural isolates of *Escherichia coli* O157:H7. *Appl. Environ. Microbiol.* **67:**4901–4907.

111. **Rodriguez-Romo, L., and A. Yousef.** 2005. Cross-protective effects of bacterial stress, p. 128–151. *In* M. Griffiths (ed.), *Understanding Pathogen Behaviour.* Woodhead Publishing Ltd., Cambridge, England.

112. **Rosche, T. M., D. J. Smith, E. E. Parker, and J. D. Oliver.** 2005. RpoS involvement and requirement for exogenous nutrient for osmotically induced cross protection in *Vibrio vulnificus*. *FEMS Microbiol. Lett.* **53:**455–462.

113. **Rychlik, I., and P. A. Barrow.** 2005. *Salmonella* stress management and its relevance to behaviour during intestinal colonisation and infection. *FEMS Microbiol. Rev.* **29:**1021–1040.

114. **Samelis, J., J. S. Ikeda, and J. N. Sofos.** 2003. Evaluation of the pH-dependent, stationary-phase acid tolerance in *Listeria monocytogenes* and *Salmonella* Typhimurium DT104 induced by culturing in media with 1% glucose: a comparative study with *Escherichia coli* O157: H7. *J. Appl. Microbiol.* **95:**563–575.

115. **Schellhorn, H. E., and V. L. Stones.** 1992. Regulation of *katF* and *katE* in *Escherichia coli* K-12 by weak acids. *J. Bacteriol.* **174:**4769–4776.

116. **Schlesinger, M. J.** 1990. Heat shock proteins. *J. Biol. Chem.* **265:**12111–12114.

117. **Schulz, A., and W. Schumann.** 1996. *hrcA*, the first gene of the *Bacillus subtilis dnaK* operon encodes a negative regulator of class I heat shock genes. *J. Bacteriol.* **178:**1088–1093.

118. **Shin, S., and C. Park.** 1995. Modulation of flagellar expression in *Escherichia coli* by acetyl phosphate and the osmoregulator OmpR. *J. Bacteriol.* **177:**4696–4702.

119. **Sleator, R. D., C. G. M. Gahan, and C. Hill.** 2003. A postgenomic appraisal of osmotolerance in *Listeria monocytogenes*. *Appl. Environ. Microbiol.* **69:**1–9.

120. **Smith, D. K., T. Kassam, B. Singh, and J. F. Elliot.** 1992. *Escherichia coli* has two homologous glutamate decarboxylase genes that map to distinct loci. *J. Bacteriol.* **174:**5820–5826.

121. **Stewart, C. M., M. B. Cole, J. D. Legan, L. Slade, and D. W. Schaffner.** 2005. Solute-

specific effects of osmotic stress on *Staphylococcus aureus*. *J. Appl. Microbiol.* **98:**193–202.

122. **Stim-Herndon, K. P., T. M. Flores, and G. N. Bennett.** 1996. Molecular characterization of *adiY*, a regulatory gene which affects expression of the biodegradative acid-induced arginine decarboxylase gene (*adiA*) of *Escherichia coli*. *Microbiology* **142:**1311–1320.
123. **Storz, G., and M. Zheng.** 2000. Oxidative stress, p. 47–59. *In* G. Storz and R. Hengge-Aronis (ed.), *Bacterial Stress Responses*. ASM Press, Washington, DC.
124. **Strøm, A. R., and I. Kaasen.** 1993. Trehalose metabolism in *Escherichia coli*: stress protection and stress regulation of gene expression. *Mol. Microbiol.* **8:**205–210.
125. **Tetsch, L., C. Koller, I. Haneburger, and K. Jung.** 2008. The membrane-integrated transcriptional activator CadC of *Escherichia coli* senses lysine indirectly via the interaction with the lysine permease LysP. *Mol. Microbiol.* **67:** 570–583.
126. **Tucker, D. L., N. Tucker, Z. Ma, J. W. Foster, R. L. Miranda, P. S. Cohen, and T. Conway.** 2003. Genes of the GadX-GadW regulon in *Escherichia coli*. *J. Bacteriol.* **185:**3190–3201.
127. **Typas, A., G. Becker, and R. Hengge.** 2007. The molecular basis of selective promoter activation by the $\sigma^S$ subunit of RNA polymerase. *Mol. Microbiol.* **63:**1296–1306.
128. **Ulmer, H. M., M. G. Gänzle, and R. F. Vogel.** 2000. Effects of high pressure on survival and metabolic activity of *Lactobacillus plantarum* TMW1.460. *Appl. Environ. Microbiol.* **66:** 3966–3973.
129. **Vaknin, A., and H. C. Berg.** 2005. Osmotic stress mechanically perturbs chemoreceptors in *Escherichia coli*. *Proc. Natl. Acad. Sci. USA* **103:** 592–596.
130. **Vasudevan, P., and K. Venkitanarayanan.** 2006. Role of the *rpoS* gene in the survival of *Vibrio parahaemolyticus* in artificial seawater and fish homogenate. *J. Food Prot.* **69:**1438–1442.
131. **Vijaranakul, U., M. J. Nadakavukaren, D. O. Bayles, B. J. Wilkinson, and R. K. Jayaswal.** 1997. Characterization of an NaCl-sensitive *Staphylococcus aureus* mutant and rescue of the NaCl-sensitive phenotype by glycine betaine but not by other compatible solutes. *Appl. Environ. Microbiol.* **63:**1889–1897.
132. **Vilhelmsson, O., and K. J. Miller.** 2002. Synthesis of pyruvate dehydrogenase in *Staphylococcus aureus* is stimulated by osmotic stress. *Appl. Environ. Microbiol.* **68:**2353–2358.
133. **Wang, G., and M. P. Doyle.** 1998. Heat shock response enhances acid tolerance of *Escherichia coli* O157:H7. *Lett. Appl. Microbiol.* **26:** 31–34.
134. **Waterman, S. R., and P. L. C. Small.** 1996. Characterization of the acid resistance phenotype and *rpoS* alleles of Shiga-like toxin-producing *Escherichia coli*. *Infect. Immun.* **64:** 2808–2811.
135. **Weber, H., T. Polen, J. Heuveling, V. F. Wendisch, and R. Hengge.** 2005. Genome-wide analysis of the general stress response network in *Escherichia coli*: $\sigma^S$-dependent genes, promoters, and sigma factor selectivity. *J. Bacteriol.* **187:**1591–1603.
136. **Welch, T. J., and D. H. Bartlett.** 1998. Identification of a regulatory protein required for pressure-responsive gene expression in the deep-sea bacterium *Photobacterium* species strain SS9. *Mol. Microbiol.* **27:**977–985.
137. **Welch, T. J., A. Farewell, F. C. Neidhardt, and D. H. Bartlett.** 1993. Stress response of *Escherichia coli* to elevated hydrostatic pressure. *J. Bacteriol.* **175:**7170–7177.
138. **Wemekamp-Kamphuis, H. H., A. K. Karatzas, J. A. Wouters, and T. Abee.** 2002. Enhanced levels of cold shock proteins in *Listeria monocytogenes* LO28 upon exposure to low temperature and high hydrostatic pressure. *Appl. Environ. Microbiol.* **68:**456–463.
139. **Wemekamp-Kamphuis, H. H., J. A. Wouters, P. P. L. A. de Leeuw, T. Hain, T. Chakraborty, and T. Abee.** 2004. Identification of sigma factor $\sigma^B$-controlled genes and their impact on acid stress, high hydrostatic pressure, and freeze survival in *Listeria monocytogenes* EGD-e. *Appl. Environ. Microbiol.* **70:** 3457–3466.
140. **White, D.** 2007. *The Physiology and Biochemistry of Prokaryotes,* 3rd ed. Oxford University Press, New York, NY.
141. **Wiedmann, M., T. J. Arvik, R. J. Hurley, and K. J. Boor.** 1998. General stress transcription factor $\sigma^B$ and its role in acid tolerance and virulence of *Listeria monocytogenes*. *J. Bacteriol.* **180:**3650–3656.
142. **Yuk, H.-G., and D. L. Marshall.** 2003. Heat adaptation alters *Escherichia coli* O157:H7 membrane lipid composition and verotoxin production. *Appl. Environ. Microbiol.* **69:**5115–5119.
143. **Zheng, M., X. Wang, L. J. Templeton, D. R. Smulski, R. A. LaRossa, and G. Storz.** 2001. DNA microarray-mediated transcriptional profiling of the *Escherichia coli* response to hydrogen peroxide. *J. Bacteriol.* **183:** 4562–4570.

# USING MICROBIAL SUCCESSION TO THE PROCESSOR'S ADVANTAGE: FOOD FERMENTATION AND BIOCONTROL

*Trevor G. Phister*

# 9

Food fermentation is one of humankind's oldest methods of preservation. It is used to preserve, enhance, and add flavor to many different types of foods. The involvement of fermentation in some foods, such as beer, wine, and cheese, is obvious, while other foods involving fermentation may be surprising, such as coffee or chocolate (15, 77). Other fermented foods, such as the ancient Mayan food pozol or the Chinese food sufu, may seem exotic to those of us in the United States and Europe (2, 8). While these foods may involve the fermentation of different substrates such as milk, grapes, or soybeans and have different primary fermentation organisms such as yeasts *(Saccharomyces cerevisiae),* bacteria *(Lactococcus lactis),* or molds *(Rhizopus oligosporus),* they all have similarities in the underlying principles of their fermentations. Fundamentally, growth of the fermenting organism(s) causes a change in the substrate, often enhancing flavor, occasionally removing toxic compounds, and always preserving the food, generally through both the production of lactic and acetic acids or ethanol and often through removal of substrates useful for the growth of spoilage or pathogenic microbes.

Early fermentations were somewhat magical occurrences. In fact, humans were using fermentation long before they had an understanding of the process; for example, Egyptian bakers or medieval monks brewing beer gave little thought to the fundamental mechanism of fermentation even while they were creating processes that are still used today. They simply knew that the product would be more palatable than the grain they used or would be safer to drink than the village water. The process of fermentation would continue to be a mystery for quite some time. Even though van Leeuwenhoek examined tiny animalcules in beer during the 17th century, most scientists thought that fermentation was a chemical process (36). It was not until the elegant experiments of Pasteur in the 19th century that the role of microbes in fermentation became accepted (6). Pasteur was hired by the French wine industry to study wine spoilage. He observed differences between good fermentations and bad fermentations under the microscope and found different microbial populations in the wines (6). He went on to demonstrate the role of microbes in both fermentation and spoilage. It was this work that eventually led to the development, by Emil

*Trevor G. Phister,* Department of Food, Bioprocessing and Nutrition Sciences, North Carolina State University, Raleigh, NC 27695.

*Food-Borne Microbes: Shaping the Host Ecosystem,* Edited by L.-A. Jaykus, H. H. Wang, and L. S. Schlesinger, © 2009 ASM Press, Washington, DC

Christian Hansen of the Carlsberg brewery along with other microbiologists of the time, of the starter culture systems used today. While early civilizations practiced the use of a simple type of starter culture by back-slopping, i.e., using material from a successful fermentation to conduct the succeeding fermentation, the modern system of starter cultures allows processors more control of the fermentation results than the Egyptian bakers and medieval monks possessed. Some say that this control leads to a loss of regional styles, which has stimulated the growth in the farmhouse cheese industry and in microbreweries producing lambic-style beers. These processes often use natural fermentations and achieve some unique flavors at the cost of consistency and occasional loss of product.

Most microbiologists and food scientists learn in their introductory classes that *Saccharomyces cerevisiae* is used to make bread and wine or that *Lactococcus lactis* is the bacterium used in making cheese. And while these statements are correct—these organisms are generally the starter cultures used for these processes—a large number of other microorganisms play both positive and negative roles in these fermentations. These other organisms can be as obvious as the succession of microbes seen during sauerkraut fermentation, in which *Leuconostoc mesenteroides* subsp. *mesenteroides* gives way to *Lactobacillus plantarum* and other lactic acid bacteria (LAB) as conditions in the fermentation change, or more subtle changes noticeable only since the advent of molecular biology techniques such as random amplified polymorphic DNA PCR. For example, Urso et al. (111) demonstrated that in wine fermentation, the inoculated *Saccharomyces cerevisiae* strain is not always the *Saccharomyces* strain that is isolated at the end of the fermentation. In this chapter, I examine the roles these microbes play in fermentation and the general principles involved in fermentations, including how the primary fermentation organisms interact with the other microbes present in the fermentation and the roles these other microbes have in both product quality and safety.

## GENERAL PRINCIPLES OF FOOD FERMENTATION

According to *Brock Biology of Microorganisms*, the definition of fermentation is anaerobic catabolism in which an organic compound serves as an electron donor and another organic compound serves as an electron acceptor, with ATP being produced by substrate level phosphorylation (73). In the case of food fermentations, the organic compound is often a sugar such as lactose in milk, which is taken up by *L. lactis* and metabolized by the Embden-Meyerhof-Parnas (EMP) pathway (Fig. 1) to create pyruvate. *L. lactis,* a homofermenting LAB, further reduces pyruvate (the electron acceptor) to two molecules of lactic acid. In heterofermentative LAB, such as the *Leuconostoc* species found at the start of sauerkraut fermentation, the pyruvate is reduced to lactic acid, ethanol, and carbon dioxide (Fig. 2). The yeast *S. cerevisiae,* under the conditions of wine fermentation, would metabolize glucose from grapes using the EMP pathway and reduce the resulting pyruvate to ethanol, although *Saccharomyces* is also capable of respiration under the right conditions. However, other food fermentations do not fit neatly into this definition, as the primary fermentation microbe may need oxygen for growth, as is the case in the *Acetobacter aceti* transformation of ethanol to acetic acid in vinegar fermentation, or the transformation of the substrate may be completed by an enzyme (2, 56).

It is the job of the winemaker, cheesemaker, or brewer to control the growth of these microbes or the activity of the enzymes responsible for the fermentation and to ensure that their conversion of the substrate is beneficial, not producing off-flavors or other faults such as $H_2S$ (hydrogen sulfide) in wine. Further, they want to ensure that the substrate is converted to the desired end products (such as lactic acid and ethanol), as these are often responsible for the control of other spoilage and

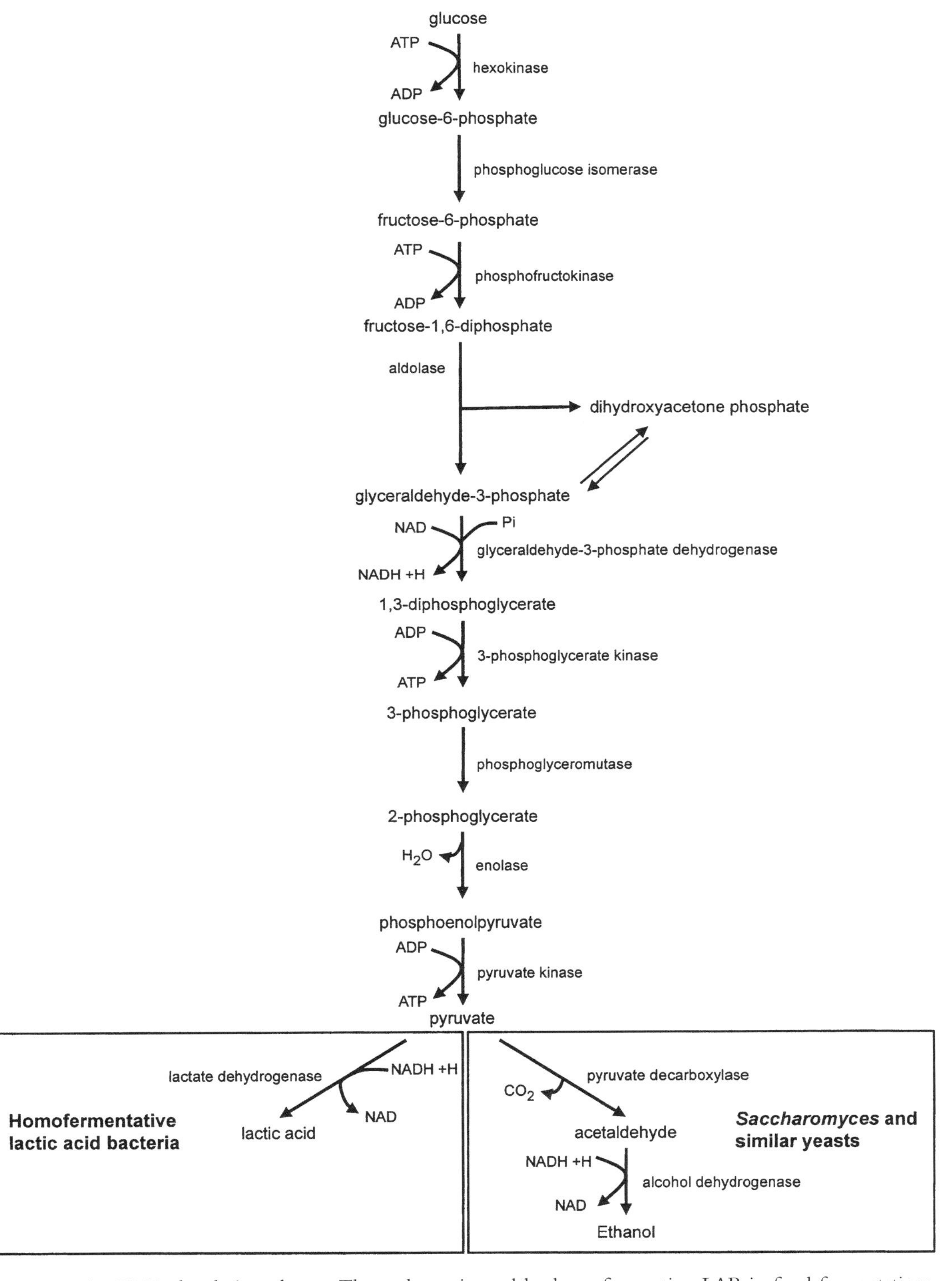


**FIGURE 1** EMP glycolysis pathway. The pathway is used by homofermenting LAB in food fermentations. Lactate dehydrogenase reduces pyruvate to lactic acid. *Saccharomyces* and other yeasts convert pyruvate to ethanol during fermentation. It is important to note that *Saccharomyces* is a Crabtree-positive yeast and can conduct respiration in the presence of oxygen. However, when high levels of glucose are available it uses the EMP pathway to produce ethanol, even in the presence of oxygen. Adapted from reference 55 with permission.

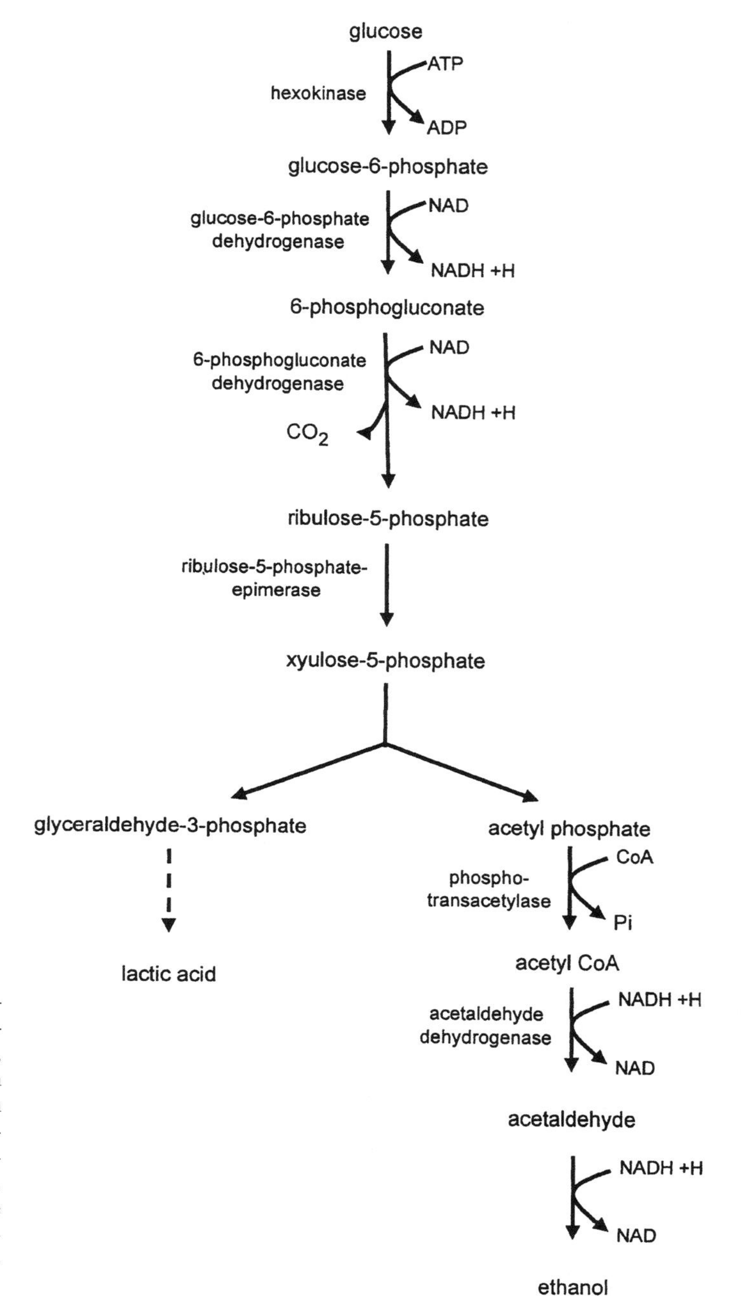


**FIGURE 2** Phosphoketolase pathway. The pathway is used by heterofermenting LAB and produces ethanol, $CO_2$, and lactic and acetic acids in equamolar amounts. Certain LAB can be facultative heterofermentative bacteria and contain both the EMP pathway and the phosphoketolase pathway, while others lacking the aldolase of the EMP pathway are obligate heterofermentative bacteria. Used from reference 55 with permission.

pathogenic microbes. Additionally, the fermenting microbes will produce an array of antimicrobial compounds such as bacteriocins, antimicrobial peptides which function against closely related species, short-chain fatty acids, $H_2O_2$, and $CO_2$, which help to control pathogens and spoilage organisms alike (104).

This control can be a difficult task, as the fermentation substrate originating from the field, dairy farm, or elsewhere will contain many different microbes even before the fermentation begins. For example, on a typical grape berry, the yeast populations range from $10^3$ to $10^5$ CFU/ml at harvest, with *S. cerevisiae* being very difficult, if not impossible, to find (41). However, there are many tools (Table 1) which the processor can use to control these fermentations and ensure that the correct organisms come to dominate. In fact, in many fermentations, early steps may reduce or control the microbial population arriving on the substrate, such as boiling of the wort before pitching of the yeast in brewing. These tools are often just manipulations of the intrinsic and extrinsic factors that affect microbial growth in foods (pH, oxidation-reduction potential, water activity, and temperature) in order to select for one microbial population over another (82). They may range from traditional techniques such as salting, a method of manipulating water activity, or controlling the fermentation temperature, to more recent advances like the addition of chemicals (for example, the use of lysozyme in wine) (31).

As in any naturally occurring ecosystem, the microbes in food fermentations, whether brought in on the substrate or used as a starter culture, interact with one another. In fermentations one can find classic examples of antagonism, such as acid and bacteriocin pro-

**TABLE 1** Processor tools for controlling a fermentation

| Factor affecting microbial growth | Examples of control in a fermentation | Reference(s) |
|---|---|---|
| Temp | Heating: controlling the cooking temp during cheesemaking to control acid production by the starter | 63 |
| | Cooling: conducting sauerkraut fermentation at 18°C helps to ensure *Leuconostoc*'s domination | 13 |
| pH | Acid addition: the must in winemaking can have acid adjustments before fermentation to help the wine's balance and to limit growth of spoilage microbes | 12 |
| | Starter culture: the cultures for dairy fermentations are often chosen for the amount of acid produced | 63 |
| Oxygen (oxidation-reduction potential) | The $CO_2$ produced in a wine fermentation creates an anaerobic environment that allows *S. cerevisiae* to outcompete other yeasts | 12 |
| Water activity | Additions: salt in fermenting vegetables lowers the water activity and influences microbial succession | 13 |
| | Drying: barley is dried for storage before malting | 3 |
| Physical interventions | Filtration: used as a final step in a number of beverage fermentations to remove spoilage agents | 3, 12 |
| Antimicrobials | Sulfur dioxide: primary antimicrobial added to control wild yeasts in wine | 12 |
| | Natamycin: used in cheese production to prevent yeast and mold growth | 63 |
| Nutrient availability | Milk standardization: cheesemakers may standardize milk to create a constant product | 63 |
| | DAP[a]: winemakers may add nutrients such as DAP, a nitrogen source, to ensure successful fermentation | 12 |

[a]DAP, diammonium phosphate.

duction by *L. lactis* in cheese. While the antagonism of pathogens and spoilage microbes is often the goal of any fermentation, examples of mutualism, commensalism, amensalism, predation, parasitism, competition, and synergism can all be found in most food fermentations (10, 40). Additionally, food fermentations even provide some classic examples of microbial succession, such as those found in the well-studied sauerkraut fermentation (13).

The following sections do not describe in detail the processing steps constituting each fermentation, other than to comment on how a particular step may affect the microbial succession during that fermentation. For a wonderful overview of the processing steps involved in individual fermentations, please see references 2 and 56. Instead, the discussion focuses on how and why the succession of microbes in each type of fermentation may occur.

## ALCOHOLIC FERMENTATIONS

Fermentation of grains and fruits resulting in the production of ethanol has been used as a form of preservation for thousands of years (19, 79). The primary microbes involved in ethanol production are typically yeasts, often strains of *S. cerevisiae,* as it has a high ethanol tolerance. In fact, *Saccharomyces* and other Crabtree-positive yeasts use the production of ethanol as a method to limit competition in high-sugar environments. Crabtree-positive yeasts undergo fermentative metabolism to produce ethanol from six-carbon sugars such as glucose, even in the presence of oxygen (30). While this is not as metabolically efficient as the Crabtree-negative yeasts' production of $CO_2$ through respiration, the ethanol produced is used to eliminate competition. After the competition is removed and the glucose supply is exhausted, the Crabtree-positive yeast can utilize the ethanol it produced as a substrate (97). Ethanol production is not, however, limited to yeasts, as bacteria such as *Zymomonas mobilis* may also be primary fermentation organisms (4).

While the types of substrate and fermented products can vary greatly, from fermented milk products that contain ethanol such as koumiss to the production of distilled beverages such as whiskey, this section focuses on beer and wine, two of today's most popular fermented products. Beer and wine were some of the earliest industrialized fermentations. In fact, it was the growth of these industries and the problems involved in large-scale production, such as microbial spoilage, that led to a number of early advances in the field of microbiology.

### Wine

According to McGovern et al. (78), wine has been produced since at least 5400 B.C.E. While it has only been within the last 150 years or so that we have studied the role of microbes, both good and bad, in fermentation, the techniques used in making wine involve some of the earliest advances in food processing, such as the use of sulfur dioxide by the Romans and Egyptians (56). While the interactions among the many organisms present throughout winemaking are important for the stability and quality of the final product, the whole process begins in the vineyard.

The microbial ecology on the grape berry is varied and includes bacteria, yeasts, and even molds. Bacteria such as lactobacilli and acetic acid bacteria may cause wine spoilage, while molds such as *Aspergillus* and *Penicillium* may produce off-flavors and even ochratoxin, which has been found in wine produced from infected grapes (103). The principal benefit provided by a grape berry's microbial community is its contribution to terroir, with the primary contributors being the non-*Saccharomyces* yeasts present on the grape skin.

These organisms are thought to contribute to a winemaking region's terroir, as they differ from region to region, as well as among grape varieties (5, 41, 76, 95). The role of these grape-associated microbes at the start of wine fermentation and their impact on flavor production are increasingly active areas of research. It is thought that specific strains of non-*Saccharomyces* yeasts such as *Hanseniaspora uvarum* (anamorph, *Kloeckera apiculata*) or *Metschnikowia pulcherrima* may be associated with a particular region or even vineyard and may impact wine flavor through production of various enzymes such as β-glycosidase, an enzyme which releases grape flavor compounds which are bound to sugar molecules (21). Interestingly, the principal wine fermentation yeasts, *S. cerevisiae* and *Saccharomyces bayanus,* are difficult to isolate from grapes, and their origin in wine production is still somewhat controversial (41, 84, 103). Mortimer and Polsinelli (84) suggested that they were inoculated onto berries by flies and bees, and they found that *Saccharomyces* was isolated only from damaged berries. In a recent study, Comitini and Ciani (24) also demonstrated that *S. cerevisiae* inoculated onto either healthy or damaged grapes was not able to persist, suggesting that the source of *Saccharomyces* in natural fermentations remains to be discovered.

The yeast populations on grapes typically range from $10^3$ to $10^5$ CFU/ml at harvest (41). *Hanseniaspora, Metschnikowia,* and *Candida* are the yeast genera most commonly isolated from the berry at the time of harvest, although recent work suggests that these yeasts may come to dominate due to damaged berries and that *Aureobasidium pullulans* is the dominant yeast in undamaged berries (5, 41, 76, 90, 101, 103).

While this observation does not appear to be a regional variation in microflora, the effects of *A. pullulans* on the microbial ecology of the grape and on wine fermentation remain to be studied (90, 101). One possible advantage of encouraging this yeast for the grower is its suppression of fungi such as *Botrytis cinerea* and *Penicillium expansum* (18). Other yeasts commonly isolated from grapes include *Cryptococcus, Rhodotorula, Pichia, Torulopsis,* and *Zygosaccharomyces,* although typically at lower numbers (41, 101–103).

A number of factors such as vineyard location, climate, and viticultural practices may impact the size and diversity of the yeast populations present on a grape berry (41, 47, 69, 76, 90, 95, 101–103). A classic study by Longo et al. (69) demonstrated how the difference in yeast populations among vineyards (three coastal and three inland) and between vintages (two years: the first rainy and the second a typical year in Spain) might occur. They found that the main difference among regions was in yeasts with an oxidative metabolism (different species of *Hansenula, Cryptococcus, Candida,* and *Pichia*) and that the diversity of these yeasts decreased in drier years. This suggested that the increase in diversity was due to the sanitary conditions of the grapes, with the proliferation of fungi (molds) such as *Botrytis* and *Aspergillus* being favored by the humidity and causing an imbalance in the yeast populations.

While this effect of fungal infection was only a suggestion by Longo et al. (69) and others (103), a study by Nisiotou and Nychas (90) comparing the yeast populations on healthy and *Botrytis*-infected grapes found not only that the total yeast populations were higher but also that there was a greater diversity of both fermentative and spoilage species present on the *Botrytis*-infected grapes. They further found that in a survey of the yeast community in another vineyard, the communities were different in uninfected grapes but similar in *Botrytis*-infected grapes, suggesting that the presence of *Botrytis* plays a role in structuring the yeast community, most likely by increasing the availability of nutrients.

Because of the large number of organisms on the grape berry which inoculate the fermentation itself, wine provides an ideal opportunity to study yeast-yeast, yeast-mold, and yeast-bacterium interactions throughout the process. To start, the juice (must) is a naturally selective environment possessing a low pH and typically a high sugar concentration, leading to a higher osmotic pressure. Often, the winemaker also adds sulfur dioxide during harvest or crush. Ultimately, these pressures determine which microbes play a role during the course of the fermentation. The interactions among the yeasts, molds, and bacteria create a final product which prevents growth of both pathogens and spoilage agents through low pH, limitation of nutrients, ethanol, an anaerobic environment created by the $CO_2$ produced during the fermentation, and, of course, the preservative sulfur dioxide, used throughout winemaking to control wild microbial populations and prevent spoilage (103). Additionally, since this interplay among organisms varies from year to year, this contributes to differences between vintages.

At the beginning of alcoholic fermentation, *Hanseniaspora, Candida,* and *Metschnikowia* predominate and reach populations of $10^6$ to $10^7$ CFU/ml (42). At midfermentation, these yeasts die off and *Saccharomyces* becomes the dominant organism. While the ethanol tolerance of each yeast species is suspected to play a primary role as in the succession of the various strains of each species during fermentation, a number of interactions do take place (40, 111). First, yeasts may produce several compounds which inhibit growth of other yeasts, including killer factors, short- to medium-chain fatty acids, and even ethanol (97, 105, 113). Further, they compete with each other for nutrients. For example, *H. uvarum,* one of the dominant yeasts on grapes, is known to strip the juice of thiamine and limit the growth of *Saccharomyces* (83). Aside from competing for nutrients to limit growth of other yeasts, this competition may also affect the finished wine. If nitrogen is deficient, which could happen for a number of reasons, including microbial competition, it may not only lead to one yeast dominating the fermentation at a particular point but also could lead to production of off-flavor compounds such as $H_2S$, which produces a rotten-egg odor (68). During the course of normal growth, *Saccharomyces* produces $H_2S$ to form methionine and cysteine; however, when nitrogen levels are low, instead of producing thiol-containing amino acids, the free $H_2S$ accumulates in the cell and diffuses into the wine, creating off-odors (59).

Recently, work by Nissen and Arneborg (91) suggests that other interactions among yeasts may also be important in determining the microbial succession during fermentation. They demonstrated that the presence of a high concentration ($10^7$ CFU/ml) of viable *Saccharomyces* was needed to cause an early growth arrest of *Kluyveromyces thermotolerans* and *Torulaspora delbrueckii.* The presence of ethanol and nutrient depletion were ruled out by growth of each yeast in media taken from a previous mixed culture at the time of early growth arrest (91). The need for live *Saccharomyces* cells was demonstrated by the continued growth of *K. thermotolerans* and *T. delbrueckii* after addition of $10^7$ CFU of heat-inactivated *Saccharomyces* per ml (91). The investigators reported that in addition to viable *Saccharomyces,* cell-cell contact was needed to observe growth arrest (91). However, they also found that the dialysis tubing used to separate the cultures was fouled by *Saccharomyces* biofilm formation, and this affected the growth of *Saccharomyces* inside the dialysis tube. And even though *Saccharomyces* still reached concentrations of $10^7$ CFU/ml inside the tube, the biofilm did affect the diffusion of molecules across the membrane (91), leaving open the possibility of a soluble factor being responsible for the interaction. One possibility, which has yet to be explored, is the presence of quorum sensing compounds.

To date, quorum sensing has been demonstrated or suspected to regulate gene expression in a number of yeasts, including *Histoplasma capsulatum, Ceratocystis ulmi,* and *S. cerevisiae* (53, 54, 88, 106). However, the greatest depth of research has been on *Candida albicans* and the regulation of its yeast-to-mycelium conversion by farnesol (88).

In *Saccharomyces,* farnesol may be produced at high levels (88). When added exogenously, it produces a growth-inhibitory effect through the creation of reactive oxygen species, possibly caused by inhibition of a phosphatidylinositol-type signaling pathway (71, 72). Farnesol, however, does not appear to be a quorum sensing molecule in *Saccharomyces,* as its effects are not cell density dependent. Currently, the only quorum sensing molecules identified in *Saccharomyces* are the aromatic alcohols tryptophol and phenylethyl alcohol (22). These alcohols stimulate the transition from yeast to filamentous form in response to low nitrogen levels (22). As quorum sensing-induced behaviors in *Saccharomyces* are a recent observation, their role in interactions among yeasts during wine fermentation or any other fermentation has yet to be explored (40).

After primary fermentation, wine (especially red wine) may under go a secondary fermentation, known as malolactic fermentation, which is normally conducted by the bacterium *Oenococcus oeni* (35, 64). This is somewhat dependent on the pH of the wine; below pH 3.5, only strains of *O. oeni* predominate, while above this pH, other less desirable LAB, such as *Pediococcus* and heterofermentative lactobacilli, constitute the majority of the remaining bacterial population (116). Since these bacteria may cause a number of defects, such as undesirable buttery, metallic, and/or earthy odors, as well as an excess of acetic acid, it is to a winemaker's advantage to ensure that *O. oeni* conducts malolactic fermentation, if it is desired.

The purpose of malolactic fermentation is primarily to convert malic acid to lactic acid. This conversion raises the pH of the wine by 0.1 to 0.3 unit, which on the surface would appear to make the product less microbiologically stable; however, the removal of malic acid eliminates a substrate which spoilage bacteria could use for growth (46). As an added benefit, it may often make a wine smoother, increasing the flavor complexity through production of buttery, nutty, and vanilla flavors, and it helps stabilize the color in red wines.

Microbiologists have debated for quite some time as to the advantage conferred on the bacterium *O. oeni* or another LAB by the conversion of malic acid to lactic acid. In the end, it appears to come down to the reaction allowing the cell to expel a proton without expending any energy. Malic acid is taken into the cell with a charge of $-2$, and lactic acid with a $-1$ charge is expelled (62).

While malolactic fermentation generally occurs unless a winemaker takes steps to prevent it (such as keeping the pH low and the ethanol concentration high and maintaining $SO_2$ levels), the interactions between the *Saccharomyces* organism responsible for alcoholic fermentation and *O. oeni* play a large role in its success or failure (1). These interactions may vary based on the strains used to inoculate each fermentation and may run the gamut from inhibition to neutralism and even stimulation (1). Work by Nehme et al. (86) nicely illustrates these interactions. Using synthetic grape juice media, these investigators conducted seven sequential fermentations. First, a series of four fermentations used four different *Saccharomyces* strains and the same *O. oeni* strain, while the last three fermentations used only three of the *Saccharomyces* strains and a different strain of *O. oeni*. Growth of the first *O. oeni* strain was inhibited to some extent by all four yeast strains (86). This could be due to a number of factors ranging from production of ethanol, short-chain fatty acids, and $SO_2$ to competition for nutrients (7, 16, 17, 37). Additionally, two recent reports suggest production of antibacterial peptides by *Saccharomyces* (25, 93). The fermentations conducted with the second *O. oeni* strain using three of the four yeasts demonstrated both neutralism and stimulation, with one yeast strain causing *O. oeni* to grow and consume malic acid approximately 2.5 and 2.6 times faster, respectively, than the control (86). While the reasons for this stimulation were not explored by Nehme et al. (86), it is known that release of nitrogen compounds and mannoproteins can stimulate *O. oeni* growth, again dependent upon the yeast strain, bacterial strain, and winemaking conditions involved (1).

## Beer

There are only four ingredients needed to make beer, i.e., water, hops, malt, and yeast. By the time processing is over, the growth of spoilage microbes is prevented in beers by a number of hurdles: low pH (3.8 to 4.0), the antimicrobial properties of hops, anaerobic conditions created through generation of $CO_2$ during fermentation, and, of course, ethanol. The manufacture of beer can be separated into four stages, those being malting, mashing, fermentation, and postfermentation (2, 56). Each stage has its own microbial concerns and controls. From a microbial ecology standpoint, the most interesting is the malting process, discussed below. Before fermentation, the malted barley is mashed (mixed with water and slowly heated by either infusion or deconcoction) to extract starches and other carbohydrates. The slow heating allows each enzyme present in the mash to work at its optimum temperature. Once the wort reaches 75°C, the hops are added and the wort is transferred to the brew kettle, where it is boiled, essentially creating a sterile environment containing antimicrobial compounds (hop acids). While a few problematic bacteria and yeasts may survive the kettle boil, most of them are found in the brewhouse environment, and if good sanitation is not practiced, they may reinfect the fermenting wort or beer. The last point of control is often a pasteurization or filtration step (2, 56).

The malting of the barley is conducted in three steps: steeping, germination, and kilning (for a review of the process, see reference 3). The primary goal of malting is to produce a number of stably maintained enzymes such as amylases and proteinases, which will break down starches and other macromolecules so they are available to the yeast during fermentation, if the malt is to be used for brewing (3). The malting process has a complex ecology containing yeasts, bacteria, and fungi; additionally, it includes the metabolically active barley grains themselves (39, 66).

The initial step, steeping, provides ideal conditions for the growth of microbes and is thus considered, from a microbial point of view, the most important step in the malting process, as the microbial population developed in this step impacts the finished beer. The barley is steeped in cool water (10 to 20°C for 2 or 3 days), resulting in a moisture content of 45 to 50%, and during this time the microbiota undergoes a change. In fact, steeping activates dormant microorganisms, and their numbers are almost always higher in the finished malt than in the starting barley. Work by O'Sullivan et al. (94) demonstrated that the microbial population of the malt may be over 2,000 times greater than that of the barley. In one case, the LAB increased from about 40 CFU/g of malt to over $10^8$ CFU/g (94). Other microbes which increase during steeping include aerobic bacteria, such as the members of the *Enterobacteriaceae* family and pseudomonads, and both yeasts and molds (39). Even though the kilning (which can be considered a pasteurization step eliminating 1 to 2 logs) and the boiling of the wort during mashing eliminate most of the microbes present at this stage, the interactions between microbes during steeping and germination still have an impact on the finished beer. For example, growth of *Fusarium* during the malting process can lead to gushing in the finished bottled beer. In this defect, the beer spontaneously overfoams upon opening of the finished product (39). While the cause of this problem has not been completely identified, it is thought that hydrophobins, or small cysteine-rich proteins produced by *Fusarium,* stabilize the $CO_2$ bubbles (65).

One of the newer tools to control the microbiota in malting is the use of starter cultures during steeping (11, 65, 66). A number of microbes have been used as malting starter cultures, including the mold *Geotrichum candidum,* the yeast *Pichia anomala,* and the bacteria *Lactobacillus plantarum* and *Pediococcus pentosaceus.* Laitila et al. (66) examined the use of *L. plantarum* and *P. pentosaceus* during steeping and found that their addition, while delaying complete germination of the barley by about a day, actually enhanced the filterability of the wort, a major bottleneck in postfermentation processing.

Use of starter cultures limited the growth of aerobic bacteria during steeping by up to 3 logs, especially *Pseudomonas* spp. (66). This effect was primarily due to the production of lactic acid, as culture on acidified MRS without glucose exhibited the same response (66). The authors thought that this suppression of the aerobic bacterial population limited the amount of exopolysaccharide produced, thus leading to the increase in the filterability of the wort. Use of the starter culture also inhibited the growth of *Fusarium,* which, aside from creating gushing problems, has been implicated in production of mycotoxins during malting (reviewed in reference 118). While the effect of *L. plantarum* on the mold during malting was primarily due to acid production, this particular lactobacillus has demonstrated production of antifusarial compounds [benzoic acid, methyl-2,4-imidazolidione, mevalonolactone, and cyclo(glycyl-L-leucyl)] which worked synergistically with lactic acid in the laboratory to completely inhibit growth of *Fusarium* (89).

The starter cultures also increased growth of yeasts during the malting, which the authors attributed to a decline in the competitive microbiota, creating more opportunities for growth (66). This increase in both the yeast and LAB populations led to an increase in the production of both microbial β-glucanase and xylanase, which help to break down the malt cell wall, resulting in a higher-quality final product (66). While this malt can be used for beer production, it may have other uses as well, including production of the fermented products malt whiskey and malt vinegar.

## BREAD AND SOURDOUGH FERMENTATION

Cereals represent the largest food crop in the world, accounting for approximately 60% of the total global food crop production (51). The majority of these crops are used in fermentations. The type of fermentation can vary from beer fermentation, described above, to production of leavened bread; in fact, the development of brewing and the production of leavened bread are intertwined (51). The fermentation process in the case of cereals differs from the processes involved in vegetable, wine, and dairy fermentations in that it is conducted in order to create a more functional product, whereas the other fermentations are primarily conducted to increase the shelf life of the substrate (56).

A traditional bread fermentation using an *S. cerevisiae* inoculum also contains other microbes, including LAB, gram-negative bacteria, molds, and yeasts. Surprisingly, *S. cerevisiae* is not present in flour or on cereal grains (34). This microbial consortium contributes to the leavening of the bread, due primarily to $CO_2$ and ethanol production by *S. cerevisiae,* although the LAB present contribute both $CO_2$ and acids as well. During baking, the yeasts are killed and the $CO_2$ and ethanol are expelled, leading to a final product with a pH between 5.0 and 6.0 (56).

In sourdough bread, the LAB and occasionally a yeast, although not necessarily *S. cerevisiae,* are intentionally inoculated into the dough, often, though not always, through back-slopping. The final product in this case typically has a pH between 4 and 4.5. Sourdough breads produced in this fashion across the world, ranging from the famous San Francisco, CA, sourdough to panettone in Italy and rye breads throughout Europe, are considered to be type I or traditional sourdoughs. In this case, the starter is maintained by a regimented addition of flour and water, which produces a very stable microbiota consisting of yeasts and LAB. Interestingly, it is thought that the $CO_2$ formation responsible for leavening in type I sourdoughs comes from heterofermentative LAB instead of yeasts (48, 80). There are also type II sourdoughs, which are semifluid silo preparations, often used as dough souring supplements, and type III preparations, which are dried sourdoughs used as acidifier supplements (29, 34). The $CO_2$ for leavening of these products is produced by the yeasts used in the starter cultures.

Sourdough fermentations typically contain *Lactobacillus* spp., with over 50 identified (34). However, the lactobacilli most commonly associated with sourdough fermentations around the world is *L. sanfranciscensis* (61). Others which are typically found in sourdough environments include *Lactobacillus pontis, Lactobacillus panis, L. paralimentarius, Lactobacillus frumenti,* and *Lactobacillus mindensis* (14, 38, 85, 115, 117). Other LAB, such as *L. plantarum, L. brevis,* and *L. reuteri,* are also common in sourdoughs around the world, although they are not, to date, found exclusively associated with cereal fermentations (34). While *L. plantarum* and *L. brevis* are considered ubiquitous,

*L. reuteri* and *L. pontis* are often associated with intestinal sources, and it is thought that they enter the dough through the feces of mice in the grain elevator, although according to Vogel and Ehrmann (114), legend says that the Italians used to add cow and horse dung to the traditional panettone fermentation.

The yeasts found in sourdoughs are typically *S. cerevisiae, Candida krusei, Candida milleri, Saccharomyces exiguus,* and *Candida humilis* (34, 114). While a number of studies have examined the yeast population in a completed fermentation, Meroth et al. (81) used denaturing gradient gel electrophoresis (DGGE) to examine yeast population dynamics during the course of a sourdough fermentation. They found that a number of yeasts were present at the start of the fermentation, including *C. humilis, Debaryomyces hansenii, S. cerevisiae, Saccharomyces uvarum, C. krusei,* a *Cladosporium* sp., *Saccharomyces servazii, Dekkera bruxellensis, Epicoccum nigrum,* and an unidentified yeast. These populations, however, were eventually outcompeted by the yeast thought of as typical of rye sourdough fermentations in the case of the type I fermentation, *C. humilis*. The other fermentations were type II fermentations, meaning that they used a starter culture instead of the traditional maintenance of the sponge with regular additions of flour. Furthermore, they used either rye flour or rye bran and were held at either 30 or 40°C. Both of these parameters played a role in selecting the yeast that would become dominant. *S. cerevisiae* dominated in the type II rye flour sourdough at 30°C and was identified by random amplified polymorphic DNA PCR as belonging to the commercial starter. In the rye flour fermentation conducted at 40°C, no yeast was able to dominate. However, in the type II sourdough fermentation using rye bran instead of rye flour at 40°C, *C. krusei* became dominant.

The interactions between yeasts and LAB in traditional type I sourdough fermentations have been well studied. The yeast-to-LAB ratio is typically 1:100 (50). In traditional sourdoughs such as those found in San Francisco, there is a strict association between *L. sanfranciscensis* and the yeast *S. exiguus* (108). Stable cometabolism of carbohydrates, adaptation to the environment, and production of antimicrobial compounds all contribute to the maintenance of a stable consortium. In fact, type I continuous-use sourdoughs are reputed to last for centuries and have been maintained in the laboratory for more than seven decades (34). The primary driver of this relationship is the ability of *L. sanfranciscensis* to ferment maltose, one of the four carbohydrates present in flour; sucrose, glucose, and fructose are the others (49).

Maltose is produced by amylases in the flour (75). The heterofermentative *L. sanfranciscensis* preferentially uses maltose through a constitutively expressed maltose phosphorylase (107). This energetically favorable reaction creates a glucose-1-phosphate and a glucose molecule without utilizing an ATP. The glucose-1-phosphate is further metabolized by the phosphoketolase pathway (Fig. 2) to lactic acid, acetic acid, and ethanol (49). The other glucose molecule is excreted from the cell (87).

This excreted glucose molecule plays an important role in establishing the dominant *S. exiguus* yeast population. First, it initiates glucose repression of maltose utilization in other microbes capable of using maltose. One of those microbes is *S. cerevisiae* (58). However, *S. cerevisiae* preferentially utilizes the sucrose that is present, causing its rapid depletion in the fermentation. Thus, a point is reached where the glucose produced by *L. sanfranciscensis* represses the ability of *S. cerevisiae* to use the primary carbohydrate, maltose, and the non-maltose-fermenting microbes present have consumed the other carbohydrates. Furthermore, as *L. sanfranciscensis* continues to grow on the maltose, it produces acetic acid, which lowers the pH of the fermentation. As the pH of the fermentation approaches 4, the acetic acid favors the growth of *S. exiguus* (a maltose-negative yeast) over the more sensitive maltose-utilizing *S. cerevisiae* (109). At this point, *S. exiguus* is able to outcompete *S.*

*cerevisiae* for the glucose, allowing it to become the dominant yeast in the fermentation. For its part in this mutually beneficial relationship, *S. exiguus* stimulates the growth of *L. sanfranciscensis* through the production of amino acids, including a peptide (Asp-Cys-Glu-Gly-Lys) that specifically stimulates the growth of *L. sanfranciscensis* (9).

The dough is further protected from other LAB and spoilage bacteria such as *Bacillus subtilis* through production of bacteriocins by strains of *L. sanfranciscensis* (28). This bacterium also produces a mixture of acids (acetic, caproic, formic, propionic, butyric, and *n*-valeric acids) that act synergistically to inhibit bread spoilage molds such as *Penicillium* and *Fusarium,* thus accounting for the longer shelf life of sourdough breads (27).

## VEGETABLE FERMENTATIONS

Vegetable fermentations were thought to have originated in Asia and the Middle East as early as the 3rd century B.C.E.. (96, 100). The practice developed from the dry-salting of vegetables, which was done for thousands of years as a method of preservation (96). In fact, modern vegetable fermentations typically do not use a starter culture but instead rely upon the development of ideal conditions, primarily created by brining, to initiate the proper microbial succession. In the United States, there are three main types of fermented vegetables: sauerkraut, pickles, and olives. However, the majority of the pickle and olive products are produced simply by pickling or packing the fresh vegetables in vinegar and not through fermentation (56).

Sauerkraut fermentation is conducted by LAB and is considered the classic example of microbial succession in food fermentations. In fact, it is taught in most food fermentation classes. The fermentation takes up to 3 weeks and can be separated into two stages. The first is the heterolactic stage, conducted primarily by *L. mesenteroides*. The second is the homolactic stage, conducted primarily by *L. plantarum*. The cabbage itself contains between 4 and 5% carbohydrate, primarily glucose and fructose, which the bacteria use as a substrate (43).

To start, the cabbage is shredded and salt is added. It is important that the salt concentration be 2 to 3%, typically 2.25%, as a salt concentration which is too low will allow spoilage bacteria to grow and higher salt concentrations favor the growth of spoilage yeasts (45). The salt addition really performs three functions. First, it draws water out of the cabbage, which contains sugars and other nutrients. Second, while the salt concentration is not that high, it provides just enough of a selective advantage to the LAB, which are not the dominant microbial population on the cabbage, for them to dominate the fermentation (92). Finally, it contributes to the flavor of the final product.

The second step the processor takes to ensure a successful fermentation is to place a plastic tarp over the top of the fermentation vessel. This tarp is weighted down with water, which both weighs down the cabbage and excludes air from the fermentation. Air is rapidly removed from the fermentation through the metabolism of both the resident microbial population and the respiration of the cabbage (13, 56). This change in the redox potential of the substrate, combined with the salt addition, allows the LAB (which are typically present at levels anywhere between a low of 7 CFU/g of cabbage and a high of $10^4$ CFU/g of cabbage) to outcompete the larger aerobic bacterial population ($10^4$ to $10^6$ CFU/g), consisting of *Pseudomonas, Flavobacterium, Staphylococcus,* and other bacteria, as well as the facultative anaerobes, such as *Escherichia coli* and yeasts, that are present.

In the heterolactic stage of fermentation, *L. mesenteroides* is able to grow faster than other salt-tolerant microbes at the reduced temperature of 18°C. As *L. mesenteroides* grows, it produces lactic and acetic acids, ethanol, and $CO_2$. It creates up to 0.8% lactic acid, which is inhibitory to the other anaerobic microbes and favors growth of more acid-tolerant LAB. As the acid levels increase, the environment becomes inhibitory to *L. mesenteroides,* which has done its part by eliminating the non-LAB competition, and *Pediococcus pentosaceus, L. brevis,* and *L. plantarum* continue the fermentation, reaching levels of around $10^7$ CFU/g. *L. plantarum* in particular dominates the later stages of fermentation, and the facultatively homolactic microbes are traditionally the only isolates found after the pH is reduced to below 4.0 (13, 56).

This succession, however, was developed using traditional culturing and biochemical identification methods. Recent work by Plengvidhya et al. (99), using DNA fingerprinting methods, has demonstrated a much greater diversity in both the heterolactic and homolactic fermentations than originally thought. Aside from *L. mesenteroides,* the heterolactic population in all four commercial fermentations studied also included *Weissella* and *Leuconostoc citreum.* While *L. plantarum* was still the dominant bacterium in the homolactic stage of the fermentation (accounting for between 56.7 and 100% of the day 9 isolates), neither *L. brevis* nor *P. pentosaceus,* which were typically associated with this stage, were recovered at high numbers. Instead, *Leuconostoc argentinum,* an organism previously unassociated with sauerkraut, was isolated at higher levels, up to 10% in one of the fermentation studies. While this study has changed the current view of sauerkraut fermentation, the role that these new organisms play in both the microbial succession of the fermentation and the final quality of the sauerkraut itself is still under investigation, with the hope that a better understanding of the microbial ecology will lead to the development of a reduced-salt starter culture system (60).

Another factor, demonstrated by the U.S. Department of Agriculture Agricultural Research Service Vegetable Fermentations Unit, that plays a role in the microbial succession of sauerkraut fermentation is bacteriophage (119). Interestingly, it was found that as with the bacterial populations, there were two separate phage-host populations, with phage from the heterolactic segment of the fermentation unable to infect bacteria from the succeeding homolactic fermentation (70). The phage populations increased with the corresponding bacterial population, and as that microbial population died out, so did its phage population. The authors speculated that the phage could contribute significantly to the rapid decline of each bacterial population. For example, 12% of the heterolactic (primarily *Leuconostoc*) population on days 1 and 3 of the fermentation were found to be susceptible to phage infection (70). While the bacteriophage ecology of sauerkraut fermentation may currently be of interest only to microbiologists, as low-salt starter culture systems are developed for vegetable fermentations, the role of phage in fermentation will take on greater industrial importance.

## DAIRY FERMENTATIONS

One of the great examples of the effects that processing can have on the outcome of any fermentation can be seen in a simple trip down the dairy aisle at the grocery store, or when spending some time trying to decide which cheese to have with one's glass of wine. Fermented dairy products all start with the same basic ingredient, milk. In general, cow's milk consists of 87% water, 5% lactose, approximately 3.5% fat, and around 3.4% protein (56). This simple ingredient, through the use of a number of different fermentation steps, can result in products ranging from yogurt and buttermilk to a variety of different cheeses. While there are many interesting microbial interactions in yogurt and other dairy fermentations, this discussion is limited to the interactions found among *Streptococcus thermophilus, Lactobacillus helveticus,* and *Propionibacterium freudenreichii* subsp. *shermanii* during Swiss cheese fermentation. The interactions between *L. helveticus* and *S. thermophilus* can be found in a number of other cheeses and cultured dairy products as well.

In the production of any cheese, there are several steps that affect the microbiota and the microbial succession during production. The first of these is the quality of the milk. The presence of antibiotics in the milk can have a definite impact on the course of fermentation;

this, however, is now strictly controlled in the United States and rarely presents a problem. The second is the pasteurization of the milk. From a safety standpoint this should not be debated, as it eliminates potential pathogens such as *Listeria monocytogenes* and others which have been linked to production of cheeses using unpasteurized milk (20, 110). However, from a quality and aging standpoint, the use of unpasteurized milk is still a hotly debated issue, as some claim that unpasteurized milk produces a cheese that develops better flavors during aging.

The next step in processing is the addition of the starter culture. The primary role of the starter is to produce acid, which not only acts as a preservative but also helps in precipitation of the casein proteins to form the curd. One interesting point regarding dairy starter cultures is that while starter cultures in most industries have moved to a single microbe in order to create a more consistent fermentation, the dairy industry, which may be slightly more advanced than other industries, has evolved to using multiple species and even strains (typically LAB) in their starter cultures. This helps not only to achieve the desired effects but also to protect the fermentation from bacteriophage, which can easily wipe out an entire fermentation if based upon the use of a single bacterium.

The first trait a cheesemaker has to consider in the choice of an appropriate starter culture is the optimum growth temperature. Both mesophiles (such as *Lactococcus lactis* subsp. *lactis*) and thermophiles (such as *S. thermophilus*) can be used, depending upon the type of cheese to be made and the cooking temperature. In fact, the difference in temperature used to cook the curd does not have to be very great to make a big difference. For instance, the growth and acid production by *Lactococcus lactis* subsp. *cremoris* are inhibited by slightly lower temperatures (38°C) than is the growth of *L. lactis* subsp. *lactis* (40°C). Although both grow at about 38°C, *L. lactis* subsp. *cremoris* does so more slowly, thus creating little or no acid in comparison to *L. lactis* subsp. *lactis*. This type of effect means that cheesemakers must pay close attention to the cooking temperature when the preferred *L. lactis* subsp. *cremoris* is used.

The next step in the process is the addition of chymosin, which is often done at the same time as the addition of starter culture. Chymosin is an acid protease that, together with the acid produced by the starter culture, creates the curd. At this point the microbial population is no longer growing planktonically but instead is entrapped within the forming casein network and growing on a solid surface. After coagulation, the curd is cut, cooked, molded or handled, and then salted and aged. Each of these steps can affect the microbiota and impact the final product (26, 63).

In Swiss-type cheeses, a mixed starter culture consisting of *S. thermophilus, L. helveticus,* and *P. freudenreichii* subsp. *shermanii* is used. The cooking temperature of Swiss cheese is higher than that of cheddar, thus the use of thermophilic starters, although during the 24 h of the fermentation after the forming and cutting of the curd, the temperature is around 35°C (63). All three starter cultures are added to the milk, with the concentration of *S. thermophilus* being 10 times greater than that of *L. helveticus* and *P. freudenreichii,* which are only present at around 100 CFU/ml.

Due to the greater number of *S. thermophilus* cells in the milk, it is the first microbe to grow. *S. thermophilus* transports the lactose present in the milk into the cell using a secondary transport system (LacS); this system acts as either a galactoside-proton symport system or a lactose-galactose antiport system (44). Once the lactose is in the cytoplasm, β-galactosidase hydrolyzes the molecule into glucose and galactose (57). The glucose is immediately phosphorylated to glucose-6-phosphate, which is the first step in the EMP pathway (Fig. 1). While the glucose is converted into lactic acid, the galactose is used by the LacS system to bring another molecule of lactose into the cell.

Surprisingly, *S. thermophilus* contains the enzymes of the Leloir pathway, which theoretically allow it to use the galactose (57). These enzymes are a galactokinase, a galactose-1-phosphate uridylyltransferase, a UDP-glucose 4-epimerase, and a mutarotase, and they convert galactose into glucose-1-phosphate, which is typically fed into the EMP pathway by phosphoglucomutase (33). However, in most *S. thermophilus* strains, the enzymes are poorly expressed, specifically galactokinase *(galK)* (112). Among a number of mutations in the enzymes themselves, galactose-fermenting strains of *S. thermophilus* contain mutations in the promoter region between the divergently transcribed regulator, *galR,* and *galK* (32). After about 8 h, as the pH of the fermentation is reduced, *L. helveticus* begins to grow, at first competing for the lactose and then consuming the galactose after the lactose is depleted (56). Upon completion of the primary fermentation, the pH is around 5.2 (67).

At this point the cheese is placed in brine for about 3 days, after which it is moved to rooms with a temperature of 20 to 25°C (63, 67). It is during the incubation in these rooms that *P. freudenreichii* begins to grow. While *P. freudenreichii* is capable of utilizing lactose, after the primary fermentation no lactose is left. Instead, it metabolizes the lactic acid produced by *S. thermophilus* and *L. helveticus*. Using the propionic acid pathway (Fig. 3), it produces acetic acid, propionic acid, and the $CO_2$ that forms the holes characteristic of the cheese (52, 55).

## FUTURE PROSPECTS

The study of food fermentations is currently undergoing a revolution. With the genome sequence available for representatives of most of the major starter cultures, scientists are well equipped to study the interactions between organisms in their favorite fermentations using various -omics techniques (74, 80, 98). Additionally, using culture-independent population analysis such as denaturing gradient gel electrophoresis, we are rapidly developing a

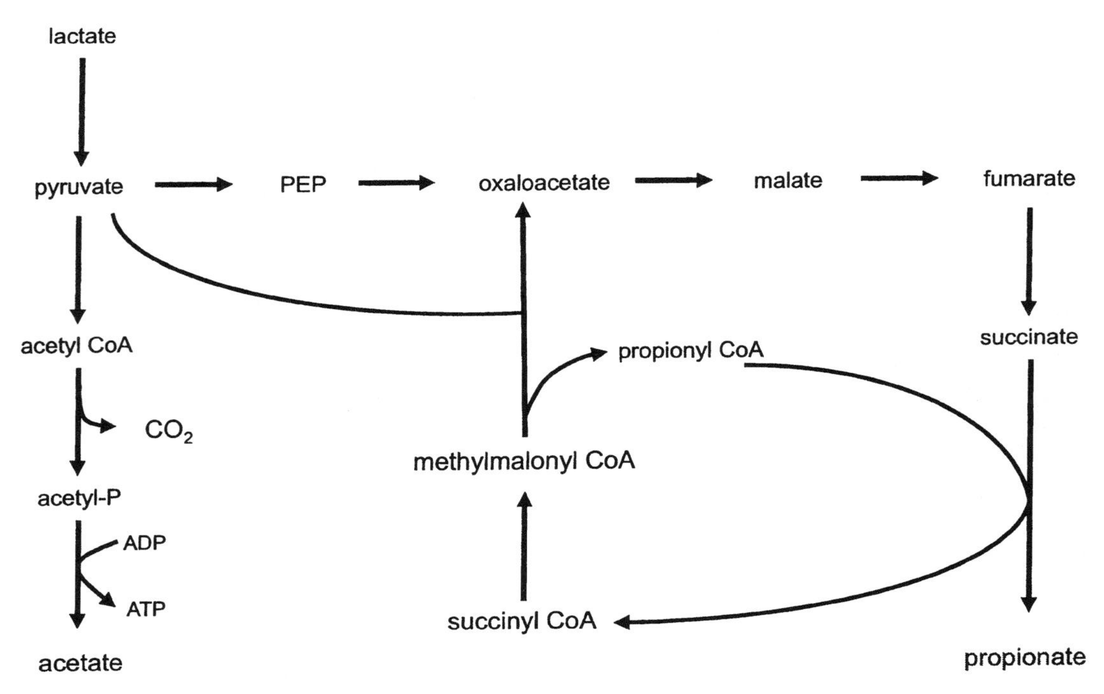


**FIGURE 3** The propionic acid pathway. Used from reference 55 with permission.

better understanding of the microbial ecology of nearly all fermentations (23). As we gain a better understanding of both the ecology of food fermentation and how the microbes in that fermentation interact to affect the quality of the final product, we are developing better systems (such as use of multiple starter cultures in the wine industry) to consistently create better and safer products (12).

## REFERENCES

1. **Alexandre, H., P. Costello, F. Remize, J. Guzzo, and M. Guilloux-Benatier.** 2004. *Saccharomyces cerevisiae-Oenococcus oeni* interactions in wine: current knowledge and perspectives. *Int. J. Food Microbiol.* **93:**141–154.
2. **Bamforth, C.** 2005. *Food, Fermentation and Microorganisms.* Blackwell Publishing, Ames, IA.
3. **Bamforth, C., and A. Barclay.** 1993. Malting technology and the uses of malt, p. 297–354. *In* A. MacGregor and R. Bhatty (ed.), *Barley: Chemistry, Technology.* American Association of Cereal Chemists, St. Paul, MN.
4. **Baratti, J. C., and J. D. Bullock.** 1986. *Zymomonas mobilis*: a bacterium for ethanol production. *Biotechnol. Adv.* **4:**95–115.
5. **Barnett, J., M. Delaney, E. Jones, A. Magson, and B. Winch.** 1972. The numbers of yeasts associated with wine grapes of Bordeaux. *Arch. Mikrobiol.* **83:**52–55.
6. **Barnett, J. A.** 2000. A history of research on yeasts 2: Louis Pasteur and his contemporaries, 1850–1880. *Yeast* **16:**755–771.
7. **Beelman, R., R. Keen, M. Banner, and S. King.** 1982. Interactions between wine yeast and malolactic bacteria under wine conditions. *Dev. Ind. Microbiol.* **23:**107–121.
8. **ben Omar, N., and F. Ampe.** 2000. Microbial community dynamics during production of the Mexican fermented maize dough pozol. *Appl. Environ. Microbiol.* **66:**3664–3673.
9. **Berg, R. W., W. E. Sandine, and A. W. Anderson.** 1981. Identification of a growth stimulant for *Lactobacillus sanfrancisco. Appl. Environ. Microbiol.* **42:**786–788.
10. **Boddy, L., and J. W. T. Wimpenny.** 1992. Ecological concepts in food microbiology. *J. Appl. Bacteriol.* **73:**23S–38S.
11. **Bolvin, P., and M. Malanda.** 1997. Improvement of malt quality and safety by adding starter culture during the malting process. *Tech. Q. Master Brew Assoc. Am.* **34:**96–101.
12. **Boulton, R. B., V. L. Singleton, L. F. Bisson, and R. E. Kunkee.** 1999. *Principles and Practices of Winemaking.* Springer, New York, NY.
13. **Breidt, F., Jr., R. F. McFeeters, and I. Díaz-Muñiz.** 2007. Fermented vegetables, p. 783–793. *In* M. P. Doyle and L. R. Beuchat (ed.), *Food Microbiology: Fundamentals and Frontiers,* 3rd ed. ASM Press, Washington, DC.
14. **Cai, Y., H. Okada, H. Mori, and T. N. Y. Benno.** 1999. *Lactobacillus paralimentarius* sp. nov. isolated from sourdough. *Int. J. Syst. Bacteriol.* **49:** 1451–1455.
15. **Camu, N., A. Gonzalez, T. De Winter, A. Van Schoor, K. De Bruyne, P. Vandamme, J. S. Takrama, S. K. Addo, and L. De Vuyst.** 2008. Influence of turning and environmental contamination on the dynamics of populations of lactic acid and acetic acid bacteria involved in spontaneous cocoa bean heap fermentation in Ghana. *Appl. Environ. Microbiol.* **74:**86–98.
16. **Capucho, I., and M. S. Romao.** 1994. Effect of ethanol and fatty acids on malolactic activity of *Leuconostoc oenos. Appl. Microbiol. Biotechnol.* **42:** 391–395.
17. **Carrete, R., M. Teresa Vidal, A. Bordons, and M. Constanti.** 2002. Inhibitory effect of sulphur dioxide and other stress compounds in wine on the ATPase activity of *Oenococcus oeni. FEMS Microbiol. Lett.* **211:**155–159.
18. **Castoria, R., F. Curtis, G. Lima, S. Caputo, S. Pacifico, and V. DeCicco.** 2001. *Aureobasidium pullulans* (LS-30) an antagonist of postharvest pathogens of fruits: study on its modes of action. *Postharvest Biol. Technol.* **22:**7–17.
19. **Cavalieri, D., P. E. McGovern, D. L. Hartl, R. Mortimer, and M. Polsinelli.** 2003. Evidence for *S. cerevisiae* fermentation in ancient wine. *J. Mol. Evol.* **57**(Suppl. 1):S226–S232.
20. **Centers for Disease Control.** 1985. Listeriosis outbreak associated with Mexican-style cheese—California. *MMWR Morb. Mortal. Wkly. Rep.* **34:** 357–359.
21. **Charoenchai, C., G. Fleet, P. Henschke, and B. Todd.** 1997. Screening of non-*Saccharomyces* wine yeasts for the presence of extracellular hydrolytic enzymes. *Aust. J. Grape Wine Res.* **3:**2–8.
22. **Chen, H., and G. R. Fink.** 2006. Feedback control of morphogenesis in fungi by aromatic alcohols. *Genes Dev.* **20:**1150–1161.
23. **Cocolin, L., and D. Ercolini (ed.).** 2008. *Molecular Techniques in the Microbial Ecology of Fermented Foods.* Springer, New York, NY.
24. **Comitini, F., and M. Ciani.** 2006. Survival of inoculated *Saccharomyces cerevisiae* strain on wine grapes during two vintages. *Lett. Appl. Microbiol.* **42:**248–253.
25. **Comitini, F., R. Ferretti, F. Clementi, I. Mannazzu, and M. Ciani.** 2005. Interactions

between *Saccharomyces cerevisiae* and malolactic bacteria: preliminary characterization of a yeast proteinaceous compound(s) active against *Oenococcus oeni*. *J. Appl. Microbiol.* **99:**105–111.

26. **Coppola, S., G. Blaiotta, and D. Ercolini.** 2008. Dairy products, p. 31–90. *In* L. Cocolin and D. Ercolini (ed.), *Molecular Techniques in the Microbial Ecology of Fermented Foods.* Springer, New York, NY.
27. **Corsetti, A., M. Gobbetti, J. Rossi, and P. Damiani.** 1998. Antimould activity of sourdough lactic acid bacteria: identification of a mixture of organic acids produced by *Lactobacillus sanfrancisco* CB1. *Appl. Microbiol. Biotechnol.* **50:** 253–256.
28. **Corsetti, A., M. Gobbetti, and E. Smacchi.** 1996. Antibacterial activity of sourdough lactic acid bacteria: isolation of a bacteriocin-like inhibitory substance from *Lactobacillus sanfrancisco* C57. *Food Microbiol.* **13:**447–456.
29. **Corsetti, A., and L. Settanni.** 2007. Lactobacilli in sourdough fermentation. *Food Res. Int.* **40:** 539–558.
30. **De Deken, R. H.** 1966. The Crabtree effect: a regulatory system in yeast. *J. Gen. Microbiol.* **44:** 149–156.
31. **Delfini, C., M. Cersosimo, V. Del Prete, M. Strano, G. Gaetano, A. Pagliara, and S. Ambro.** 2004. Resistance screening essay of wine lactic acid bacteria on lysozyme: efficacy of lysozyme in unclarified grape musts. *J. Agric. Food Chem.* **52:**1861–1866.
32. **De Vin, F., P. Radstrom, L. Herman, and L. De Vuyst.** 2005. Molecular and biochemical analysis of the galactose phenotype of dairy *Streptococcus thermophilus* strains reveals four different fermentation profiles. *Appl. Environ. Microbiol.* **71:** 3659–3667.
33. **De Vos, W. M.** 1996. Metabolic engineering of sugar catabolism in lactic acid bacteria. *Antonie van Leeuwenhoek* **70:**223–242.
34. **De Vuyst, L., and P. Neysens.** 2005. The sourdough microflora: biodiversity and metabolic interactions. *Trends Food Sci. Technol.* **16:**43–56.
35. **Dicks, L. M. T., F. Dellaglio, and M. D. Collins.** 1995. Proposal to reclassify *Leuconostoc oenos* as *Oenococcus oeni* [corrig.] gen. nov., comb. nov. *Int. J. Syst. Bacteriol.* **45:**395–397.
36. **Dobell, C.** 1932. *Antonie van Leeuwenhoek and His "Little Animals."* Staples Press, London, England.
37. **Edwards, C. G., and R. B. Beelman.** 1987. Inhibition of the malolactic bacterium, *Leuconostoc oenos* (PSU-1), by decanoic acid and subsequent removal of the inhibition by yeast ghosts. *Am. J. Enol. Vitic.* **38:**239–242.
38. **Ehrmann, M. A., M. R. A. Müller, and R. F. Vogel.** 2003. Molecular analysis of sourdough reveals *Lactobacillus mindensis* sp. nov. *Int. J. Syst. Evol. Microbiol.* **53:**7–13.
39. **Flannigan, B.** 2003. The microbiota of barley and malt, p. 113–180. *In* F. G. Priest and I. Campbell (ed.), *Brewing Microbiology,* 3rd ed. Kluwer Academic/Plenum Publishers, New York, NY.
40. **Fleet, G. H.** 2003. Yeast interactions and wine flavour. *Int. J. Food Microbiol.* **86:**11–22.
41. **Fleet, G. H., C. Prakitchaiwattana, A. I. Beh, and G. Heard.** 2002. The yeast ecology of wine grapes, p. 1–17. *In* M. Ciani (ed.), *Biodiversity and Biotechnology of Wine Yeast.* Research Signpost, Kerala, India.
42. **Fleet, G. H., and G. Heard.** 1993. Yeast-growth during fermentation, p. 27–54. *In* G. H. Fleet (ed.), *Wine Microbiology and Biotechnology.* Harwood Academic Publishers, Chur, Switzerland.
43. **Fleming, H. P., R. F. McFeeters, and E. G. Humphries.** 1988. A fermentor for study of sauerkraut fermentation. *Biotechnol. Bioeng.* **31:**189–197.
44. **Foucaud, C., and B. Poolman.** 1992. Lactose transport system of *Streptococcus thermophilus.* Functional reconstitution of the protein and characterization of the kinetic mechanism of transport. *J. Biol. Chem.* **267:**22087–22094.
45. **Fred, E. B., and W. H. Peterson.** 1922. The production of pink sauerkraut by yeasts. *J. Bacteriol.* **7:**257–269.
46. **Fugelsang, K. C., and C. G. Edwards.** 2007. *Wine Microbiology Practical Applications and Procedures,* 2nd ed. Springer, New York, NY.
47. **Ganga, M., and C. Martinez.** 2004. Effect of wine yeast monoculture practice on biodiversity of non-*Saccharomyces* yeasts. *J. Appl. Microbiol.* **96:** 76–83.
48. **Ganzle, M. G., S. Hausle, and W. P. Hammes.** 1997. Wechselwirkungen zwischen Laktobazillen und Hefen. *Getreide Mehl Brot* **51:**209–215.
49. **Ganzle, M. G., N. Vermeulen, and R. F. Vogel.** 2007. Carbohydrate, peptide and lipid metabolism of lactic acid bacteria in sourdough. *Food Microbiol.* **24:**128–138.
50. **Gobbetti, M., A. Corsetti, J. Rossi, F. LaRosa, and S. DeVincenzi.** 1994. Identification and clustering of lactic acid bacteria and yeast from wheat sourdoughs of central Italy. *Ital. J. Food Sci.* **1:**85–94.
51. **Hammes, W. P., M. J. Brandt, K. L. Francis, J. Rosenheim, M. F. H. Seitter, and S. A. Vogelmann.** 2005. Microbial ecology of cereal fermentations. *Trends Food Sci. Technol.* **16:** 4–11.
52. **Hettinga, D. H., and G. W. Reinbold.** 1972. The propionic acid bacteria—a review. II. Metabolism. *J. Milk Food Technol.* **35:**358–372.

53. **Hogan, D. A.** 2006. Quorum sensing: alcohols in a social situation. *Curr. Biol.* **16:**R457–R458.

54. **Hogan, D. A.** 2006. Talking to themselves: autoregulation and quorum sensing in fungi. *Eukaryot. Cell* **5:**613–619.

55. **Hutkins, R.** 2001. Metabolism of starter cultures, p. 207–241. *In* E. H. Marth and J. L. Steele (ed.), *Applied Dairy Microbiology*. Marcel Dekker, Inc., New York, NY.

56. **Hutkins, R.** 2006. *Microbiology and Technology of Fermented Foods.* Blackwell Publishing, Ames, IA.

57. **Hutkins, R., and H. A. Morris.** 1987. Carbohydrate metabolism by *Streptococcus thermophilus*: a review. *J. Food Prot.* **50:**876–884.

58. **Jiang, H., I. Medintz, B. Zhang, and C. A. Michels.** 2000. Metabolic signals trigger glucose-induced inactivation of maltose permease in *Saccharomyces*. *J. Bacteriol.* **182:**647–654.

59. **Jiranek, V., P. Langridge, and P. A. Henschke.** 1995. Regulation of hydrogen sulfide liberation in wine-producing *Saccharomyces cerevisiae* strains by assimilable nitrogen. *Appl. Environ. Microbiol.* **61:**461–467.

60. **Johanningsmeier, S., R. F. McFeeters, H. P. Fleming, and R. L. Thompson.** 2007. Effects of *Leuconostoc mesenteroides* starter culture on fermentation of cabbage with reduced salt concentrations. *J. Food Sci.* **72:**M166–M172.

61. **Kline, L., and T. T. Sugihara.** 1971. Microorganisms of the San Francisco sour dough bread process. II. Isolation and characterization of undescribed bacterial species responsible for the souring activity. *Appl. Microbiol.* **21:**459–465.

62. **Konings, W. N.** 2002. The cell membrane and the struggle for life of lactic acid bacteria. *Antonie van Leeuwenhoek* **82:**3–27.

63. **Kosikowski, F., and V. V. Mistry.** 1997. *Cheese and Fermented Milk Foods,* 3rd ed. F.V. Kosikowski Llc, Westport, CT.

64. **Kunkee, R. E.** 1967. Malolactic fermentation. *Adv. Appl. Microbiol.* **9:**235–279.

65. **Laitila, A., T. Sarlin, E. Kotaviita, T. Huttunen, S. Home, and A. Wilhelmson.** 2007. Yeast isolated from industrial maltings can suppress *Fusarium* growth and formation of gushing factors. *J. Ind. Microbiol. Biotechnol.* **34:**701–713.

66. **Laitila, A., H. Sweins, A. Vilpola, E. Kotaviita, J. Olkku, S. Home, and A. Haikara.** 2006. *Lactobacillus plantarum* and *Pediococcus pentosaceus* starter cultures as a tool for microflora management in malting and for enhancement of malt processability. *J. Agric. Food Chem.* **54:**3840–3851.

67. **Langsrud, T., and G. W. Reinbold.** 1973. Flavor development and microbiology of Swiss cheese—a review. II. Starters, manufacturing process and procedures. *J. Milk Food Technol.* **36:**531–542.

68. **Linderholm, A. L., C. L. Findleton, G. Kumar, Y. Hong, and L. F. Bisson.** 2008. Identification of genes affecting hydrogen sulfide formation in *Saccharomyces cerevisiae*. *Appl. Environ. Microbiol.* **74:**1418–1427.

69. **Longo, E., J. Cansado, D. Agrelo, and T. G. Villa.** 1991. Effect of climatic conditions on yeast diversity in grape musts from northwest Spain. *Am. J. Enol. Vitic.* **42:**141–144.

70. **Lu, Z., F. Breidt, V. Plengvidhya, and H. P. Fleming.** 2003. Bacteriophage ecology in commercial sauerkraut fermentations. *Appl. Environ. Microbiol.* **69:**3192–3202.

71. **Machida, K., and T. Tanaka.** 1999. Farnesol-induced generation of reactive oxygen species dependent on mitochondrial transmembrane potential hyperpolarization mediated by F0F1-ATPase in yeast. *FEBS Lett.* **462:**108–112.

72. **Machida, K., T. Tanaka, Y. Yano, S. Otani, and M. Taniguchi.** 1999. Farnesol-induced growth inhibition in *Saccharomyces* by a cell cycle mechanism. *Microbiology* **145:**293–299.

73. **Madigan, M. T., and J. M. Martinko.** 2006. *Brock Biology of Microorganisms,* 11th ed. Prentice Hall, Upper Saddle River, NJ.

74. **Makarova, K., A. Slesarev, Y. Wolf, A. Sorokin, B. Mirkin, E. Koonin, A. Pavlov, N. Pavlova, V. Karamychev, N. Polouchine, V. Shakhova, I. Grigoriev, Y. Lou, D. Rohksar, S. Lucas, K. Huang, D. M. Goodstein, T. Hawkins, V. Plengvidhya, D. Welker, J. Hughes, Y. Goh, A. Benson, K. Baldwin, J. H. Lee, I. Diaz-Muniz, B. Dosti, V. Smeianov, W. Wechter, R. Barabote, G. Lorca, E. Altermann, R. Barrangou, B. Ganesan, Y. Xie, H. Rawsthorne, D. Tamir, C. Parker, F. Breidt, J. Broadbent, R. Hutkins, D. O'Sullivan, J. Steele, G. Unlu, M. Saier, T. Klaenhammer, P. Richardson, S. Kozyavkin, B. Weimer, and D. Mills.** 2006. Comparative genomics of the lactic acid bacteria. *Proc. Natl. Acad. Sci. USA* **103:**15611–15616.

75. **Martinez-Anaya, M. A.** 1996. Enzymes and bread flavor. *J. Agric. Food Chem.* **44:**2469–2480.

76. **Martini, A., M. Ciani, and G. Scorzetti.** 1996. Direct enumeration and isolation of wine yeasts from grape surfaces. *Am. J. Enol. Vitic.* **47:**435–440.

77. **Masoud, W., L. B. Cesar, L. Jespersen, and M. Jakobsen.** 2004. Yeast involved in fermentation of *Coffea arabica* in East Africa determined by genotyping and by direct denaturating gradient gel electrophoresis. *Yeast* **21:**549–556.

78. **McGovern, P. E., D. L. Glusker, L. J. Exner, and M. M. Voigt.** 1996. Neolithic resinated wine. *Nature* **381:**480–481.

79. **McGovern, P. E., J. Zhang, J. Tang, Z. Zhang, G. R. Hall, R. A. Moreau, A. Nu-

**nez, E. D. Butrym, M. P. Richards, C. S. Wang, G. Cheng, Z. Zhao, and C. Wang.** 2004. Fermented beverages of pre- and protohistoric China. *Proc. Natl. Acad. Sci. USA* **101:** 17593–17598.

80. **Merico, A., P. Sulo, J. Piskur, and C. Compagno.** 2007. Fermentative lifestyle in yeasts belonging to the *Saccharomyces* complex. *FEBS J.* **274:**976–989.
81. **Meroth, C. B., W. P. Hammes, and C. Hertel.** 2003. Identification and population dynamics of yeasts in sourdough fermentation processes by PCR-denaturing gradient gel electrophoresis. *Appl. Environ. Microbiol.* **69:**7453–7461.
82. **Montville, T.** 1997. Principles which influence microbial growth, survival, and death in foods, p. 13–29. *In* M. P. Doyle, L. R. Beuchat, and T. J. Montville (ed.), *Food Microbiology: Fundamentals and Frontiers*. ASM Press, Washington, DC.
83. **Mortimer, R.** 2000. *Kloeckera apiculata* controls the rates of natural fermentation. *Rev. Vitic. Enol.* **53:**61–68.
84. **Mortimer, R., and M. Polsinelli.** 1999. On the origins of wine yeast. *Res. Microbiol.* **150:**199–204.
85. **Müller, M. R. A., M. A. Ehrmann, and R. F. Vogel.** 2000. *Lactobacillus frumenti* sp. nov., a new lactic acid bacterium isolated from rye-bran fermentations with a long fermentation period. *Int. J. Syst. Bacteriol.* **50:**2127–2133.
86. **Nehme, N., F. Matieu, and P. Taillandier.** 4 March 2008. Quantitative study of interactions between *Saccharomyces cerevisiae* and *Oenococcus oeni* strains. *J. Ind. Microbiol. Biotechnol.* [Epub ahead of print.] doi:10.1007/s10295-008-0328-7.
87. **Neubauer, H., E. Glaasker, W. P. Hammes, B. Poolman, and W. N. Konings.** 1994. Mechanism of maltose uptake and glucose excretion in *Lactobacillus sanfrancisco*. *J. Bacteriol.* **176:** 3007–3012.
88. **Nickerson, K. W., A. L. Atkin, and J. M. Hornby.** 2006. Quorum sensing in dimorphic fungi: farnesol and beyond. *Appl. Environ. Microbiol.* **72:**3805–3813.
89. **Niku-Paavola, M. L., A. Laitila, T. Mattila-Sandholm, and A. Haikara.** 1999. New types of antimicrobial compounds produced by *Lactobacillus plantarum*. *J. Appl. Microbiol.* **86:**29–35.
90. **Nisiotou, A. A., and G.-J. E. Nychas.** 2007. Yeast populations residing on healthy or *Botrytis*-infected grapes from a vineyard in Attica, Greece. *Appl. Environ. Microbiol.* **73:**2765–2768.
91. **Nissen, P., and N. Arneborg.** 2003. Characterization of early deaths of non-*Saccharomyces* yeast in mixed cultures with *Saccharomyces cerevisiae*. *Arch. Microbiol.* **180:**257–263.
92. **Nout, M. R. J., and F. M. Rombouts.** 1992. Fermentative preservation of plant foods. *J. Appl. Bacteriol.* **73:**136S–147S.
93. **Osborne, J. P., and C. G. Edwards.** 2007. Inhibition of malolactic fermentation by a peptide produced by *Saccharomyces cerevisiae* during alcoholic fermentation. *Int. J. Food Microbiol.* **118:**27–34.
94. **O'Sullivan, T. F., Y. Walsh, A. O'Mahony, G. Fitzgerald, and D. V. Sideren.** 1999. A comparative study of malthouse and brewhouse microflora. *J. Inst. Brew.* **105:**55–61.
95. **Parish, M., and D. Carroll.** 1985. Indigenous yeasts associated with Muscadine *(Vitis rotundifolia)* grapes and musts. *Am. J. Enol. Vitic.* **36:** 165–169.
96. **Pederson, C.** 1979. *Microbiology of Food Fermentations,* 2nd ed. AVI, Westport, CT.
97. **Piskur, J., E. Rozpedowska, S. Polakova, A. Merico, and C. Compagno.** 2006. How did *Saccharomyces* evolve to become a good brewer? *Trends Genet.* **22:**183–186.
98. **Pizarro, F., F. A. Vargas, and E. Agosin.** 2007. A systems biology perspective of wine fermentations. *Yeast* **24:**977–991.
99. **Plengvidhya, V., F. Breidt, Jr., Z. Lu, and H. P. Fleming.** 2007. DNA fingerprinting of lactic acid bacteria in sauerkraut fermentations. *Appl. Environ. Microbiol.* **73:**7697–7702.
100. **Prajapati, J., and B. M. Nair.** 2003. History of fermented foods, p. 1–27. *In* E. R. Farnworth (ed.), *Handbook of Fermented Functional Foods.* CRC Press, New York, NY.
101. **Prakitchaiwattana, C., G. Fleet, and G. Heard.** 2004. Application and evaluation of denaturing gradient gel electrophoresis to analyse the yeast ecology of wine grapes. *FEMS Yeast Res.* **4:**865–877.
102. **Raspor, P., D. Milek, J. Polanc, S. Mozina, and N. Cadez.** 2006. Yeasts isolated from three varieties of grapes cultivated in different locations of Dolenjska vine-growing region, Slovenia. *Int. J. Food Microbiol.* **109:**97–102.
103. **Ribereau-Gayon, P., D. Dubourdieu, B. Doneche, and A. Lonvaud.** 2006. *Handbook of Enology: the Microbiology of Wine and Vinification.* Wiley & Sons, Hoboken, NJ.
104. **Ross, P., S. Morgan, and C. Hill.** 2002. Preservation and fermentation: past, present and future. *Int. J. Food Microbiol.* **79:**3–16.
105. **Schmitt, M. J., and J. Reiter.** 2008. Viral induced yeast apoptosis. *Biochim. Biophys. Acta* **1783:**1413–1417.
106. **Sprague, G. F., Jr., and S. C. Winans.** 2006. Eukaryotes learn how to count: quorum sensing by yeast. *Genes Dev.* **20:**1045–1049.
107. **Stolz, P., W. P. Hammes, and R. F. Vogel.** 1996. Maltose-phosphorylase and hexokinase

activity in lactobacilli from traditionally prepared sourdoughs. *Adv. Food Sci.* **18:**1–6.

108. **Sugihara, T., L. Kline, and M. McGready.** 1970. Nature of San Francisco sour dough in French bread process. II. Microbial aspects. *Baker's Dig.* **44:**51–57.

109. **Suihko, M. L., and V. Makinen.** 1984. Tolerance of acetate, propionate and sorbate by *Saccharomyces cerevisiae* and *Torulopsis holmii*. *Food Microbiol.* **1:**105–110.

110. **Tham, W. A., and V. M. Danielsson-Tham.** 1988. *Listeria monocytogenes* isolated from soft cheese. *Vet. Rec.* **122:**540.

111. **Urso, R., K. Rantsiou, P. Dolci, L. Rolle, G. Comi, and L. Cocolin.** 11 March 2008. Yeast biodiversity and dynamics during sweet wine production as determined by molecular methods. *FEMS Yeast Res.* [Epub ahead of print.] doi:10.1111/j.1567-1364.2008.00364.x.

112. **Vaughan, E. E., P. T. C. van den Bogaard, P. Catzeddu, O. P. Kuipers, and W. M. de Vos.** 2001. Activation of silent *gal* genes in the *lac-gal* regulon of *Streptococcus thermophilus*. *J. Bacteriol.* **183:**1184–1194.

113. **Viegas, C. A., M. F. Rosa, I. Sa-Correia, and J. M. Novais.** 1989. Inhibition of yeast growth by octanoic and decanoic acids produced during ethanolic fermentation. *Appl. Environ. Microbiol.* **55:**21–28.

114. **Vogel, R., and M. Ehrmann.** 2008. Sourdough fermentations, p. 119–144. *In* L. Cocolin and D. Ercolini (ed.), *Molecular Techniques in Microbial Ecology of Fermented Foods*. Springer, New York, NY.

115. **Vogel, R. F., G. Böcker, P. Stolz, M. Ehrmann, D. Fanta, W. Ludwig, B. Pot, K. Kersters, K. H. Schleifer, and W. P. Hammes.** 1994. Identification of lactobacilli from sourdough and description of *Lactobacillus pontis* sp. nov. *Int. J. Syst. Bacteriol.* **44:**223–229.

116. **Wibowo, D., R. Eschenbruch, C. R. Davis, G. H. Fleet, and T. H. Lee.** 1985. Occurrence and growth of lactic acid bacteria in wine: a review. *Am. J. Enol. Vitic.* **36:**302–313.

117. **Wiese, B. G., W. Strohmar, F. A. Rainey, and H. Diekmann.** 1996. *Lactobacillus panis* sp. nov., from sourdough with a long fermentation period. *Int. J. Syst. Bacteriol.* **46:**449–453.

118. **Wolf-Hall, C. E.** 2007. Mold and mycotoxin problems encountered during malting and brewing. *Int. J. Food Microbiol.* **119:**89–94.

119. **Yoon, S. S., R. Barrangou-Poueys, F. Breidt, Jr., T. R. Klaenhammer, and H. P. Fleming.** 2002. Isolation and characterization of bacteriophages from fermenting sauerkraut. *Appl. Environ. Microbiol.* **68:**973–976.

# THE INTERACTION OF BILE SALTS WITH PATHOGENIC AND NONPATHOGENIC INTESTINAL BACTERIA

*Robert W. Crawford and John S. Gunn*

# 10

Normal physiology of healthy individuals requires emulsification and solubilization of lipids as well as elimination of substances that cannot be efficiently excreted in urine. This is accomplished by bile, a complex digestive secretion produced by the liver at nearly 1 liter per day that in general is lipid rich and protein poor (58). Bile acids are the primary constituents of bile, act as detergents due to their surface-active, amphipathic properties, and comprise 12% of the weight and approximately 50% of the organics in bile. In addition to aiding the dispersal of lipids, bile acids serve as potent antimicrobial agents. Thus, bile contributes to the harsh environment of the gastrointestinal tract, which also includes variations in pH and oxygen tension, the presence of commensal bacteria, and limitations in nutrients. These and other factors combine to be a significant barrier for invading enteric pathogens (9).

Gastrointestinal commensal bacteria and pathogens have developed mechanisms to resist the deleterious effects of bile and subsequently colonize human intestines. The manner in which microbes sense and respond to bile has been a subject of recent interest. This chapter begins with a brief overview of bile composition and conjugation mediated by the normal flora and its aforementioned antimicrobial effects and follows with a discussion about known molecular mechanisms behind bile resistance and the role bile has in altering the virulence of enteric pathogens.

## BILE AND BILE SALTS

Along with bile acids, other lipid-containing molecules found in bile include cholesterol and phospholipids (e.g., phosphatidylcholine). One of the major protein constituents is immunoglobulin, while the pigment biliverdin provides bile its characteristic yellowish green color (58, 59, 81). Immunoglobulin A and mucus in bile contribute to making it a sterile compound by preventing bacterial growth and adhesion to the biliary and small intestinal epithelium (59). Synthesis of bile occurs in pericentral hepatocytes of the liver and is followed by secretion into bile canaliculi and drainage into bile ducts that merge to form hepatic ducts. The common hepatic duct meets the cystic duct from the gallbladder, and the two channels merge into the common bile duct to provide an exit route for bile from the

*Robert W. Crawford and John S. Gunn,* Center for Microbial Interface Biology, Department of Molecular Virology, Immunology and Medical Genetics, The Ohio State University, Columbus, OH 43210.

*Food-Borne Microbes: Shaping the Host Ecosystem,* Edited by L.-A. Jaykus, H. H. Wang, and L. S. Schlesinger, © 2009 ASM Press, Washington, DC

liver. From there, the sphincter of Oddi regulates the flow of bile into the duodenum (9).

Bile is concentrated up to 10-fold by removal of water and electrolytes and can be acidified by ion exchange. This process occurs in the gallbladder, the storage organ for bile between meals (59, 66). About half the hepatic bile produced is diverted to the gallbladder until fats stimulate the enteric hormones secretin and cholecystokinin to induce bile secretion and flow. They cause the gallbladder to contract and the sphincter of Oddi to relax so that bile may reach the duodenum and aid digestion.

Bile acids are derived from cholesterol in the liver through a multienzyme process that includes an important oxidative step which converts the molecule from an eight-carbon to a five-carbon acid moiety. This mechanism permits conjugation of bile acids through a peptide linkage with either glycine or taurine in a 3:1 ratio depending on diet (53, 110). Cholic acid and chenodeoxycholic acid are the primary salts of human bile. In their unconjugated form, bile acids are only slightly soluble at physiologically relevant pH. Therefore, amidation of bile acids is critical because it lowers their $pK_a$ to provide solubility over a wide range of ionic strengths, calcium concentrations, and pH values (60, 61).

Additionally, conjugation of bile acids discourages passive absorption in the biliary tract and small intestine, so bile acids are maintained at between 0.2 and 2% in the intestine and approximately 8% in the gallbladder (58). Fasting and malnourishment have been shown to decrease the amount of bile in the intestine and, consequently, leave individuals vulnerable to bacterial pathogens (90, 105, 109). As mentioned above, bile acids are amphipathic and therefore self-associate in water to form polymolecular micelles that solubilize other lipids. It has been suggested that these aggregates decrease the toxicity individual bile salts exude upon biliary epithelial cells (59).

De novo synthesis of bile acids from cholesterol is not always required, because up to 95% of the bile acid pool is recycled on a daily basis in a process called enterohepatic circulation (9). Absorption of conjugated and unconjugated bile acids occurs along the entire length of the gut mucosa to facilitate transfer back to the liver for reconjugation and delivery into bile (57). The constant flow of bile through the gastrointestinal tract is critical for bile homeostasis. Excess intraluminal bile acids can cause diarrhea, whereas suboptimal levels may result in cholesterol-supersaturated bile, which is responsible for gallstone formation (59, 91).

## THE NORMAL FLORA AND BILE

The resident, bile-resistant microbial flora of the cecum and colon can alter the properties of bile acids that were bypassed for enterohepatic circulation. The organisms of the intestinal microbiota, primarily anaerobes, deconjugate bile acids and stimulate mechanisms that convert the primary bile acids cholate and chenodeoxycholate into secondary bile acids deoxycholate and lithocholate, respectively (42). As outlined by Bortolini et al., the categories of microbially induced biotransformations of bile acids include epimerization, oxidation, hydroxylation, reduction, and dehydroxylation (17). Some of these natural reactions are pharmaceutically relevant, such as ursodeoxycholic acid, the 7-hydroxy epimer of chenodeoxycholic acid that can be synthesized industrially and prescribed as a treatment for cholestatic liver diseases (28).

Bacteria may alter the composition of intestinal bile acids via epimerization of the hydroxyl groups at specific carbon sites. Inversions of chenodeoxycholate stereochemistry at C-3, C-7, and C-12 by clostridia and eubacteria can enhance bacterial growth in an in vitro coculture model (79). Oxidation reactions of bile acids follow formal expulsion or insertion of $H_2$ and are conducted in the presence of commensal bacteria such as *Bacteroides uniformis* and *Lactobacillus plantarum* (34, 35). A variety of organisms of the microbiota, including colonic species of fusobacteria, catalyze replacement of hydrogen with hydroxyl groups to change bile acid structure in a process termed hydroxylation (17). The opposite of this reaction is dehydroxylation, in which hydroxyl groups of primary bile acids are converted to secondary bile acids.

Deconjugation cleaves amino acid side chains and is an important mechanism used by bacteria to impact the properties of bile. This reaction may be carried out by intracellular bile salt hydrolases (BSHs) belonging to the choloylglycine hydrolase family of enzymes (9). Genera of the intestinal microflora expressing BSH activity include *Clostridium*, *Bifidobacterium*, and *Enterococcus* (27, 47, 68). Although the physiological significance of BSH activity remains uncharacterized, key studies have advanced our understanding of their proposed roles in enteric microorganisms. Van Eldere et al. demonstrated that clostridium energetics utilize taurine released from bile acids, suggesting that deconjugation by hydrolytic strains provides nutritional sources of carbon and nitrogen (121). It has also been postulated that BSHs increase tensile strength of bacterial membranes residing in the intestine and serve as a detoxification system mediating bile tolerance (15, 46). Similar to commensal enteric bacteria, pathogenic *Listeria monocytogenes* contains BSH genes thought to confer bile resistance and successful colonization and disease manifestation. This mechanism is described below in the section on *Listeria*.

The inherent cost of altering the properties of bile acids to cultivate a healthy microflora may be at the expense of the human host. Unconjugated bile acids emulsify dietary lipids less effectively than their conjugated counterparts, leaving lipid digestion and fatty acid removal impaired (30). The products of dehydroxylation reactions are potentially toxic and have been associated with an increased risk of colon cancer (84). On the other hand, probiotics researchers encourage consumption of BSH-active bacteria. They suggest that subsequent limitations to enterohepatic circulation in the presence of unconjugated bile acids cause enhanced fecal loss of bile salts. The result may be manifested in lower serum cholesterol levels due to an increased demand by the liver for cholesterol for de novo bile acid synthesis (30). The role of the intestinal flora and pathogens in altering bile acid structure and function is an ongoing area of investigation.

## BILE AND ITS EFFECT ON PATHOGENIC BACTERIA

### Bile Sensing

Adaptability to the harsh effects of bile acids is a critical component of survival for gastrointestinal pathogens (Fig. 1). However, bacterial components that sense and respond to environmental concentrations of bile are currently unknown. Enteric bacteria are known to alter protein production and increase mechanisms of resistance to survive the deleterious effects of bile, but whether this is a generalized stress response or a specific sensory phenomenon is unclear (69). A comparison of bile and sodium dodecyl sulfate stress responses in *Enterococcus faecalis* revealed differential protein regulation, suggesting a unique pathway employed by the bacteria when exposed to bile (37). In addition, incubation of *E. faecalis* and *Salmonella enterica* serovar Typhimurium with a sublethal bile dose facilitates growth in extremely high bile concentrations (37, 122).

The presence of bile in the extracellular milieu may be sensed directly by regulatory factors or through regulatory systems such as two-component systems. The result is an adaptive response typically mediated by a change in gene expression. Regarding two-component regulatory systems, more than 4,000 two-component signal transduction systems have been identified in 145 sequenced bacterial genomes accounting for diverse response regulation, including nutrient acquisition, energy metabolism, and virulence (83). While evidence exists linking several regulators to bile-mediated effects, few direct bile-protein interactions have been demonstrated (51).

Alternatively, indirect effects of bile salts may result in phenotypic changes that enhance colonization and persistence. The detergent-

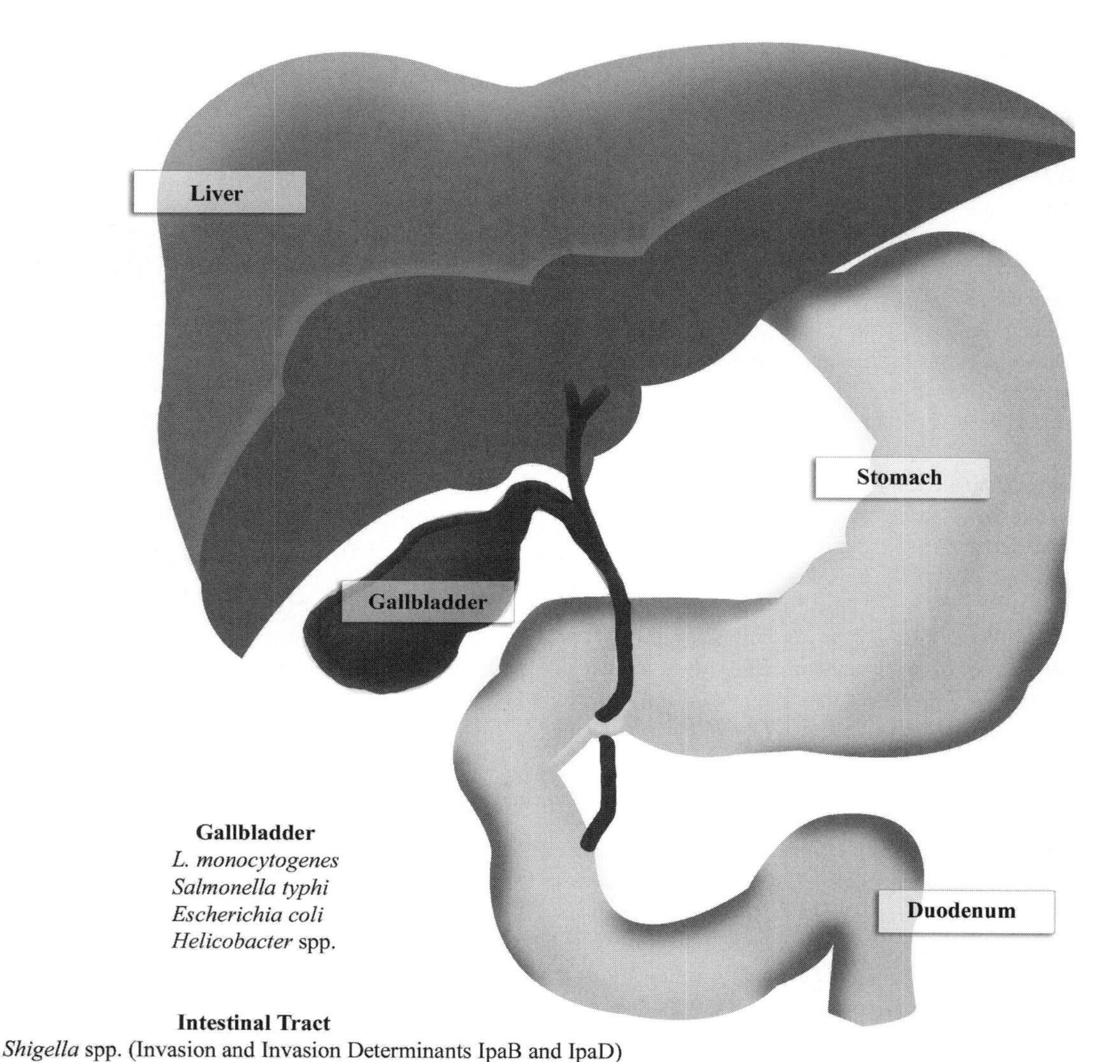


**FIGURE 1** Bacterial intestinal pathogens discussed in this chapter that are affected by bile. Listed are microbes known to cause acute or chronic disease in the gallbladder or the intestinal tract. Interactions between invading microorganisms and bile, which is produced in the liver and found in the intestine and gallbladder, affect various genes and virulence properties (listed) that enhance colonization and persistence.

like activity of bile acids may hinder bacterial membrane integrity such that the corresponding damage induces transcription of genes involved in stress responses. The *Escherichia coli* Cpx and RcsCB systems sense just this type of perturbation at the cell membrane (31). It could also be the case that gene regulation during bile adaptation occurs through sigma factors associated with stress-induced regulons or through regulation by small RNAs (25). Further explanation of these and other possibilities comprises the remainder of this section.

### *Salmonella*

*Salmonella* is a pathogen of global concern, particularly in nonindustrialized nations where food and water sanitation systems remain subpar. The two most common clinical manifestations are gastroenteritis and enteric (typhoid) fever, both of which result in large yearly financial burdens to hospitals and patients. During the course of infection, *Salmonella* spp. interact with bile and as a result have evolved resistance to its detergent-like properties. *S. enterica* serovar Typhimurium colonizes the distal ileum of the small intestine and was shown in vitro to have a bile minimal bactericidal concentration of more than 60% (122). Approximately 3 to 5% of patients suffering from typhoid fever develop an asymptomatic chronic carrier state thought to account for much of the human-to-human transmission of *S. enterica* serovar Typhi (100). The main site of carriage is the bile-rich gallbladder, suggesting unique or enhanced bile resistance mechanisms for salmonellae leading to colonization of and persistence in this organ.

Bile affects the expression of important *Salmonella* proteins necessary for bile tolerance, some of which are thought to be regulated by the two-component system PhoPQ. Van Velkinburgh and Gunn observed that mutant strains constitutively expressing this regulon survived prolonged incubation with bile at concentrations exceeding 60%, whereas a *phoP*-null strain was susceptible to significantly lower bile concentrations than wild-type strains. However, no PhoPQ-regulated genes responsible for bile resistance were identified (122). Therefore, PhoPQ-mediated bile resistance may occur via genes that overlap this regulon.

The lipopolysaccharide (LPS) of the outer membrane of gram-negative bacteria protects cells from damaging agents like amphipathic bile acids. Loss of the O-antigen component of LPS, which protrudes from the membrane to serve as its main barrier, creates a rough colony phenotype resulting in decreased resistance to bile (67, 99). Interestingly, LPS modifications (aminoarabinose addition to the lipid A anchor portion of LPS and ethanolamine changes to the core and lipid A regions) did not alter the high tolerance of *Salmonella* to bile, eliminating the possibility that electrostatic interactions between bile and serovar Typhimurium contributed to resistance (122). Furthermore, activation of the PmrAB system, which mediates these modifications, actually sensitizes both *E. coli* and *Salmonella* to killing by bile salts (J. Gunn, unpublished observation), suggesting that genes other than those mediating LPS alteration are likely to be possible.

One of the most well-characterized mechanisms through which bacteria exhibit bile tolerance is efflux pump activity. In addition to mediating resistance to a variety of antibiotics and oxidative stress agents, these systems expel out of the cell bile salts that have traversed the inner membrane. Salmonellae express the bile-activated AcrAB efflux pump, which has a proton motive force-dependent drug efflux transporter, AcrB, and an element to bridge the inner and outer membranes, AcrA (78, 127). The *acrAB* efflux pump genes are absolutely required for growth of serovar Typhimurium in the presence of bile. Similarly, the *marRAB* operon of serovar Typhimurium is involved in resistance to structurally unrelated antimicrobials, including chloramphenicol, tetracycline, and bile. DNA microarray and transcriptional assays performed by Prouty et al. showed activation of *marRAB* in the presence of bile. Deoxycholate, but not other bile acids or a nonionic detergent, interacted directly with the *marRAB* repressor

MarR to interfere with its ability to bind to the *mar* operator. Interestingly, growth in sublethal amounts of bile enhanced chloramphenicol resistance by both *mar*-dependent and *mar*-independent pathways (98).

Tol proteins function in the uptake of colicins and filamentous phage DNA, and transport systems in many gram-negative bacteria require a Tol homologue at the outer membrane to serve as a pore (49). In this regard, TolC contributes to the AcrAB efflux pump system and is an essential component of bile resistance. Several bile-sensitive mutations in serovar Typhimurium isolated from a MudJ transposon experiment were located in the *tolQRA* region (99). Mutations in these genes compromise membrane integrity, possibly allowing an increase in bile acid traffic into the cell. The enterobacterial common antigen of the serovar Typhimurium outer membrane also provides bile protection. Strains carrying mutations in the enterobacterial common antigen biosynthesis and assembly genes *wecD* and *wecA* were 40-fold more sensitive to deoxycholate in vitro and highly attenuated in oral infections in vivo compared to the wild-type strain (104).

Bacterially mediated endocytosis of serovar Typhimurium into nonphagocytic eukaryotic cells utilizes a type III secretion system (TTSS) located in *Salmonella* pathogenicity island 1 (SPI-1). Invasion gene regulators, including *sirC* and *invF,* are transcriptionally repressed by bile, leading to repression of protein secretion by the SPI-1 TTSS and a dramatic reduction in epithelial cell invasion (95, 97). The presence of bile also reduced expression of flagellar biosynthesis genes and motility. Therefore, successful host invasion may rely on temporal sensing of bile by salmonellae. High concentrations of bile encountered in the intestinal lumen may halt invasive properties until bacteria traverse the epithelial mucous layer to the underlying surface. It is there, where bile is likely at a low concentration and invasion is needed for disease progression, that SPI-1 TTSS repression is alleviated.

Individuals afflicted with gallstones are at a significantly higher risk of carrying *Salmonella* asymptomatically in the gallbladder (100). The formation of biofilms on gallstones is hypothesized to enhance colonization and persistence in this hostile environment. A biofilm is defined as a community of microorganisms that adhere to each other and to a biotic or abiotic surface, mediating protection from antibiotics through a self-initiated exopolysaccharide matrix. Interestingly, bile has been shown to promote biofilm formation in an in vitro human gallstone test tube assay (100). In addition, the support matrix seemed to provide signals that determined *Salmonella* biofilm components, and those differed for gallstones and glass surfaces (96). Characterizing the exopolysaccharide matrix of these biofilms may provide a novel therapeutic target for eliminating chronic carriage in the gallbladder in patients with bile-induced *Salmonella* gallstone biofilms.

### *E. coli* and *Shigella*

The ubiquitous, gram-negative bacterium *Escherichia coli* facilitates waste processing, vitamin K production, and food absorption in a healthy functioning gastrointestinal microenvironment. However, it is in this niche that commensal *E. coli* plays destructive roles implicated in gastroenteritis, urinary tract infections, and peritonitis. Despite advances in sterilization techniques aimed at killing active bacteria, *E. coli* also remains an important food-borne pathogen. Recent outbreaks of contamination of beef and vegetables have caused numerous cases of severe diarrhea (65).

Like *Salmonella* and other gram-negative bacteria, *E. coli* actively removes bile acids that have traversed the membranes into the cytoplasm as a mechanism of bile resistance. Thanassi et al. demonstrated that bile acids in the unconjugated, lipophilic form entered cells through the OmpF porin of the outer membrane and were subsequently expelled via energy-dependent efflux (117). The *mdtABCD* multidrug resistance cluster encodes three different transporter proteins and has been linked to enhanced resistance to deoxycholate (7). *E. coli* possesses two additional efflux systems, AcrAB and EmrAB, required for tolerance to antibiotics such as tetracycline and chloramphenicol and toxic stressors such as bile. As mentioned above, the constituents of AcrAB are well characterized, whereas EmrAB is not as well described outside of its negative regulator EmrR (75, 76). *E. coli* strains carrying *acrA emrB* double mutations demonstrated growth inhibition in the presence of bile and intracellular accumulation of chenodeoxycholate.

Enteropathogenic *E. coli* (EPEC) strains are significant causes of infantile diarrhea. Virulence factors encoded by the chromosomal locus of enterocyte effacement region and adherence factors allow EPEC attachment to the intestinal mucosa (120). The ability of bile to mediate EPEC adhesion was demonstrated by de Jesus et al. in experiments showing that bile acids, including chenodeoxycholate and deoxycholate, indeed increased adherence to Hep-2 cells in tissue culture (29). Although specific EPEC adhesins responsible for the bile acid stress-induced adherence were not elucidated, the virulence factors intimin and flagella were ruled out. This issue was similarly addressed by Torres et al. using strains of atypical EPEC (aEPEC). Bile acids enhanced aEPEC adhesion to in vitro-cultured HeLa cells by regulating expression of the afimbrial locus for diffuse adherence (LDA) adhesin (119). Other detergents tested could not achieve the same induction of this locus, suggesting a specific mechanism by which bile promotes disease manifestation by increasing the binding of aEPEC to intestinal epithelial cells.

Contaminated beef is a source of human infection, implying that cattle serve as primary reservoirs of *E. coli* (85). Several studies using bovine host models have isolated strain O157: H7, capable of growth in 15% bile, throughout the intestinal tract and from the gallbladder following experimental inoculation (115). The hypothesis that *E. coli* exists in a persistent state is difficult to prove because shedding from cattle is generally sporadic and bacteria are recovered in low numbers (92). Jeong et al. detected three types of *E. coli* in the gallbladder up to 36 days after oral inoculation, suggesting an essential bile tolerance phenotype during animal-to-animal or environmental transmission (65).

*Shigella flexneri,* closely related to enteroinvasive *E. coli,* is a facultative intracellular pathogen and the etiologic agent of bacillary dysentery (52). Invasion of colonic epithelial cells is mediated by a TTSS and triggers a strong inflammatory response. Disease onset is characterized by diarrhea, abdominal cramps, and fever (13, 38). Interestingly, natural infections in humans require only 10 to 100 organisms of *S. flexneri,* whereas simulated infection of tissue cultured cells is relatively inefficient and needs a centrifugation step to initiate contact between bacteria and target cells. This observation led to the hypothesis that bile acids encountered in vivo act as specific environmental factors that enhance *Shigella* invasion and disease progression (93).

Indeed, Pope et al. demonstrated that exposure to physiological concentrations of deoxycholate and chenodeoxycholate increased attachment to and entry of *Shigella* species into HeLa cells in an in vitro model system (93). Interestingly, this phenomenon did not occur in enteroinvasive *E. coli.* The presence of bile up-regulates TTSS machinery and invasion in *Shigella* but has the opposite effect in *Salmonella,* further suggesting a unique mechanism contributing to bile signaling and pathogenesis. This group also observed the ability of bile acids to enhance secretion of Ipa proteins released during TTSS-dependent cell entry. Olive et al. elaborated upon this finding, reporting that deoxycholate stimulated recruitment of IpaB to the *S. flexneri* surface. Once there, IpaB colocalized with IpaD at the tip of the type III secretion needle. The authors speculate that IpaD is an environmental sensor capable of changing conformation in the presence of bile acids to permit type III-mediated secretion (89).

### *Campylobacter*

Since the discovery in the late 1970s that *Campylobacter jejuni* could induce human illness, this gram-negative pathogen has become the leading bacterial cause of enteritis in the United States (112). Symptoms of *Campylo-*

*bacter* infection vary in severity from mild diarrhea to extensive cramping and abdominal pain (71). A small percentage of patients develop Guillain-Barré syndrome, an autoimmune disorder of the peripheral nervous system characterized by weak respiratory muscles, neurologic deficits, or death (88). Epidemiological studies implicate consumption of undercooked, contaminated poultry as the most frequent cause of sporadic *C. jejuni* illness in humans (112). Effective control and prevention of *Campylobacter* infection remain elusive, as the pathogen is becoming increasingly resistant to fluoroquinolone treatment and innate host defense (116).

Bile salts present an important antimicrobial barrier for *Campylobacter* in the intestine. Accordingly, the pathogen expresses a multidrug resistance efflux pump called CmeABC which is essential for adaptation to the harsh gastrointestinal environment (77). This resistance-nodulation-division (RND)-type multidrug efflux system is capable of mediating the bile resistance phenotype *C. jejuni* requires for successful host colonization (48, 103).

CmeABC is encoded by a three-gene operon and is constitutively expressed at moderate levels in wild-type *Campylobacter* strains cultured in conventional media (72). Lin et al. showed expression of *cmeABC* to be controlled by the transcriptional repressor CmeR, which, when inactivated, leads to overexpression of the pump (73). Researchers have also demonstrated direct interaction of bile salts with CmeR much like that described previously for *Salmonella* MarR (72). Both conjugated and unconjugated bile salts increased expression of the CmeABC efflux pump, while other antimicrobials, including chloramphenicol, ethidium bromide, and erythromycin, did not affect transcription of *cmeABC*. Surface plasmon resonance provided evidence that bile salts were capable of inhibiting binding of CmeR to the *cmeABC* promoter, leading to increased pump expression and elevated bile resistance. Interestingly, the presence of bile salts in culture media enhanced the resistance of *Campylobacter* to multiple antibiotics, including cefotaxime, novobiocin, and fusidic acid (72). Mutations in *cmeABC* drastically reduced resistance to bile salts, as inactivation of the CmeABC pump led to impaired growth in the presence of 2 mM choleate culture media, whereas wild-type bacteria displayed no growth defect under the same conditions (74).

*Campylobacter* also requires the CmeABC efflux pump for growth in the animal intestinal tract, where bile resistance is crucial for colonization and persistence. In experiments using young chickens, wild-type *C. jejuni* colonized the birds as early as 2 days postinoculation, while *cmeABC* mutants failed to do so at any point during the 20-day study (74). Expression of the CmeABC efflux pump was also shown to be significantly up-regulated in rabbit ileal loops using real-time reverse transcription-PCR (114). These data offer the CmeABC efflux pump as a therapeutic target for *Campylobacter* pathogenesis. Efflux pump inhibitors present a new avenue of intervention, and towards this goal, Lin et al. recently showed that oral administration of two different efflux pump inhibitors reduced colonization of *C. jejuni* in chickens at 2 to 4 days postinoculation (71).

Bile salts are proposed to act as a stimulatory signal for synthesis of virulence-associated *Campylobacter* invasion antigens (Cia) present upon coculture with epithelial cells (106). These invasion-related proteins were not induced by altering the pH, calcium concentration, osmolarity, or temperature of the culture media, suggesting a specific effect for the bile salts sodium deoxycholate, cholate, and chenodeoxycholate. The flagellum of *C. jejuni* is another important virulence determinant whose expression is known to be enhanced by bile. Experiments done by Allen and Griffiths showed that bile positively induced the flagellar *flaA* $\sigma^{28}$ promoter and that subsequent chemotaxis led to colonization of the mucous layer and pathogenesis (2).

### *Vibrio*

The presence of bile causes pleiotropic responses that affect production of virulence factors, motility, and other phenotypes in the pathogenic bacterium *Vibrio cholerae*. Vibrios are noninvasive, gram-negative enteric bacteria that can survive diverse environmental conditions. The diarrheal disease cholera is acquired following consumption of food or water contaminated with *V. cholerae*. Developing countries are particularly susceptible to devastating outbreaks, in which mortality rates may exceed 20% in affected populations (23).

Production of cholera toxin (CT) and toxin-coregulated pilus is typically required for colonization of the intestinal epithelium (67). Expression of these virulence factors, as well as others, is controlled by a cascade of regulatory proteins comprising the ToxR regulon. ToxR and TcpP are inner membrane DNA binding proteins that activate ToxT, an important transcriptional regulator that controls a variety of factors necessary for colonization and disease in the host, including *ctxAB* and *tcpA* (23).

As evidenced by several pieces of conflicting data, the relationship between *Vibrio* and bile is quite complex. It was initially reported that the production of CT and toxin-coregulated pilus is dramatically diminished when *V. cholerae* is grown in the presence of crude ox bile extract (50). This finding was extended by Schuhmacher and Klose, who used a plasmid-based heterologous $P_{lac}$ promoter to show that sodium choleate addition resulted in a significantly reduced ability of ToxT to activate transcription of regulated genes. In sum, these results suggested that ToxT interacts directly with bile or with a factor induced in the presence of bile to alter its ability to regulate target genes (108).

Furthering the studies of the Klose laboratory, Chatterjee et al. examined bile-regulated virulence factor expression by elaborately fractionating crude bile to identify the components mediating virulence gene repression and enhancement of motility in *V. cholerae* (23). They isolated the unsaturated fatty acids arachidonic, linoleic, and oleic acids from bile and showed them to repress expression of the *ctxAB* and *tcpA* genes independently of ToxT. These fatty acids also increased expression of *flrA*, the first gene in a cascade controlling motility. It was proposed that the ToxR-ToxT system that negatively regulates motility is nonfunctional in the presence of unsaturated fatty acids. The resulting increase in motility was also observed upon addition of cholesterol. However, Hung and Mekalanos later reported that purified bile acids, unlike crude bile extracts, do not inhibit CT or TcpA expression. Interestingly, this work shows that bile acids can activate CT expression in a ToxR-dependent, ToxT-independent manner (63).

*Vibrio cholerae* also expresses outer membrane proteins that are affected by bile. ToxR activity, independently of ToxT, leads to reciprocal expression of OmpU and OmpT, two general diffusion porins that affect bile resistance (20, 21). Gupta and Chowdhury reported that crude ox bile extracts increased *V. cholerae* motility in a ToxR-independent manner but had no effect on OmpU/OmpT expression (50). However, Duret and Delcour (32) and Provenzano et al. (101, 102) later demonstrated that bile increased OmpU expression in a ToxR-dependent manner to confer enhanced bile resistance on *V. cholerae*.

Like other enteric gram-negative pathogens, *V. cholerae* expresses an efflux system contributing to antimicrobial resistance. Bina et al. recently characterized two RND family efflux systems called *vexAB* and *vexCD* that provide *V. cholerae* resistance to bile (12). Microarrays were used previously to show that the two open reading frames, now known to contain *vexAB*, were up-regulated in stools of cholera patients and in rabbit ileal loops (11, 82, 126). Experiments by Bina et al. showed that expression of the *vexAB* and *vexCD* efflux systems is regulated by bile acids and that these pathways have overlapping roles in resistance to bile (12). Surprisingly, though, they demonstrated that *vexB*, *vexD*, and *vexB-vexD* deletion strains efficiently colonized the infant mouse, suggesting the dispensability of these systems in vivo or the involvement of other, as-yet-uncharacterized efflux systems.

As seen with *Salmonella*, bile-induced biofilm formation occurs in *V. cholerae* and may enhance colonization of the host. Hung et al. found enhanced *Vibrio* biofilm formation in the presence of bile acids using a glass test tube assay (64). The observed biofilm induction occurred independently of known biofilm regulatory pathways, acting instead through the *vps* (*Vibrio* polysaccharide synthesis) genes and their transcriptional activator VpsR. Interestingly, they showed that biofilms provided more resistance to harsh bile acids than did planktonic cells. These data suggest that bile-induced biofilms protect *Vibrio* from the antimicrobial effects of bile during infection to allow persistence in the intestinal environment.

### *Helicobacter*

Species of the gram-negative, helix-shaped *Helicobacter* belonged to the genus *Campylobacter* until it was classified on its own in 1989. The majority of research over the past two decades has centered on conditions related to *Helicobacter pylori* infection such as chronic gastritis, peptic ulceration, gastric cancer, and, more rarely, malignant tumors of the mucosa-associated lymphoid tissue (3). Recently, other members of the *Helicobacter* genus have been discovered living in the gastrointestinal systems of humans, other mammals, and birds, meriting consideration alongside *H. pylori* by the scientific community. Their presence in the intestines and spleens of infected individuals suggests that bile tolerance mechanisms contribute to the pathogenesis of gastroenteritis, hepatitis, and chronic liver disease (4).

One such microorganism, *Helicobacter pullorum,* was first isolated from the livers and intestinal contents of laying hens afflicted with vibrionic hepatitis. Interestingly, studies showed that asymptomatic poultry transmitted *H. pullorum* via contaminated feces (6, 18, 113). For humans, PCR was used to detect this strain in bile isolated from patients with chronic cholecystitis (40). This report concluded that, reminiscent of *Salmonella* carriage and its associated risk factors, *Helicobacter* was present in the bile and gallbladder tissue of a high proportion of Chilean patients with gallbladder cancer.

*Helicobacter bilis,* originally identified from inbred mice with chronic hepatitis, is also resistant to bile. Fox et al. sequenced PCR-amplified 16S rRNA fragments of DNA and showed that *H. bilis* was found in the gallbladders of five of eight Chilean patients with chronic cholecystitis (40, 41). Another bile-tolerant species, *H. hepaticus,* was isolated from livers and intestinal mucosal scrapings from mice suffering from hepatocellular carcinoma (39, 124). Collectively, *H. pullorum, H. bilis,* and *H. hepaticus* are implicated in disease states requiring their ability to survive harsh concentrations of bile in infected hosts. Therefore, bile may represent a signal in *Helicobacter* to both increase resistance and alter virulence factor expression when the organism is within the gastrointestinal system.

Several studies have evaluated the correlation between chronic cholecystitis and the presence of *H. pylori* in the gallbladder. Its existence in this organ is inherently intriguing because *H. pylori* is susceptible to bile in vitro (118). Ananieva et al. (3) and Apostolov et al. (5) used immunohistochemistry with *H. pylori*-specific CagA and VacA antigens and observed a high prevalence of this strain in gallbladder epithelial cells of Ukrainian patients needing cholecystectomy. They suggest that a spiral-to-coccoid shape conversion may allow *H. pylori* the fitness required to counteract the normally lethal effects of bile and enhance bacteremia in patients with gallstones. On the other hand, Bohr et al. draw little analogy between the presence of *Helicobacter* and cholecystitis (16). They investigated gallbladder tissues from German patients with gallstones and found a low prevalence of *Helicobacteraceae* upon immunohistochemical analysis. In addition, Maurer et al. used C57L/J mice, a strain in which gallstone formation can be induced by a lithogenic diet, to conclude that *H. pylori* infection plays no role in murine cholesterol gallstone formation (80). The limited amount of contradictory literature suggests that the cause-and-effect relationship between the presence of *Helicobacter* and induction of various hepatobiliary diseases, as influenced by bile during infection, needs further investigation.

### *Listeria*

As is the case with the majority of food-borne pathogens, *Listeria* requires mechanisms to sense and respond to a variety of stressful conditions. Food processing and storage expose the gram-positive bacterium *Listeria monocytogenes* to growth challenges, including cold temperatures, acidic pH, and high osmolarity (123). The surviving bacteria contaminate food and subsequently can infect a human host, in whom temperatures are higher, pH is lower, and osmolarity shifts during transit through the gastrointestinal tract (43). Fatty acids, limited amounts of oxygen and nutrients, and bile further enhance the volatility of the gastrointestinal tract as a barrier to *L. monocytogenes* pathogenesis (26).

Similar to the case with *S. enterica* serovar Typhimurium described above, preexposure of *Listeria* to sublethal concentrations of bile prepares cells for subsequent growth in otherwise lethal amounts of bile. Begley et al. mimicked experiments performed by Flahaut et al. in *E. faecalis* (36) and showed this adaptation phenomenon in low levels of bile to occur in as little as 5 s (8). Although bile sensing yields coordinate changes to cellular physiology through as-yet-unknown mechanisms, it has been suggested that adaptation to other stresses also confers cross-protection to bile. Indeed, exposing exponential-phase *L. monocytogenes* to acidic pH, high temperature, 5% NaCl, or 0.01% sodium dodecyl sulfate increased bile tolerance (8).

*Listeria* expresses BSHs thought to be important for bile resistance. These enzymes catalyze hydrolysis of the amide bond between the C-24 position of the steroid moiety and the amino acid side chain of bile acids. It has been proposed that deconjugation of bile salts by BSHs represents a detoxification mechanism that promotes *Listeria* persistence in the gastrointestinal tract (9). A *bsh* gene deletion in *L. monocytogenes* created by Dussurget et al. decreased the MICs of both bile and bile salts and yielded reduced bacterial fecal carriage after oral infection of guinea pigs (33). In addition, the *bsh* gene is activated by PrfA, the major virulence gene regulator of *L. monocytogenes*. Interestingly, when wild-type cells carried the *bsh* gene on a plasmid, researchers observed an increase in intestinal multiplication.

Begley et al. corroborated these results using a rifampin-marked strain in a murine model of infection, showing that *bsh* mutants could not be found in the intestine 3 days after oral inoculation (10). They also showed involvement of the general stress responsive sigma factor $\sigma^B$ in *bsh* gene regulation. This indicates that anaerobic conditions of the intestine trigger the sigma B regulon to prepare *Listeria* for environmental stressors, including bile. Genomic sequences revealed *bsh* to be absent from nonpathogenic *Listeria,* further implicating BSH activity as a specific mediator of bile resistance for successful colonization of the host.

Another virulence determinant in *L. monocytogenes* is the recently described bile tolerance locus BilE, which contains two distinct ATP-dependent bile acid permease signature sequences common to multidrug efflux pumps (111). Sleator et al. revealed that a *bilE* mutant retained approximately 1.5 times the amount of radiolabeled bile retained by the wild type, rendering it bile sensitive. Intriguingly, they showed that bile susceptibility is more pronounced in a *bilE* deletion mutant than in a *bsh* mutant, suggesting that active extrusion of bile salts is more effective than detoxification of internalized bile. The authors also present data showing that *bilE* transcription is coordinately controlled by $\sigma^B$ and *prfA,* much like BSH enzymes mentioned above (111).

The bile resistance phenotype of *Listeria* is likely important in the gallbladder environment as well as the intestine. Hardy et al. performed elaborate in vivo bioluminescence imaging experiments indicating that the gallbladder was the primary site for *L. monocytogenes* extracellular replication during murine infection (54). The bacteria trafficked to this organ in asymptomatic as well as diseased animals, and it has been estimated that up to 10% of healthy adult humans shed *Listeria* (107). Therefore, it is possible that growth in the gallbladder leads to secretion via bile into the intestine and transmission of *L. monocytogenes* to unsuspecting hosts in a manner similar to that of typhoid fever.

### Viruses

In addition to altering the pathogenic course of invasive bacteria, bile is known to affect disease outcome for a variety of viral types. Human immunodeficiency virus type 1 (HIV-1) transmission and pathogenic progression require infection or propagation of viral replication in $CD4^+$ T lymphocytes. This mechanism has been suggested to rely upon an important interaction between the gp120 envelope glycoprotein of HIV-1 with dendritic-cell (DC)-specific ICAM-3 grabbing nonintegrin (DC-SIGN) expressed on DCs and a subset of B cells (44, 45, 62, 94). Naarding et al. demonstrated that Lewis X ($Le^X$) found in human milk could inhibit DC-SIGN-dependent transfer of HIV-1 to $CD4^+$ T lymphocytes by binding to DC-SIGN and blocking the interaction between the virus and receptor (87). They further showed that bile salt-stimulated lipase, a $Le^X$-carrying glycoprotein secreted by the pancreas and activated by bile salts in the intestine, was the primary compound in human milk binding to DC-SIGN and inhibiting HIV-1 transfer (86). Bile salt-stimulated lipase variations caused alterations in $Le^X$ expression that limited viral activity, suggesting a potential role for bile derivatives against HIV-1 transmission. In another study investigating the effects of bile on HIV-1 activity, Al-Jabri et al. showed that tyrosine-conjugated tri- and dihydroxy bile salt derivatives exhibited high surfactant activity, low cellular toxicity to human tissues, and anti-HIV-1 virucidal activity (1).

The viral hepatitides, common causes of liver disease globally, have been shown to encounter bile during pathogenesis. Bile duct damage is a characteristic histopathological consequence of the massive immune response triggered by chronic hepatitis C virus (HCV) infection (55, 70). Haruna et al. obtained bile by needle aspiration from the gallbladders of HCV-infected patients and found that viral RNA levels were as high in bile as they were in serum. They provided immunohistochemical evidence that HCV could infect bile duct epithelial cells, implicating the biliary system and bile in the pathobiology of HCV (55). Similarly, the virion and DNA-free particles of hepatitis B virus (HBV) surface antigen have been found in bile. Blum et al. hybridized a radiolabeled probe specific for HBV nucleotide sequences to liver tissue of infected patients and observed the virus in both hepatocytes and bile duct epithelial cells (14). These findings have led to the use of bile acids in clinical trials as therapeutic agents for patients with viral hepatitis. It has been noted that administration of bile acids, such as ursodeoxycholic acid, induced significant improvements in serum transaminase activities in hepatitis B and hepatitis C (24).

On the other hand, the presence of bile acids seems to be required for the growth of porcine enteric calicivirus (PEC), a small, nonenveloped virus responsible for a high proportion of nonbacterial gastroenteritis outbreaks in the United States each year (22). The PEC strain Cowden can be propagated by serial passage in a continuous cell line (LLC-PK), but only with addition of intestinal content fluid filtrate obtained from uninfected gnotobiotic pigs. Chang et al. elucidated the bile acids cholic acid, glycocholic acid, chenodeoxycholic acid, and glycochenodeoxycholic acid in intestinal contents as the essential active factors mediating growth of PEC in cultured cells. The presence of bile during PEC replication triggered the protein kinase A cell signaling pathway and induced increased cyclic AMP levels in LLC-PK cells, leading to a down-regulation of innate immunity associated with decreased interferon-mediated signal transducer and activator of transcription 1 phosphorylation. Interestingly, addition of cyclic AMP/protein kinase A pathway inhibitors to infected LLC-PK cells suppressed bile acid-mediated PEC replication. The dependence of PEC growth on the presence of bile suggests that bile salts may actively suppress innate immune function in susceptible target cells, so enteric viruses such as PEC may utilize this mechanism to establish gut infections (22).

## CONCLUSIONS

The normal flora of the gastrointestinal microenvironment has evolved to resist the deleterious effects of bile while still allowing nutrient absorption and providing an immune barrier in healthy individuals. However, recent work by a number of laboratories has shown that bile is a temporal and spatial signal sensed by invading enteric organisms that regulates expression of bile resistance and virulence factors during infection. Whereas adaptive responses of bacteria to physiological stresses, including osmolarity and pH, have been well characterized, the mechanisms contributing to bile resistance and subsequent virulence factor production remain poorly understood.

Colonization of the intestines or gallbladder relies on a high degree of bile resistance, and such residence results in major public health concerns. For example, the *Salmonella* asymptomatic carrier state is the number one risk factor for carcinomas of the gallbladder and has also been implicated in cancers of the pancreas and lung (19, 125). Therefore, discovery of antibacterial countermeasures related to bile resistance properties has direct correlations with cancer prevention. Other bacterial targets include biofilm-related factors and efflux pumps. Compounds could also be identified that target the gallbladder.

Clinical administration of bile itself has therapeutic activity. Indeed, bile acids have been shown to be efficacious as topical agents countering sexually transmitted diseases and intravenously for viral hepatitis (24, 56). Manipulation of microbial bile resistance genes may also have benefits. Enhancing the BSH properties of probiotics could increase the demand for cholesterol in the liver for de novo bile acid synthesis, thereby removing it from the blood to provide a treatment for patients with hypercholesterolemia (9). In malnourished individuals, who are susceptible to enteric infections, increasing the bile acid levels in the intestine via oral administration or by an increase in hormones that result in increased bile production or secretion would bolster their defense arsenal against invading pathogens.

As long as the integrity of the normal microbial flora is maintained, compounds such as those discussed above could be manufactured to target known factors contributing to bile resistance. It is likely that future bile research will target temporal and spatial sensing mechanisms of bacteria, as bile-mediated virulence factor expression leading to colonization and persistence must be better understood.

## REFERENCES


1. **Al-Jabri, A. A., M. D. Wigg, E. Elias, R. Lamkin, C. O. Mills, and J. S. Oxford.** 2000. *In vitro* anti-HIV-1 virucidal activity of tyrosine-conjugated tri- and dihydroxy bile salt derivatives. *J. Antimicrob. Chemother.* **45:**617–621.
2. **Allen, K. J., and M. W. Griffiths.** 2001. Effect of environmental and chemotactic stimuli on the activity of the *Campylobacter jejuni flaA* $\sigma^{28}$ promoter. *FEMS Microbiol. Lett.* **205:**43–48.
3. **Ananieva, O., I. Nilsson, T. Vorobjova, R. Uibo, and T. Wadstrom.** 2002. Immune responses to bile-tolerant *Helicobacter* species in patients with chronic liver diseases, a randomized population group, and healthy blood donors. *Clin. Diagn. Lab. Immunol.* **9:**1160–1164.
4. **Anderson, L. P.** 2001. New *Helicobacter* species in humans. *Dig. Dis.* **19:**112–115.
5. **Apostolov, E., W. A. Al-Soud, I. Nilsson, I. Kornilovska, V. Usenko, V. Lyzogubov, Y. Gaydar, T. Wadstrom, and A. Ljungh.** 2005. *Helicobacter pylori* and other *Helicobacter* species in gallbladder and liver of patients with chronic cholecystitis detected by immunological and molecular methods. *Scand. J. Gastroenterol.* **40:**96–102.
6. **Atabay, H. I., J. E. Corry, and S. L. On.** 1998. Identification of unusual *Campylobacter*-like isolates from poultry products as *Helicobacter pullorum*. *J. Appl. Microbiol.* **84:**1017–1024.
7. **Baranova, N., and H. Nikaido.** 2002. The BaeSR two-component regulatory system activates transcription of the *yegMNOB (mgtABCD)*

transporter gene cluster in *Escherichia coli* and increases its resistance to novobiocin and deoxycholate. *J. Bacteriol.* **184:**4168–4176.

8. **Begley, M., C. G. M. Gahan, and C. Hill.** 2002. Bile stress response in *Listeria monocytogenes* L028: adaptation, cross-protection and identification of genetic loci involved in bile resistance. *Appl. Environ. Microbiol.* **68:**6005–6012.
9. **Begley, M., C. G. M. Gahan, and C. Hill.** 2005. The interaction between bacteria and bile. *FEMS Microbiol. Rev.* **29:**625–651.
10. **Begley, M., R. D. Sleator, C. G. M. Gahan, and C. Hill.** 2005. Contribution of three bile-associated loci, *bsh, pva,* and *btlB,* to gastrointestinal persistence and bile tolerance of *Listeria monocytogenes. Infect. Immun.* **73:**894–904.
11. **Bina, J., J. Zhu, M. Dziejman, S. Faruque, S. Calderwood, and J. J. Mekalanos.** 2004. ToxR regulon of *Vibrio cholerae* and its expression in vibrios shed by cholera patients. *Proc. Natl. Acad. Sci. USA* **101:**2801–2806.
12. **Bina, J. E., D. Provenzano, C. Wang, X. R. Bina, and J. J. Mekalanos.** 2006. Characterization of the *Vibrio cholerae vexAB* and *vexCD* efflux systems. *Arch. Microbiol.* **186:**171–181.
13. **Blocker, A., K. Komoriya, and S. Aizawa.** 2003. Type III secretion systems and bacterial flagella: insights into their function from structural similarities. *Proc. Natl. Acad. Sci. USA* **100:**3027–3030.
14. **Blum, H. E., L. Stowring, A. Figus, C. K. Montgomery, A. T. Haase, and G. N. Vyas.** 1983. Detection of hepatitis B virus DNA in hepatocytes, bile duct epithelium, and vascular elements by *in situ* hybridization. *Proc. Natl. Acad. Sci. USA* **80:**6685–6688.
15. **Boggs, J. M.** 1987. Lipid intermolecular hydrogen bonding: influence on structural organization and membrane function. *Biochim. Biophys. Acta* **906:**353–404.
16. **Bohr, U. R. M., D. Kuester, F. Meyer, T. Wex, M. Stillert, A. Csepregi, H. Lippert, A. Roessner, and P. Malfertheiner.** 2007. Low prevalence of Helicobacteraceae in gallstone disease and gallbladder carcinoma in the German population. *Clin. Microbiol. Infect.* **13:**525–531.
17. **Bortolini, O., A. Medici, and S. Poli.** 1997. Biotransformations on steroid nucleus of bile acids. *Steroids* **62:**564–577.
18. **Burnens, A. P., J. Stanley, and J. Nicolet.** 1994. Possible association of *Helicobacter pullorum* with lesions of vibrionic hepatitis in poultry, p. 1259–1274. *In* D. G. Newell, J. M. Ketley, and R. A. Feldman (ed.), *Campylobacters, Helicobacters, and Related Organisms.* Plenum Press, New York, NY.
19. **Caygill, C. P., M. J. Hill, M. Braddick, and J. C. Sharp.** 1994. Cancer mortality in chronic typhoid and paratyphoid carriers. *Lancet* **343:**83–84.
20. **Chakrabarti, S. R., K. Chaudhuri, K. Sen, and J. Das.** 1996. Porins of *Vibrio cholerae:* purification and characterization of OmpU. *J. Bacteriol.* **178:**524–530.
21. **Champion, G. A., M. N. Neely, M. A. Brennan, and V. J. DiRita.** 1997. A branch in the ToxR regulatory cascade of *Vibrio cholerae* revealed by characterization of *toxT* mutant strains. *Mol. Microbiol.* **23:**323–331.
22. **Chang, K., S. V. Sosnovtsev, G. Belliot, Y. Kim, L. J. Saif, and K. Y. Green.** 2004. Bile acids are essential for porcine enteric calicivirus replication in association with down-regulation of signal transducer and activator of transcription 1. *Proc. Natl. Acad. Sci. USA* **101:**8733–8738.
23. **Chatterjee, A., P. K. Dutta, and R. Chowdhury.** 2007. Effect of fatty acids and cholesterol present in bile on expression of virulence factors and motility of *Vibrio cholerae. Infect. Immun.* **75:**1946–1953.
24. **Chen, W., J. Liu, and C. Gluud.** 2003. Bile acids for viral hepatitis. *Cochrane Database Syst. Rev.* **2:**CD003181.
25. **Cheville, A. M., K. W. Arnold, C. Buchreiser, C. M. Cheng, and C. W. Casper.** 1996. *rpoS* regulation of acid, heat, and salt tolerance in *Escherichia coli* O157:H7. *Appl. Environ. Microbiol.* **62:**1822–1824.
26. **Chowdhury, R., G. K. Sahu, and J. Das.** 1996. Stress response in pathogenic bacteria. *J. Biosci.* **21:**149–160.
27. **Coleman, J. P., and L. L. Hudson.** 1995. Cloning and characterization of a conjugated bile acid hydrolase gene from *Clostridium perfringens. Appl. Environ. Microbiol.* **61:**2514–2520.
28. **Crosignani, A., K. D. Setchell, P. Invernizzi, A. Larghi, C. M. Rodrigues, and M. Podda.** 1996. Clinical pharmacokinetics of therapeutic bile acids. *Clin. Pharmacokinet.* **30:**333–358.
29. **de Jesus, M. C., A. A. Urban, M. E. Marasigan, and D. E. Barnett Foster.** 2005. Acid and bile-salt stress of enteropathogenic *Escherichia coli* enhances adhesion to epithelial cells and alters glycolipid receptor binding specificity. *J. Infect. Dis.* **192:**1430–1440.
30. **De Smet, I., L. Van Hoorde, N. De Sayer, M. Vande Woestyne, and W. Verstraete.** 1994. *In vitro* study of bile salt hydrolase (BSH) activity of BSH isogenic *Lactobacillus plantarum* 80 strains and estimation of cholesterol lowering through enhanced BSH activity. *Microb. Ecol. Health Dis.* **7:**315–329.
31. **DiGiuseppe, P. A., and T. J. Silhavy.** 2003. Signal detection and target gene induction by the

CpxRA two-component system. *J. Bacteriol.* **185:** 2432–2440.

32. **Duret, G., and A. H. Delcour.** 2006. Deoxycholic acid blocks *Vibrio cholerae* OmpT but not OmpU porin. *J. Biol. Chem.* **281:**19899–19905.
33. **Dussurget, O., D. Cabanes, P. Dehoux, M. Lecuit, C. Buchrieser, P. Glaser, and P. Gossart.** 2002. *Listeria monocytogenes* bile salt hydrolase is a *prfA*-regulated virulence factor involved in the intestinal and hepatic phases of listeriosis. *Mol. Microbiol.* **45:**1095–1106.
34. **Edenharder, R.** 1984. Dehydroxylation of cholic acid at C12 and epimerization at C5 and C7 by *Bacteroides* species. *J. Steroid Biochem.* **21:** 413–420.
35. **Edenharder, R., and K. Mielek.** 1984. Epimerization, oxidation and reduction of bile acids by *Eubacterium lentum. Syst. Appl. Microbiol.* **5:** 287–298.
36. **Flahaut, S., A. Hartke, J. C. Giard, A. Benachour, P. Boutibonnes, and A. Auffray.** 1996. Relationship between stress response towards bile salts, acid and heat treatment in *Enterococcus faecalis. FEMS Microbiol. Lett.* **138:**49–54.
37. **Flahaut, S., J. Frere, P. Boutibonnes, and Y. Auffray.** 1996. Comparison of the bile salts and sodium dodecyl sulfate stress responses in *Enterococcus faecalis. Appl. Environ. Microbiol.* **62:**2416–2420.
38. **Formal, S. B., T. L. Hale, and P. J. Sansonetti.** 1983. Invasive enteric pathogens. *Rev. Infect. Dis.* **5:**S702–S707.
39. **Fox, J. G., F. E. Dewhirst, J. G. Tully, B. J. Paster, L. Yan, N. S. Taylor, M. J. Collins, Jr., P. L. Gorelick, and J. M. Ward.** 1994. *Helicobacter hepaticus* sp. nov., a microaerophilic bacterium isolated from livers and intestinal mucosal scrapings from mice. *J. Clin. Microbiol.* **32:** 1238–1245.
40. **Fox, J. G., F. E. Dewhirst, Z. Shen, Y. Feng, N. S. Taylor, B. J. Paster, R. L. Ericson, C. N. Lau, P. Correa, J. C. Araya, and I. Roa.** 1998. Hepatic *Helicobacter* species identified in bile and gallbladder tissue from Chileans with chronic cholecystitis. *Gastroenterology* **114:**755–763.
41. **Fox, J. G., L. L. Yan, F. E. Dewhirst, B. J. Paster, B. Shames, J. C. Murphy, A. Hayward, J. C. Belcher, and E. N. Mendes.** 1995. *Helicobacter bilis* sp. nov., a novel *Helicobacter* species isolated from bile, livers, and intestines of aged, inbred mice. *J. Clin. Microbiol.* **33:**445–454.
42. **Franklund, C. V., S. F. Baron, and P. B. Hylemon.** 1993. Characterization of the *baiH* gene encoding a bile acid-inducible NADH:flavin oxidoreductase from *Eubacterium* sp. strain VPI 12708. *J. Bacteriol.* **175:**3002–3012.
43. **Gahan, C. G. M., and C. Hill.** 1999. The relationship between acid stress responses and virulence in *Salmonella typhimurium* and *Listeria monocytogenes. Int. J. Food Microbiol.* **50:**93–100.
44. **Geijtenbeek, T. B., D. Kwon, R. Torensma, S. van Vliet, G. van Duijnhoven, J. Middel, I. Cornelissen, H. Nottet, V. KewalRamini, and D. Littman.** 2000. DC-SIGN, a dendritic cell-specific HIV-1-binding protein that enhances trans-infection of T cells. *Cell* **100:**587–597.
45. **Geijtenbeek, T. B., R. Torensma, S. van Vliet, G. van Duijnhoven, G. Adema, Y. van Kooyk, and C. Figdor.** 2000. Identification of DC-SIGN, a novel dendritic cell-specific ICAM-3 receptor that supports primary immune responses. *Cell* **100:**578–585.
46. **Grill, J. P., C. Cayuela, J. M. Antoine, and F. Schneider.** 2000. Isolation and characterization of a *Lactobacillus amylovorus* mutant depleted in conjugated bile salt hydrolase activity: relation between activity and bile salt resistance. *J. Appl. Microbiol.* **89:**553–563.
47. **Grill, J. P., S. Perrin, and F. Schneider.** 2000. Bile toxicity to some bifidobacteria strains: role of conjugated bile salt hydrolase and pH. *Can. J. Microbiol.* **46:**878–884.
48. **Grkovic, S., M. H. Brown, and R. A. Skurray.** 2002. Regulation of bacterial drug export systems. *Microbiol. Mol. Biol. Rev.* **66:**671–701.
49. **Gunn, J. S.** 2000. Mechanisms of bacterial resistance and response to bile. *Microbes Infect.* **2:** 907–913.
50. **Gupta, S., and R. Chowdhury.** 1997. Bile affects production of virulence factors and motility of *Vibrio cholerae. Infect. Immun.* **65:**1131–1134.
51. **Haghjoo, E., and J. E. Galan.** 2007. Identification of a transcriptional regulator that controls intracellular gene expression in *Salmonella* Typhi. *Mol. Microbiol.* **64:**1549–1561.
52. **Hale, T. L.** 1991. Genetic basis of virulence in *Shigella* species. *Microbiol. Rev.* **55:**206–224.
53. **Hardison, W. G.** 1978. Hepatic taurine concentration and dietary taurine as regulators of bile acid conjugation with taurine. *Gastroenterology* **75:** 71–75.
54. **Hardy, J., K. P. Francis, M. DeBoer, P. Chu, K. Gibbs, and C. H. Contag.** 2004. Extracellular replication of *Listeria monocytogenes* in the murine gallbladder. *Science* **303:**851–853.
55. **Haruna, Y., T. Kanda, M. Honda, T. Takao, and N. Hayashi.** 2001. Detection of hepatitis C virus in the bile and bile duct epithelial cells of hepatitis C virus-infected patients. *Hepatology* **33:**977–980.
56. **Herold, B. C., R. Kirkpatrick, D. Marcellino, A. Travelstead, V. Pilipenko, H. Krasa,**

**J. Bremer, L. J. Dong, and M. D. Cooper.** 1999. Bile salts. Natural detergents for the prevention of sexually transmitted diseases. *Antimicrob. Agents Chemother.* **43:**745–751.

57. **Hofmann, A. F.** 1989. Overview of bile secretion, p. 549–566. *In Handbook of Physiology,* section 6, vol. 3. *The Gastrointestinal System. Salivary, Gastric, Pancreatic, and Hepatobiliary Secretion.* American Physiological Society, Bethesda, MD.
58. **Hofmann, A. F.** 1998. Bile secretion and the enterohepatic circulation of bile acids, p. 937–948. *In* M. Feldman, B. F. Scharschmidt, and M. H. Sleisenger (ed.), *Sleisenger and Fordtran's Gastrointestinal and Liver Disease,* 6th ed. W. B. Saunders, Co., Philadelphia, PA.
59. **Hofmann, A. F.** 1999. Bile acids: the good, the bad, and the ugly. *News Physiol. Sci.* **14:**24–29.
60. **Hofmann, A. F., and A. Roda.** 1984. Physicochemical properties of bile acids and their relationship to biological properties: an overview of the problem. *J. Lipid Res.* **25:**1477–1489.
61. **Hofmann, A. F., and K. J. Mysels.** 1992. Bile acid solubility and precipitation *in vitro* and *in vivo:* the role of conjugation, pH and $Ca^{2+}$ ions. *J. Lipid Res.* **33:**617–626.
62. **Hu, J., M. B. Gardner, and C. J. Miller.** 2000. Simian immunodeficiency virus rapidly penetrates the cervicovaginal mucosa after intravaginal inoculation and infects intraepithelial dendritic cells. *J. Virol.* **74:**6087–6095.
63. **Hung, D. T., and J. J. Mekalanos.** 2005. Bile acids induce cholera toxin expression in *Vibrio cholerae* in a ToxT-independent manner. *Proc. Natl. Acad. Sci. USA* **102:**3028–3033.
64. **Hung, D. T., J. Zhu, D. Sturtevant, and J. J. Mekalanos.** 2006. Bile acids stimulate biofilm formation in *Vibrio cholerae. Mol. Microbiol.* **59:** 193–201.
65. **Jeong, K. C., M. Y. Kang, C. Heimke, J. A. Shere, I. Erol, and C. W. Kaspar.** 2007. Isolation of *Escherichia coli* O157:H7 from the gallbladder of inoculated and naturally-infected cattle. *Vet. Microbiol.* **119:**339–345.
66. **Johnson, L. R.** 1998. Bile secretion and gallbladder function, p. 465–471. *In* L. R. Johnson (ed.), *Essential Medical Physiology,* 2nd ed. Lippincott–Raven, Philadelphia, PA.
67. **Kaper, J. B., J. G. Morris, and M. M. Levine.** 1995. Cholera. *Clin. Microbiol. Rev.* **8:**48–86.
68. **Knarreborg, A., R. M. Engberg, S. K. Jensen, and B. B. Jensen.** 2002. Quantitative determination of bile salt hydrolase activity in bacteria isolated from the small intestine of chickens. *Appl. Environ. Microbiol.* **68:**6425–6428.
69. **Lacroix, F. J., C. Ayoyne, C. Pinault, M. Y. Popoff, and P. Pardon.** 1995. *Salmonella typhimurium* TnphoA mutants with increased sensitivity to biological and chemical detergents. *Res. Microbiol.* **146:**659–670.
70. **Lefkowitch, J. H.** 1993. Pathological diagnosis of chronic hepatitis C: a multicenter comparative study with chronic hepatitis B. *Gastroenterology* **104:**595–603.
71. **Lin, J., and A. Martinez.** 2006. Effect of efflux pump inhibitors on bile resistance and *in vivo* colonization of *Campylobacter jejuni. J. Antimicrob. Chemother.* **58:**966–972.
72. **Lin, J., C. Cagliero, B. Guo, Y. Barton, M. Maurel, S. Payot, and Q. Zhang.** 2005. Bile salts modulate expression of the CmeABC multidrug efflux pump in *Campylobacter jejuni. J. Bacteriol.* **187:**7417–7424.
73. **Lin, J., M. Akiba, O. Sahin, and Q. Zhang.** 2005. CmeR functions as a transcriptional repressor for the multidrug efflux pump CmeABC in *Campylobacter jejuni. Antimicrob. Agents Chemother.* **49:**1067–1075.
74. **Lin, J., O. Sahin, L. Overbye Michel, and Q. Zhang.** 2003. Critical role of multidrug efflux pump CmeABC in bile resistance and in vivo colonization of *Campylobacter jejuni. Infect. Immun.* **71:**4250–4259.
75. **Lomovskaya, O., and K. Lewis.** 1992. Emr, an *Escherichia coli* locus for multidrug resistance. *Proc. Natl. Acad. Sci. USA* **89:**8938–8942.
76. **Lomovskaya, O., K. Lewis, and A. Matin.** 1995. EmrR is a negative regulator of the *Escherichia coli* multidrug resistance pump EmrAB. *J. Bacteriol.* **177:**2328–2334.
77. **Luo, N., O. Sahin, J. Lin, L. O. Michel, and Q. Zhang.** 2003. In vivo selection of *Campylobacter* isolates with high levels of fluoroquinolone resistance associated with *gyrA* mutations and the function of the CmeABC efflux pump. *Antimicrob. Agents Chemother.* **47:**390–394.
78. **Ma, D., D. N. Cook, J. E. Hearst, and H. Nikaido.** 1994. Efflux pumps and drug resistance in gram-negative bacteria. *Trends Microbiol.* **2:** 489–493.
79. **Macdonald, I. A., and D. M. Hutchison.** 1982. Epimerization versus dehydroxylation of the 7α-hydroxyl-group of primary bile acids: competitive studies with *Clostridium absonum* and 7α-dehydroxylating bacteria (*Eubacterium* sp.). *J. Steroid Biochem.* **17:**295–303.
80. **Maurer, K. J., A. B. Rogers, Z. Ge, A. J. Wiese, M. C. Carey, and J. G. Fox.** 2006. *Helicobacter pylori* and cholesterol gallstone formation in C57L/J mice: a prospective study. *Am. J. Physiol. Gastrointest. Liver Physiol.* **290:**G175–G182.
81. **McPhee, M. S., and N. J. Greenberger.** 1987. Diseases of the gallbladder and bile duct, p. 1358–1362. *In* R. T. Harrison (ed.), *Harrison's*

*Principles of Internal Medicine,* 11th ed. McGraw-Hill, New York, NY.

82. **Merrell, D. S., S. M. Butler, F. Qadri, N. A. Dolganov, A. Alam, M. B. Cohen, S. B. Calderwood, G. K. Schoolnik, and A. Camilli.** 2002. Host-induced epidemic spread of the cholera bacterium. *Nature* **417:**642–645.
83. **Mohapatra, N. P., S. Soni, B. L. Bell, R. Warren, R. K. Ernst, A. Muszynski, R. W. Carlson, and J. S. Gunn.** 2007. Identification of an orphan response regulator required for the virulence of *Francisella* spp. and transcription of pathogenicity island genes. *Infect. Immun.* **75:** 3305–3314.
84. **Mower, H. F., R. M. Ray, R. Shoff, G. N. Stemmermann, A. Nomura, G. A. Glober, S. Kamiyama, A. Shimada, and H. Yamakawa.** 1979. Fecal bile acids in two Japanese populations with different colon cancer risks. *Cancer Res.* **39:**328–331.
85. **Moxley, R. A.** 2004. *Escherichia coli* O157:H7: an update on intestinal colonization and virulence mechanisms. *Anim. Health Res. Rev.* **5:**15–33.
86. **Naarding, M. A., A. M. Dirac, I. S. Ludwig, D. Speijer, S. Lindquist, E. L. Vestman, M. J. Stax, T. B. Geijenbeek, G. Pollakis, O. Hernell, and W. A. Paxton.** 2006. Bile salt-stimulated lipase from human milk binds DC-SIGN and inhibits human immunodeficiency virus type 1 transfer to CD4$^+$ T cells. *Antimicrob. Agents Chemother.* **50:**3367–3374.
87. **Naarding, M. A., I. S. Ludwig, F. Groot, B. Berkhout, T. B. Geijenbeek, G. Pollakis, and W. A. Paxton.** 2005. Lewis X component in human milk binds DC-SIGN and inhibits HIV-1 transfer to CD4$^+$ T lymphocytes. *J. Clin. Investig.* **115:**3256–3264.
88. **Nachamkin, I., B. M. Allos, and T. Ho.** 1998. *Campylobacter* species and Guillain-Barré syndrome. *Clin. Microbiol. Rev.* **11:**555–567.
89. **Olive, A. J., R. Kenjale, M. Espina, D. S. Moore, W. L. Picking, and W. D. Picking.** 2007. Bile salts stimulate recruitment of IpaB to the *Shigella flexneri* surface, where it colocalizes with IpaD at the tip of the type III secretion needle. *Infect. Immun.* **75:**2626–2629.
90. **Opleta, K., J. D. Butzner, E. A. Shaffer, and D. G. Gall.** 1988. The effect of protein-calorie malnutrition on the developing liver. *Pediatr. Res.* **23:**505–508.
91. **Paumgartner, G., and T. Sauerbruch.** 1991. Gallstones: pathogenesis. *Lancet* **338:**1117–1121.
92. **Pearce, M. C., C. Jenkins, L. Vali, A. W. Smith, H. I. Knight, T. Cheasty, H. R. Smith, G. J. Gunn, M. E. J. Woolhouse, S. G. B. Amyes, and G. Frankel.** 2004. Temporal shedding pattern and virulence factors of *Escherichia coli* O157 serogroups O26, O103, O111, O145, and O157 in a cohort of beef calves and their dams. *Appl. Environ. Microbiol.* **70:**1708–1716.
93. **Pope, L. M., K. E. Reed, and S. M. Payne.** 1995. Increased protein secretion and adherence to HeLa cells by *Shigella* spp. following growth in the presence of bile salts. *Infect. Immun.* **63:** 3642–3648.
94. **Pope, M., S. Gezelter, N. Gallo, L. Hoffman, and R. M. Steinman.** 1995. Low levels of HIV-1 infection in cutaneous dendritic cells promote extensive viral replication upon binding to memory CD4$^+$ T cells. *J. Exp. Med.* **182:** 2045–2056.
95. **Prouty, A. M., and J. S. Gunn.** 2000. *Salmonella enterica* serovar Typhimurium invasion is repressed in the presence of bile. *Infect. Immun.* **68:**6763–6769.
96. **Prouty, A. M., and J. S. Gunn.** 2003. Comparative analysis of *Salmonella enterica* serovar Typhimurium biofilm formation on gallstones and on glass. *Infect. Immun.* **71:**7154–7158.
97. **Prouty, A. M., I. E. Brodsky, J. Manos, R. Belas, S. Falkow, and J. S. Gunn.** 2004. Transcriptional regulation of *Salmonella enterica* serovar Typhimurium genes by bile. *FEMS Immunol. Med. Microbiol.* **41:**177–185.
98. **Prouty, A. M., I. E. Brodsky, S. Falkow, and J. S. Gunn.** 2004. Bile-salt-mediated induction of antimicrobial and bile resistance in *Salmonella typhimurium*. *Microbiology* **150:**775–783.
99. **Prouty, A. M., J. C. Van Velkinburgh, and J. S. Gunn.** 2002. *Salmonella enterica* serovar Typhimurium resistance to bile: identification and characterization of the *tolQRA* cluster. *J. Bacteriol.* **184:**1270–1276.
100. **Prouty, A. M., W. H. Schwesinger, and J. S. Gunn.** 2002. Biofilm formation and interaction with the surfaces of gallstones by *Salmonella* spp. *Infect. Immun.* **70:**2640–2649.
101. **Provenzano, D., and K. E. Klose.** 2000. Altered expression of the ToxR-regulated porins OmpU and OmpT diminishes *Vibrio cholerae* bile resistance, virulence factor expression, and intestinal colonization. *Proc. Natl. Acad. Sci. USA* **97:**10220–10224.
102. **Provenzano, D., D. A. Schuhmacher, J. L. Barker, and K. E. Klose.** 2000. The virulence regulatory protein ToxR mediates enhanced bile resistance in *Vibrio cholerae* and other pathogenic *Vibrio* species. *Infect. Immun.* **68:**1491–1497.
103. **Putman, M., H. W. van Veen, and W. N. Konings.** 2000. Molecular properties of bacterial multidrug transporters. *Microbiol. Mol. Biol. Rev.* **64:**672–693.

**104. Ramos-Morales, F., A. I. Prieto, C. R. Beuzon, D. W. Holden, and J. Casadesus.** 2003. Role for *Salmonella enterica* enterobacterial common antigen in bile resistance and virulence. *J. Bacteriol.* **185:**5328–5332.

**105. Reed, R. P., F. O. Wegerhoff, and A. D. Rothberg.** 1996. Bacteraemia in malnourished African children. *Ann. Trop. Paediatr.* **16:**61–68.

**106. Rivera-Amill, V., B. J. Kim, J. Seshu, and M. E. Konkel.** 2001. Secretion of the virulence-associated *Campylobacter* invasion antigens from *Campylobacter jejuni* requires a stimulatory signal. *J. Infect. Dis.* **183:**1607–1616.

**107. Schlech, W. F.** 2000. Foodborne listeriosis. *Clin. Infect. Dis.* **31:**770–775.

**108. Schuhmacher, D. A., and K. E. Klose.** 1999. Environmental signals modulate ToxT-dependent virulence factor expression in *Vibrio cholerae*. *J. Bacteriol.* **181:**1508–1514.

**109. Shimeles, D., and S. Lulseged.** 1994. Clinical profile and pattern of infection in Ethiopian children with severe protein-energy malnutrition. *East Afr. Med. J.* **71:**264–267.

**110. Sjovall, J.** 1959. Dietary glycine and taurine on bile acid conjugation on man; bile acids and steroids 75. *Proc. Soc. Exp. Biol. Med.* **100**(4): 676–678.

**111. Sleator, R. D., H. H. Wemekamp-Kamphuis, C. G. M. Gahan, T. Abee, and C. Hill.** 2005. A PrfA-regulated bile exclusion system (BilE) is a novel virulence factor in *Listeria monocytogenes*. *Mol. Microbiol.* **55:**1183–1195.

**112. Slutsker, L., S. F. Altekruse, and D. L. Swerdlow.** 1998. Foodborne diseases. Emerging pathogens and trends. *Infect. Dis. Clin. N. Am.* **12:**199–216.

**113. Stanley, J.** 1994. *Helicobacter pullorum* sp. nov.—genotype and phenotype of a new species isolated from poultry and from human patients with gastroenteritis. *Microbiology* **140:** 3441–3449.

**114. Stintzi, A., D. Marlow, K. Palyada, H. Naikare, R. Panciera, L. Whitworth, and C. Clarke.** 2005. Use of genome-wide expression profiling and mutagenesis to study the intestinal lifestyle of *Campylobacter jejuni*. *Infect. Immun.* **73:**1797–1810.

**115. Stoffregen, W. C., J. F. Pohlenz, and E. A. Dean-Nystrom.** 2004. *Escherichia coli* O157:H7 in the gallbladders of experimentally infected calves. *J. Vet. Diagn. Investig.* **16:**79–83.

**116. Tauxe, R. V.** 2002. Emerging foodborne pathogens. *Int. J. Food Microbiol.* **78:**31–41.

**117. Thanassi, D. G., L. W. Cheng, and H. Nikaido.** 1997. Active efflux of bile salts by *Escherichia coli*. *J. Bacteriol.* **179:**2512–2518.

**118. Tiwari, S. K., A. A. Khan, M. Ibrahim, M. A. Habeeb, and C. M. Habibullah.** 2006. *Helicobacter pylori* and other *Helicobacter* species DNA in human bile samples from patients with various hepato-biliary diseases. *World J. Gastroenterol.* **12:**2181–2186.

**119. Torres, A. G., C. B. Tutt, L. Duval, V. Popov, A. B. Nasr, J. Michalski, and I. C. Scaletsky.** 2007. Bile salts induce expression of the afimbrial LDA adhesin of atypical enteropathogenic *Escherichia coli*. *Cell. Microbiol.* **9:** 1039–1049.

**120. Trabulsi, L. R., R. Keller, and T. A. Tardelli Gomes.** 2002. Typical and atypical enteropathogenic *Escherichia coli*. *Emerg. Infect. Dis.* **8:**508–513.

**121. Van Eldere, J., P. Celis, G. De Pauw, E. Lesaffre, and H. Eyssen.** 1996. Tauroconjugation of cholic acid stimulates 7α-dehydroxylation by fecal bacteria. *Appl. Environ. Microbiol.* **62:**656–661.

**122. Van Velkinburgh, J. C., and J. S. Gunn.** 1999. PhoP-PhoQ-regulated loci are required for enhanced bile resistance in *Salmonella* spp. *Infect. Immun.* **67:**1614–1622.

**123. Vazquez-Boland, J. A., M. Kuhn, P. Berche, T. Chakraborty, G. Dominguez-Bernal, W. Goebel, B. Gonzalez-Zorn, J. Wehland, and J. Kreft.** 2001. *Listeria* pathogenesis and molecular virulence determinants. *Clin. Microbiol. Rev.* **14:**584–640.

**124. Ward, J. M., J. G. Fox, M. R. Anver, D. C. Haines, C. V. George, M. J. Collins, P. L. Gorelick, K. Nagashima, M. A. Gonda, and R. V. Gilden.** 1994. Chronic active hepatitis and associated liver tumors in mice caused by a persistent bacterial infection with a novel *Helicobacter* species. *J. Natl. Cancer Inst.* **86:**1222–1227.

**125. Welton, J. C., J. S. Marr, and S. M. Friedman.** 1979. Association between hepatobiliary cancer and typhoid carrier state. *Lancet* **i:**791–794.

**126. Xu, Q., M. Dziejman, and J. J. Mekalanos.** 2003. Determination of the transcriptome of *Vibrio cholerae* during intraintestinal growth and midexponential phase *in vitro*. *Proc. Natl. Acad. Sci. USA* **100:**1286–1291.

**127. Zgurskaya, H. I., and H. Nikaido.** 1999. Bypassing the periplasm: reconstitution of the AcrAB multidrug efflux pump of *Escherichia coli*. *Proc. Natl. Acad. Sci. USA* **96:**7190–7195.

# THE INFLUENCE OF HELMINTHS ON IMMUNOLOGICAL DISEASES

*Joel V. Weinstock and David M. Elliott*

# 11

Ethnic and geographic variations in immunological diseases like inflammatory bowel disease (IBD) and multiple sclerosis suggest that environmental factors are important for disease pathogenesis. Improved hygiene leading to a lack of exposure to parasitic worms (helminths) may be one factor causing the increased prevalence of these diseases of immune dysregulation. Helminths are powerful modulators of host mucosal and systemic immunity. Helminth exposure protects animals from experimental colitis and other immunological diseases and can be used to treat ongoing murine and human diseases. Helminths induce mucosal T cells to make Th2 and regulatory cytokines that participate in the protective process. Helminths affect pathways of innate immunity and induce various regulatory T-cell subsets that limit immune reactivity. This chapter reviews the various concepts outlined above and explores the implications of expanding industrialization on the demographics of immunological diseases worldwide.

*Joel V. Weinstock,* Division of Gastroenterology-Hepatology, Department of Internal Medicine, Tufts Medical Center, Boston, MA 02111. *David M. Elliott,* Division of Gastroenterology-Hepatology, Department of Internal Medicine, University of Iowa, Iowa City, IA 52240.

## THE NATURE OF HELMINTHS

Helminths are multicellular parasitic worms that live in humans, other animals, or plants to derive nourishment from their host. The parasitic helminths of humans belong to three major groups. The annelids are the leeches, which attach to the skin to feed. Worms that invade the body include the nematodes, which are nonsegmented roundworms that range in size from a few millimeters to over 30 cm in length. They usually live in the small bowel or colon (e.g., *Trichuris trichiura* [whipworm], *Enterobius vermicularis* [pinworm], *Ascaris lumbricoides* [roundworm], *Necator americanus* and *Ancylostoma duodenale* [hookworms], and *Strongyloides* [threadworm]). The other major group is called the platyhelminths and comprises the cestodes and trematodes. The cestodes live in the small intestine and have flat, ribbon-like bodies, and they are thus usually called tapeworms. Common examples of human tapeworms include *Taenia saginata* (beef tapeworm), *Taenia solium* (pig tapeworm), *Diphyllobothrium latum* (fish tapeworm), and *Hymenolepis nana.* Trematodes comprise the blood (schistosomes), liver *(Fasciola, Clonorchis,* and *Opisthorchis),* intestinal *(Fasciolopsis* and *Echinostoma),* and lung *(Paragonimus)* flukes. They are flat, leaf-shaped organisms.

*Food-Borne Microbes: Shaping the Host Ecosystem,* Edited by L.-A. Jaykus, H. H. Wang, and L. S. Schlesinger, © 2009 ASM Press, Washington, DC

Fossilized remains of dinosaurs suggest that they harbored intestinal worms, showing that helminths have colonized other animals for millions of years. Helminth ova have been identified within petrified human waste dating back more than 10,000 years (27), revealing that humans and helminths have coexisted for at least many thousands of years. Many helminths have complex life cycles, but all require successful colonization of a host. Most helminths are quite selective in host species requirements. Thus, humans and their helminths have coexisted and undergone mutual adaptation over the millennia.

Helminth colonization is most common in children who live in warm climates and are subject to poor sanitation. Even today, more than one-third of people around the world harbor helminths. The infective forms of both animal and human helminthic parasites are spread through contact with contaminated soil, water, or food. Children living on farms may readily encounter animal helminths (e.g., *Trichuris suis*) whose ova are in the feces of the farm animals. Household pets also are a source of helminth exposure, such as to the dog worm *Toxocara canis* (47, 48). Helminth exposure was nearly universal until the development of modern-day sanitation. In the United States, helminth colonization has steadily declined except in people living in medically underserved regions (29) and in recent immigrants from less developed countries (55).

The human immune system frequently displays distinct patterns of immune reactivity when challenged by various invading microorganisms. The Th1 response most commonly is associated with vigorously activated T cells and macrophages that produce cytokines, such as gamma interferon (IFN-γ) and interleukin-12 (IL-12), to eradicate bacteria and viruses. The Th2 response is an immunological reaction commonly associated with eosinophilia and production of immunoglobulin E and cytokines like IL-4, IL-5, IL-9, and IL-13. Most helminthic organisms stimulate the host to vigorously and selectively produce Th2 cytokines while blocking Th1 cytokine responses. This usually leads to strong immunoglobulin E, mast cell, and eosinophil responses. A Th2, as opposed to Th1, cytokine response (4) is host protective against intestinal worms. IL-4 and IL-13 in particular help expel worms, although their relative importance in worm expulsion varies among different worm species. With regard to the worm, a "healthy" host response is one that only limits worm colonization, thus preventing overcolonization and injury to the host. This is mutually beneficial for the helminth and the host.

## THE RISE OF IMMUNOLOGICAL DISEASES

The prevalence of immunological diseases like IBD, type 1 diabetes, multiple sclerosis, and asthma has risen substantially in industrialized countries over the last 60 years (3). From 3 to 5% of the population in the United States suffers from one or more of these diseases. Less developed countries (like India) with improving living standards are now experiencing a rapid rise in these diseases. Profound increases in such diseases in stable populations suggest that environmental factors underlie these increases.

The growth of IBDs like Crohn's disease and ulcerative colitis illustrates the point. IBDs probably result from loss of immune tolerance to the normal intestinal flora (17). The genetic influence on IBD was first demonstrated in twin studies in which monozygotic as opposed to dizygotic twin pairs separated at birth were shown to have a higher concordance rate for Crohn's disease and, to a lesser extent, ulcerative colitis (28, 65). Specific gene variations modulate the risk for developing IBD, but none of these genes are the underlying cause of these conditions. For instance, people with selective defects in the *NOD2 (CARD15)* gene, which encodes an important intracellular signaling molecule of the innate immune system, are more likely to get Crohn's disease (1, 14, 51), whereas people with dysfunctional coding variants of the receptor for the proinflammatory cytokine IL-23 are relatively protected from this disease. However, most people with a *NOD2* defect, which is common in people from Western nations, never acquire Crohn's disease. Other genes also influence disease susceptibility, but none account for the rapid worldwide rise in IBD frequency in the 20th century.

Environmental factors affect the risk for IBD (37). Appendicitis followed by appendectomy lowers the incidence of ulcerative colitis (2, 16, 52), whereas cigarette smoking enhances chances for Crohn's disease (9). In some patients, an immune response to pathogenic organisms like *Entamoeba histolytica* or cytomegalovirus provokes IBD.

There are striking geographic variations in IBD frequency. Early studies showed that IBD was more common in northern than in southern regions of the United States and Europe (58, 61). Migration studies suggest that children born to individuals who relocate from regions with a low frequency of Crohn's disease or ulcerative colitis to areas with a high disease prevalence acquire greater disease risk (10, 31, 50). IBD prevalence has increased in Western Europe and North America, and IBD now affects about 1 in 250 people in some affluent regions (7, 38). The original association of IBD mostly with affluent Jewish populations in the United States and Europe now is much less striking, since IBD now is commonly diagnosed in people of African-American and Hispanic origin. Jewish immigrants to Israel from Western Europe and North America have a higher frequency of IBD than do Jewish immigrants from Mediterranean and Middle Eastern countries (46).

The "hygiene hypothesis" was proposed more than 20 years ago to explain the rise of immunological diseases in industrial countries. The hypothesis states that immunological diseases may be an unanticipated consequence of improvements in public health and hygiene. Epidemiological data related to IBD support this hypothesis. Patients with Crohn's disease are likely to live in more hygienic environments and have houses with indoor plumbing and hot water (18, 26). IBD is more common in urban areas than in rural areas (19) and less common in people having jobs exposing them to dirt (60). Past prisoner-of-war status or combat in tropical regions lowers the risk of IBD (15). Similar epidemiological findings have been reported for multiple sclerosis and other immunological diseases.

## THE CASE FOR DEWORMING AS A RISK FACTOR

People living in environments with poor sanitation frequently harbor worms. Deworming of children and adults, associated with improved public hygiene, could be one of the factors leading to the rise in immunological diseases. Humans and helminths have coexisted throughout our evolution. It is well appreciated that helminths strongly modulate the host's immune system to promote worm survival. However, the immune response to helminths also alters host immunity to immunologic stimuli unrelated to the parasite. Human studies suggest that exposure to various helminths decreases T-cell signal transduction, lowers expression of CD28 with increased expression of CTLA4, reduces proliferation to recall antigens and mitogens, decreases delayed-type hypersensitivity responses (8), and raises IL-10 levels (6). In mice, the Th2 response to helminths can deviate Th1 antigenic immunity toward Th2 (35, 49, 54). People carrying helminths can show immune bias away from the Th1 response normally elicited with tetanus vaccination (53) or in vitro mitogen stimulation (6). Even maternal helminth infection during pregnancy skews the normal Th1 immune reactivity of newborn children (39). Maternal helminth infection appears to protect against infantile eczema (20). Thus, deworming of the population after thousands of years of coevolution may have negative consequences for the development of the human immune system, particularly during childhood.

## HELMINTHS PREVENT AND REVERSE IMMUNOLOGICAL DISEASE IN ANIMAL MODELS

Animal experimentation shows that helminths can prevent the onset of various immunological diseases and reverse ongoing pathology. A Th1-type colitis can be induced in rodents by placing trinitrobenzene sulfonic acid (TNBS) in alcohol in the colon. Rodents are protected from TNBS-induced colitis by infection with the intestinal roundworm *Heligmosomoides polygyrus* or *Trichuris muris* or the vascular fluke *Schistosoma mansoni* (43), or by treatment with schistosome ova (21). Other helminths, like the intestinal roundworm *Trichinella spiralis* (33) or the rat intestinal tapeworm *Hymenolepis diminuta,* also protect. *T. muris* or *H. polygyrus* infection or exposure to nonviable schistosome ova can prevent or reverse the chronic Th1-type colitis that develops in mice deficient in IL-10 (22, 23). Helminth infection also protects rodents with immunological diseases that simulate conditions such as experimental autoimmune encephalomyelitis (36, 57), asthma (34, 68), rheumatoid arthritis (41), type 1 diabetes (69), food allergy (5), and Graves' hyperthyroidism (45).

## MORE EPIDEMIOLOGICAL AND CLINICAL DATA SUPPORTING THE PREMISE

Growing data support a link between worm carriage and protection from multiple sclerosis. *Trichuris trichiura* is a common human helminth with global prevalence. The rising prevalence of multiple sclerosis worldwide correlates with the decreasing prevalence of *T. trichiura* in various regions (24). Multiple sclerosis patients acquiring helminthic infection have fewer disease exacerbations and magnetic resonance imaging changes than do uninfected patients. Also, their peripheral blood cells make more regulatory cytokines (IL-10 and transforming growth factor β [TGF-β]) and less of the Th1 cytokine IL-12, and they display higher clonal frequencies of peripheral blood T cells expressing regulatory phenotypes (12). The latter provides a potential mechanism for this protection.

Epidemiological data suggest that helminth infections correlate with a lower risk of allergic disorders and sensitization. Infection with *Schistosoma haematobium* stimulates a parasite-specific IL-10 response associated with suppression of atopic reactivity in children living in Gabon (66). A case control study in Ethiopia showed a lower frequency of asthma in people infected with hookworm than in uninfected people (56). A randomized, controlled prospective study noted that repeated antihelminthic treatment of chronically helminth-infected Gabonese schoolchildren resulted in increased allergic sensitivity to house dust mites in the active-treatment group. The latter study provides evidence that helminths mediate the suppression of atopy (67). A study in Vietnam suggested that hookworm was protective against allergic sensitization (25).

Exposure to helminths has therapeutic potential for people with IBD and other immunological disorders. While presently there are no epidemiological data directly linking worm carriage with prevention of IBD, clinical improvement was noted in a double-blind clinical study on ulcerative colitis and an open-label study on Crohn's disease that used live ova from the porcine whipworm *(Trichuris suis)* as an oral therapeutic intervention (62–64). Another open-label study on Crohn's disease using human hookworm cutaneously administered showed efficacy (13). These investigations collectively suggest that exposure to helminths can modulate disease activity in IBD and that natural exposure to animal as well as human helminths can affect disease susceptibility.

## PROPOSED IMMUNOLOGICAL MECHANISMS OF PROTECTION

Various effector cytokines can drive immunological diseases. These include the Th1 (e.g., IL-12 and IFN-γ), Th2 (e.g., IL-4 and IL-13), and Th17 (IL-17A) pathways of inflammation. The Th17 pathway is a newly described proinflammatory branch of the immune system implicated in the immunopathology of IBD, multiple sclerosis, asthma, and various animal models of such diseases.

Extensive experimentation conducted in murine models of IBD has elucidated some of the mechanisms of worm-induced immunoregulation. Worm infestation dampens Th1, Th2, and/or Th17-type reactivity and modulates lymphocyte responsiveness to various antigens and mitogens. This suggests that worms trigger various immunoregulatory circuits to regulate or prevent intestinal inflammation.

Colonization with intestinal helminths protects mice from TNBS-induced colitis, which is a Th1-driven disease. Intestinal helminths inhibit the intestinal mucosa from mounting a Th1-type cytokine response (IFN-γ and IL-12 p40) in the distal bowel. Mucosal Th1 pathways also are blocked in healthy wild-type mice without colitis. Several mechanisms mediate this process of protection/regulation. Worm infection induces mucosal lamina propria mononuclear cells to produce Th2 cytokines (e.g., IL-4 and IL-13) and large quantities of regulatory factors (e.g., IL-10 and TGF-β). In TNBS colitis, abrogation of the Th2 pathway blocks worm protection, showing the importance of Th2 cytokines in this IBD animal model (21). In worm-infected mice, blocking the IL-10 receptor partly restores IFN-γ and IL-12 production in lamina propria mononuclear cells from mice with TNBS colitis or from healthy wild-type mice, suggesting that IL-10 is important in regulation of the Th1 pathway. However, IL-10 is not necessarily essential for control of the intestinal Th1 pathway of inflammation or colitis, since mice lacking IL-10 get colitis, and worm colonization curtails intestinal IFN-γ and IL-12 p40 production and prevents or reverses the mucosal inflammation in this disease (22).

Colitis resulting from IL-10 deficiency is driven in part through the Th17 pathway. IL-12 p40 is a molecule shared by IL-12 and IL-23. IL-23 and other cytokines induce and sustain the mucosal IL-17 response. Worm infection blocks mucosal IL-23 and IL-17 secretion, suggesting that regulation of IL-17 is another important mechanism of action (21a).

Regulatory T cells are functionally important in limiting inflammation in various animal models of IBD (44). There are several regulatory T-cell phenotypes. Helminths induce regulatory T cells in their hosts. CD8 T cells help limit colitis in IL-10 KO colitis. T cells from the mesenteric lymph nodes of *H. polygyrus*-colonized IL-10-deficient mice, unlike mesenteric lymph node cells from worm-free IL-10 littermates, stop colitis when transferred into IL-10-deficient recipients (22). Lamina propria T cells from *H. polygyrus*-infected mice, in contrast to lamina propria T cells from uninfected controls, strongly suppress proliferation of both $CD4^+$ and $CD8^+$ effector T cells. This regulatory activity lies exclusively within the lamina propria $CD8^+$ T-cell subset, which must contact the responder T cells to mediate their effect (42). This regulatory activity requires class I interactions, but neither regulatory cell development nor function requires IL-10 or TGF-β. $CD8^+$ T-cell-mediated suppression has a role in several animal models of inflammation, like in the disease model of multiple sclerosis (32).

Lipopolysaccharides (LPS) are components of most gram-negative bacteria. LPS interact with a specific receptor (Toll-like receptor 4), usually to drive secretion of proinflammatory molecules from various cell types. The intestinal microflora releases huge amounts of LPS, but the normal intestinal mucosa is unresponsive to this molecule. *H. polygyrus* infection induces a small subset of mucosal T cells to express Toll-like receptor 4, which, in turn, produces TGF-β and, to a lesser extent, IL-10, upon LPS exposure. LPS does not stimulate these cells to make proinflammatory cytokines (30). Mucosal injury allows LPS from commensal bacteria to enter the mucosa. In the presence of worms, LPS may help limit activation of adaptive immunity through driving TGF-β-producing regulatory T cells.

Mice with a T-cell-specific defect in TGF-β signaling develop colitis and are unable to appropriately limit either mucosal Th1 or Th2 reactivity. Worms cannot limit Th1 responsiveness or control the colitis in this animal model of IBD, suggesting that TGF-β signaling via the mucosal T cells is critical in the protective process. TGF-β, in conjunction with some other cytokines, is important for driving development of peripherally induced regulatory T cells, which have a vital role in worm protection and reversal of colitis.

Dextran sodium sulfate administered to rodents induces intestinal inflammation through a T-cell-independent process. This is a somewhat different model of IBD, since the inflammation is not driven by adaptive immunity. Schistosomiasis protects BALB/c mice, but not C57BL/6 mice, from this form of IBD. The schistosome worms, not the eggs, mediate the protection. Resistance to colitis is strictly macrophage dependent and is independent of regulatory T cells or the usual regulatory cytokines (e.g., IL-10 or TGF-β) (59).

Rodents infected with various nematode parasites have decreased susceptibility to allergic responses (5, 34, 40, 68). Protection in animal models of asthma also may involve IL-10 and regulatory-type T cells. Thus, it appears that helminths trigger various immunoregulatory pathways to control aberrant inflammation (Fig. 1).

## POTENTIAL NEGATIVE CONSEQUENCES OF WORM COLONIZATION

Helminth colonization can have negative consequences. Some worms can cause disease. For instance, *Schistosoma mansoni* can induce liver fibrosis and portal hypertension that can lead to gastrointestinal bleeding. *Ascaris lumbricoides* can migrate up the biliary tree, causing stone formation, duct obstruction, and infection. Other worms, like hookworms, have been associated with iron deficiency anemia. However, the vast majority of mild to moderate worm infestations rarely manifest as disease.

Recent experiments using murine and larger animal models suggest that worm infection can interfere with vaccine efficacy. This

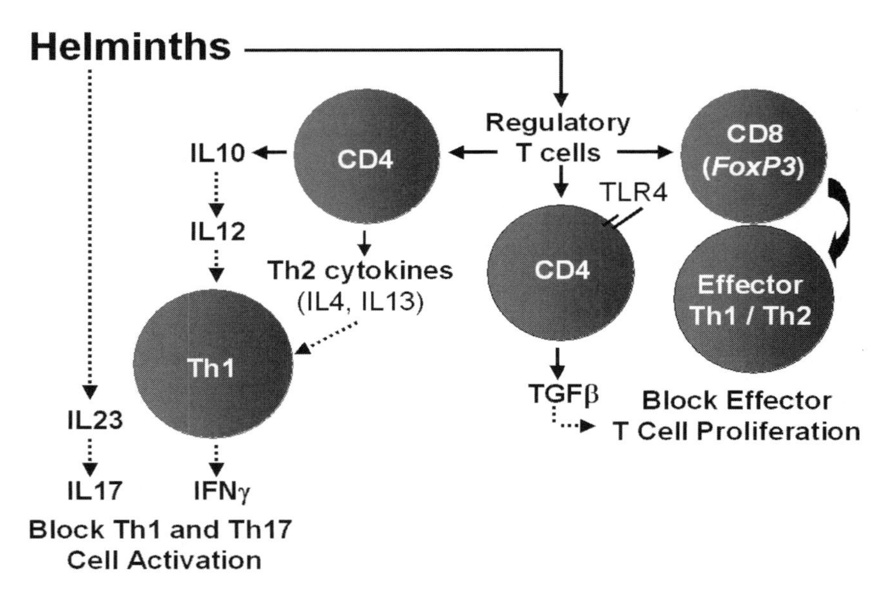


**FIGURE 1** Helminths induce the host's immune system to produce various regulatory T-cell subtypes. These work through several mechanisms to limit the production and function of Th1, Th2, and Th17 effector T cells that drive immunopathology. Helminths also induce changes in innate immunity that can limit inflammation independently of the adoptive immunity (not shown). TLR4, Toll-like receptor 4.

could have important implications for both veterinary livestock health and human disease (53). An underlying helminth infection also may interfere with the accurate diagnosis of infectious status or leave some animals more susceptible to some serious pathogen-induced enteric infections (11). However, the relative risk to humans of using helminths therapeutically in our present hygienic societies may be negligible.

## CONCLUSION

The human race has lived with helminths throughout its evolution. This close relationship has resulted in bilateral immune adaptation to their presence. Various helminths can modulate the host immune response, damping responsiveness to various antigens. Helminths trigger many immunoregulatory/immunomodulatory pathways in the host's immune system to promote their survival. It is possible that the sudden removal of helminthic infestation, particularly from children in our hygienic industrialized countries, is leading to immune dysregulation and the increase in immunological diseases. Reexposure to helminths may help to at least partly correct this imbalance and reverse the processes leading to disease.

## REFERENCES


1. **Abreu, M. T., K. D. Taylor, Y. C. Lin, T. Hang, J. Gaiennie, C. J. Landers, E. A. Vasiliauskas, L. Y. Kam, M. Rojany, K. A. Papadakis, J. I. Rotter, S. R. Targan, and H. Yang.** 2002. Mutations in NOD2 are associated with fibrostenosing disease in patients with Crohn's disease. *Gastroenterology* **123:**679–688.
2. **Andersson, R. E., G. Olaison, C. Tysk, and A. Ekbom.** 2001. Appendectomy and protection against ulcerative colitis. *N. Engl. J. Med.* **344:**808–814.
3. **Bach, J. F.** 2002. The effect of infections on susceptibility to autoimmune and allergic diseases. *N. Engl. J. Med.* **347:**911–920.
4. **Bancroft, A. J., K. J. Else, J. P. Sypek, and R. K. Grencis.** 1997. Interleukin-12 promotes a chronic intestinal nematode infection. *Eur. J. Immunol.* **27:**866–870.
5. **Bashir, M. E., P. Andersen, I. J. Fuss, H. N. Shi, and C. Nagler-Anderson.** 2002. An enteric helminth infection protects against an allergic response to dietary antigen. *J. Immunol.* **169:** 3284–3292.
6. **Bentwich, Z., Z. Weisman, C. Moroz, S. Bar-Yehuda, and A. Kalinkovich.** 1996. Immune dysregulation in Ethiopian immigrants in Israel: relevance to helminth infections? *Clin. Exp. Immunol.* **103:**239–243.
7. **Bernstein, C. N., J. F. Blanchard, P. Rawsthorne, and A. Wajda.** 1999. Epidemiology of Crohn's disease and ulcerative colitis in a central Canadian province: a population-based study. *Am. J. Epidemiol.* **149:**916–924.
8. **Borkow, G., Q. Leng, Z. Weisman, M. Stein, N. Galai, A. Kalinkovich, and Z. Bentwich.** 2000. Chronic immune activation associated with intestinal helminth infections results in impaired signal transduction and anergy. *J. Clin. Investig.* **106:**1053–1060.
9. **Calkins, B. M.** 1989. A meta-analysis of the role of smoking in inflammatory bowel disease. *Dig. Dis. Sci.* **34:**1841–1854.
10. **Carr, I., and J. F. Mayberry.** 1999. The effects of migration on ulcerative colitis: a three-year prospective study among Europeans and first- and second-generation South Asians in Leicester (1991–1994). *Am. J. Gastroenterol.* **94:**2918–2922.
11. **Chen, C. C., S. Louie, B. A. McCormick, W. A. Walker, and H. N. Shi.** 2006. Helminth-primed dendritic cells alter the host response to enteric bacterial infection. *J. Immunol.* **176:**472–483.
12. **Correale, J., and M. Farez.** 2007. Association between parasite infection and immune responses in multiple sclerosis. *Ann. Neurol.* **61:**97–108.
13. **Croese, J., J. O'Neil, J. Masson, S. Cooke, W. Melrose, D. Pritchard, and R. Speare.** 2006. A proof of concept study establishing *Necator americanus* in Crohn's patients and reservoir donors. *Gut* **55:**136–137.
14. **Cuthbert, A. P., S. A. Fisher, M. M. Mirza, K. King, J. Hampe, P. J. Croucher, S. Mascheretti, J. Sanderson, A. Forbes, J. Mansfield, S. Schreiber, C. M. Lewis, and C. G. Mathew.** 2002. The contribution of NOD2 gene mutations to the risk and site of disease in inflammatory bowel disease. *Gastroenterology* **122:** 867–874.
15. **Delco, F., and A. Sonnenberg.** 1998. Military history of patients with inflammatory bowel disease: an epidemiological study among U.S. veterans. *Am. J. Gastroenterol.* **93:**1457–1462.
16. **Derby, L. E., and H. Jick.** 1998. Appendectomy protects against ulcerative colitis. *Epidemiology* **9:**205–207.
17. **Duerr, R. H., K. D. Taylor, S. R. Brant, J. D. Rioux, M. S. Silverberg, M. J. Daly,**

**A. H. Steinhart, C. Abraham, M. Regueiro, A. Griffiths, T. Dassopoulos, A. Bitton, H. Yang, S. Targan, L. W. Datta, E. O. Kistner, L. P. Schumm, A. T. Lee, P. K. Gregersen, M. M. Barmada, J. I. Rotter, D. L. Nicolae, and J. H. Cho.** 2006. A genome-wide association study identifies IL23R as an inflammatory bowel disease gene. *Science* **314:**1461–1463.

18. **Duggan, A. E., I. Usmani, K. R. Neal, and R. F. Logan.** 1998. Appendicectomy, childhood hygiene, Helicobacter pylori status, and risk of inflammatory bowel disease: a case control study. *Gut* **43:**494–498.
19. **Ekbom, A., C. Helmick, M. Zack, and H. O. Adami.** 1991. The epidemiology of inflammatory bowel disease: a large, population-based study in Sweden. *Gastroenterology* **100:**350–358.
20. **Elliott, A. M., H. Mpairwe, M. A. Quigley, M. Nampijja, L. Muhangi, J. Oweka-Onyee, M. Muwanga, J. Ndibazza, and J. A. Whitworth.** 2005. Helminth infection during pregnancy and development of infantile eczema. *JAMA* **294:**2032–2034.
21. **Elliott, D. E., J. Li, A. Blum, A. Metwali, K. Qadir, J. F. Urban, Jr., and J. V. Weinstock.** 2003. Exposure to schistosome eggs protects mice from TNBS-induced colitis. *Am. J. Physiol. Gastrointest. Liver Physiol.* **284:**G385–G391.

21a. **Elliott, D. E., A. Metwali, J. Leung, T. Setiawan, A. M. Blum, M. N. Ince, L. E. Bazzone, M. J. Stadecker, J. F. Urban, Jr., and J. V. Weinstock.** 2008. Colonization with *Heligmosomoides polygyrus* suppresses mucosal IL-17 production. *J. Immunol.* **181:**2414–2419.

22. **Elliott, D. E., T. Setiawan, A. Metwali, A. Blum, J. F. Urban, Jr., and J. V. Weinstock.** 2004. *Heligmosomoides polygyrus* inhibits established colitis in IL-10-deficient mice. *Eur. J. Immunol.* **34:**2690–2698.
23. **Elliott, D. E., J. F. Urban, Jr., C. K. Argo, and J. V. Weinstock.** 2000. Does the failure to acquire helminthic parasites predispose to Crohn's disease? *FASEB J.* **14:**1848–1855.
24. **Fleming, J. O., and T. D. Cook.** 2006. Multiple sclerosis and the hygiene hypothesis. *Neurology* **67:**2085–2086.
25. **Flohr, C., L. N. Tuyen, S. Lewis, R. Quinnell, T. T. Minh, H. T. Liem, J. Campbell, D. Pritchard, T. T. Hien, J. Farrar, H. Williams, and J. Britton.** 2006. Poor sanitation and helminth infection protect against skin sensitization in Vietnamese children: a cross-sectional study. *J. Allergy Clin. Immunol.* **118:**1305–1311.
26. **Gent, A. E., M. D. Hellier, R. H. Grace, E. T. Swarbrick, and D. Coggon.** 1994. Inflammatory bowel disease and domestic hygiene in infancy. *Lancet* **343:**766–767.
27. **Goncalves, M. L., A. Araujo, and L. F. Ferreira.** 2003. Human intestinal parasites in the past: new findings and a review. *Mem. Inst. Oswaldo Cruz* **98**(Suppl. 1)**:**103–118.
28. **Halfvarson, J., L. Bodin, C. Tysk, E. Lindberg, and G. Jarnerot.** 2003. Inflammatory bowel disease in a Swedish twin cohort: a long-term follow-up of concordance and clinical characteristics. *Gastroenterology* **124:**1767–1773.
29. **Healy, G. R., N. N. Gleason, R. Bokat, H. Pond, and M. Roper.** 1969. Prevalence of ascariasis and amebiasis in Cherokee Indian school children. *Public Health Rep.* **84:**907–914.
30. **Ince, M. N., D. E. Elliott, T. Setiawan, A. Blum, A. Metwali, Y. Wang, J. F. Urban, Jr., and J. V. Weinstock.** 2006. Heligmosomoides polygyrus induces TLR4 on murine mucosal T cells that produce TGFβ after lipopolysaccharide stimulation. *J. Immunol.* **176:**726–729.
31. **Jayanthi, V., C. S. Probert, D. Pinder, A. C. Wicks, and J. F. Mayberry.** 1992. Epidemiology of Crohn's disease in Indian migrants and the indigenous population in Leicestershire. *Q. J. Med.* **82:**125–138.
32. **Jiang, H., S. I. Zhang, and B. Pernis.** 1992. Role of CD8+ T cells in murine experimental allergic encephalomyelitis. *Science* **256:**1213–1215.
33. **Khan, W. I., P. A. Blennerhasset, A. K. Varghese, S. K. Chowdhury, P. Omsted, Y. Deng, and S. M. Collins.** 2002. Intestinal nematode infection ameliorates experimental colitis in mice. *Infect. Immun.* **70:**5931–5937.
34. **Kitagaki, K., T. R. Businga, D. Racila, D. E. Elliott, J. V. Weinstock, and J. N. Kline.** 2006. Intestinal helminths protect in a murine model of asthma. *J. Immunol.* **177:**1628–1635.
35. **Kullberg, M. C., E. J. Pearce, S. E. Hieny, A. Sher, and J. A. Berzofsky.** 1992. Infection with Schistosoma mansoni alters Th1/Th2 cytokine responses to a non-parasite antigen. *J. Immunol.* **148:**3264–3270.
36. **La Flamme, A. C., K. Ruddenklau, and B. T. Backstrom.** 2003. Schistosomiasis decreases central nervous system inflammation and alters the progression of experimental autoimmune encephalomyelitis. *Infect. Immun.* **71:**4996–5004.
37. **Loftus, E. V., Jr., and W. J. Sandborn.** 2002. Epidemiology of inflammatory bowel disease. *Gastroenterol. Clin. N. Am.* **31:**1–20.
38. **Loftus, E. V., Jr., P. Schoenfeld, and W. J. Sandborn.** 2002. The epidemiology and natural history of Crohn's disease in population-based

patient cohorts from North America: a systematic review. *Aliment. Pharmacol. Ther.* **16:**51–60.

39. **Malhotra, I., P. Mungai, A. Wamachi, J. Kioko, J. H. Ouma, J. W. Kazura, and C. L. King.** 1999. Helminth- and Bacillus Calmette-Guerin-induced immunity in children sensitized in utero to filariasis and schistosomiasis. *J. Immunol.* **162:**6843–6848.
40. **Mangan, N. E., N. van Rooijen, A. N. McKenzie, and P. G. Fallon.** 2006. Helminth-modified pulmonary immune response protects mice from allergen-induced airway hyperresponsiveness. *J. Immunol.* **176:**138–147.
41. **McInnes, I. B., B. P. Leung, M. Harnett, J. A. Gracie, F. Y. Liew, and W. Harnett.** 2003. A novel therapeutic approach targeting articular inflammation using the filarial nematode-derived phosphorylcholine-containing glycoprotein ES-62. *J. Immunol.* **171:**2127–2133.
42. **Metwali, A., T. Setiawan, A. M. Blum, J. Urban, D. E. Elliott, L. Hang, and J. V. Weinstock.** 2006. Induction of CD8+ regulatory T cells in the intestine by Heligmosomoides polygyrus infection. *Am. J. Physiol. Gastrointest. Liver Physiol.* **291:**G253–G259.
43. **Moreels, T. G., R. J. Nieuwendijk, J. G. De Man, B. Y. De Winter, A. G. Herman, E. A. Van Marck, and P. A. Pelckmans.** 2004. Concurrent infection with Schistosoma mansoni attenuates inflammation induced changes in colonic morphology, cytokine levels, and smooth muscle contractility of trinitrobenzene sulphonic acid induced colitis in rats. *Gut* **53:**99–107.
44. **Mottet, C., H. H. Uhlig, and F. Powrie.** 2003. Cutting edge: cure of colitis by CD4+CD25+ regulatory T cells. *J. Immunol.* **170:**3939–3943.
45. **Nagayama, Y., K. Watanabe, M. Niwa, S. M. McLachlan, and B. Rapoport.** 2004. *Schistosoma mansoni* and α-galactosylceramide: prophylactic effect of Th1 immune suppression in a mouse model of Graves' hyperthyroidism. *J. Immunol.* **173:**2167–2173.
46. **Odes, H. S., D. Fraser, and J. Krawiec.** 1989. Inflammatory bowel disease in migrant and native Jewish populations of southern Israel. *Scand. J. Gastroenterol. Suppl.* **170:**36–38.
47. **Overgaauw, P. A.** 1997. Aspects of Toxocara epidemiology: human toxocarosis. *Crit. Rev. Microbiol.* **23:**215–231.
48. **Overgaauw, P. A.** 1997. Aspects of Toxocara epidemiology: toxocarosis in dogs and cats. *Crit. Rev. Microbiol.* **23:**233–251.
49. **Pearlman, E., J. W. Kazura, F. E. Hazlett, Jr., and W. H. Boom.** 1993. Modulation of murine cytokine responses to mycobacterial antigens by helminth-induced T helper 2 cell responses. *J. Immunol.* **151:**4857–4864.
50. **Probert, C. S., V. Jayanthi, A. O. Hughes, J. R. Thompson, A. C. Wicks, and J. F. Mayberry.** 1993. Prevalence and family risk of ulcerative colitis and Crohn's disease: an epidemiological study among Europeans and south Asians in Leicestershire. *Gut* **34:**1547–1551.
51. **Radlmayr, M., H. P. Torok, K. Martin, and C. Folwaczny.** 2002. The c-insertion mutation of the NOD2 gene is associated with fistulizing and fibrostenotic phenotypes in Crohn's disease. *Gastroenterology* **122:**2091–2092.
52. **Russel, M. G., E. Dorant, R. J. Brummer, M. A. van de Kruijs, J. W. Muris, J. M. Bergers, J. Goedhard, and R. W. Stockbrügger.** 1997. Appendectomy and the risk of developing ulcerative colitis or Crohn's disease: results of a large case-control study. *Gastroenterology* **113:** 377–382.
53. **Sabin, E. A., M. I. Araujo, E. M. Carvalho, and E. J. Pearce.** 1996. Impairment of tetanus toxoid-specific Th1-like immune responses in humans infected with Schistosoma mansoni. *J. Infect. Dis.* **173:**269–272.
54. **Sacco, R., M. Hagen, M. Sandor, J. V. Weinstock, and R. G. Lynch.** 2002. Established $T_{H1}$ granulomatous responses induced by active *Mycobacterium avium* infection switch to $T_{H2}$ following challenge with *Schistosoma mansoni*. *Clin. Immunol.* **104:**274–281.
55. **Salas, S. D., R. Heifetz, and E. Barrett-Connor.** 1990. Intestinal parasites in Central American immigrants in the United States. *Arch. Intern. Med.* **150:**1514–1516.
56. **Scrivener, S., H. Yemaneberhan, M. Zebenigus, D. Tilahun, S. Girma, S. Ali, P. McElroy, A. Custovic, A. Woodcock, D. Pritchard, A. Venn, and J. Britton.** 2001. Independent effects of intestinal parasite infection and domestic allergen exposure on risk of wheeze in Ethiopia: a nested case-control study. *Lancet* **358:**1493–1499.
57. **Sewell, D., Z. Qing, E. Reinke, E. Elliott, J. V. Weinstock, M. Sandor, and Z. Fabry.** 2003. Immunomodulation of experimental autoimmune encephalomyelitis by helminth ova immunization. *Int. Immunol.* **15:**59–69.
58. **Shivananda, S., J. Lennard-Jones, R. Logan, N. Fear, A. Price, L. Carpenter, and M. van Blankenstein.** 1996. Incidence of inflammatory bowel disease across Europe: is there a difference between north and south? Results of the European Collaborative Study on Inflammatory Bowel Disease (EC-IBD). *Gut* **39:**690–697.
59. **Smith, P., N. E. Mangan, C. M. Walsh, R. E. Fallon, A. N. J. McKenzie, N. van Rooijen, and P. G. Fallon.** 2007. Infection with a helminth parasite prevents experimental

colitis via a macrophage-mediated mechanism. *J. Immunol.* **178:**4557–4566.

60. **Sonnenberg, A.** 1990. Occupational distribution of inflammatory bowel disease among German employees. *Gut* **31:**1037–1040.

61. **Sonnenberg, A., D. J. McCarty, and S. J. Jacobsen.** 1991. Geographic variation of inflammatory bowel disease within the United States. *Gastroenterology* **100:**143–149.

62. **Summers, R. W., D. E. Elliott, J. F. Urban, Jr., R. Thompson, and J. V. Weinstock.** 2005. Trichuris suis therapy in Crohn's disease. *Gut* **54:**87–90.

63. **Summers, R. W., D. E. Elliott, J. F. Urban, Jr., R. A. Thompson, and J. V. Weinstock.** 2005. Trichuris suis therapy for active ulcerative colitis: a randomized controlled trial. *Gastroenterology* **128:**825–832.

64. **Summers, R. W., D. E. Elliott, and J. V. Weinstock.** 2005. Is there a role for helminths in the therapy of inflammatory bowel disease? *Nat. Clin. Pract. Gastroenterol. Hepatol.* **2:**62–63.

65. **Tysk, C., E. Lindberg, G. Jarnerot, and B. Floderus-Myrhed.** 1988. Ulcerative colitis and Crohn's disease in an unselected population of monozygotic and dizygotic twins. A study of heritability and the influence of smoking. *Gut* **29:** 990–996.

66. **van den Biggelaar, A. H., C. Lopuhaa, R. van Ree, J. S. van der Zee, J. Jans, A. Hoek, B. Migombet, S. Borrmann, D. Luckner, P. G. Kremsner, and M. Yazdanbakhsh.** 2001. The prevalence of parasite infestation and house dust mite sensitization in Gabonese schoolchildren. *Int. Arch. Allergy Immunol.* **126:**231–238.

67. **van den Biggelaar, A. H., L. C. Rodrigues, R. van Ree, J. S. van der Zee, Y. C. Hoeksma-Kruize, J. H. Souverijn, M. A. Missinou, S. Borrmann, P. G. Kremsner, and M. Yazdanbakhsh.** 2004. Long-term treatment of intestinal helminths increases mite skin-test reactivity in Gabonese schoolchildren. *J. Infect. Dis.* **189:**892–900.

68. **Wilson, M. S., M. D. Taylor, A. Balic, C. A. Finney, J. R. Lamb, and R. M. Maizels.** 2005. Suppression of allergic airway inflammation by helminth-induced regulatory T cells. *J. Exp. Med.* **202:**1199–1212.

69. **Zaccone, P., Z. Fehervari, F. M. Jones, S. Sidobre, M. Kronenberg, D. W. Dunne, and A. Cooke.** 2003. Schistosoma mansoni antigens modulate the activity of the innate immune response and prevent onset of type 1 diabetes. *Eur. J. Immunol.* **33:**1439–1449.

# ANTIBIOTIC RESISTANCE

# THE EVOLUTION OF ANTIBIOTIC-RESISTANT MICROBES IN FOODS AND HOST ECOSYSTEMS

*Marilyn C. Roberts*

# 12

## INTRODUCTION

The modern antibiotic era began with the discovery of penicillin in 1928 and its use in the 1940s. During the next 40 years, a number of antibiotics were identified and introduced into the marketplace. The term antibiotic was first used in 1941 to include natural substances made by one microbe which inhibit or kill other microbes (52). Antibiotics represent one of the major achievements in medicine during the 20th century. Antibiotic therapy has led to a reduction in morbidity and mortality of infectious diseases, and these drugs have been a major contributor to increased life expectancy in the last 60 years (52). In the years between the introduction of antibiotic therapy and the mid-1970s, antibiotic-resistant bacteria were generally restricted to the hospital setting and epidemic diseases which were not major issues in much of the industrialized world. As a result, treatment failure due to antibiotic-resistant bacteria was rare, and clinicians usually did not need to identify the pathogen causing the disease or worry about what antibiotic to use for therapy. That all changed in the mid-1970s with the isolation of ampicillin-resistant *Haemophilus influenzae,* the cause of childhood meningitis, followed by isolation of penicillin-resistant *Neisseria gonorrhoeae,* the cause of gonorrhea. Both of these diseases are community acquired, and the bacteria were resistant to the first-line therapies. This was the start of a new era in which clinicians would need to identify the pathogen and determine whether the pathogen was resistant to commonly used therapies (73, 74).

We now know that the use of antibiotics affects all the bacteria in an individual and not just the pathogen which is the therapeutic target. Data have shown that given enough time and antibiotic exposure, antibiotic-resistant bacteria, which are able to grow in increasing concentrations of the antibiotic, will develop. This may happen quickly, as with penicillin-resistant staphylococci, which appeared shortly after the introduction of penicillin, or may take years, as occurred with the development of vancomycin-resistant *Staphylococcus aureus* (52). Often, clinically important bacterial resistance is due to acquisition of new genes which reside on mobile elements. These elements can be readily transferred to related and nonrelated bacteria within the same environment and between different ecosystems. This allows for rapid spread of bacterial antibiotic

*Marilyn C. Roberts,* Department of Pathobiology and Department of Environmental and Occupational Health Sciences, University of Washington, Seattle, WA 98195.

*Food-Borne Microbes: Shaping the Host Ecosystem,* Edited by L.-A. Jaykus, H. H. Wang, and L. S. Schlesinger, © 2009 ASM Press, Washington, DC

resistance not seen in other types of organisms and has been important in the spread of these genes around the world (21, 74, 75).

Today antibiotics are used for treatment and prevention of bacterial diseases and protozoan parasitic diseases as well as a number of noninfectious diseases of humans, animals, and plants (21, 29). In addition, the overuse and misuse of antibiotics also contribute to the increase in bacterial resistance. Diseases caused by antibiotic-resistant bacterial pathogens can lead to a decrease in therapeutic options, which may require more expensive alternative therapies, an increase in the length of time needed for effective treatment, an increase in hospital stays, and an increase in mortality. Ultimately this situation may lead to a time when particular bacterial diseases will no longer have viable options for treatment available because the pathogen is multiresistant and/or the effective drug is not available to the patient because of the high cost (52).

## Antibiotics and Food Animals

In the early 1950s, antibiotic use was introduced into agriculture and food production (52). Traditionally prescriptions were not needed to treat animals, and even today antibiotics for farm use can be obtained without veterinary prescription. Antibiotics were given to domesticated animals for treatment and prevention of infectious diseases as in human medicine. In addition, early studies using low subtherapeutic levels as growth promoters in animal feed seemed to demonstrate that the animals had improved weight gain and improved feed efficiency compared to animals without the added antibiotics. This became the basis for the use of subtherapeutic-level antibiotics as growth promoters (66). The association between the use of antibiotics for growth promotion in animals and the occurrence of antibiotic-resistant animal bacteria has been demonstrated (4, 9, 10). In some industrialized countries, it has been estimated that 50% of the antibiotics consumed yearly by the country was for human use and 50% for animal use, though this varies, especially in countries where food animals outnumber humans (52).

The majority of antibiotics used for animal food production are of the same classes as those used in human medicine (42). The result is that the same antibiotic resistance genes found in animal bacteria and their environment are found in human bacteria (1, 3, 28). Antibiotic resistance genes are often associated with mobile elements, which allow for horizontal transfer of these resistance genes to related and unrelated species and genera. Antibiotic-resistant animal bacteria may transfer their antibiotic resistance genes to human-adapted bacteria, even when the animal bacteria are not well adapted for long-term survival in humans (51, 74). In addition, most U.S. antibiotics used in animal food production are available over the counter (90).

The use of subtherapeutic doses of antibiotics as growth promoters readily selects for high-level antibiotic-resistant bacteria which can be identified in the animals, food, associated environment, and farm workers (1, 4, 9, 32, 43–45, 52, 60). Antibiotic-resistant bacteria may contaminate meat during slaughtering and processing, and many studies have shown that retail meat products bought by the consumer may be important sources of human exposure to antibiotic-resistant bacterial pathogens in addition to resistant bacteria not normally associated with human disease (28, 32, 34, 38). Antibiotic-resistant bacteria have also been identified in a number of different processed foods as well as imported foods (3, 79).

Recently, the recovery of multidrug-resistant *Enterococcus* spp. and *Streptococcus* spp. from the indoor air of large-scale swine feeding operations has been reported (19, 35, 82). The results of the studies suggest that airborne gram-positive bacteria from swine operations may be an important contributor to environmental reservoirs of resistant bacteria and antibiotic resistance genes, which not only affects the farm workers but also has the potential to reach individuals living or working in close proximity to these facilities, as well as soils, crops, and surface waters in the surrounding area due to the bacteria being carried by air currents. Other studies have identified antibiotic-resistant bacteria contaminating water and soil surrounding animal production facilities (20, 44). In a study by Gibbs et al. (35), multidrug-resistant bacteria were recovered 150 m downwind of a large-scale swine feeding operation, while another study was able to detect airborne antibiotic-resistant bacteria up to 200 m away (http://www.ces.purdue.edu/pork/house/conf.htm) and within nearby residences (83).

Enterococci have been isolated from milk, cheese, meat, and raw produce. A recent study also found that antibiotic-resistant *Enterococcus* spp. (88% *E. faecalis*) could be found in the digestive tracts of houseflies collected from five restaurants in Kansas; of the 176 *E. faecalis* isolates tested, 66% were resistant to tetracycline and 24% to erythromycin (57).

Antibiotics have been used for ~50 years in the aquaculture industry to control bacterial infections found in salmon, catfish, and lobster (25, 42, 52). The number of products produced by aquaculture continues to increase, and the industry is an important supplier for many of the seafood and fish products consumed worldwide. In later sections of this chapter, the development of tetracycline-resistant ($Tc^r$) fish pathogens and bacteria associated with aquaculture environments is discussed.

## Antibiotics, Resistance Genes, and Plants

Antibiotics have also been used since the 1950s to control for bacterial diseases of high-value fruits, vegetables, and ornamental plants, with oxytetracycline and streptomycin being the most commonly used (62). However, unlike antibiotics used for food animals, with which usage can be ≥50% of a country's consumption, plant usage normally accounts for <0.5% of the total antibiotic use of a country (62). Antibiotics are used to treat plant infections due to the bacteria *Erwinia amylovora* (fire blight) and mycoplasma (lethal yellow) and to control infection of seeds by the bacterium *Xanthomonas campestris* (black rot) (52). Treatments of fruit trees are often done by airplane, and thus the antibiotic application also contaminates the surrounding soil, water, animals, other plants, and humans under the spray zone. $Tc^r$ orchard bacteria were shown to carry the same *tet* genes on the same transposons as described for $Tc^r$ bacteria in the environment or isolated from animals and humans, suggesting that gene exchange between the various ecosystems has been occurring (47, 84). However, more studies are needed to determine if the situation will continue to change over time. Antibiotics have also been used to treat honeybees for foulbrood disease caused by gram-positive bacteria *Bacillus larvae* and *Melissococcus plutonius* (formerly *Streptococcus pluton*) (52).

Food imported from around the world is commonly available in North America and Europe. These imports come from countries which may use antibiotics to manage plant bacterial diseases without knowledge of appropriate dosages, clear national policies, or monitoring programs. Because of the limited testing done upon entry, food with antibiotic residue does make it to the local store. Recently a group of investigators examined the occurrence of oxytetracycline- and gentamicin-resistant gram-negative bacteria from lettuce collected at one organic farm and nine conventional farms located in Costa Rica (81). The lettuce studied contained between $10^4$ and $10^7$ CFU of antibiotic-resistant bacteria/g (fresh weight), and the proportions of antibiotic-resistant bacteria ranged from 10 to 100% and certainly could be a potential source for dissemination of antibiotic-resistant bacteria and resistance genes. Interestingly, the organic farm had high levels of gentamicin-resistant bacteria. A variety of different tetracycline resistance genes, usually associated with mobile elements, were identified. The investigators also found fecal bacteria in some of the irrigation water samples, although the level was in compliance with national and international water quality standards. Other reports have previously implicated lettuce and other fresh vegetables as sources of antibiotic-resistant bacteria in Europe (81).

Another potential source of antibiotic resistance genes can be found in transgenic plants. Bacterial antibiotic resistance genes were first used as markers of the transgenic plant, though currently there is a move away from use of these genes as indicators of transformation. In some plant systems the bacterial resistance genes were engineered to be expressed in the eukaryotic cells, which allowed the companies to directly select for transformed plant cells. These genes are not likely to function even if transformed into the bacterial host because of the differences between eukaryotic and prokaryotic machinery. However, in other plant systems the bacterial resistance genes were not modified, so they are not expressed in the plant cell but could theoretically be expressed if transformed into a suitable bacterial host. In one experiment, five chickens were fed transgenic maize. No transformation was found in the lower gut; however, given the sheer number of chickens raised each year, the possibility of this happening either within the chicken or in environmental bacteria is higher than zero (18). The unmodified antibiotic resistance genes are the most likely to act as a source of transforming DNA which could potentially be incorporated into environmental bacteria or oral or intestinal bacteria of animals or humans. One study demonstrated that transfer from transgenic tobacco plants to bacteria was dependent on the gene copy number, bacterial competence, and the presence of homologous sequences (47). Studies have shown that transgenic DNA released can persist in the soil-containing litter from transgenic tobacco, potato, and sugar beet plants, and some of the DNA was extracellular (26). In another study with tobacco, the authors found that DNA recovered from roots and pollen was of high molecular weight and could be used to transform *Acinetobacter* under laboratory conditions. The transforming activity was maintained over the winter and during 4 years of storage in the laboratory. To date, transfer of recombinant DNA from transgenic plants to microbes in soil, animals, or humans under natural conditions has not yet been identified (26). However, the longer-term consequences of using unaltered bacterial antibiotic resistance genes in transgenic bacteria remain controversial.

### Antibiotic Residues

Antibiotic residues may be found in a variety of foods produced around the world. Currently, unlabeled but high levels of cephalosporins (ceftiofur) are allowed in some foods (22). Rules exist which aim to minimize the level of antibiotic residues found in food products; however, not all foods are tested, nor is it clear that the standards required by farms in North America and Europe are followed when the food is produced for overseas consumption. This may also be true for ingredients purchased overseas and made into food products for humans or animals, as the recent description of pigs eating contaminated pet food illustrates (http://petnblog.preciouspets.org/?p=510, http://www.msnbc.msn.com/id/18210224/). Ingestion of antibiotic residues will select for more resistant bacteria and could change the composition of the commensal microflora of the person or animal consuming the food (89).

### DNA from Antibiotic Producers

Another potential source of antibiotic resistance genes is from the producers, which has been identified in the antibiotics used in animal feeds. In one study, DNA extracted from crude avoparcin preparations intended for animal use was from *Amycolatopsis coloradensis,* the producer of avoparcin. In addition, genes whose products had >50% amino acid identity to the products of various *van* genes were also identified from the same DNA preparation (54). The authors suggest that the use of crude avoparcin preparations in animal feeds from 1975 to 1996 in Europe was the origin of the *vanHAX* genes and the source of the genes now found in vancomycin-resistant enterococci (VRE) (10). Lu and colleagues (54) suggested that crude antibiotic preparations used for growth promoters provided another delivery system for bacterial antibiotic resistance genes to directly contaminate the environment as well as potentially the bacteria associated with food animals acquiring these genes.

### Probiotic and Food Bacteria

Probiotic bacteria such as *Lactobacillus* spp. are used in food animal production. Recently, probiotic bacteria, including *Lactobacillus, Lactococcus,* and *Pediococcus* strains, which are used for large-scale starter cultures in the food industry and/or as probiotics in animal food production, were examined for their resistances to a variety of antibiotics. Among the *Lactobacillus* isolates recently tested for antibiotic susceptibility, the following were resistant to ≥1 of the antibiotics examined: 9.7% (11 of 113) of isolates used in probiotic products; 8.7% (2 of 23) of isolates used for probiotic or nutritional purposes, used for research, under research for inclusion in probiotic products, and/or used in food products as starter cultures; 7.7% (2 of 26) of isolates used for nutritional purposes or used in food products as starter cultures; 20% (1 of 5) of isolates which came from animals; but only 1.2% (3 of 249) of isolates from human samples. However, duplicate cultures for the probiotic or nutritional isolates could not be ruled out (49). Some of the isolates were shown to carry the *erm*(B), *tet*(M), and/or *tet*(W) genes; all were known to be associated with conjugative elements (21, 72, 74, 77). Use of these antibiotic-resistant strains for starter cultures or probiotics could provide additional avenues for these mobile bacterial antibiotic resistance genes to be introduced into the food chain (49). Clearly, other bacteria associated with production of food products should be characterized for potential antibiotic resistance genes with the aim of replacing these antibiotic-resistant strains with susceptible strains.

### Miscellaneous

As mentioned above, Macovei and Zurek (57) examined houseflies collected from local food settings and found them to carry antibiotic-resistant bacteria. These flies could easily contaminate food as well as the local environment. Contamination of food may occur during processing or at the final store due to contaminated knives or equipment.

## MECHANISM OF RESISTANCE

Antibiotic-resistant bacteria have developed in response to the use of antibiotics worldwide (52). Initially, resistant bacteria were identified because of treatment failure; however, the setup of surveillance studies can lead to the identification of emerging resistant bacteria before they cause treatment failure (27). Bacteria have used a variety of methods to adapt to increasing exposure to antibiotics. All bacteria are able to respond with random changes in genes which code for specific proteins. Some mutations can lead to changes in the amino acid sequence of the protein which alter the protein's function. Other mutations may arise which prevent the protein's production or shorten the peptide, resulting in either no protein or a nonfunctional protein. Mutation may also occur in regulator genes or regulatory regions, affecting the expression of functional proteins directly responsible for the phenotypic changes. In each situation the end result is that the bacterium's susceptibility to one or more antibiotics is reduced. Multiple changes are usually needed for the development of clinical resistance (53, 58, 59, 74–76). Mutations are important in the development of resistance in certain bacteria and to some classes of antibiotics; however, generally their impact on clinical disease is moderate (see below).

A limited number of naturally transformable species, such as *Acinetobacter, Haemophilus, Neisseria,* and *Streptococcus,* are able to take up external free DNA and integrate it within their genome. Transformation is sensitive to DNase in the environment and does not require cell-to-cell contact. Transformation can lead to the addition of new genes or to chimeric (mosaic) genes. Mosaic genes include some sequences which correspond to regions from more resistant species and part from the host bacteria (17, 86). This change may result in a functional protein which has slightly different properties from the parental protein.

For example, clinically relevant low-affinity penicillin binding proteins (PBPs) from *Neisseria meningitidis* and *Streptococcus pneumoniae* have been characterized and have been shown to have sections that have the same sequences as found in commensal *Neisseria* or oral *Streptococcus* spp. (15, 86). Normally most of the foreign DNA found in these genes is from closely related species. Even before the DNA molecule was known, early research demonstrated in vivo transformation in *S. pneumoniae* (37).

Mobile elements, including plasmids, transposons, conjugative transposons, and integrons, are primarily responsible for the lateral transfer of most antibiotic resistance genes from one cell to another. Most bacteria can participate in this exchange, which allows for gene transfer across natural bacterial and ecological barriers and between very different ecosystems (73–77, 79, 91, 92). It has been hypothesized that horizontal gene exchange is the major method of spreading new antibiotic resistance genes through the local environment and throughout the microbial population. This is why antibiotic resistance may spread so quickly. As more bacterial genomes are completely sequenced, it is clear that many of these genomes have regions which differ in G+C content from the majority of the chromosome. It has been hypothesized that these regions represent sequences that have originated in other species and been laterally transferred into the current isolate (24). These mobile elements carry new genes, which allow for the production of new proteins, some of which may confer resistance to one or more antibiotics (31, 52). Mobile elements can carry a variety of genes, which are usually not required for the bacteria's normal growth but provide them with flexibility which allows them to respond to changes in the environment. One study examined an epidemic methicillin-resistant *Staphylococcus aureus* strain genome versus the genome from methicillin-susceptible *S. aureus* and found that the resistant strain carried ~6% novel genes (40). The majority of the novel genes were mobile elements carrying antibiotic resistance and virulence genes. The authors suggest that their data illustrate the critical role mobile elements have played in the rapid evolution of this bacterium, and it has been hypothesized that these elements play a similar role in many other bacteria.

## Mutations

Random DNA mutations which lead to changes in the amino acid or premature termination of the protein chain can lead to moderate changes in susceptibility to one or more antibiotics. These chromosomal changes are usually transferred to daughter cells during cell division or to the same or closely related species by natural transformation, which is a process that allows certain bacterial species to acquire DNA from the surrounding environment, or transduction, in which bacterial DNA is packaged into a phage capsid and injected into a new host cell after this phage particle attaches to another bacterium-specific phage receptor (24, 92). Bacterial mutations have often been identified because of treatment failure or, more recently, due to active surveillance and screening.

High-level resistance usually requires multiple changes to one or more genes, although exceptions can be found, as demonstrated by the selection to high-level streptomycin resistance (75). Resistance to antibiotics used for treatment of mycobacterial diseases of humans and animals is due to chromosomal mutations, although acquired genes have been identified in human isolates (70). In general, resistance to quinolones has been associated with mutation in the DNA gyrase genes, *gyrA* and/or *gyrB,* and topoisomerase IV genes, *parC* and/or *parE,* although more recently, plasmid-associated *qnr* genes have been identified in gram-negative species (80).

Various mutations have been identified in the 23S rRNA at either the A2058 or A2059 position (*Escherichia coli* numbering) (59, 60) which confer increased resistance to macrolide-lincosamide-streptogramin B, macrolide-lincosamide, telithromycin, and/or linezolid antibiotics. In the last 10 years, the role of efflux pumps in conferring antibiotic resistance has been elucidated for both gram-positive and gram-negative species. Many innate efflux pumps transport substrates including clinically relevant antibiotics from within the cell into the external environment (93). Mutations may result in overexpression of efflux pumps, which may result in resistance to antibiotics of multiple classes, some dyes, detergents, and disinfectants (23, 93). Mutations are also responsible for changing the spectrum of activity of β-lactamases and account for most of the differences in the various classical and extended-spectrum β-lactamases (15), and they may have significant impact on the pathogens' susceptibility to normal therapy. For example, the extended-spectrum OXA-10 and OXA-2 enzymes, when found in multidrug-resistant *Pseudomonas aeruginosa,* confer resistance to all β-lactams except carbapenems, making therapy an issue (53). Mutations can also influence the expression of PBPs or innate and acquired β-lactamases, resulting in resistance (15, 69).

### Chimeric (Mosaic) Genes

Naturally transformable species, e.g., *Streptococcus pneumoniae, Streptococcus* species, *Neisseria meningitidis,* and *Haemophilus* species, can develop chimeric (mosaic) genes which have reduced affinity for β-lactam antibiotics by interspecies recombination. These chimeric PBPs develop by the replacement of parts of the PBP-encoding genes by transformation with corresponding regions from more resistant species (86). These altered genes are transferred to daughter cells during cell division or to the same or closely related species by transformation or transduction (where chromosomal DNA is packaged into a phage and injected into a new host). These chimeric genes can be found in a limited number of gram-positive *(Bacillus, Enterococcus, Staphylococcus,* and *Streptococcus)* and gram-negative *(Acinetobacter, Haemophilus, Helicobacter, Neisseria,* and *Pseudomonas)* genera which are all naturally transformable (17, 86).

### Acquisition of New Genes

With most bacteria, resistance is associated with the acquisition of new genes associated with mobile elements (plasmids, transposons, and/or integrons) (14, 31, 33, 51, 52, 55, 56, 64, 68, 71–76, 78, 79, 84, 87, 88). Most clinically relevant bacteria with reduced susceptibility to antibiotics are due to their acquisition of new genes conferring antibiotic resistance. Plasmids have been identified in human *E. coli* isolates collected between 1917 and 1954, but the lack of antibiotic resistance genes in these isolates suggests that the mobile elements were associated with bacteria before they were associated with the carriage of antibiotic resistance genes (41). A few studies have looked at antibiotic resistance of particular species over time and in general found that the number of isolates carrying acquired antibiotic resistance genes increased with time (8, 23). The data suggested that the accumulation of these antibiotic resistance genes is linked to the increased use of antibiotics worldwide in both humans and agriculture (52).

## MOBILE ELEMENTS ASSOCIATED WITH ACQUIRED RESISTANCE GENES

Mobile elements were first suggested by McClintock in the 1940s, when she described elements which could move from place to place within the maize genome (61). Plasmids, transposons, integrons, and gene cassettes play major roles in horizontal transfer of antibiotic resistance genes (24). Each of these elements is composed of double-stranded DNA. Plasmids were first identified in the 1950s, with the first review of these elements done in 1963 on plasmids in gram-negative enteric bacteria (5, 91). In 1960, Akiba et al. showed that multidrug resistance was transmissible. Within 8 years of the first detection of plasmids in Japan, similar plasmids in gram-negative bacteria were found in England and Germany (31). In the 1970s conjugative plasmids, which conferred ampicillin resistance in clinical strains of *Haemophilus influenzae* and *Neisseria gonorrhoeae,* both associated with community-acquired diseases, were identified (73, 74). Prior to this time both organisms were considered uniformly susceptible to ampicillin, which was the drug of choice for treatment of diseases associated with these bacteria. This period ushered in a new era of antibiotic resistance, because from then on antibiotic-resistant pathogens were no longer limited to just the hospital and epidemic diarrheal diseases. Clinicians needed to start worrying about what was causing the disease and if the strain was resistant to the commonly used first-line antibiotics (74).

Many antibiotic resistance genes are associated with transposons and/or conjugative transposons. A transposon is a discrete set of DNA sequences that can move to different positions within the genome of a single cell and can be found in the chromosome and on plasmids. Conjugative transposons in gram-positive bacteria and gram-negative *Bacteroides* spp. were first described in the 1980s. These elements carry their own genes which allow for conjugal transfer from one bacterium to another, from one species to another, and between unrelated genera (72–74, 80). Integrons were first described in 1989 for enteric bacteria and *Staphylococcus* spp. (87). Integrons carry gene cassettes that confer resistance to antibiotics and disinfectants. They have been found on chromosomes, transposons, and plasmids, which allow for their transfer through the bacterial population (12).

## Plasmids

The term plasmid was proposed by Lederberg to mean "any extrachromosomal hereditary determinant" (50). Plasmids are autonomously replicating extrachromosomal elements, are typically double stranded and circular, and usually carry nonessential genes (67). Plasmids exist in many types of bacteria and come in a wide range of sizes, from 1 to >400 kb, and may exist within the bacteria as one or multiple copies. Plasmids were first identified because they carried genes that conferred antibiotic resistance phenotypes on their hosts (5, 91). Plasmids provide the bacterial host with flexibility to survive in a changing environment and may carry a gene(s) that confers resistance to heavy metals, UV light, and disinfectants as well as a gene(s) encoding virulence factors, toxins, adhesins, type III secretion systems, insertion sequences, metabolic pathways for degradation of specific compounds, and genes for replication. In conjugative plasmids, genes for transfer are also present (24, 31). Today plasmids often contain discrete transposons, integrons, and regions which differ in their G+C contents, suggesting acquisition from different hosts (24). Many plasmids have limited host ranges and replicate in a single genus, while other plasmids, such as RSF1010, are able to replicate in a number of different unrelated genera. This is regulated by the genes coding for replication and maintenance functions. A number of different plasmid incompatibility groups have been identified (46). Plasmids may also integrate into the bacterial host chromosome (40). Most plasmids have incompatibility genes which prevent superinfection of a second plasmid, from the same incompatibility group that is already present in the bacteria. Plasmids from other incompatibility groups can become established (67). Natural plasmids are rarely able to cross from gram-positive to gram-negative strains or vice versa and continue to exist as independent replicating units.

Plasmids were first identified in the 1950s, and initially it was thought that plasmids were found in a limited number of enteric bacteria. However, in the 1970s ampicillin resistance due to the carriage of plasmids was identified in *H. influenzae, Haemophilus parainfluenzae, Haemophilus ducreyi,* and *N. gonorrhoeae*. All of these isolates carried the same TEM β-lactamase gene first described for *E. coli* (73). Both large and small plasmids were identified in *Haemophilus* spp., but the small β-lactamase plasmids were found in the different *Haemophilus* spp. and *N. gonorrhoeae*. These small plasmids were genetically related and most likely originated from a small plasmid from *H. parainfluenzae*. This was the first indication that plasmid exchange has a much wider host range than within enteric species. A second β-lactamase plasmid called ROB-1, which encoded a different β-lactamase, was identified in a few *H. influenzae* isolates, but the same plasmid was also found in the pig pathogen *Actinobacillus pleuropneumoniae* and in human isolates of *Pasteurella multocida* and *Pasteurella haemolytica* from Europe and the United States. In each case the bacterium was specific for one host, yet it was clear that the plasmid was moving between different ecosystems and different parts of the world (73). Clearly, a number of other intermediary bacteria were needed to make the transfer from humans to pigs and/or vice versa.

One study found a correlation between the plasmid incompatibility group and the particular *tet* genes the plasmid carried (46). These authors suggested that *tet* genes may become genetically linked to specific incompatibility and/or replication genes which could influence the distribution to specific genera or even species. This relationship was not shown in an earlier study by Mendez et al. (63).

### Transposons

Transposons are discrete entities which are incapable of autonomous replication and thus, to be maintained, must be integrated into the host chromosome or on host plasmids. Transposons are normally flanked by terminal inverted or direct repeats and carry the genes needed for movement to different locations (transposition) on the host's DNA. Conjugative transposons are self-transmissible integrating elements that carry additional genes which allow for movement from one bacterium to another by excision from the donor, circularization of the transposon, transfer to the recipient bacterial cell, and incorporation into the recipient's DNA (72). The host range of the conjugative transposons is very broad, with movement between gram-positive, gram-negative, anaerobic, and aerobic species readily demonstrated in the laboratory and the same conjugative transposon found in all of these types of bacteria in nature (21, 72, 74). The *tet*(M) gene is normally associated with conjugative chromosomal elements, and since these elements have less host specificity than do plasmids, this may explain why the *tet*(M) gene has been found naturally in 30 different genera (77). My laboratory has proposed that the location of antibiotic resistance genes such as *tet*(M) primarily on conjugative transposons may explain their wide distribution in nature (21). Transposons vary in size, with some being ≥150 kb. In addition, composite transposons, in which one transposon is inserted within a second transposon, have been described, and new variants carrying different combinations of antibiotic resistance and other genes continue to be characterized (72).

### Integrons

Integrons are gene expression elements that accumulate promoterless open reading frames known as gene cassettes. Integrons have an integrase *(int)* gene, a nearby recombination site *(attI)*, and a promoter ($P_{ANT}$) (87). All captured genes are read from the single promoter region, and the genes closest to the promoter will have the highest level of expression. The genes (cassettes) can be switched. Five classes of integrons associated with cassettes that contain antibiotic resistance genes, with three types of integrons (classes 1, 2, and 3), have been found in gram-negative bacteria, including multidrug-resistant *Salmonella enterica* serovar Typhimurium DT104, and they carry integrases that share 40 to 58% identity (12). Their importance to food production was shown when Nandi et al. (65) demonstrated that gram-positive bacteria are a major reservoir of class 1 antibiotic resistance integrons in poultry litter. Integrons vary in size, and a 126-kb superintegron has been identified in *Vibrio cholerae* with 179 cassettes, while more recently, a multidrug-resistant *Acinetobacter baumannii* strain was partially sequenced and shown to carry an 86-kb mosaic genomic island which contained most of the antibiotic resistance genes associated with multiple integrons and transposons. In contrast, a susceptible strain appeared to have the ends of the 86-kb region but no antibiotic resistance genes, integrons, or transposons in the region (78).

Over 40 different genes, coding for resistance to aminoglycosides, β-lactams, chloramphenicol, macrolides, sulfonamides, and trimethoprim and resistance to quaternary ammonium compounds and/or heavy metals, have all been found in integrons (24). Very recently, the first *tet* gene, *tet*(C), has been identified within *intl1* of the integron in Tn*1404** (88). This is the first time a *tet* gene has been associated with an integron and illustrates how diversity continues to evolve over time. How this will impact the spread of the *tet*(C) gene, which is currently found in 12 genera, is unclear. It is also of interest that the *tet*(C) gene became integrated rather than the more common *tet*(B) gene, which is found in 25 genera (77).

## SPECIFIC EXAMPLES

### *Enterococcus*

In the past 20 years, enterococci, formerly thought to be of minimal clinical impact, have emerged as important hospital-acquired pathogens in immunosuppressed patients and in intensive care units (36). Enterococci are innately resistant to a number of antibiotics and are able to survive in the hospital environment. Enterococci are commonly associated with nosocomial disease, and ~10% of hospital-acquired infections in the United States during the 1990s were due to enterococcal disease. They are heat tolerant and survive under adverse environmental conditions and, as a result, are widely dispersed. Enterococci are normal inhabitants of the intestinal flora of most mammals, birds, and humans, as well as from soil, surface waters, plants, vegetables, raw foods such as milk and meat, and fermented meats such as Italian salami or raw sausage (36). Enterococci may also be found in spoiled cooked and processed meats, may also be isolated from natural milk cultures, and also have been identified in probiotics used for animal feed (2). They are opportunistic pathogens which are able to cause a range of diseases in animals and humans.

Virulence in enterococci is enhanced by both innate and acquired antibiotic resistance genes. Enterococci are intrinsically resistant to cephalosporins, β-lactams, and low levels of aminoglycosides. Acquired resistance genes, which confer resistance to aminoglycosides, chloramphenicol, tetracycline, glycopeptides, and some β-lactam antibiotics, are also found in enterococci. Many of these genes are associated with mobile elements and can be transferred by conjugation to other enterococci as well as unrelated genera (72).

Enterococcal animal infections are rarely treated with antibiotics, but because they are part of the normal intestinal flora, they are exposed to antibiotic selection every time animals are given antibiotic therapy to treat disease or fed food with antibiotic growth promoters, leading to selection of antibiotic-resistant enterococci (2). Much work has been done with resistance to vancomycin used in human medicine and enterococcal resistance associated with the use of avoparcin, a related glycopeptide, as a growth promoter in food animals in Europe. VRE were first detected in 1986. In the United Kingdom, some of the VRE from human and nonhuman sources had the same ribotype, suggesting that the human and nonhuman isolates were related. This has led to the hypothesis that the use of avoparcin as a feed additive in food production selected for VRE in the animals (9). In the 1990s in Europe, VRE were associated with production animals and nonhospitalized humans, while in North America, where avoparcin has never been used, VRE isolates were not found in food production but were found in hospitalized U.S. patients. The difference was attributed to the extensive use of vancomycin in the hospital. Thus, the difference in distribution of VRE isolates is thought to be directly related to the different uses in Europe and North America (2, 10, 11).

In Europe, a number of studies have focused on vancomycin-resistant *Enterococcus faecium* and the use of avoparcin, and streptogramin-resistant *E. faecium* and the use of virginiamycin, a streptogramin, both used in animal husbandry. Work linked the high prevalence of resistance to one or both of these drugs used on farms. These resistant isolates were found in the poultry, in meat at food stores, in feces of nonhospitalized Europeans (including vegetarians), and in hospitalized patients (9, 10, 38). Numerous studies have shown that these genes are associated with mobile elements and are able to be transferred to recipient enterococci in human volunteers without antibiotic treatment (51). In addition, the apparent transfer of a plasmid carrying vancomycin resistance genes from *Enterococcus faecalis* to *Staphylococcus aureus* within a patient highlights the concerns about increased resistance and the ability of commensal or opportunistic bacteria to transfer their resistance genes to a more robust pathogen (33).

To counteract this trend, the European Union has banned the use of avoparcin, virginiamycin, spiramycin, tylosin, and bacitracin as growth promoters as of 1 July 1999 (30). The rationale was that these antibiotics select for bacterial resistance to important human antibiotics and thus are potential sources of bacterial resistance. In one study, the prevalence of antibiotic-resistant *Enterococcus* spp. rapidly dropped following the removal of growth promoters in pigs and chickens from multiple countries in the European Union. In contrast, in the United States and Canada, a variety of antibiotics continue to be used as food additives with the assumption that they will improve growth and feed conversion in the food animals. Antibiotics used as food additives are given at subtherapeutic doses, which select for antibiotic-resistant commensal, opportunistic, and pathogenic bacteria within the animal, as well as bacteria in the surrounding environment (52).

The two clinically important vancomycin resistance genes are *vanA,* which confers high-level inducible resistance to vancomycin and teicoplanin, and *vanB,* which confers variable levels of inducible vancomycin resistance. Other *van* genes *(vanC, vanD, vanE,* and *vanG)* have more recently been described and confer lower levels of vancomycin resistance and no or little resistance to teicoplanin (11). As stated above, crude avoparcin has been shown to carry genes related to the *vanA* enterococcal genes, and these authors feel that the drug avoparcin is the actual original source for the *vanA* cluster of enterococcal genes (54).

## Tetracycline Resistance Genes in Aquaculture Environments and Public Health Implications

Oxytetracycline has been used to treat and prevent bacterial disease of fish for more than 50 years. The antimicrobial agents used for aquaculture often contaminate the surrounding environment. Studies have shown that the level of bacteria underneath fish pens generally increases due to an increase in nutrient levels, by food and fish waste products, once the fish are in place. Various studies suggest that the environmental sites impacted by fish farms have increased numbers of antibiotic-resistant bacteria with increased carriage of antibiotic-resistant plasmids compared to background levels of nonimpacted sediments (13, 39). The limited data available indicate that like for human bacterial pathogens, the level of antibiotic resistance in most of the important fish pathogens has increased over time in a number of countries (7), while there has been much less investigation of commensal and environmental bacteria associated with aquaculture environments (16, 39, 48, 64, 85). Studies have shown that fish foods, even unlabeled ones, may contain antibiotic-resistant bacteria and/or antibiotic residues. In either case the food can then be a source of contamination for the fish farm even when antibiotics are not knowingly used (25, 39, 64).

An association between the tetracycline resistance genes *tet*(D) and *tet*(E) in bacterial fish pathogens and in aquaculture ponds has been described. In a 1995 study, we found that 69% of tetracycline-resistant *Aeromonas hydrophila* isolates from catfish ponds in the United States carried the *tet*(E) gene, while this gene is uncommon in gram-negative bacteria in the fecal flora of terrestrial animals (http://faculty.washington.edu/marilynr/). How the *tet*(E) gene has been disseminated around the world is still a mystery, since it is usually associated with large nonconjugative plasmids. In contrast, the *tet*(D) gene is associated with conjugative plasmids which are readily transferred in vitro.

More recently, my laboratory found a more diverse set of tetracycline resistance *(tet)* genes associated with Chilean freshwater salmon farms with no history of recent antibiotic use than previously described from other geographical locations (64). This included tetracycline resistance *tet*(H) genes found in *Moraxella* and *Acinetobacter radioresistens.* These two *tet*(H) genes were indistinguishable from Tn*5706* found previously in *Pasteurella multocida* from food animals, suggesting that horizontal gene exchange has occurred between salmon farms in Chile and land animal farms in Germany, although the Chilean genes could not be transferred in the laboratory (64). Not only were a number of new bacterial genera found to carry known *tet* genes but also approximately 40% of the tetracycline-resistant isolates may carry novel resistance genes, suggesting that environmental bacteria associated with food production may have a wider array of antibiotic resistance genes than have been identified in animal or human samples (64).

Relevance for human bacteria was illustrated by the study of Rhodes et al. (71). The authors found that closely related incompatibility plasmids containing the IncU replicon previously associated only with fish farm environments were also common in hospital sewage and patients, suggesting exchange between these groups of bacteria. It is also of interest that more diversity of *tet* genes has been found in water-associated bacteria than in all other species outside of *E. coli* (10 *tet* genes). For instance, nine different *tet* efflux genes have been indentified in *Aeromonas* and *Pseudomonas* and nine different *tet* efflux genes have been identified in *Vibrio* spp. (http://faculty.washington.edu/marilynr/).

Another example of horizontal gene exchange between fish farms and land animal farms was identified in the work of Anderson et al. (6). These authors found that a *tet*(C) gene associated with the chromosome of the obligate intercellular pig pathogen *Chlamydia suis* was part of a 10.1-kb fragment which shared 99% identity with plasmid pRAS3.2, commonly found in the fish pathogen *Aeromonas salmonicida.* In addition, the *C. suis tet*(C) gene carried an ISCS605 element, similar to IS elements identified in *Helicobacter.* This marks the first description of any acquired antibiotic resistance gene identified in obligate intracellular bacteria. One can only speculate how the *tet*(C) gene came to be linked to a genetic island that has components from two distantly unrelated genera that inhabit distinctive ecological niches. However, the farm that the resistant *C. suis* was isolated from had used tetracyclines as growth promoters over an extended period. It is unlikely that there was direct gene exchange between the three species, because the optimal growth temperature for *C. suis* and *Helicobacter,* found in warm-blooded animals, is >10°C higher than for *A. salmonicida,* with a growth range of 4 to 25°C. This example illustrates how antibiotic resistance genes are able to spread from one bacterial environment to other distinct bacterial environments across the globe, and to bacteria which have no apparent direct link. It also demonstrates that once antibiotic resistance genes become established in a bacterial community, regardless of whether such establishment is in humans, animals, food crops, or the environment, these genes are able to establish themselves in unrelated bacteria and can ultimately pose a threat for treatment of bacterial diseases.

## CONCLUSION

Antibiotic use and bacterial resistance are global issues, especially as changing technology and practices allow food, goods, and people to travel freely and quickly across national borders and bring their associated bacteria with them. A growing number of bacterial diseases which were once treatable with antibiotics are becoming untreatable because the bacterial pathogens have become resistant to currently available antibiotics. In addition, reemergence of diseases and "new diseases" continue to complicate public health measures. Food and supplements that go into a variety of processed food products may have antibiotic-resistant bacteria, antibiotic residues, and/or DNA from antibiotic producers within them. The impact of the use of concentrated animal feeding operations on the surrounding environment, with higher levels of air, water, and environmental contamination, and the subsequent public health consequences are just being recognized (90). Imported foods have become increasingly available to consumers, yet how safe these are is not always clear. Unfortunately, even the careful consumer who never buys imported fruits and vegetables is unlikely to know if some components in processed food were imported. A good example is the melamine contamination in 2007, which primarily affected cats and dogs. However, this contaminated product was also fed to pigs and farm-raised salmon destined for human consumption. Instead of melamine, an antibiotic residue or antibiotic-resistant bacteria could have been the contaminating ingredient. In the last few years it has become clearer that food production and practices anywhere in the world can impact the safety of the food available locally. The long-term impact on public health is still not clear.

This chapter has discussed a few examples in which specific antibiotic resistance genes, regions of DNA, and/or plasmids have been found in bacteria from very different ecosystems, from different parts of the world, and in bacteria that are host species specific. These examples illustrate that once antibiotic resistance genes develop in any bacteria, horizontal gene transfer events across geographical and bacterial host boundaries do occur. Therefore, we must now think of the world as a single connected system where changes in bacterial resistance at one location may lead to changes in distant locations in unexpected ways. Antibiotic-resistant commensal and environmental bacteria can act as reservoirs for antibiotic resistance genes and are often able to transfer these genes to pathogenic bacteria. Diseases caused by antibiotic-resistant pathogens often have consequences not seen in those due to susceptible pathogens. Changes to a more centralized food production, larger livestock and aquaculture facilities, and more importation of food all contribute to the spread of potential pathogens and antibiotic-resistant bacteria. Other factors which impact treatment of antibiotic-resistant diseases include the extremely limited number of new classes of antibiotics which have been introduced and the reduction of research and development of antibiotics by the major pharmaceutical companies. Unless significant changes are made at all levels, the number and level of antibiotic-resistant bacteria will continue to increase, ultimately leading to most bacterial diseases becoming untreatable, as was the case in the preantibiotic era.

## REFERENCES

1. **Aarestrup, F. M., Y. Agerso, P. Gerner-Smidt, M. Madsen, and L. B. Jensen.** 2000. Comparison of antimicrobial resistance phenotypes and resistance genes in *Enterococcus faecalis* and *Enterococcus faecium* from humans in the community, broilers, and pigs in Denmark. *Diagn. Microbiol. Infect. Dis.* **37:**127–137.
2. **Aarestrup, F. M., P. Butaye, and W. Witte.** 2002. Nonhuman reservoirs of enterococci, p. 55–99. *In* M. S. Gilmore, D. B. Clewell, P. Courvalin, G. M. Dunny, B. E. Murray, and L. B. Rice (ed.), *The Enterococci: Pathogenesis, Molecular Biology, and Antibiotic Resistance.* ASM Press, Washington, DC.
3. **Aarestrup, F. M., R. S. Hendriksen, J. Lockett, K. Gay, K. Teates, P. F. McDermott, D. G. White, H. Hasman, G. Sorenen, A. Bangtrakulonth, S. Pornreongwong, C. Puylsrikarn, F. J. Angulo, and P. Garner-Smidt.** 2007. International spread of multidrug-resistant *Salmonella* Schwarzengrund in food products. *Emerg. Infect. Dis.* **13:**726–731.
4. **Aarestrup, F. M., H. Kruse, E. Tast, A. M. Hammerum, and L. B. Jensen.** 2000. Associations between the use of antimicrobial agents for growth promotion and the occurrence of resistance among *Enterococcus faecium* from broilers and pigs in Denmark, Finland, and Norway. *Microb. Drug Resist.* **6:**63–70.
5. **Akiba, T., L. Koyama, Y. Ishiki, S. Kimura, and T. Rukushima.** 1960. On the mechanism

of the development of multiple-drug-resistant clones of *Shigella. Jpn. J. Microbiol.* **4:**219–227.

6. **Anderson, A. A., J. Dugan, L. Jones, and D. Rockey.** 2004. Stable chlamydial tetracycline resistance associated with a *tet*(C) resistance allele. *Antimicrob. Agents Chemother.* **48:**3989–3995.
7. **Aoki, T.** 1988. Drug-resistant plasmids from fish pathogens. *Microbiol. Sci.* **5:**219–223.
8. **Atkinson, B. A., A. Abu-Al-Jaibat, and D. J. LeBlanc.** 1997. Antibiotic resistance among enterococci isolated from clinical specimens between 1953 and 1954. *Antimicrob. Agents Chemother.* **41:**1598–1600.
9. **Bager, F., M. Madsen, J. Christensen, and F. M. Aarestrup.** 1997. Avoparcin used as a growth promoter is associated with the occurrence of vancomycin-resistant *Enterococcus faecium* on Danish poultry and pig farms. *Prev. Vet. Med.* **31:**95–112.
10. **Bates, J.** 1997. Epidemiology of vancomycin-resistant enterococci in the community and the relevance of farm animals to human infection. *J. Hosp. Infect.* **37:**507–514.
11. **Beltrametti, F., A. Consolandi, L. Carrano, F. Bagatin, R. Rossi, L. Leoni, E. Zennaro, E. Selva, and F. Marinelli.** 2007. Resistance to glycopeptide antibiotics in the teicoplanin producer is mediated by *van* gene homologue expression directing the synthesis of a modified cell wall peptidoglycan. *Antimicrob. Agents Chemother.* **51:**1135–1141.
12. **Bonomo, R. A., A. M. Hujer, and K. M. Hujer.** 2007. Integrons and superintegrons, p. 331–338. *In* R. A. Bonomo and M. Tolmasky (ed.), *Enzyme-Mediated Resistance to Antibiotics: Mechanisms, Dissemination, and Prospects for Inhibition.* ASM Press, Washington, DC.
13. **Brazil, G., D. Curley, F. Gannon, and P. Smith.** 1986. Persistence and acquisition of antibiotic resistance plasmids in *Aeromonas salmonicida,* p. 107–114. *In* S. B. Levy and R. P. Novick (ed.), *Antibiotic Resistance Genes: Ecology, Transfer, and Expression.* Cold Spring Harbor Laboratory, Cold Spring Harbor, NY.
14. **Brisson-Noel, A., M. Arthur, and P. Courvalin.** 1988. Evidence for natural gene transfer from gram-positive cocci to *Escherichia coli. J. Bacteriol.* **170:**1739–1745.
15. **Bush, K., and P. A. Bradford.** 2007. β-Lactamases: historical perspectives, p. 67–79. *In* R. A. Bonomo and M. Tolmasky (ed.), *Enzyme-Mediated Resistance to Antibiotics: Mechanisms, Dissemination, and Prospects for Inhibition.* ASM Press, Washington, DC.
16. **Casas, C., E. C. Anderson, K. K. Ojo, I. Keith, D. Whelan, D. Rainnie, and M. C. Roberts.** 2005. Characterization of pRAS1-like plasmids from atypical North American psychrophilic *Aeromonas salmonicida. FEMS Microbiol. Lett.* **242:**59–63.
17. **Chalkley, L. J., and H. J. Koornhof.** 1990. Intra- and inter-specific transformation of *Streptococcus pneumoniae* to penicillin resistance. *J. Antimicrob. Chemother.* **26:**21–28.
18. **Chambers, P. A., P. S. Duggan, J. Heritage, and J. M. Forbes.** 2002. The fate of antibiotic resistance marker genes in transgenic plant feed material fed to chickens. *J. Antimicrob. Chemother.* **49:**161–164.
19. **Chapin, A., A. Rule, K. Gibson, T. Buckley, and K. Schwab.** 2005. Airborne multidrug-resistant bacteria isolated from a concentrated swine feeding operation. *Environ. Health Perspect.* **113:**137–142.
20. **Chee-Sanford, J. C., R. I. Aminov, I. J. Krapac, N. Garrigues-Jeanjean, and R. I. Mackie.** 2001. Occurrence and diversity of tetracycline resistance genes in lagoons and groundwater underlying two swine production facilities. *Appl. Environ. Microbiol.* **67:**1494–1502.
21. **Chopra, I., and M. Roberts.** 2001. Tetracycline antibiotics: mode of action, applications, molecular biology, and epidemiology of bacterial resistance. *Microbiol. Mol. Biol. Rev.* **65:**232–260.
22. **Collignon, P., and F. M. Aarestrup.** 2007. Extended-spectrum β-lactamases, food, and cephalosporin use in food animals. *Clin. Infect. Dis.* **44:**1391–1392.
23. **Cousin, S. L., Jr., W. L. Whittington, and M. C. Roberts.** 2003. Acquired macrolide resistance genes in pathogenic *Neisseria* spp. isolated between 1940 and 1987. *Antimicrob. Agents Chemother.* **47:**3877–3880.
24. **Craig, M. L., R. Craigie, M. Gellert, and A. M. Lambowitz (ed.).** 2002. *Mobile DNA II.* American Society for Microbiology, Washington, DC.
25. **DePaola, A., P. A. Flynn, R. M. McPhearson, and S. B. Levy.** 1988. Phenotypic and genotypic characterization of tetracycline- and oxytetracycline-resistant *Aeromonas hydrophila* from cultured channel catfish *(Ictalurus punctatus)* and their environments. *Appl. Environ. Microbiol.* **54:**1861–1863.
26. **deVries, J., M. Heine, K. Harms, and W. Wackernagel.** 2003. Spread of recombinant DNA by roots and pollen of transgenic potato plants, identified by highly specific biomonitoring using natural transformation of an *Acinetobacter* sp. *Appl. Environ. Microbiol.* **69:**4455–4462.
27. **Doern, G. V., A. Brueggemann, H. P. Holley, and A. M. Rauch.** 1996. Antimicrobial resistance of *Streptococcus pneumoniae* recovered from outpatients in the United States during the winter

months of 1994 to 1995: results of a 30-center national surveillance study. *Antimicrob. Agents Chemother.* **40:**1208–1213.

28. **Donabedian, S. M., L. A. Thal, E. Hershberger, M. B. Perri, J. W. Chow, P. Bartlett, R. Jones, K. Joyce, S. Rossiter, K. Gay, J. Johnson, C. Mackinson, E. DeBess, J. Madden, F. Angulo, and M. J. Zervos.** 2003. Molecular characterization of gentamicin-resistant enterococci in the United States: evidence of spread from animals to humans through food. *J. Clin. Microbiol.* **41:**1109–1113.
29. **Edlind, T. D.** 1991. Protein synthesis as a target for antiprotozoal drugs, p. 569–586. *In* G. Coombs and M. North (ed.), *Biochemical Protozoology.* Taylor and Francis, London, England.
30. **Emborg, H., A. K. Ersboll, O. E. Heuer, and H. C. Wegener.** 2001. The effect of discontinuing the use of antimicrobial growth promoters on the productivity in the Danish broiler production. *Prev .Vet. Med.* **50:**53–70.
31. **Falkow, S.** 1977. *Infectious Multiple Drug Resistance.* Pion Limited, London, England.
32. **Fey, P. D., T. J. Safranek, M. E. Rupp, E. F. Dunne, E. Ribot, P. C. Iwen, P. A. Bradford, F. J. Angulo, and S. H. Hinrichs.** 2000. Ceftriaxone-resistant salmonella infection acquired by a child from cattle. *N. Engl. J. Med.* **342:**1242–1249.
33. **Flannagan, S. E., J. W. Chow, S. M. Donabedian, W. J. Brown, M. B. Perri, M. J. Zervos, Y. Ozawa, and D. B. Clewell.** 2003. Plasmid content of a vancomycin-resistant *Enterococcus faecalis* isolate from a patient also colonized by *Staphylococcus aureus* with a VanA phenotype. *Antimicrob. Agents Chemother.* **47:**3954–3959.
34. **Gambarotto, K., M. C. Ploy, F. Dupron, M. Giangiobbe, and F. Denis.** 2001. Occurrence of vancomycin-resistant enterococci in pork and poultry products from a cattle-rearing area of France. *J. Clin. Microbiol.* **39:**2354–2355.
35. **Gibbs, S. G., C. F. Green, P. M. Tarwater, and P. V. Scarpino.** 2004. Airborne antibiotic resistant and nonresistant bacteria and fungi recovered from two swine herd confined animal feeding operations. *J. Occup. Environ. Hyg.* **1:** 699–706.
36. **Giraffa, G.** 2002. Enterococci from foods. *FEMS Microbiol. Rev.* **26:**163–171.
37. **Griffith, F.** 1928. The significance of pneumococcal types. *J. Hyg.* **27:**113–159. [Reprint, *J. Hyg.* **64:**129–175, 1966.]
38. **Hayes, J. R., L. L. English, P. J. Carter, T. Proescholdt, K. Y. Lee, D. D. Wagner, and D. G. White.** 2003. Prevalence and antimicrobial resistance of *Enterococcus* species isolated from retail meats. *Appl. Environ. Microbiol.* **69:**7153–7160.
39. **Herwig, R. P., J. P. Gray, and D. P. Weston.** 1997. Antibacterial resistant bacteria in surficial sediments near salmon net-cage farms in Puget Sound, Washington. *Aquaculture* **149:**263–283.
40. **Holden, M. T. G., E. J. Feil, J. A. Lindsay, S. J. Peacock, N. P. J. Day, M. C. Enright, T. J. Foster, C. E. Moore, L. Hurst, R. Atkin, A. Barron, N. Bason, S. D. Bentley, C. Chillingworth, T. Chillingworth, C. Churcher, L. Clark, C. Corton, A. Cronin, J. Doggett, L. Dowd, T. Feltwell, Z. Hance, B. Harris, H. Hauser, S. Holroyd, K. Jagels, K. D. James, N. Lennard, A. Line, R. Mayes, S. Mould, K. Mungall, D. Ormond, M. A. Quail, E. Rabbinowitch, K. Rutherford, M. Sanders, S. Sharp, M. Simmonds, K. Stevens, S. Whitehead, G. G. Barrell, B. G. Spratt, and J. Parkhill.** 2004. Complete genomes of two clinical *Staphylococcus aureus* strains: evidence for the rapid evolution of virulence and drug resistance. *Proc. Natl. Acad. Sci. USA* **101:**9786–9791.
41. **Hughes, V. M., and N. Datta.** 1983. Conjugative plasmids in bacteria of the "pre-antibiotic" era. *Nature* **302:**725–726.
42. **Institute of Medicine, Division of Health Promotion and Disease Prevention.** 1998. *Report of a Study. Human Health Risks With the Subtherapeutic Use of Penicillin or Tetracyclines in Animal Feed.* National Academy Press, Washington, DC.
43. **Jackson, C. R., P. J. Fedorka-Cray, J. B. Barrett, and S. R. Ladely.** 2004. Effects of tylosin use on erythromycin resistance in enterococci isolated from swine. *Appl. Environ. Microbiol.* **70:**4205–4210.
44. **Jensen, L. B., Y. Agerso, and G. Sengelov.** 2002. Presence of *erm* genes among macrolide-resistant Gram-positive bacteria isolated from Danish farm soil. *Environ. Int.* **28:**487–491.
45. **Jensen, L. B., A. M. Hammerum, F. Bager, and F. M. Aarestrup.** 2002. Streptogramin resistance among *Enterococcus faecium* isolated from production animals in Denmark in 1997. *Microb. Drug Resist.* **8:**369–374.
46. **Jones, C. S., D. J. Osborne, and J. Stanley.** 1992. Enterobacterial tetracycline resistance in relation to plasmid incompatibility. *Mol. Cell. Probes* **6:**313–317.
47. **Kay, E., T. M. Vogel, F. Bertolla, R. Natlin, and P. Simonet.** 2002. In situ transfer of antibiotic resistance genes from transgenic (transplastomic) tobacco plants to bacteria. *Appl. Environ. Microbiol.* **68:**3345–3351.
48. **Kim, E. H., and T. Aoki.** 1994. The transposon-like structure of IS*26*-tetracycline,

and kanamycin resistance determinant derived from transferable R plasmids of fish pathogen, *Pasteurella piscicida. Microbiol. Immunol.* **38:**31–38.

49. **Klare, I., C. Konstabel, G. Werner, G. Huys, V. Vankerckhoven, G. Kahlmeter, B. Hildebrandt, S. Muller-Bertling, W. Witte, and H. Goossens.** 2007. Antimicrobial susceptibilities of *Lactobacillus, Pediococcus* and *Lactococcus* human isolates and cultures intended for probiotic or nutritional use. *J. Antimicrob. Chemother.* **59:**900–912.
50. **Lederberg, J.** 1952. Cell genetics and hereditary symbiosis. *Physiol. Rev.* **32:**403–430.
51. **Lester, C. H., N. Frimodt-Moller, T. L. Sorensen, D. L. Monnet, and A. M. Hammerum.** 2006. In vivo transfer of the *vanA* gene between *Enterococcus faecium* strains and the intestine of humans. *Antimicrob. Agents Chemother.* **50:** 596–599.
52. **Levy, S. B.** 2002. *The Antibiotic Paradox: How the Misuse of Antibiotics Destroys Their Curative Powers,* 2nd ed. Plenum Press, New York, NY.
53. **Livermore, D. M.** 2002. Multiple mechanisms of antimicrobial resistance in *Pseudomonas aeruginosa:* or worst nightmare? *Clin. Infect. Dis.* **34:** 634–640.
54. **Lu, K., R. Asano, and J. Davies.** 2004. Antimicrobial resistance gene delivery in animal feeds. *Emerg. Infect. Dis.* **10:**679–683.
55. **Luna, V. A., S. Cousin, W. L. Whittington, and M. C. Roberts.** 2000. Identification of the conjugative *mef* gene in clinical *Acinetobacter junii* and *Neisseria gonorrhoeae* isolates. *Antimicrob. Agents Chemother.* **44:**2503–2506.
56. **Luna, V. A., M. Heiken, K. Judge, C. Ulep, C., N. Van Kirk, H. Luis, M. Bernardo, J. Leitao, and M. C. Roberts.** 2002. Distribution of *mef*(A) in gram-positive bacteria from healthy Portuguese children. *Antimicrob. Agents Chemother.* **46:**2513–2517.
57. **Macovei, L., and L. Zurek.** 2006. Ecology of antibiotic resistance genes: characterization of enterococci from houseflies collected in food settings. *Appl. Environ. Microbiol.* **72:**4028–4035.
58. **Malbruny, B., A. Canu, B. Bozdogan, B. Fantin, V. Zarrouk, S. Dutka-Malen, C. Feger, and R. Leclercq.** 2002. Resistance to quinupristin-dalfopristin due to mutations of L22 ribosomal protein in *Staphylococcus aureus. Antimicrob. Agents Chemother.* **46:**2200–2207.
59. **Malbruny, B., K. Nagai, M. Coquemont, B. Bozdogan, A. T. Andrasevic, H. Hupkova, R. Leclercq, and P. C. Appelbaum.** 2002. Resistance to macrolides in clinical isolates of *Streptococcus pyogenes* due to ribosomal mutations. *J. Antimicrob. Chemother.* **49:**935–939.
60. **Martel, A., L. A. Devriese, A. Decostere, and F. Haesebrouck.** 2003. Presence of macrolide resistance genes in streptococci and enterococci isolated from pigs and pork carcasses. *Int. J. Food Microbiol.* **84:**27–32.
61. **McClintock, B.** 1950. The origin and behavior of mutable loci in maize. *Proc. Natl. Acad. Sci. USA* **36:**344–349.
62. **McManus, P. S., V. O. Stockwell, G. W. Sundin, and A. L. Jones.** 2002. Antibiotic use in plant agriculture. *Annu. Rev. Phytopathol.* **40:** 443–465.
63. **Mendez, B., C. Tachibana, and S. B. Levy.** 1980. Heterogeneity of tetracycline resistance determinants. *Plasmid* **3:**99–108.
64. **Miranda, C. D., C. Kehrenberg, C. Ulep, S. Schwarz, and M. C. Roberts.** 2003. Diversity of tetracycline resistance genes in bacteria from Chilean salmon farms. *Antimicrob. Agents Chemother.* **47:**883–888.
65. **Nandi, S., J. J. Maurer, C. Hofacre, and A. O. Summers.** 2004. Gram-positive bacteria are a major reservoir of Class 1 antibiotic resistance integrons in poultry litter. *Proc. Natl. Acad. Sci. USA* **101:**7118–7122.
66. **National Research Council.** 1999. *The Use of Drugs in Food Animals: Benefits and Risks.* National Academy Press, Washington, DC.
67. **Novick, R. P.** 1987. Plasmid incompatibility. *Microbiol. Rev.* **51:**381–395.
68. **Ojo, K. K., C. Ulep, N. Van Kirk, H. Luis, M. Bernardo, J. Leitao, and M. C. Roberts.** 2004. The *mef*(A) gene predominates among seven macrolide resistance genes identified in gram-negative strains representing 13 genera, isolated from healthy Portuguese children. *Antimicrob. Agents Chemother.* **48:**3451–3456.
69. **Page, M. G.** 2007. Resistance mediated by penicillin-binding proteins, p. 81–99. *In* R. A. Bonomo and M. Tolmasky (ed.), *Enzyme-Mediated Resistance to Antibiotics: Mechanisms, Dissemination, and Prospects for Inhibition.* ASM Press, Washington, DC.
70. **Pang, Y., B. A. Brown, V. A. Steingrube, R. J. Wallace, Jr., and M. C. Roberts.** 1994. Tetracycline resistance determinants in *Mycobacterium* and *Streptomyces* species. *Antimicrob. Agents Chemother.* **38:**1408–1412.
71. **Rhodes, G., G. Huys, J. Swings, P. McGann, M. Hiney, P. Smith, and R. W. Pickup.** 2000. Distribution of oxytetracycline resistance plasmids between aeromonads in hospital and aquaculture environments: implication of Tn*1721* in dissemination of the tetracycline resistance determinant Tet A. *Appl. Environ. Microbiol.* **66:**3883–3890.
72. **Rice, L. B.** 2007. Conjugative transposons, p. 271–284. *In* R. A. Bonomo and M. Tolmasky (ed.). *Enzyme-Mediated Resistance to Antibiotics: Mechanisms, Dissemination, and Prospects for Inhibition.* ASM Press, Washington, DC.

73. **Roberts, M. C.** 1989. Plasmids of *Neisseria gonorrhoeae* and other *Neisseria* species. *Rev. Clin. Microbiol.* **2:**S18–S23.

74. **Roberts, M. C.** 1997. Genetic mobility and distribution of tetracycline resistance determinants. *Ciba Found. Symp.* **207:**206–218.

75. **Roberts, M. C.** 1998. Antibiotic resistance in oral/respiratory bacteria. *Crit. Rev. Oral Biol. Med.* **9:**522–540.

76. **Roberts, M. C.** 2002. Antibiotic toxicity, interactions and resistance development. *Periodontology 2000.* **28:**280–297.

77. **Roberts, M. C.** 2005. Update on acquired tetracycline resistance genes. *FEMS Microbiol. Lett.* **245:**195–203.

78. **Roberts, M. C.** 2006. Multidrug resistant genes associated with a 86 kb island in *Acinetobacter baumannii. Trends Microbiol.* **14:**375–378.

79. **Roberts, M. C., B. Facinelli, E. Giovanetti, C. Casolari, and P. E. Varaldo.** 1996. Transferable erythromycin resistance in food isolates of *Listeria. Appl. Environ. Microbiol.* **62:**269–270.

80. **Robicsek, A., G. A. Jacoby, and D. C. Hooper.** 2006. The worldwide emergence of plasmid-mediated quinolone resistance. *Lancet Infect. Dis.* **6:**629–640.

81. **Rodríguez, C., L. Lang, A. Wang, K. Altendorf, F. García, and A. Lipski.** 2006. Lettuce for human consumption collected in Costa Rica contains complex communities of culturable oxytetracycline- and gentamicin-resistant bacteria. *Appl. Environ. Microbiol.* **72:**5870–5876.

82. **Sapkota, A. R., K. K. Ojo, M. C. Roberts, and K. J. Schwab.** 2006. Antibiotic resistance genes in multidrug-resistant *Enterococcus* spp. and *Streptococcus* spp. recovered from indoor air of a concentrated swine feeding operation. *Lett. Appl. Microbiol.* **43:**534–540.

83. **Scarpino, P. V., and H. Quinn.** 1998. Bioaerosol distribution patterns adjacent to two swine-growing-finishing housed confinement units in the American Midwest. *J. Aerosol Sci.* **29:**S553.

84. **Schnabel, E. L., and A. L. Jones.** 1999. Distribution of tetracycline resistance genes and transposons among phylloplane bacteria in Michigan apple orchards. *Appl. Environ. Microbiol.* **65:** 4898–4907.

85. **Sorum, H., M. C. Roberts, and J. H. Crosa.** 1992. Identification and cloning of a tetracycline resistance gene from the fish pathogen *Vibrio salmonicida. Antimicrob. Agents Chemother.* **36:**611–615.

86. **Spratt, B. G., Q. Y. Zhang, D. M. Jones, A. Hutchison, J. A. Brannigan, and C. G. Dowson.** 1989. Recruitment of a penicillin-binding protein gene from *Neisseria flavescens* during the emergence of penicillin resistance in *Neisseria meningitidis. Proc. Natl. Acad. Sci. USA* **86:** 8988–8992.

87. **Stokes, H. W., and R. M. Hall.** 1989. A novel family of potentially mobile DNA elements encoding site-specific gene-integration functions: integrons. *Mol. Microbiol.* **3:**1669–1683.

88. **Stokes, H. W., L. D. H. Elbourne, and R. M. Hall.** 2007. Tn*1403,* a multiple-antibiotic resistance transposon made up of three distinct transposons. *Antimicrob. Agents Chemother.* **51:** 1827–1829.

89. **Summers, A. O.** 2002. Generally overlooked fundamentals of bacterial genetics and ecology. *Clin. Infect. Dis.* **34:**S85–S92.

90. **Thorne, P. S.** 2007. Environmental health impacts of concentrated animal feeding operations: anticipating hazards—searching for solutions. *Environ. Health Perspect.* **115:**296–297.

91. **Watanabe, T.** 1963. Infective heredity of multiple drug resistance in bacteria. *Bacteriol. Rev.* **27:** 87–115.

92. **Waters, V. L.** 2007. The dissemination of antibiotic resistance by bacterial conjugation, p. 285–312. *In* R. A. Bonomo and M. Tolmasky (ed.), *Enzyme-Mediated Resistance to Antibiotics: Mechanisms, Dissemination, and Prospects for Inhibition.* ASM Press, Washington, DC.

93. **Webber, M. A., and L. J. V. Piddock.** 2003. The importance of efflux pumps in bacterial antibiotic resistance. *J. Antimicrob. Chemother.* **51:**9–11.

# ANTIMICROBIAL RESISTANCE IN FOOD-BORNE PATHOGENS

*David G. White and Patrick F. McDermott*

# 13

Antimicrobial-resistant (AMR) enteric zoonotic bacteria usually reach humans and cause disease following ingestion of contaminated food or water (8, 114, 115). The environmental dissemination of AMR food-borne pathogens, and the magnitude of the resulting public health burden, is a complex issue (Fig. 1) that differs by country and region. It is influenced by a number of variables such as antimicrobial use practices in farming, process control at slaughter, the integrity of storage and distribution systems, the availability of clean water, farm sanitation practices (biocontainment, composting, manure fertilization of crops, etc.), and proper cooking and home hygiene, among others (203). In areas where these variables are loosely controlled and public health infrastructure is lacking, indigenous exposure levels to multidrug-resistant (MDR) food-borne bacteria may be high (210, 211).

Three scenarios have been previously proposed by which the use of antimicrobials in food animals could affect treatment of food-borne diarrheal diseases in humans. In the first, AMR zoonotic bacterial pathogens are selected, and food is contaminated during slaughter and/or preparation. After consumption of the food, these pathogens cause an infection that requires antimicrobial treatment, and therapy is compromised. The second scenario occurs when AMR bacteria nonpathogenic to humans are selected in the animal. When the contaminated food is ingested, the bacteria transfer resistance determinants to commensal and opportunistic pathogens in the human gut. Finally, there is the instance in which antimicrobials remain as residues in food products, which allows the selection of antibiotic-resistant bacteria after the food is consumed (141).

While a large body of evidence shows that food animals are a reservoir of AMR food-borne pathogens that can reach humans via the food supply (14, 51, 64, 76, 167, 189, 193, 214, 216, 220), other studies have demonstrated the spread of resistant fecal bacteria to humans living in close proximity to farm animals. One of the earliest prospective studies showed the movement of AMR *Escherichia coli* from poultry to farm personnel following treatment with tetracycline (98). Another study profiled the emergence of both

*David G. White,* Office of Research, Center for Veterinary Medicine, U.S. Food and Drug Administration, Laurel, MD 20708. *Patrick F. McDermott,* National Antimicrobial Resistance Monitoring System, Center for Veterinary Medicine, U.S. Food and Drug Administration, 8401 Muirkirk Rd., Mod 2, Laurel, MD 20708.

*Food-Borne Microbes: Shaping the Host Ecosystem,* Edited by L.-A. Jaykus, H. H. Wang, and L. S. Schlesinger, © 2009 ASM Press, Washington, DC

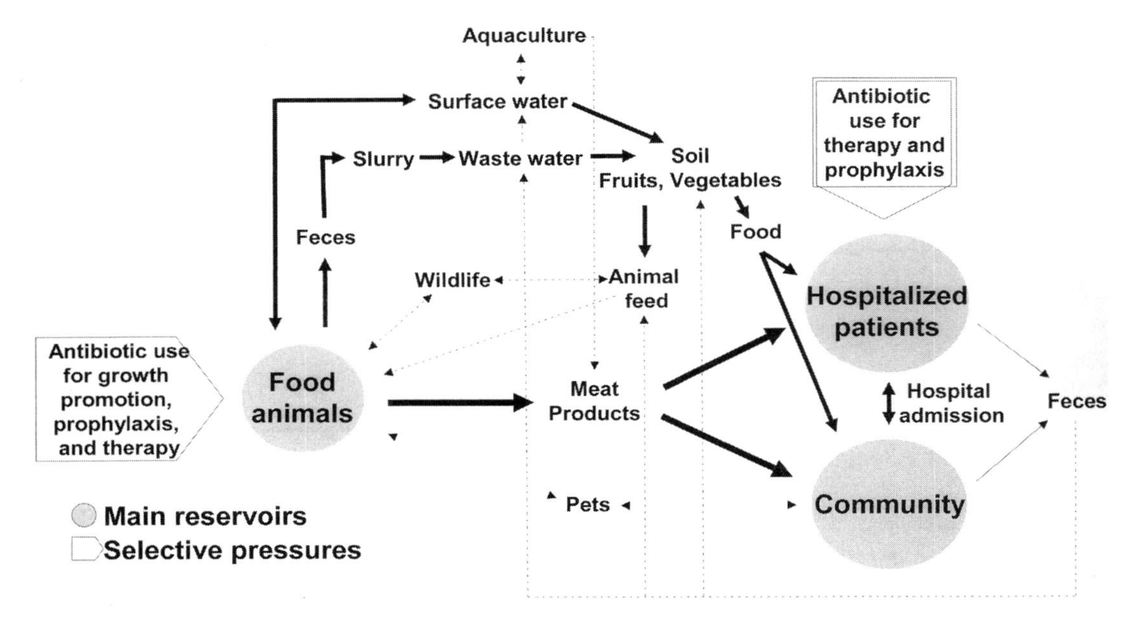


**FIGURE 1** Routes of dissemination of AMR bacteria and determinants. Adapted from reference 197a.

gentamicin- and apramycin-resistant *E. coli* and *Salmonella* spp. following the use of these antibiotics in food animals, prompting the dissemination of a common plasmid carrying the resistance gene (the gene for aminoglycoside-3-*N*-acetyltransferase) among bacteria from both the animals and the workers caring for them (31). In the United States, Fey et al. (51) demonstrated that a ceftriaxone-resistant isolate of *Salmonella enterica* serotype Typhimurium recovered from a child with diarrhea was indistinguishable from a cattle isolate and was likely contracted by direct contact. More recently, methicillin-resistant *Staphylococcus aureus* of clonal lineage ST398 has been recovered from humans and animals in Central Europe (41, 78, 198) and from meat products in The Netherlands (179), raising questions about the dynamics of its spread in the community.

While ingestion of contaminated food products and contact with farm animals are well-established modes of transmission, less attention has been paid to the impact of fertilizing agricultural fields for crop production using untreated wastewater or manure, and the subsequent contamination of produce by resistant fecal pathogens (Fig. 1). Wastewater is a known source of contamination in areas where it is used (3). As animal wastes are contaminated with significant levels of zoonotic pathogens (81), the practice may provide a pathway for resistant enteric pathogens, selected in the farm environment, to contaminate uncooked food commodities. The finding of AMR *E. coli* from broiler feces and litters also suggests the potential for contamination of the environment when litters are applied as fertilizer (44). Additionally, the use of antimicrobials in plant agriculture through spraying over large areas of land may influence the prevalence of AMR bacteria in fresh fruit and vegetables (e.g., apples, pears, and tomatoes) (113). The public health significance of these routes of AMR dissemination has yet to be fully explored.

AMR emerges in bacteria by an acquired ability to inactivate the drugs or their targets or by decreasing access to the cytoplasm. Acquired resistance typically arises as the result of structural gene mutations or by the horizontal transfer of preexisting resistance genes that may impart resistance to one or more antimicrobial classes. A number of these resistance genes have been associated with large transferable plasmids on which may reside other DNA mobile elements, such as transposons and integrons. Sequence data show that many different resistance determinants can amass in linked clusters on plasmids or the chromosome, such that antimicrobials of a different class, including substances such as disinfectants or heavy metals, may select for MDR bacteria (42, 68, 171, 189). Though resistance, in particular, MDR, appears to be most serious in certain bacterial species, this situation may be shifting as large mobile MDR elements spread to new hosts in different environs (189). Thus, there is a continuing need for global surveillance of AMR in pathogenic and commensal bacteria of both animal and human origin. In addition, more quantitative information is necessary to describe the magnitude and nature of the contribution of antimicrobial use in food and companion animals and its impact on the severity of illness worldwide (52, 63, 80, 100, 140, 161, 162, 165, 166). Furthermore, there is an urgent need for empirical studies and risk models that evaluate interventions aimed at minimizing the negative repercussions associated with specific antimicrobial uses in animals (63, 163). This chapter focuses on the occurrence of AMR phenotypes among selected food-borne bacteria, with an emphasis on isolates recovered from foods of animal origin and the potential public health consequences. Current risk assessment and risk management strategies to mitigate the spread of AMR food-borne bacteria are also discussed.

## USE OF ANTIMICROBIAL AGENTS IN FOOD ANIMALS

Antimicrobials are used primarily in food animal production for either (i) therapy in animals that show overt signs of clinical illness, (ii) prophylaxis in healthy animals at risk for contracting an infection, (iii) infection control (metaphylaxis) in a herd or flock that displays elevated morbidity or mortality, and (iv) improved feed efficiency (growth promotion) in healthy animals. It was the unexpected discovery in the 1950s that antimicrobials increased the growth rates and improved the efficiency of feed utilization in animals that led to their use as growth promoters (88). The use of antimicrobials in healthy animals for improved growth performance is the most controversial practice, since it involves the continuous administration of antimicrobials to healthy animals. In addition, many licensed compounds may be administered for both growth-promoting and prophylactic purposes, depending on the dosage. Growth-promoting dosages are given to healthy cattle, pigs, and poultry, usually at low concentrations (2.5 to 50 ppm) for extended periods. Past studies have shown that this practice is a potentially significant driving force in accelerating the emergence of resistance in some bacterial species that can infect humans (45, 91, 110, 187). The European Union has banned the use of growth promoter antibiotics in feed because of their ability to select for resistance to antimicrobials used for treating human infections (139, 177). As with prophylactic and feed efficiency practices, entire groups of food-producing animals (e.g., fish and poultry) may be treated for therapeutic purposes, which is accomplished via medicated feed and/or water as the only practical means of administration (110). Herd or flock treatment is often initiated when illness is first detected in a small fraction of the animals. It is aimed at treating sick animals while exposing others in the herd or flock to prevent disease (86, 110). Other prophylactic antimicrobial treatments are also typically implemented during high-risk or high-stress periods when illness is likely to be observed, such as during transport or after weaning.

The extensive use of antimicrobials in livestock, poultry, and farmed seafood has had considerable beneficial effects on food animal health and production efficiency. A list of animal antimicrobials used in the United States is shown in Table 1. The principal diseases requiring the most extensive use of antimicrobial drugs for treatment or prophylaxis in veterinary medicine are respiratory and enteric diseases in cattle and pigs, colibacillosis in poultry, and mastitis in dairy cattle. In fact, one of the earliest reports of antimicrobial use in veterinary medicine was for the treatment of mastitis in dairy cows in the mid-1940s (53).

Most classes of antimicrobials used in food animals have human analogues and therefore have the potential to select for resistance to human antimicrobials. These include penicillins, cephalosporins, macrolides, lincosamides, streptogramins, aminoglycosides, fluoroquinolones, tetracyclines, folate pathway inhibitors (sulfonamides and potentiated sulfonamides, e.g., trimethoprim-sulfamethoxazole), phenicols, and polypeptides (Table 1). Other antimicrobial agents, which are mostly administered in feeds for growth promotion and prophylaxis, belong to classes used only in veterinary medicine, such as the arsenical compounds, pleuromutilins, quinoxalines (e.g., olaquindox), phosphoglycolipids (e.g., flavomycin) and ionophores (e.g., monensin, salinomycin, and lasalocid).

Antimicrobials are also frequently used in small animal practice to treat a variety of illnesses, including infections of the urinary tract, respiratory tract, wounds, skin, and ears (Table 1). Similar to the situation in food animals, antimicrobial use in companion animals will, over time, select for resistant strains. Several longitudinal studies in veterinary hospitals have documented the emergence of MDR zoonotic pathogens, such as *Staphylococcus intermedius, Enterococcus, E. coli,* and *Salmonella* (132, 133, 145, 155). These isolates tend to be recovered from the worst cases, often representing repeated treatment failures, for which cultures are more likely to be submitted to veterinary diagnostic laboratories (145). There is a growing body of evidence showing that companion animals, in particular, cats and dogs, may serve as sources of AMR zoonotic bacteria in close proximity to humans (8, 63). The hazard to human health associated with the use of antimicrobials in companion animals has yet to be fully determined and warrants further investigation.

## DISSEMINATION OF AMR

Numerous studies have demonstrated an association between antimicrobial use in animals and the subsequent isolation of resistant bacteria from the same animals (6, 11, 19, 31, 37, 103, 109, 110, 126). Once an AMR determinant has become widely disseminated in different ecosystems, it is difficult to conduct trace studies to identify particular origins. It may be possible, however, to track the spread of resistant organisms after the introduction of a new animal antimicrobial drug. One example which showcased the emergence and dissemination of a particular AMR genotype/phenotype after introduction of a unique antimicrobial in animal husbandry occurred in the early 1980s, when the streptothricin antimicrobial nourseothricin replaced oxytetracycline as a growth-promoting agent in pigs in the former German Democratic Republic (178, 197). Streptothricin antimicrobials do not demonstrate cross-resistance with other antimicrobials and had not previously been used in human medicine (178). No resistance was detected prior to the agent's use. Within 2 years of nourseothricin use, resistance was observed in *E. coli* isolates from the gut flora of treated pigs. The resistance mechanism was identified as a streptothricin acetyltransferase *(sat-1)* and was mapped to the transposon Tn*1825* (178). Nourseothricin use was discontinued upon German reunification in 1990; however, by that time, resistance had disseminated to *E. coli* recovered from the gut flora of pig farmers and their family members, as well as *E. coli* from urban residents and *E. coli* causing urinary tract infections (197). The *sat-1* gene was later identified in *Salmonella* and *Shigella* spp. associated with diarrhea, the latter being most interesting, as *Shigella* does not have an animal reservoir (79). Since then, Tn*1825* and the *sat-1* gene have been identified in several other zoonotic pathogens, in-

**TABLE 1** Examples of antimicrobials approved for use in animals in the United States[a]

| Indication | Drugs approved for use in: | | | | |
|---|---|---|---|---|---|
| | Aquaculture | Companion animals[b] | Cattle | Poultry | Swine |
| Treatment | Florfenicol<br>Formalin<br>Hydrogen peroxide<br>Ormetoprim-sulfadimethoxine<br>Oxytetracycline<br>Sulfamerazine | Amikacin<br>Amoxicillin<br>Amoxicillin-clavulanic acid<br>Ampicillin<br>Cefadroxil<br>Ceftiofur<br>Chloramphenicol<br>Clindamycin<br>Difloxacin<br>Enrofloxacin<br>Erythromycin<br>Gentamicin<br>Lincomycin<br>Marbofloxacin<br>Neomycin<br>Orbifloxacin<br>Ormetoprim-sulfadimethoxine<br>Penicillin<br>Spectinomycin<br>Sulfonamides<br>Tetracyclines<br>Trimethoprim-sulfadiazine<br>Tylosin | Amoxicillin<br>Ampicillin<br>Ceftiofur<br>Cephapirin<br>Danofloxacin<br>Enrofloxacin<br>Erythromycin<br>Florfenicol<br>Gentamicin<br>Neomycin<br>Novobiocin<br>Penicillin<br>Pirlimycin<br>Spectinomycin<br>Sulfonamides<br>Tetracyclines<br>Tilmicosin<br>Tylosin | Ceftiofur<br>Erythromycin<br>Gentamicin<br>Lincomycin<br>Neomycin<br>Ormetoprim-sulfadimethoxine<br>Penicillin<br>Spectinomycin<br>Sulfonamides<br>Tetracyclines<br>Tylosin<br>Virginiamycin | Amoxicillin<br>Ampicillin<br>Ceftiofur<br>Florfenicol<br>Gentamicin<br>Lincomycin<br>Neomycin<br>Penicillin<br>Spectinomycin<br>Sulfonamides<br>Tetracyclines<br>Tiamulin<br>Tilmicosin<br>Tylosin<br>Virginiamycin |
| Growth promotion and feed efficiency[c] | | | Bacitracin<br>Bambermycin<br>Chlortetracycline<br>Erythromycin<br>Tylosin<br>Virginiamycin | Bacitracin<br>Bambermycin<br>Chlortetracycline<br>Erythromycin<br>Lincomycin<br>Ormetoprim-sulfadimethoxine<br>Penicillin<br>Spectinomycin<br>Tylosin<br>Virginiamycin | Bacitracin<br>Bambermycin<br>Chlortetracycline<br>Erythromycin<br>Penicillin<br>Tiamulin<br>Tylosin<br>Virginiamycin |

[a]FDA database of approved animal drug products (http://www.fda.gov/cvm/greenbook.html).
[b]Includes horses, dogs and cats; not all antimicrobials are approved in each animal species.
[c]Does not incude ionophores (e.g., monensin, lasalocid, and salinomycin).

cluding Shiga toxin-producing *E. coli* from numerous countries across the globe (137, 138, 156, 164, 191, 214, 219). The nourseothricin example illustrates how a particular resistance determinant carried on a mobile genetic element can undergo international expansion in multiple pathogens long after the specific selection pressure has been removed.

While it is often assumed that continual selection pressure is needed to enrich and maintain resistance traits in a population, there is no guarantee that bacteria recovered from animals or animal meats "raised without antibiotics" will be more susceptible than isolates obtained from conventionally reared animals for some drugs. A study by LeJeune and Christie compared the microbiological quality of 77 samples of ground beef from conventionally raised cattle with 73 samples of ground beef from cattle raised without antimicrobial agents (97). Resistance to ceftiofur and chloramphenicol was more prevalent in beef samples from conventionally raised cattle, at 18 and 30%, respectively, compared to 5 and 12% prevalence in beef from cattle raised without antimicrobial agents. However, the rates of prevalence of *E. coli* resistant to ampicillin (39%), amoxicillin-clavulanic acid (23%), ceftriaxone (5%), tetracycline (19%), streptomycin (19%), kanamycin (11%), trimethoprim-sulfamethoxazole (2%), and gentamicin (1%) were similar in the two groups. Bischoff et al. (16) found that 53% of diagnostic swine *E. coli* isolates exhibited resistance to chloramphenicol, an antibiotic whose use in food animals has been prohibited in the United States since the mid-1980s. Comparing *Campylobacter* isolates from conventional and organic facilities, Luangtongkum et al. (102) found considerably higher rates of resistance to most agents used in chickens raised with antibiotics, but similar resistance levels for tetracycline. While it is clear that use leads to resistance, these data indicate that some resistance determinants can circulate in animal hosts in the absence of corresponding selection pressures. Evidence suggests that some resistant strains of *E. coli* and *Salmonella* have become acclimated to the energetic costs of carrying chromosomal resistance genes and have adapted to the intestinal environment of their animal hosts and that independent fitness attributes allow resistant strains to persist in the absence of selection pressure (40, 92, 93).

More attention is being paid to the coselection of linked genes carried by extrachromosomal elements and their role in the spread of MDR phenotypes. A few examples suggest that antimicrobials used only in food animal production can coselect for resistance to antimicrobials used in human medicine. Food animal slaughtering facilities in the United States apply carcass sanitization or decontamination treatments immediately before and after hide removal, at the end of the dressing process, and, in some cases, after carcass chilling (8). Many of these treatments involve the use of antimicrobial solutions as rinses or spray washes designed to reduce bacterial loads on animal hides (169). For example, the U.S. Food and Drug Administration (FDA) approved cetylpyridinium chloride (CPC) for use as an antimicrobial treatment for use on raw poultry carcasses in 2004 (http://www.fda.gov/ohrms/dockets/98fr/04-7399.htm). CPC is also being evaluated as an antimicrobial intervention to reduce microbial contamination on beef cattle hides (18). Resistance to CPC has been infrequently reported. Chung and Saier (35) reported that overexpression of the *E. coli sugE* gene (a member of the small MDR family of drug efflux pumps) conferred high-level resistance to a narrow range of quaternary ammonium compounds, including CPC. Of food safety significance is the recent identification of the *sugE* gene on large MDR plasmids in human and poultry isolates of *Salmonella* from the United States and *Salmonella* and uropathogenic *E. coli* from Taiwan (58, 170, 189, 208). These investigators have also reported the close physical location of the *sugE* gene with the $bla_{CMY-2}$ β-lactamase gene, which confers resistance to extended-spectrum cephalosporins (e.g., ceftriaxone). According to a report by Welch et al. (189), two copies of the *sugE* gene flanking two $bla_{CMY-2}$ β-lactamase genes were found on a large IncA/C MDR plasmid from *Salmonella enterica* serovar Newport (Fig. 2), which also possessed genes conferring resistance to tetracycline, gentamicin, streptomycin, sulfa drugs, florfenicol, and mercury. This IncA/C plasmid was common among *Salmonella* isolates recovered from retail meats obtained from the U.S. National Antimicrobial Resistance Monitoring System (NARMS) between 2002 and 2005 (189). The presence of *sugE* on MDR plasmids in *Salmonella* suggests the possibility that carcass treatments might disproportionately enrich for MDR *Salmonella* in cases where they fail to decontaminate the meat surface.

It is thought that ionophore antimicrobials have little to do with therapeutic AMR, since they have a distinctly different mode of action (25, 151). However, the recent identification of *Campylobacter* isolates recovered from retail poultry for which the MICs of roxarsone (an arsenical compound) were decreased raised the possibility that arsenic resistance determinants could become linked with other AMR genotypes on extrachromosomal elements in *Campylobacter,* analogous to the situation in *Enterococcus faecium,* in which copper resistance is linked with macrolide and glycopeptide resistance genotypes (69, 157).

Another example of coresistance to agents of animal and human medicine is documented in the recent report describing the first identified genetic basis for resistance to the swine growth enhancer olaquindox (OQX) in an *E. coli* isolate recovered from swine manure (66). The gene *(oqxAB)* involved in resistance to OQX encoded a multidrug efflux pump and

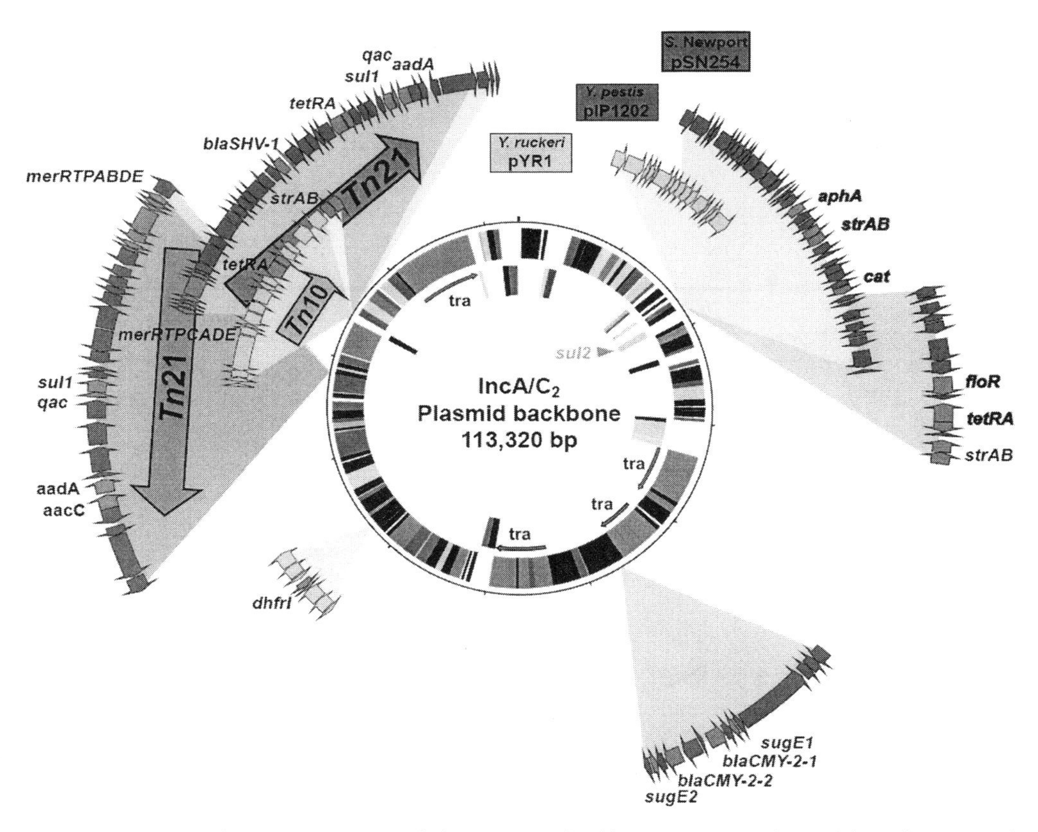


**FIGURE 2** Circular representation of the IncA/C backbone (inner circle) and laterally acquired regions (outer circles) on R plasmids from *Yersinia ruckeri, Yersinia pestis,* and *Salmonella* serovar Newport. Printed with permission of W. Florian Fricke, University of Maryland School of Medicine, Baltimore.

was localized on a conjugative plasmid. Plasmids carrying *oqxAB* conferred high-level resistance to OQX (MIC > 128 μg/ml) as well as to chloramphenicol (>64 μg/ml) and ampicillin (>32 μg/ml) in *E. coli* (65). This raises some concern that the use of OQX could select for the proliferation of a conjugative plasmid that also carries resistance to ampicillin and chloramphenicol. The extent of genetic linkage and the degree to which coresistance is maintained by different selection pressures constitute an interesting area of investigation and an important consideration in assessing risks associated with antimicrobial use.

## MONITORING AMR IN FOOD-BORNE BACTERIA

Continuous monitoring of antimicrobial susceptibility/resistance profiles in food-borne pathogens is needed in order to identify important trends that signify a need to amend antimicrobial use practices. Monitoring is also necessary to gauge the effect of different mitigation steps. The World Health Organization (WHO), the Food and Agriculture Organization (FAO), and the World Organization for Animal Health (OIE) recommend that countries implement programs to monitor the use of antimicrobials in animals, as well as the occurrence of AMR in bacteria from animals and from foods of animal origin (http://whqlibdoc.who.int/hq/2004/WHO_CDS_CPE_ZFK_2004.7.pdf). The WHO, FAO, and OIE recognize that data obtained by such monitoring may be used to (i) document the usage of antimicrobials and the occurrence of resistance, (ii) identify epidemiological trends, (iii) compare the occurrence of resistance between countries or regions over specific periods, (iv) develop risk assessments, (v) evaluate the effectiveness of any control measures implemented, (vi) identify focused and targeted research, and (vii) develop policies for the containment of AMR. Considering the magnitude of international travel and trade in agricultural products, global surveillance of antimicrobial susceptibility of food-borne pathogens has become increasingly important to assist veterinarians, physicians, and public health officials in determining appropriate chemotherapeutic treatment regimens as well as setting national policy.

Over the past decade, several countries have made significant efforts in establishing surveillance systems for AMR in food-borne bacteria. All of these systems continue to adapt new strategies and techniques as they work to improve upon data collection, analysis, integration, and global harmonization. Continuous surveillance for AMR among bacteria isolated from food animals was first established in Denmark in 1995 (1). The Danish Integrated Antimicrobial Resistance Monitoring and Research Program, DANMAP, was established on the initiative of the Danish Ministry of Health and the Danish Ministry of Food, Agriculture and Fisheries, as a coordinated national surveillance and research program for antimicrobial consumption and AMR in bacteria from animals, foods, and humans. Three categories of bacteria, namely, indicator bacteria, zoonotic bacteria, and animal pathogens, are continuously isolated from broilers, cattle, and pigs and tested for susceptibility to antimicrobial agents used for therapy and growth promotion (1). Annual reports are published (http://www.danmap.org/) describing the annual consumption of antimicrobial agents and the occurrence of resistance in different reservoirs over time. Similar AMR surveillance programs exist in Norway (NORM-VET) and Sweden (SVARM).

In North America, three monitoring programs exist that focus on the dissemination of AMR food-borne bacteria in farm animals, foods, and humans: (i) the Canadian Integrated Program for Antimicrobial Resistance Surveillance (CIPARS), (ii) the NARMS program in the United States, and (iii) the ResistVet program in Mexico. CIPARS was formally recommended in 1997 under the auspices of Health Canada and the Canadian Infectious Diseases Society. The NARMS program began in 1996 and is a collaboration between the FDA Center for Veterinary Medicine (FDA-CVM), the Centers for Disease

Control and Prevention (CDC), and the U.S. Department of Agriculture (USDA). ResistVet began in 2000 as a collaborative effort between the FDA, the USDA, and the O'Horan General Hospital in Yucatan with a plan to monitor AMR in five states in Mexico. Currently, annual reports are published by CIPARS (http://www.phac-aspc.gc.ca/cipars-picra/index_e.html) and NARMS (http://www.fda.gov/cvm/narms_pg.html).

The NARMS program monitors antimicrobial susceptibility/resistance among two categories of enteric bacteria recovered from food animals, humans, and retail meats: zoonotic bacterial pathogens *(Salmonella* and *Campylobacter)* and commensal bacteria *(E. coli* and *Enterococcus).* Bacterial isolates recovered from ill people are sent to the CDC Coordinating Center for Infectious Diseases in Atlanta, GA, by participating state and local health departments. The animal isolates are gathered from healthy farm animals, animal clinical specimens, carcasses of food animals at slaughter, and ground products obtained from processing plants. These isolates are tested by the USDA Agricultural Research Service (USDA-ARS), Bacterial Epidemiology and Antimicrobial Resistance Research Unit, in Athens, GA. Animal isolates also come from federally inspected slaughter and processing facilities, the USDA's animal health monitoring studies on farms, and veterinary diagnostic laboratories. Retail meat samples are collected from grocery stores in states participating in the Foodborne Diseases Active Surveillance Network (FoodNet). Participating laboratories from FoodNet states isolate the bacteria of interest and forward the isolates to the FDA-CVM, Office of Research, laboratory in Laurel, MD, for further analysis. All three arms (CDC, USDA-ARS, and FDA-CVM) also characterize *Salmonella* and *Campylobacter* through the use of pulsed-field gel electrophoresis (PFGE) in an effort to determine genetic relatedness between isolates. Epidemiological and microbiological research studies are conducted within each agency or between agencies on isolates of special interest, such as those of a particular serotype or expressing a particular resistance pattern.

The most recent data from NARMS on the number and percentage of retail meat samples positive for *Salmonella, Campylobacter, Enterococcus,* and *E. coli* are presented in Table 2. Approximately 20,000 retail meat samples were tested for the presence of *Campylobacter* and *Salmonella* between 2002 and 2006 through NARMS. *Salmonella* was more often recovered from chicken breast (9 to 13%) and ground turkey (11 to 15%) samples, whereas *Campylobacter* was recovered almost exclusively from chicken breasts (46 to 60%). Only a single *Campylobacter* isolate was obtained from 5,100 ground beef samples and just 18 isolates were obtained from 5,076 pork chop samples tested between 2002 and 2006 (Table 2), confirming other reports from North America, Europe, Asia, and New Zealand that these retail meats are rarely contaminated with *Campylobacter* (4, 12, 17, 57, 77, 200, 213). In retail chicken breast samples, *Campylobacter jejuni* was more often isolated than *Campylobacter coli,* while *C. coli* predominated among pork chops, albeit at a very low prevalence (≤1%).

Enterococci were readily isolated from all retail meats tested, ranging, on average, from 88% of pork chops to 98% of all chicken breasts sampled, which is consistent with other studies from the United States (70, 112). Interestingly, *Enterococcus faecalis* was the predominant species recovered from pork chop, ground beef, and ground turkey samples, whereas *E. faecium* was most common among chicken breasts (data not shown). These data contrast somewhat with those reported by Hayes et al. (70), who showed *E. faecium* as the predominant species recovered from ground turkey, ground beef, and chicken breasts from a retail study in Iowa. Additionally, the majority of *Enterococcus hirae* isolates were recovered from ground beef samples over time, a finding consistent with prior studies (70, 95). Non-type-specific (NTS) *Escherichia coli* was also consistently recovered,

**TABLE 2** Number and percentage of retail meats positive for *Campylobacter, Salmonella, Enterococcus,* and *E. coli* from NARMS, 2002 to 2006

| Meat type | Yr | *Campylobacter* | | | *Salmonella* | | | *Enterococcus* | | | *E. coli* | | |
|---|---|---|---|---|---|---|---|---|---|---|---|---|---|
| | | No. of samples | No. of isolates | % Positive | No. of samples | No. of isolates | % Positive | No. of samples | No. of isolates | % Positive | No. of samples | No. of isolates | % Positive |
| Chicken breast | 2002 | 616 | 288 | 46.8 | 616 | 60 | 9.7 | 390 | 381 | 97.7 | 390 | 282 | 72.3 |
| | 2003 | 897 | 469 | 52.3 | 897 | 83 | 9.3 | 477 | 466 | 97.7 | 477 | 396 | 83.0 |
| | 2004 | 1,172 | 706 | 60.2 | 1,172 | 157 | 13.4 | 476 | 466 | 97.9 | 476 | 400 | 84.0 |
| | 2005 | 1,190 | 554 | 46.6 | 1,194 | 153 | 12.8 | 470 | 457 | 97.2 | 468 | 393 | 84.0 |
| | 2006 | 1,196 | 585 | 48.9 | 1,196 | 149 | 12.5 | 478 | 471 | 98.5 | 478 | 431 | 90.2 |
| Ground turkey | 2002 | 642 | 4 | 0.6 | 642 | 74 | 11.5 | 395 | 387 | 98.0 | 395 | 304 | 77.0 |
| | 2003 | 857 | 5 | 0.6 | 857 | 114 | 13.3 | 447 | 418 | 93.5 | 447 | 333 | 74.5 |
| | 2004 | 1,165 | 12 | 1.0 | 1,165 | 142 | 12.2 | 466 | 437 | 93.8 | 466 | 376 | 80.7 |
| | 2005 | 1,195 | 20 | 1.7 | 1,195 | 183 | 15.3 | 470 | 452 | 96.2 | 470 | 396 | 84.3 |
| | 2006 | 1,184 | 29 | 2.4 | 1,184 | 159 | 13.4 | 466 | 435 | 93.3 | 466 | 395 | 84.8 |
| Ground beef | 2002 | 642 | 0 | 0.0 | 642 | 9 | 1.4 | 399 | 383 | 96.0 | 399 | 295 | 73.9 |
| | 2003 | 880 | 1 | 0.1 | 880 | 10 | 1.1 | 470 | 432 | 91.9 | 470 | 311 | 66.2 |
| | 2004 | 1,186 | 0 | 0.0 | 1,186 | 14 | 1.2 | 480 | 448 | 93.3 | 480 | 338 | 70.4 |
| | 2005 | 1,196 | 0 | 0.0 | 1,196 | 8 | 0.7 | 470 | 447 | 95.1 | 468 | 316 | 67.5 |
| | 2006 | 1,196 | 0 | 0.0 | 1,196 | 20 | 1.7 | 480 | 454 | 94.6 | 480 | 328 | 68.3 |
| Pork chop | 2002 | 613 | 5 | 0.8 | 613 | 10 | 1.6 | 390 | 369 | 94.6 | 390 | 184 | 47.2 |
| | 2003 | 899 | 4 | 0.4 | 899 | 5 | 0.6 | 479 | 426 | 88.9 | 479 | 218 | 45.5 |
| | 2004 | 1,176 | 3 | 0.3 | 1,176 | 11 | 0.9 | 478 | 404 | 84.5 | 478 | 232 | 48.5 |
| | 2005 | 1,196 | 2 | 0.2 | 1,196 | 9 | 0.8 | 470 | 409 | 87.0 | 465 | 205 | 44.1 |
| | 2006 | 1,192 | 4 | 0.3 | 1,192 | 9 | 0.8 | 476 | 402 | 84.5 | 476 | 236 | 49.6 |
| Total | | 20,290 | 2,691 | 13.3 | 20,294 | 1,379 | 6.8 | 9,127 | 8,544 | 93.6 | 9,118 | 6,369 | 69.8 |

with the lowest prevalence in pork chops, as previously reported in the United States (82).

Data gathered in NARMS and other similar surveillance programs can be used to inform and prioritize science-based approaches to ensuring food safety (e.g., development of risk assessments and mathematical models) and to reduce uncertainty about the potential public health hazards posed by antimicrobial use in food animals. However, a lack of harmonization between various national integrated surveillance systems needs to be resolved in order to better compare data from different regions. For example, antimicrobials under surveillance may differ by national program, as well as by the interpretive criteria used to calculate resistance prevalence. Considering the extent of international travel and agricultural product trade, it is important that national AMR surveillance programs are standardized in their activities, so data can be reliably compared on a global basis. Global surveillance of food-borne pathogens and sharing of antimicrobial susceptibility data among developed and developing countries are also needed. The WHO began the Global Salmonella Surveillance (WHO-GSS; http://www.who.int/salmsurv/en/) program in January 2000. WHO-GSS is a global network of laboratories and individuals involved in surveillance, isolation, identification, and antimicrobial susceptibility testing of *Salmonella* and other enteric pathogens. The major goal of WHO-GSS is to expand integrated laboratory-based food-borne disease surveillance and foster collaboration among human health, veterinary, and food-related disciplines, thereby enhancing the capacity of countries to monitor the prevalence and AMR among food-borne pathogens. WHO-GSS can serve as a platform for coordination and harmonization at the international level in an effort to contain AMR in food-borne pathogens.

## AMR PHENOTYPES AMONG SELECTED FOOD-BORNE BACTERIA

AMR phenotypes of enteric organisms isolated from retail food have been examined for both commensal bacteria such as *E. coli* and *Enterococcus* and the zoonotic pathogens *Salmonella* and *Campylobacter*. Due to the value of commensal organisms as sentinels for the selection pressure of antimicrobial drug exposure, and because of the burden of enteric illness caused by the major pathogens, resistance in *E. coli* and *Salmonella* is highlighted.

### *Escherichia coli*

*Escherichia coli* is a member of the *Enterobacteriaceae* family and a normal constituent of the intestinal microflora in humans and warm-blooded animals. *E. coli* organisms were therefore long considered as having low virulence and being generally beneficial to the host. It is now clear that there are many pathogenic *E. coli* strains that can cause a variety of diseases in animals and humans. Particular pathogenic types of *E. coli* usually differ from those that predominate in the enteric floras of healthy animals and individuals in that they are more likely to express particular virulence factors (46). The enteric habitat of *E. coli* in animals provides an easy source of contamination for animal-derived meats at slaughter and at points downstream in the food production process. The variety of *E. coli* biotypes that can pass from animal-derived foods to humans is still not completely clear. In terms of human enteric infections, enterohemorrhagic *E. coli* (EHEC) O157:H7 produces one of the most severe disease syndromes when it presents as hemorrhagic colitis and/or hemolytic-uremic syndrome (89, 125).

Historically, serotyping based on the O (somatic lipopolysaccharide), F (fimbrial), K (capsular), and H (flagellar) antigens has been used to compare strains. Based on the 100 different K groups, 180 O groups, and 60 H antigens (and growing) described, nearly 700 different antigenic types can be determined (150). Research in this area shows that a relatively small number of serogroups predominate in cases of both diarrheal and extraintestinal infections (82, 134–136, 158). While useful for distinguishing strains, serotyping alone does not indicate the presence of disease-associated virulence factors. Over time, a host of specific virulence properties have been identified in distinct pathogenic strains associated with different spectra of diseases, including diarrheal illness. This has led to the current classification of pathogenic types of *E. coli* by their specific virulence attributes (e.g., toxins, adhesins, and invasiveness), serotypes, clinical symptoms, and pathogenesis. Currently, at least seven distinct classes of pathogenic *E. coli* are recognized: (i) enterotoxigenic *E. coli,* (ii) enteropathogenic *E. coli,* (iii) EHEC, (iv) enteroinvasive *E. coli,* (v) diffuse adhering *E. coli,* (vi) necrotoxigenic *E. coli,* and (vi) enteroaggregative *E. coli* (90, 125, 186). Additionally, those *E. coli* strains with the ability to cause extraintestinal disease have been grouped under the term extraintestinal pathogenic *E. coli* (ExPEC) (84).

Various types of AMR *E. coli* were recovered from retail meats (in particular, poultry products) as early as the 1970s (28, 99) and continue to be reported for a variety of foods today. In recent years, a number of surveys have looked for the presence of *E. coli,* particularly O157:H7, or total coliforms from various food commodities; however, quantitative susceptibility data on such isolates are still very limited. Nearly all of the published data describing AMR phenotypes in *E. coli* originating from foods have come from isolates originating from retail raw meat products in the past few years. With the exception of verocytotoxigenic *E. coli* (VTEC), serotype data are limited among food isolates. For example, Meng et al. (116) compared *E. coli* O157:H7 isolates from animals, food, and humans and found that 7 of 29 (24%) *E. coli* O157 isolates from ground beef in the United States were resistant to streptomycin and trimethoprim-sulfamethoxazole only, whereas 6 isolates from raw milk were susceptible to all the antimicrobials tested. In another study, Schroeder et al. (160) tested a total of 361 *E. coli* O157 isolates from cattle, swine, and humans, along with 27 isolates from foods. Eleven of these were identified as O157:H7 and were susceptible to all tested antimicrobials. Similar antimicrobial susceptibility results have been reported for *E. coli* O157:H7 isolates from other countries, including Greece (47), Malaysia (147), Brazil (30), and Ireland (184). MDR phenotypes have, however, been documented among VTEC isolates, including resistance to the extended-spectrum cephalosporins (190).

AMR in NTS strains of *E. coli* may also play an important role in the ecology of resistance. Transmission of resistance genes from normally nonpathogenic species to more virulent organisms within the animal or human intestinal tract may be an important mechanism for acquiring clinically relevant AMR organisms (196). In fact, MDR phenotypes are often more frequently reported among NTS *E. coli* than among pathogenic variants such as VTEC (122, 153, 159, 160, 184). Although AMR NTS *E. coli* isolates have been recovered from various foods, including vegetables, confectionaries, milk, and milk products, the majority of resistant strains have been recovered from retail meat and poultry samples (122, 153, 159, 160, 184). A large study conducted in Spain compared human and animal *E. coli* isolates with those obtained from ground beef, sausage, chicken breasts, turkey meats, and precooked chicken products (153). Forty-seven *E. coli* isolates were recovered from 69 diverse food samples, with the most common resistance phenotypes being resistance to nalidixic acid (53%), ampicillin (47%), kanamycin (40%), gentamicin (17%), and ciprofloxacin (13%). MDR strains were not uncommon, with six strains displaying resistance to four antimicrobials. The authors hypothesized that the high prevalence of quinolone resistance among recovered isolates was the result of extensive enrofloxacin use in animals in Spain (153).

In regions where laboratory-based surveillance is not established, only small pilot studies are available. Data are sparse from the Middle East, but one study from Lebanon tested selected foods (lahm bi'ajeen [meat pies] and shawarma, Lebanese meat sandwiches similar to gyros and donairs, containing meat, vegetables, and sesame seed oil-based sauce) for AMR *E. coli* (67). Sixty-nine percent of *E. coli* isolates were found to display resistance to at least one of the antimicrobials tested, in particular, trimethoprim-sulfamethoxazole. The results showed that meat-based fast foods in Lebanon could be a public health hazard, especially shawarma, as they may act as potential vehicles for many AMR pathogenic organisms. Similar results were observed in an earlier study in Australia which characterized AMR *E. coli* isolates from 100 samples of beef, veal, lamb, and pork at the retail level (13). Sixteen percent of the 400 samples were contaminated with resistant *E. coli,* with significantly more *E. coli* isolates recovered from retail pork products than from other meats ($P < 0.01$).

The Norwegian monitoring program for AMR, NORM (for humans) and NORM-VET (for food, feed, and animals), has been operating since the year 2000. The NORM-VET program tested 944 *E. coli* isolates from meat and meat products of poultry ($n = 359$), pork ($n = 295$), cattle ($n = 190$), and sheep ($n = 100$) during the years 2000, 2001, 2002, and 2003 for antimicrobial susceptibilities (172). The most commonly observed resistance phenotypes were resistance to streptomycin (15.1%), sulfonamides (14.3%), tetracyclines (6.6%), ampicillin (6.5%), and trimethoprim (4.3%). Resistance to quinolones, cephalosporins, neomycin, and gentamicin was rare (~1%). A total of 241 isolates were classified as resistant to at least one of the antimicrobial agents tested and were further screened for the presence of specific resistance genes. Resistance to sulfonamides and ampicillin was primarily encoded by the *sul2* gene and $bla_{TEM}$ genes, respectively, whereas tetracycline resistance was mediated by either the *tet*(A) or *tet*(B) gene (172).

The largest published U.S. survey for NTS *E. coli* in retail meats involved a total of 1,025 meat samples purchased in the greater Washington, DC, area from 1998 to 2000 (159). Two hundred samples of ground meats (50 beef, 51 chicken, 49 pork, and 50 turkey) and 825 samples of whole meats (210 beefsteaks, 212 whole chickens, 209 pork chops, and 194 turkey breasts) were cultured for *E. coli,* and recovered isolates were further subjected to antimicrobial susceptibility testing. The incidence of *E. coli* was significantly higher in ground products than in whole meats (37% versus 22%). Eighty-six percent of isolates displayed resistance to at least one antimicrobial, 26% to four or more, and 7% to six or more. Resistance was most often observed with tetracycline (59%), sulfamethoxazole (45%), streptomycin (44%), cephalothin (38%), ampicillin (35%), and, to a lesser extent, gentamicin (12%), nalidixic acid (8%), chloramphenicol (6%), ceftiofur (4%), and ceftriaxone (1%).

In the U.S. NARMS program, *E. coli* isolates were recovered from a monthly sampling of chicken breasts, ground turkey, ground beef, and pork chops purchased from selected grocery stores in six participating FoodNet sites (Connecticut, Georgia, Maryland, Minnesota, Oregon, and Tennessee) in 2002 and an additional two sites each in 2003 (California and New York) and 2004 (Colorado and New Mexico). A total of 7,218 retail meats were examined for *E. coli* from 2002 to 2005, including 1,811 chicken breasts, 1,778 ground turkey samples, 1,817 ground beef samples, and 1,812 pork chops. Overall, 69% of the retail meat samples tested ($n = 4,979$) were contaminated with *E. coli,* with the prevalence of contamination being highest for chicken breasts (81%) and ground turkey (79%). *E. coli* was recovered less frequently from pork chops than from other retail meats under surveillance between 2002 and 2005 (44 to 49%).

AMR to tetracycline (48 to 52%), streptomycin (32 to 38%), and sulfa drugs (27 to 33%) was most often observed (Fig. 3). Resistance to gentamicin and nalidixic acid among recovered *E. coli* isolates increased between 2002 and 2005 from 14 to 20% and 2 to 6%, respectively, whereas resistance to ceftiofur and trimethoprim-sulfamethoxazole fluctuated between 2 and 4% and between 2 and 5%, respectively (Fig. 3). Interestingly, the majority of gentamicin-resistant *E. coli* isolates were recovered from either chicken breasts (30.5%) or ground turkey samples (28.4%)

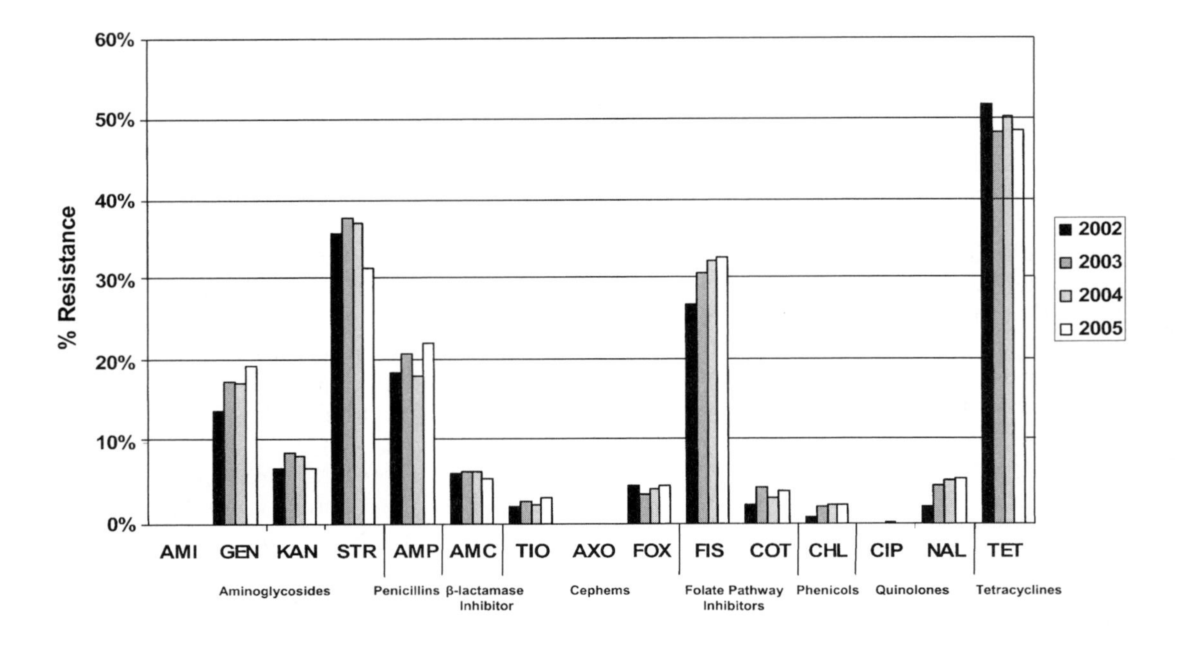


**FIGURE 3** AMR among *E. coli* isolates from NARMS retail meats, 2002 to 2005. AMI, amikacin; GEN, gentamicin; KAN, kanamycin; STR, streptomycin; AMP, ampicillin; AMC, amoxicillin-clavulanic acid; TIO, ceftiofur; AXO, ceftriaxone; FOX, cefoxitin; FIS, sulfisoxazole; COT, trimethoprim-sulfamethoxazole; CHL, chloramphenicol; CIP, ciprofloxacin; NAL, nalidixic acid; TET, tetracycline.

rather than ground beef (0.5%) or pork chops (1%). This may be due to the common practices by turkey breeders of dipping eggs in gentamicin sulfate solutions to prevent mycoplasmosis or of injecting day-old poults to prevent *E. coli*-associated illness (5, 38, 218). Likewise, the higher rate of trimethoprim-sulfamethoxazole resistance observed among poultry isolates may be due to the use of ormetoprim-sulfadimethoxine (cross-resistance) as an aid in the prevention of coccidiosis caused by *Eimeria* species in chickens and turkeys. Nalidixic acid resistance was most often observed among *E. coli* isolates recovered from either ground turkey (4 to 12%) or chicken breasts (3 to 7%), whereas ciprofloxacin resistance, albeit infrequent ($n$ = 5 isolates), was observed primarily among ground turkey *E. coli* isolates (80%). These data are somewhat in contrast to an earlier report from the Minneapolis-St. Paul, MN, area (1999 to 2000), where it was reported that 37% of retail chicken products were found to carry nalidixic acid-resistant *E. coli* (83).

Ceftiofur is the only cephalosporin approved for systemic use in food-producing animals in the United States. This drug was first approved in 1988 as an injectable therapeutic for the treatment of acute bovine respiratory disease, and subsequently it was approved for use in other food animal species, including pigs, sheep, chickens, and turkeys (220). Because ceftiofur-resistant organisms also exhibit decreased susceptibility to cephamycins and extended-spectrum cephalosporins, this antimicrobial in food animals has come under scrutiny as a selective agent responsible for the emergence and dissemination of ceftriaxone-resistant enteric pathogens such as *E. coli* and *Salmonella*. Resistance to ceftiofur was highest in *E. coli* obtained from chicken breast samples (6 to 9%) from all years of NARMS sampling, compared to other retail meats tested (Fig. 3). Additionally, the $bla_{CMY}$ β-lactamase gene was detected in all ceftiofur-resistant *E. coli* isolates, which all also showed decreased susceptibility to the extended-spectrum cephalosporin ceftriaxone (MIC $\geq$ 4 μg/ml). Extended-spectrum-cephalosporin-resistant *E. coli* isolates possessing $bla_{CMY}$ genes located on transferable plasmids have been recovered from diseased cattle in North Dakota as early as 1996 (19), retail ground meats in Maryland (220), and, more recently, from retail ground meats in Taiwan (209) and commercial broiler chicken farms in Canada (44).

*E. coli* strains with resistance or variable susceptibility to extended-spectrum cephalosporins have also been detected in cattle, swine, and poultry in France (118). These strains were shown to produce CTX-M-1 and CTX-M-15-type extended-spectrum β-lactamases, the genes for which were located on a conjugative plasmid and were linked to the insertion sequence ISEcp1, which could have contributed to dissemination of these genes. Recently, $bla_{CTX\text{-}M}$ extended-spectrum β-lactamase genes were detected in 12 *E. coli* isolates recovered over a 7-month period from the ceca of healthy poultry in seven districts in France in 2005 (59). Eleven of these strains were not clonally related and had a $bla_{CTX\text{-}M\text{-}1}$ gene located on transferable plasmids of different sizes and structures. More recently, Jouini et al. (87) detected the first *E. coli* isolates producing CTX-M and SHV extended-spectrum β-lactamases from food samples in Tunisia. The increasing occurrence of extended-spectrum β-lactamases in *E. coli* recovered from food animals as well as the community raises questions with regard to foods of animal origin serving as reservoirs for the transmission of such strains to humans (22, 59, 117, 118, 120). Further epidemiological and molecular analysis of these strains is certainly warranted to determine the reasons for the observed extended-spectrum-cephalosporin resistance phenotypes among *E. coli* isolates recovered from the different retail meats.

The 2006 Danish DANMAP program (http://www.dfvf.dk/) summarized data on susceptibility in *E. coli* strains isolated from broilers ($n$ = 123), cattle ($n$ = 93), and pigs ($n$ = 148) at abattoir and from domestic ($n$ = 534) and imported ($n$ = 550) broiler meat sold at wholesale and retail outlets. Since 2004, resistance to streptomycin-sulfonamide-tetracycline and ampicillin-streptomycin-sulfonamide-tetracycline declined in *E. coli* from pigs (in contrast to rising trends for salmonellae), in which resistance is historically most prevalent. DANMAP data also show significant differences in resistance prevalence between domestic and imported commodities, with comparatively higher rates of *E. coli* resistance observed for isolates obtained from imports. Some of the salient differences in resistance between domestic and imported broiler meat isolates, respectively, were as follows: amoxicillin-clavulanate, 0 and 5%; ceftiofur, $<$1 and 5%; nalidixic acid, 6 and 26%; and ciprofloxacin, 4 and 24%. Ceftiofur resistance was first detected in broiler meat in Denmark in 2006. In addition, ciprofloxacin resistance was significantly higher in both domestic and imported meat than in 2004.

Other studies suggest that the prevalence of AMR among food-borne *E. coli* can be low. Zhao et al. (221) tested ground beef samples purchased from retail stores from New York, NY; San Francisco, CA; Philadelphia, PA; Denver, CO; Atlanta, GA; Houston, TX; and Chicago, IL, for the presence of *Salmonella* and *E. coli*. One hundred two *E. coli* strains were cultured from 404 samples, and overall, resistance was infrequent. The most commonly observed resistance was resistance to cephalothin (31%), followed by tetracycline (13%), amoxicillin-clavulanate (5%), and ampicillin (4%); only three isolates displayed resistance to multiple antimicrobials. Similar observations were reported by Musgrove et al. (124) in their evaluation of the antibiotic susceptibility patterns for *E. coli* isolated from commercial shell eggs in the southeastern United States. In this case, 73% of the *E. coli* isolates were pansusceptible, with tetracycline (30%) and streptomycin (6%) resistance most commonly observed. Another study investigated *E. coli* strains isolated from a commercial beef packing plant in Canada (10). A total of 284 isolates from animal hides ($n$ = 71), washed carcasses ($n$ = 36), conveyers ($n$ = 55), beef trimmings ($n$ = 52), and ground beef ($n$ = 70) were analyzed. Fifty-six percent were susceptible to all 15 antimicrobial agents tested; however, resistance to one or more antimicrobial agents was observed in 49, 50, and 37% of the *E. coli* isolates recovered from conveyers, beef trimmings, and ground beef, respectively. Overall resistance was low, with tetracycline (38%), ampicillin (9%), and streptomycin (6%) resistance being the most common phenotypes observed. The authors stressed the need for improving hygienic conditions during all stages of commercial beef processing and meatpacking to avoid the risk of transfer of AMR bacteria to humans.

With the possible exception of *E. coli* O157:H7 and other toxigenic *E. coli* strains, the current data are insufficient to accurately assess the hazard and the potential public health risks associated with the presence of NTS *E. coli* in foods, regardless of their resistance traits. In addition, there is evidence that antimicrobials may not be helpful in treating some *E. coli* infections. For example, a prospective cohort study of 71 children concluded that antibiotic treatment of children infected with *E. coli* O157:H7 increased the risk of hemolytic-uremic syndrome (199). A growing body of data shows that non-O157: H7 EHEC strains may also be associated with extraintestinal human infections. For these infections, the need for antimicrobial therapy may be critical.

Recent studies suggest that food-borne *E. coli,* indirectly via the host fecal flora, may be an important cause of extraintestinal infections (82, 83, 85, 106, 123, 148). For instance, Manges et al. (106) presented early evidence indicating that an MDR clonal group of *E. coli* was responsible for community-acquired urinary tract infections in women in California, Michigan, and Minnesota between 1996 and 2000. The authors also speculated that the *E. coli* isolates may have been spread by ingestion of contaminated food products as a possible explanation for the temporal and geographical clustering of the clonal group. During the same period, Johnson et al. (83) surveyed retail chicken products for *E. coli* strains carrying virulence factors associated with extraintestinal disease (e.g., urinary tract infection, neonatal meningitis, and sepsis) and showed that 21% of 169 retail samples harbored *E. coli* which met the classification criteria for ExPEC. Additional studies by the same authors further tested the hypothesis that the food supply may disseminate ExPEC and AMR *E. coli* to humans (82, 85). In a prospective survey of 1,648 diverse food items from 10 retail markets in the Minneapolis-St. Paul area from 2001 to 2003, *E. coli* was isolated from 396 (24%) samples. Contamination with *E. coli* varied by food class, with significant differences observed when comparing miscellaneous items (produce and other nonmeat or poultry items, 9%), meats (beef or pork, 69%), and poultry (92%) (82). The prevalence of AMR and virulence-associated traits also varied significantly by food class, being lowest for miscellaneous foods and highest for poultry. Seventeen isolates (2 from miscellaneous foods, 5 from beef or pork, and 10 from poultry), all from phylogenetic group B2 or D, met molecular criteria for ExPEC and exhibited O-UTI antigens or O11/O17/O77 antigens, consistent with possible human pathogenic potential (82). More recent work has also shown that ExPEC strains of human and avian origin show similarities in virulence attributes, implying that avian strains constitute a potential zoonotic risk (123). Taken together, these data suggest that retail foods may be an important vehicle for community-wide transmission of AMR *E. coli* and ExPEC; however, further research is warranted to determine the relative risks to public health.

### Non-Serovar Typhi *Salmonella*

*Salmonella,* a genus which currently includes more than 2,500 different serotypes, can colonize and cause disease in a variety of food-producing and non-food-producing animals.

Although all serotypes may be regarded as potential human pathogens, in reality only a limited number of them have been associated with infection in humans and animals. Though a growing number of human salmonellosis cases are associated with contaminated produce, traditionally illness has been linked with consumption of contaminated food of animal origin, especially poultry and poultry products (20, 48, 54, 104, 192, 213). Although the hygienic standards of meat production are quite high in most developed countries, fecal contamination of meat products cannot be completely prevented. For example, Stopforth et al. reported that *Salmonella* incidence in broilers was higher after the immersion chiller than before, suggesting that the chilling process can serve as an important source of pathogen cross-contamination between carcasses (169). As a result, a small number of carcasses with salmonellae can play a role in spreading contamination. Recently, several investigators have suggested that processing conditions may play a significant role in promoting and influencing the selection of food-borne pathogens, including AMR variants (101, 127, 131, 154). However, the myriad factors that may contribute to this have yet to be fully evaluated.

It is known that *Salmonella* infection can be contracted through direct contact with animals (49, 51). More problematic is the fact that AMR, in particular, MDR, appears to be increasing among numerous *Salmonella* serotypes from various animal and human origins worldwide (24, 43, 56, 175, 181, 188, 218). The levels and extent of resistance vary globally and are influenced by antimicrobial use practices in humans and animals, as well as geographical variations in the epidemiology of *Salmonella* infections (216).

The recovery of *Salmonella* exhibiting resistance to clinically important antibiotics has been reported since the early 1960s, when most of the resistance was limited to a single antibiotic (23, 33, 130, 183). Since the mid-1970s, there has been an increasing worldwide trend towards MDR in *Salmonella*. Studies also suggest that infection by AMR *Salmonella* strains may result in more severe disease than infections caused by susceptible isolates (71, 73, 74, 182). From a public health perspective, the isolation of *Salmonella* exhibiting decreased susceptibility to fluoroquinolones (e.g., ciprofloxacin) and extended-spectrum cephalosporins (e.g., ceftiofur and ceftriaxone) is especially worrisome, since these two antimicrobial classes are important in treating *Salmonella* infections in adults and children, respectively (64, 121, 180, 216).

In most countries, data on antibiotic resistance of *Salmonella* strains isolated from animal-derived foods and food animals come mostly from point prevalence studies or are in the form of limited temporal surveillance data. Consistent among most of these reports is the finding that, irrespective of source (humans, foods, or food animals), resistance to older antimicrobials such ampicillin, chloramphenicol, streptomycin, sulfamethoxazole, and tetracycline (29, 34, 39, 61, 129, 142, 152; http://www.dfvf.dk/) occurs most frequently. With some regional exceptions, resistance to more recently developed compounds is less prevalent (144, 174). A survey in Spain in which 112 isolates from 691 frozen and fresh chicken meat products (75) were tested found resistance rates highest for chloramphenicol (45%), ampicillin (35%), and tetracycline (34%). A study from southern Italy (105) reported the distribution of serovars and drug resistance of 206 *Salmonella* isolates obtained between 1998 and 2000 from 172 samples of raw meats, 22 food animal fecal samples, and 12 animal feed samples. MDR was found to be more common in *Salmonella* serovar Typhimurium (52%) than in non-serovar Typhimurium (38%) strains. The characteristic serovar Typhimurium definitive phage type 104 (DT104), which is typified by resistance to ampicillin, chloramphenicol and florfenicol, streptomycin, sulfonamides, and tetracycline (R-type *ACSSuT*) (9, 21, 36, 143), was identified in 17 (8.2%) of the serovar Typhimurium isolates. In Northern Ireland, a survey of retail chicken products found resistance limited to sulfonamide (52%), streptomycin (26%), tetracycline (22%), and ampicillin

(17%), with low rates of MDR (194). A small surveillance study compared 133 *Salmonella* isolates recovered from retail meats purchased in the United States and the People's Republic of China (32). Seventy-three (82%) of the 89 *Salmonella* strains isolated from retail meats purchased in the Washington, DC, area were resistant to at least one antimicrobial, the most common ones being tetracycline (68%), streptomycin (61%), sulfamethoxazole (42%), and ampicillin (29%), with 8 isolates (6%) showing resistance to ceftriaxone and none showing resistance to quinolones. In contrast, among the 44 isolates from the People's Republic of China, 14 (32%) were resistant to nalidixic acid, and the MICs of ciprofloxacin for these were also higher. Resistance to tetracycline (43%), ampicillin (39%), and streptomycin (32%) was also observed. A large study in Austria (108) conducted over 3 years examined 922 samples of pork ($n$ = 220), beef ($n$ = 134), chicken ($n$ = 288), turkey ($n$ = 266), and minced meat ($n$ = 14) collected randomly from supermarkets, butchers, street markets, and abattoirs across the country. *Salmonella* was present in 16.4% of the chicken samples but was rare in or absent from the other meat types. In contrast to other reports, the most common resistance among the Austrian isolates was resistance to nalidixic acid (42%). This was followed by resistance to tetracycline (33%), streptomycin (27%), ampicillin (17%), and chloramphenicol (17%). Ciprofloxacin resistance was detected in five strains (9.6%).

In a 1998 U.S. pilot survey (193), *Salmonella* was recovered from 41 of 200 (21%) ground meat samples (51 chicken, 50 beef, 50 turkey, and 49 pork) purchased in the greater Washington, DC, area. *Salmonella* was isolated more frequently from poultry (33% of chicken and 24% of turkey samples) than red meats (18% of pork and 6% of beef samples), and 84% (38/45) of the total isolates were resistant to at least one antimicrobial. Again, resistance to older agents was more common: tetracycline (80%), streptomycin (73%), sulfamethoxazole (69%), and ampicillin (27%). Sixteen percent of the strains displayed resistance to amoxicillin-clavulanic acid, cephalothin, ceftiofur, and ceftriaxone. Ceftriaxone-resistant salmonellae were isolated from ground turkey, chicken, and beef. Since this initial study, the FDA conducted a larger pilot project in Iowa over 15 months (March 2001 to June 2002), sampling randomly from 300 retail outlets across the state (27). One-hundred thirty-one *Salmonella* strains were recovered from 126 of 981 (13%) meat samples, all but 5 of which originated from ground turkey and chicken breast samples. AMR was common, with 12% showing resistance to 3 antimicrobials, 5% to 7 antimicrobials, and 1% to 12 antimicrobials. Resistance to nalidixic acid ($n$ = 5) and ceftiofur ($n$ = 2) was seen only in isolates from ground turkey.

The combination of targeted point prevalence studies, along with the systematic monitoring of national trends, has suggested that recent rises in MDR *Salmonella* may be attributed to the clonal spread of MDR strains such as serovar Typhimurium DT104 and, more recently, the proliferation of serovar Newport MDR-AmpC (143, 149, 173, 176, 181, 215, 218, 219). Serovar Typhimurium is one of the most common serotypes associated with both animal and human illness (15, 40, 60, 74, 96, 143, 176). In addition to the pentaresistance phenotype (R-type *ACSSuT*), DT104 has acquired resistance to other antibiotics, including trimethoprim, trimethoprim-sulfonamides, kanamycin, nalidixic acid, ciprofloxacin (173), and extended-spectrum cephalosporins (119). The DT104 strain type first appeared in seagulls and cattle in 1984 in the United Kingdom, where it was thought to have originated from gulls and exotic birds imported from Southeast Asia (173). It has since been isolated in many countries from poultry, cattle, pigs, sheep, nondomestic birds, and other species.

While much of the rising resistance trend in *Salmonella* during the past decade was driven by the global spread of strains belonging to the DT104 clonal lineage (60, 146, 173), more recently, serovar Newport MDR-AmpC has emerged and spread in both animals and humans in the United States (49, 212). In addition to the ACSSuT resistance phenotype typical of serovar Typhimurium DT104, serovar Newport MDR-AmpC is also resistant to amoxicillin-clavulanic acid, cephalothin, cefoxitin, and ceftiofur and exhibits decreased susceptibility to ceftriaxone (MIC $\geq$ 16 $\mu$g/ml). Some Newport MDR-AmpC strains also show resistance to gentamicin, kanamycin, and trimethoprim-sulfamethoxazole. Zhao et al. (219) showed that among 87 serovar Newport isolates obtained from humans and food animals from 25 U.S. states from 2001 to 2002, 53 (60%) were identified as serovar Newport MDR-AmpC, including 16 (53%) from humans, 27 (93%) from cattle, 7 (70%) from swine, and 3 (30%) from chickens. All 53 serovar Newport MDR-AmpC isolates possessed a cephamycinase, encoded by the $bla_{CMY}$ gene. This extended-spectrum β-lactamase has been associated with resistance to narrow-, expanded-, and broad-spectrum cephalosporins and has been shown to be widespread in many other gram-negative enteric pathogens as well (189, 195, 220). Conjugation and sequencing studies have shown that $bla_{CMY}$ genes can be located on plasmids that also contain genes for resistance to chloramphenicol, streptomycin, sulfamethoxazole, tetracycline, mercury, and quaternary ammonium compounds (189). The prevalence of serovar Newport MDR-AmpC among serovar Newport isolates from humans in the United States increased from 0% during 1996 to 1997 to 26% in 2001, before declining to 14.7% in 2004. A parallel epidemic spread was observed in animals (56, 62); however, data are not yet available to ascertain whether prevalence is declining in cattle as in human isolates. At least 26 states have isolated serovar Newport MDR-AmpC either from humans, cattle, or ground beef. This particular strain of serovar Newport is thought to be domestically acquired and associated with exposure to dairy cattle or ground beef (14, 64).

The U.S. NARMS program expanded in 2002 to incorporate monitoring of retail meats for AMR salmonellae (http://www.fda.gov). Data from the first 4 years of sampling show a *Salmonella* contamination rate in chicken breast and ground turkey of 10 to 15% from 2002 to 2005, with <2% of pork chops and ground beef contaminated. The overall degree of AMR of *Salmonella* isolated from meat in the United States from 2002 to 2005 is shown in Fig. 4. Comparing ground turkey isolates from 2002 to 2005, increased resistance to streptomycin (38 to 44%), sulfamethoxazole (20 to 34%), and ampicillin (16 to 27%) was observed. Decreased resistance was seen for tetracycline (55 to 40%) and chloramphenicol (1.4 to 0.5%). Resistance to ceftiofur decreased from 8.1% in 2002 to 7.1% in 2005, and nalidixic acid resistance of ground turkey isolates decreased from 8.1 to 1.1%. For chicken breast isolates, streptomycin resistance ranged from 28 to 30%, sulfamethoxazole resistance remained at 17%, and chloramphenicol resistance remained below 3%, while ampicillin resistance increased from 17 to 27% and tetracycline resistance increased from 23 to 44% from 2002 to 2005. Ceftiofur resistance increased from 10 to 20%, and nalidixic acid resistance remained below 4%. All isolates from all sampling years were susceptible to amikacin, ceftriaxone, ciprofloxacin, and trimethoprim-sulfamethoxazole.

AMR data on *Salmonella* strains isolated from foods of animal origin imported into the United States are not included in the U.S. NARMS program, but this topic has been studied on occasion. A total of 187 *Salmonella* isolates, representing 80 serotypes, were recovered from 4,072 samples covering a wide variety of imported meat, seafood, produce, and spice products by FDA field laboratories in 2000. Fifteen isolates (8%) were resistant to at least one antimicrobial, and five (2.7%) were resistant to three or more antimicrobials (214), with four isolates resistant to nalidixic acid. A follow-up study examined the susceptibility of 208 *Salmonella* isolates obtained from more than 5,000 imported foods entering the United States (217). A wide variety of sero-

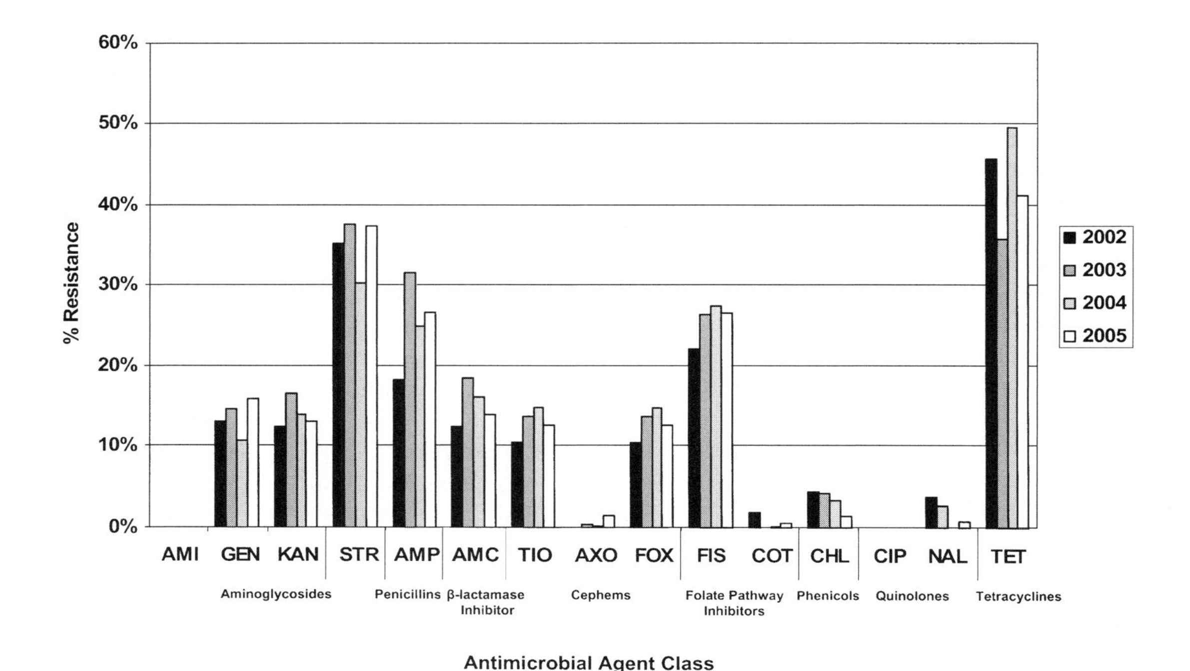


**FIGURE 4** AMR among *Salmonella* isolates from NARMS retail meats, 2002 to 2005. For abbreviations, see legend to Fig. 3.

types and susceptibility patterns were identified. Of particular note, a *Salmonella enterica* serovar Schwarzengrund isolate recovered from dehydrated chili imported from Thailand exhibited resistance to six antimicrobials, including ampicillin, chloramphenicol, ciprofloxacin, sulfamethoxazole, streptomycin, and tetracycline. Subsequent studies based on PFGE patterns showed that this strain was present in poultry and associated with human cases of salmonellosis in Thailand. It likely spread, via international trade, to cause illness in Denmark and the United States (2). This example of global dissemination emphasizes the need for international efforts to monitor the spread of food-borne pathogens and to reduce the pressures leading to the evolution of AMR strains.

Kiessling et al. (94) recently reported antimicrobial susceptibility data on *Salmonella* isolates recovered from 22,231 imported and domestic food products between 1999 and 2003 in the United States. Approximately 6% of products yielded *Salmonella* upon analysis ($n$ = 1,319), with a total of 1,382 isolates collected, as some samples yielded more than one isolate. Interestingly, 51.4% of isolates were recovered from either fresh or frozen seafood. The most common serotypes recovered included Weltevreden, Newport, Lexington, Senftenberg, and Typhimurium (94). Forty-nine percent ($n$ = 681) of *Salmonella* isolates displayed resistance to at least one antimicrobial; however, only 2% exhibited resistance to five or more antimicrobials. The most common resistance phenotypes observed were resistance to sulfisoxazole, streptomycin, tetracycline, and, to a lesser extent, nalidixic acid.

The retail meat component of Denmark's DANMAP (http://www.dfvf.dk/) program screens pork at slaughterhouses and imported pork, as well as broiler meat and imported turkey meat at retail outlets, for *Salmonella*. Isolates are also obtained from samples obtained from healthy poultry and pigs and from diagnostic samples from cattle. For *Salmonella* serotypes, antimicrobial susceptibility results are reported for *Salmonella enterica* serovars Enteritidis and Typhimurium. In contrast to the data for *E. coli,* the DANMAP data showed a significant increase in the levels of resistance to ampicillin, chloramphenicol, sulfonamide, and tetracycline in serovar Typhimurium from 1999 to 2006. As observed for *E. coli* resistance phenotypes in broiler meat, antibiotic resistance of serovar Typhimurium strains derived from imported pork products occurred more frequently than in strains originating from domestic products. Important differences between domestic and imported pork, respectively, were as follows: ciprofloxacin, 1.6 and 5.4%, and florfenicol, 10.9 and 24.3%. In addition, the first ceftiofur-resistant serovar Typhimurium strain was isolated from a Danish pig in 2006.

In Mexico, prospective laboratory surveillance for AMR in food-borne pathogens began in 2000. The first summary report of ResistVet data described the results of testing over 7,200 samples collected from healthy and ill children; chicken, pork, and beef from retail outlets; and animal intestinal samples from chickens, pigs, and cattle (210). An advantage of this program is that sampling was designed to capture both the temporal and geographical relationships of dissemination in a setting where *Salmonella* is commonly present in the meat supply. ResistVet data showed that *Salmonella* contamination was frequent in chicken (46%), pork (63%), and beef (60%), and isolates could be recovered from asymptomatic children (10%). There were some strains resistant to the extended-spectrum cephalosporins. Monitoring data showed that MDR-AmpC serovar Typhimurium ($bla_{\text{CMY-2}}^{+}$) emerged in 2002 (211) and spread rapidly in Mexico. This is in contrast to the situation in the United States, where most resistance to extended-spectrum cephalosporins is present in serovar Newport. As with serovar Newport, serovar Typhimurium MDR-AmpC strains are resistant to 8 to 14 other antimicrobial agents in addition to the extended-spectrum cephalosporins. The authors concluded that this strain type likely originated from swine production in Mexico, where serovar Typhimurium PFGE types and AMR profiles were shared in common with human strains.

## BURDEN OF ILLNESS ASSOCIATED WITH AMR FOOD-BORNE PATHOGENS

While the immediate public health impact associated with food-borne pathogens is associated with acute and frequently self-limiting disease, infection with AMR strains can potentiate more intractable and severe disease than that caused by susceptible strains (71, 74, 107, 121, 182). Diseases caused by resistant pathogens may require prolonged and more expensive medical treatments than infections caused by nonresistant pathogens, resulting in absence from work for longer periods and additional costs due to increased morbidity and mortality (55, 72). Epidemiological studies have measured the health burden of AMR in terms of excess mortality, increased duration of illness, and increased risk of invasive illness or hospitalization from food-borne disease, in particular, those caused by *Campylobacter* and *Salmonella* (43, 50, 71, 74, 107, 121, 128, 180, 182). For example, Helms et al. (71, 74) reported a higher mortality rate among patients with resistant *Salmonella* infections in Denmark, particularly quinolone-resistant infections, compared with mortality among the general population and among patients infected with susceptible *Salmonella.* In Canada, Martin et al. (107) reported a 2.3-fold-greater likelihood of hospitalization for patients who contracted serovar Typhimurium infections resistant to ampicillin, chloramphenicol, and/or kanamycin, streptomycin, sulfamethoxazole, and tetracycline than for patients with infections with organisms not resistant to these agents. In the United States, analysis of FoodNet and NARMS data by Varma et al. (182) showed a greater likelihood of hospitalization and longer hospital stays among patients who had disease caused by *Salmonella* resistant to ≥1 antimicrobial than for those individuals infected with pansensitive *Salmonella* strains. Resistant isolates were also more likely to be invasive (isolated from the bloodstream rather than stool) than were pansusceptible isolates. In Mexico, systemic serovar Typhimurium infections, with severe thrombocytopenia, hemorrhagic syndrome, and death, were only seen in children infected with MDR $bla_{CMY-2}$ strains, although there were too few cases to demonstrate a statistically significant association (210).

The human health hazards of fluoroquinolone- and macrolide-resistant *Campylobacter* have also been investigated (50, 73, 128). One of the earliest studies was conducted by Smith et al. (167), who compared cases of quinolone-resistant and quinolone-susceptible *C. jejuni* infections and showed that infection with quinolone-resistant *C. jejuni* was associated with foreign travel and with the use of a quinolone before the collection of stool specimens. Soon after, the CDC examined differences in several illness outcomes between persons infected with ciprofloxacin-resistant *Campylobacter* and persons infected with ciprofloxacin-susceptible *Campylobacter* within seven FoodNet sites from 1998 to 1999 (128). Using a multivariable analysis-of-variance model, they found that people with ciprofloxacin-resistant infections had a longer mean duration of diarrhea than those with ciprofloxacin-susceptible infections ($P = 0.01$), which was independent of foreign travel (128).

A study by Engberg et al. (50) integrated data on quinolone and macrolide susceptibility patterns with epidemiological and typing data from *Campylobacter jejuni* and *C. coli* infections in two Danish counties and showed a longer duration of illness for patients with quinolone-resistant *C. jejuni* infections (median, 13.2 days) than for patients with quinolone-susceptible *C. jejuni* infections (median, 10.3 days; $P = 0.001$). Foreign travel, eating fresh poultry other than chicken and turkey, and swimming were also associated with increased risk for quinolone-resistant *C. jejuni* infection (50). Another Danish study determined the risk of invasive illness and death associated with infection with quinolone- and erythromycin-resistant *Campylobacter* strains using data obtained from the Danish Surveillance Registry for Enteric Pathogens and Civil Registration System and National Health Registries (73). Patients infected with quinolone-resistant *Campylobacter* strains were shown to have a 6-fold-increased risk of an adverse event within 30 days of the date of receipt of samples at the microbiology laboratory compared with patients infected with quinolone- and erythromycin-susceptible *Campylobacter* strains (73). In addition, infection with erythromycin-resistant strains was associated with a >5-fold increased risk of an adverse event within 90 days of the date of receipt of samples at the microbiology laboratory (73).

Not all of the data support the viewpoint that AMR per se raises the burden or severity of food-borne campylobacteriosis and salmonellosis. Devasia et al. (43) reported no significant differences in symptoms, hospitalization, duration of illness, or other outcomes when comparing persons infected with MDR-AmpC and pansusceptible serovar Newport from a multistate population-based case control study. Data from the United Kingdom's *Campylobacter* Sentinel Surveillance Scheme also indicated no apparent differences in length of illness or hospital admission between indigenous cases of infection with ciprofloxacin-resistant *C. jejuni* strains versus susceptible counterparts (26). Additionally, a recent reanalysis of the risks attributed to ciprofloxacin-resistant *Campylobacter jejuni* infections by Wassenaar et al. (185) showed no significant differences in duration of disease between infections caused by susceptible and resistant strains. More evidence is needed to better understand the importance of AMR food-borne bacteria and the associated adverse human health outcomes.

## RISK ASSESSMENT AND RISK MANAGEMENT STRATEGIES TO MITIGATE AMR ASSOCIATED WITH THE USE OF ANTIMICROBIALS IN AGRICULTURE

The widespread use of antimicrobials in food animals, the spread of AMR zoonotic bacteria to humans via the food supply, and the potential health ramifications are ongoing topics of policy debate. Beginning with the release of the Swann report in the United Kingdom in 1969 (7), numerous expert committee reports have been published addressing this issue. There are many information gaps that prevent full assessment of the potential negative effects on human health occurring because of antimicrobial use in food animals (8, 126, 201, 203, 204). In addition to resistance to common antimicrobials such as tetracyclines, streptomycin, and sulfamethoxazole, there are increasing reports documenting decreased susceptibility to several antimicrobial classes important to animal and human health, including aminoglycosides, extended-spectrum cephalosporins, and fluoroquinolones. Resistance to the latter two antimicrobial classes is especially problematic, as they are the primary therapeutic options in cases of invasive bacterial infections in both humans and animals.

Accurately assessing the public health risks of AMR of food-borne pathogens, and the relative public health significance associated with different classes of antimicrobial agents used in animals, is a considerable challenge. Possible ecological pathways for the dissemination of AMR bacteria and resistance determinants through the food production continuum are diverse and complex, often hindering direct epidemiological measures of public health risk. As a result, microbial risk assessment is being used globally to facilitate better understanding of the risks related to the food chain, including quantification of uncertainty and prioritization of control strategies (168). Although somewhat indirect (111), microbial risk assessment efforts to date have sought to estimate the probability of negative human health effects due to AMR bacteria associated with routine antimicrobial use in animals.

In 2003, the FDA outlined a practicable risk assessment approach for answering concerns about AMR as applicable to proposed original or supplemental new animal antimicrobial drug applications (microbial food safety) in food-producing animals (177a). This document contains nonbinding recommendations and provides a paradigmatic framework to answer FDA concerns about the selection and emergence of AMR bacteria in food-producing animals as a result of the use of new antimicrobial products in different animal populations. This risk assessment is built of subcomponents addressing resistance development, human exposure to resistant bacteria, and consequences to human health. Sponsors of new antimicrobial animal drugs can follow the outline set forth in this document to estimate how their product might contribute to AMR among pathogens constituting a food-borne hazard, as well as the risk (as a probability of occurrence of the hazard) to public health. The evaluations are based on product characteristics, conditions of antimicrobial use in target food animals, and the importance of the subject drug (or drugs in the same class) in human medicine. Further, a sponsor can use the guidance to anticipate possible risk management mitigations that might be applied to their product as safeguards against AMR emergence and selection.

The WHO has also been at the forefront of this issue since the late 1990s, when medical problems arising from the use of antimicrobials in livestock production were identified, and concern was expressed that drug-resistant pathogens could be transmitted to humans via the food chain (201). In 1998, the WHO held a consultation to address concerns that the use of quinolones in food animals was leading to the emergence of quinolone-resistant zoonotic bacteria (202). The WHO later developed global principles for the containment of AMR bacteria in animals intended for food (203). In 2005, the WHO convened an expert meeting on critically important antimicrobials for human medicine, with a follow-up consultation in 2007 (205, 206). As a result of the 2007 consultation, an updated list of critical antimicrobials of AMR concern, which might merit risk management because of extensive nonhuman applications, was published (207).

The ad hoc Intergovernmental Task Force on Antimicrobial Resistance was created by the 29th Session of the Codex Alimentarius Commission (July 2006). The Task Force's objectives are to develop science-based guidelines to assess the risks to human health associated with the presence in food and feed, including aquaculture, of AMR microorganisms and resistance genes, and to develop appropriate risk management advice based on that assessment. The goal is to create an internationally accepted framework for assessing and managing the risk presented by AMR microorganisms in food, taking full account of the prior work on risk analysis principles and standards of the Codex Commission, FAO, WHO, and OIE, as well as of national and regional authorities. It is hoped that guidance developed by the task force will enable regulatory authorities to assess the potential risks to human health of AMR microorganisms, and allow these authorities to develop and apply appropriate risk management advice to reduce public health risks.

## SUMMARY

Given the increasing international demand for meat, poultry, and seafood as important sources of protein, the health hazards posed by AMR food-borne pathogens have become a global issue. Antimicrobial use and resistance in agriculture will continue to be a considerable challenge for industry and for officials charged with limiting the public health burden of food-borne illness. Over the past 2 decades, we have observed a steady loss of antimicrobial efficacy against many food-borne pathogens, with resistance to the most potent agents also appearing. The evolution and propagation of transmissible plasmids in *Salmonella* and *E. coli* carrying resistance to eight or more antimicrobials are particularly worrisome. As intensive food production systems become more common, and as expanding trade in food animals and their derived products evolve to meet global demands, national and regional surveillance systems to measure and track drug resistance in these and other enteric organisms will become more important. Developing effective ways to characterize and mitigate resistance requires continued exploitation of basic research on the ecology and epidemiology of the major food-borne pathogens and improved surveillance of retail food products, including imported products. Furthermore, continuous refinement of risk assessment methods is required to support the development of science-based risk management policies.

## ACKNOWLEDGMENT

We are grateful to M. Wekell for helpful comments and review.

## REFERENCES

1. **Aarestrup, F. M.** 2004. Monitoring of antimicrobial resistance among food animals: principles and limitations. *J. Vet. Med. Ser.* B **51:**380–388.
2. **Aarestrup, F. M., R. S. Hendriksen, J. Lockett, K. Gay, K. Teates, P. F. McDermott, D. G. White, H. Hasman, G. Sorensen, A. Bangtrakulnonth, S. Pornreongwong, C. Pulsrikarn, F. J. Angulo, and P. Gerner-Smidt.** 2007. International spread of multidrug-resistant *Salmonella* Schwarzengrund in food products. *Emerg. Infect. Dis.* **13:**726–731.
3. **Ait Melloul, A., L. Hassani, and L. Rafouk.** 2001. *Salmonella* contamination of vegetables irrigated with untreated wastewater. *World J. Microbiol. Biotechnol.* **17:**207–209.
4. **Alban, L., E. O. Nielsen, and J. Dahl.** 2008. A human health risk assessment for macrolide-resistant *Campylobacter* associated with the use of macrolides in Danish pig production. *Prev. Vet. Med.* **83:**115–129.
5. **Altekruse, S. F., F. Elvinger, K. Y. Lee, L. K. Tollefson, E. W. Pierson, J. Eifert, and N. Sriranganathan.** 2002. Antimicrobial susceptibilities of *Escherichia coli* strains from a turkey operation. *J. Am. Vet. Med. Assoc.* **221:**411–416.
6. **Angulo, F. J., V. N. Nargund, and T. C. Chiller.** 2004. Evidence of an association between use of anti-microbial agents in food ani-

mals and anti-microbial resistance among bacteria isolated from humans and the human health consequences of such resistance. *J. Vet. Med. Ser. B* **51:**374–379.

7. **Anonymous.** 1969. *Report of the Joint Committee on the Use of Antibiotics in Animal Husbandry and Veterinary Medicine.* Her Majesty's Stationery Office, London, England.
8. **Anonymous.** 2006. Antimicrobial resistance: implications for the food system. *Compr. Rev. Food Sci. Food Saf.* **5:**71–137.
9. **Arcangioli, M. A., S. Leroy-Setrin, J. L. Martel, and E. Chaslus-Dancla.** 1999. A new chloramphenicol and florfenicol resistance gene flanked by two integron structures in *Salmonella typhimurium* DT104. *FEMS Microbiol. Lett.* **174:** 327–332.
10. **Aslam, M., and C. Service.** 2006. Antimicrobial resistance and genetic profiling of *Escherichia coli* from a commercial beef packing plant. *J. Food Prot.* **69:**1508–1513.
11. **Bauer-Garland, J., J. G. Frye, J. T. Gray, M. E. Berrang, M. A. Harrison, and P. J. Fedorka-Cray.** 2006. Transmission of *Salmonella enterica* serotype Typhimurium in poultry with and without antimicrobial selective pressure. *J. Appl. Microbiol.* **101:**1301–1308.
12. **Belanger, A. E., and T. R. Shryock.** 2007. Macrolide-resistant *Campylobacter:* the meat of the matter. *J. Antimicrob. Chemother.* **60:**715–723.
13. **Bensink, J. C., and F. P. Bothmann.** 1991. Antibiotic-resistant *Escherichia coli* isolated from chilled meat at retail outlets. *N. Z. Vet. J.* **39:** 126–128.
14. **Berge, A. C., J. M. Adaska, and W. M. Sischo.** 2004. Use of antibiotic susceptibility patterns and pulsed-field gel electrophoresis to compare historic and contemporary isolates of multi-drug-resistant *Salmonella enterica* subsp. *enterica* serovar Newport. *Appl. Environ. Microbiol.* **70:**318–323.
15. **Besser, T. E., M. Goldoft, L. C. Pritchett, R. Khakhria, D. D. Hancock, D. H. Rice, J. M. Gay, W. Johnson, and C. C. Gay.** 2000. Multiresistant *Salmonella* Typhimurium DT104 infections of humans and domestic animals in the Pacific Northwest of the United States. *Epidemiol. Infect.* **124:**193–200.
16. **Bischoff, K. M., D. G. White, P. F. McDermott, S. Zhao, S. Gaines, J. J. Maurer, and D. J. Nisbet.** 2002. Characterization of chloramphenicol resistance in beta-hemolytic *Escherichia coli* associated with diarrhea in neonatal swine. *J. Clin. Microbiol.* **40:**389–394.
17. **Bohaychuk, V. M., G. E. Gensler, R. K. King, K. I. Manninen, O. Sorensen, J. T. Wu, M. E. Stiles, and L. M. McMullen.** 2006. Occurrence of pathogens in raw and ready-to-eat meat and poultry products collected from the retail marketplace in Edmonton, Alberta, Canada. *J. Food Prot.* **69:**2176–2182.
18. **Bosilevac, J. M., T. M. Arthur, T. L. Wheeler, S. D. Shackelford, M. Rossman, J. O. Reagan, and M. Koohmaraie.** 2004. Prevalence of *Escherichia coli* O157 and levels of aerobic bacteria and Enterobacteriaceae are reduced when hides are washed and treated with cetylpyridinium chloride at a commercial beef processing plant. *J. Food Prot.* **67:**646–650.
19. **Bradford, P. A., P. J. Petersen, I. M. Fingerman, and D. G. White.** 1999. Characterization of expanded-spectrum cephalosporin resistance in *E. coli* isolates associated with bovine calf diarrhoeal disease. *J. Antimicrob. Chemother.* **44:**607–610.
20. **Brandl, M. T.** 2006. Fitness of human enteric pathogens on plants and implications for food safety. *Annu. Rev. Phytopathol.* **44:**367–392.
21. **Briggs, C. E., and P. M. Fratamico.** 1999. Molecular characterization of an antibiotic resistance gene cluster of *Salmonella typhimurium* DT104. *Antimicrob. Agents Chemother.* **43:**846–849.
22. **Brinas, L., M. Zarazaga, Y. Saenz, F. Ruiz-Larrea, and C. Torres.** 2002. Beta-lactamases in ampicillin-resistant *Escherichia coli* isolates from foods, humans, and healthy animals. *Antimicrob. Agents Chemother.* **46:**3156–3163.
23. **Bulling, E., R. Stephan, and V. Sebek.** 1973. The development of antibiotics resistance among salmonella bacteria of animal origin in the Federal Republic of Germany and West Berlin. 1st communication: a comparison between the years of 1961 and 1970–71. *Zentralbl. Bakteriol. Orig. A* **225:**245–256. (Author's translation.)
24. **Butaye, P., G. B. Michael, S. Schwarz, T. J. Barrett, A. Brisabois, and D. G. White.** 2006. The clonal spread of multidrug-resistant non-*typhi Salmonella* serotypes. *Microbes Infect.* **8:** 1891–1897.
25. **Callaway, T. R., T. S. Edrington, J. L. Rychlik, K. J. Genovese, T. L. Poole, Y. S. Jung, K. M. Bischoff, R. C. Anderson, and D. J. Nisbet.** 2003. Ionophores: their use as ruminant growth promotants and impact on food safety. *Curr. Issues Intest. Microbiol.* **4:**43–51.
26. **Campylobacter Sentinel Surveillance Scheme Collaborators.** 2002. Ciprofloxacin resistance in *Campylobacter jejuni:* case-case analysis as a tool for elucidating risks at home and abroad. *J. Antimicrob. Chemother.* **50:**561–568.
27. **Carter, P. J., L. L. English, B. Cook, T. Proeschold, and D. G. White.** 2002. Prevalence and antimicrobial susceptibility profiles of

*Salmonella* and *Campylobacter* isolated from retail meats. *Abstr. 102nd Annu. Meet. Am. Soc. Microbiol.*

28. **Caudry, S. D., and V. A. Stanisich.** 1979. Incidence of antibiotic-resistant *Escherichia coli* associated with frozen chicken carcasses and characterization of conjugative R plasmids derived from such strains. *Antimicrob. Agents Chemother.* **16:**701–709.

29. **Centers for Disease Control and Prevention.** 2004. *National Antimicrobial Resistance Monitoring System for Enteric Bacteria (NARMS): 2002 Annual Report.* Centers for Disease Control and Prevention, Atlanta, GA.

30. **Cergole-Novella, M. C., L. S. Nishimura, K. Irino, T. M. Vaz, A. F. de Castro, L. Leomil, and B. E. Guth.** 2006. Stx genotypes and antimicrobial resistance profiles of Shiga toxin-producing *Escherichia coli* strains isolated from human infections, cattle and foods in Brazil. *FEMS Microbiol. Lett.* **259:**234–239.

31. **Chaslus-Dancla, E., P. Pohl, M. Meurisse, M. Marin, and J. P. Lafont.** 1991. High genetic homology between plasmids of human and animal origins conferring resistance to the aminoglycosides gentamicin and apramycin. *Antimicrob. Agents Chemother.* **35:**590–593.

32. **Chen, S., S. Zhao, D. G. White, C. M. Schroeder, R. Lu, H. Yang, P. F. McDermott, S. Ayers, and J. Meng.** 2004. Characterization of multiple-antimicrobial-resistant *Salmonella* serovars isolated from retail meats. *Appl. Environ. Microbiol.* **70:**1–7.

33. **Cherubin, C. E.** 1981. Antibiotic resistance of *Salmonella* in Europe and the United States. *Rev. Infect. Dis.* **3:**1105–1126.

34. **Chiappini, E., L. Galli, P. Pecile, A. Vierucci, and M. de Martino.** 2002. Results of a 5-year prospective surveillance study of antibiotic resistance among *Salmonella enterica* isolates and ceftriaxone therapy among children hospitalized for acute diarrhea. *Clin. Ther.* **24:**1585–1594.

35. **Chung, Y. J., and M. H. Saier, Jr.** 2002. Overexpression of the *Escherichia coli sugE* gene confers resistance to a narrow range of quaternary ammonium compounds. *J. Bacteriol.* **184:**2543–2545.

36. **Cloeckaert, A., and S. Schwarz.** 2001. Molecular characterization, spread and evolution of multidrug resistance in *Salmonella enterica* Typhimurium DT104. *Vet. Res.* **32:**301–310.

37. **Cohen, M. L., and R. V. Tauxe.** 1986. Drug-resistant *Salmonella* in the United States: an epidemiologic perspective. *Science* **234:**964–969.

38. **Constable, P. D.** 2004. Antimicrobial use in the treatment of calf diarrhea. *J. Vet. Intern. Med.* **18:**8–17.

39. Reference deleted.

40. **Davis, M. A., D. D. Hancock, and T. E. Besser.** 2002. Multiresistant clones of *Salmonella enterica:* the importance of dissemination. *J. Lab. Clin. Med.* **140:**135–141.

41. **de Neeling, A. J., M. J. van den Broek, E. C. Spalburg, M. G. Santen-Verheuvel, W. D. Dam-Deisz, H. C. Boshuizen, A. W. van de Giessen, E. van Duijkeren, and X. W. Huijsdens.** 2007. High prevalence of methicillin resistant *Staphylococcus aureus* in pigs. *Vet. Microbiol.* **122:**366–372.

42. **Depardieu, F., I. Podglajen, R. Leclercq, E. Collatz, and P. Courvalin.** 2007. Modes and modulations of antibiotic resistance gene expression. *Clin. Microbiol. Rev.* **20:**79–114.

43. **Devasia, R. A., J. K. Varma, J. Whichard, S. Gettner, A. B. Cronquist, S. Hurd, S. Segler, K. Smith, D. Hoefer, B. Shiferaw, F. J. Angulo, and T. F. Jones.** 2005. Antimicrobial use and outcomes in patients with multidrug-resistant and pansusceptible *Salmonella* Newport infections, 2002–2003. *Microb. Drug Resist.* **11:**371–377.

44. **Diarrassouba, F., M. S. Diarra, S. Bach, P. Delaquis, J. Pritchard, E. Topp, and B. J. Skura.** 2007. Antibiotic resistance and virulence genes in commensal *Escherichia coli* and *Salmonella* isolates from commercial broiler chicken farms. *J. Food Prot.* **70:**1316–1327.

45. **Dibner, J. J., and J. D. Richards.** 2005. Antibiotic growth promoters in agriculture: history and mode of action. *Poult. Sci.* **84:**634–643.

46. **Donnenberg, M. S., and T. S. Whittam.** 2001. Pathogenesis and evolution of virulence in enteropathogenic and enterohemorrhagic *Escherichia coli*. *J. Clin. Investig.* **107:**539–548.

47. **Dontorou, C., C. Papadopoulou, G. Filioussis, V. Economou, I. Apostolou, G. Zakkas, A. Salamoura, A. Kansouzidou, and S. Levidiotou.** 2003. Isolation of *Escherichia coli* O157:H7 from foods in Greece. *Int. J. Food Microbiol.* **82:**273–279.

48. **Doyle, M. P., and M. C. Erickson.** 2006. Reducing the carriage of foodborne pathogens in livestock and poultry. *Poult. Sci.* **85:**960–973.

49. **Dunne, E. F., P. D. Fey, P. Kludt, R. Reporter, F. Mostashari, P. Shillam, J. Wicklund, C. Miller, B. Holland, K. Stamey, T. J. Barrett, J. K. Rasheed, F. C. Tenover, E. M. Ribot, and F. J. Angulo.** 2000. Emergence of domestically acquired ceftriaxone-resistant *Salmonella* infections associated with AmpC beta-lactamase. *JAMA* **284:**3151–3156.

50. **Engberg, J., J. Neimann, E. M. Nielsen, F. M. Aerestrup, and V. Fussing.** 2004. Quinolone-resistant *Campylobacter* infections: risk

factors and clinical consequences. *Emerg. Infect. Dis.* **10:**1056–1063.

**51. Fey, P. D., T. J. Safranek, M. E. Rupp, E. F. Dunne, E. Ribot, P. C. Iwen, P. A. Bradford, F. J. Angulo, and S. H. Hinrichs.** 2000. Ceftriaxone-resistant *Salmonella* infection acquired by a child from cattle. *N. Engl. J. Med.* **342:**1242–1249.

**52. Finley, R., C. Ribble, J. Aramini, M. Vandermeer, M. Popa, M. Litman, and R. Reid-Smith.** 2007. The risk of salmonellae shedding by dogs fed *Salmonella*-contaminated commercial raw food diets. *Can. Vet. J.* **48:**69–75.

**53. Foley, E. G., S. W. Lee, and N. J. Hartley.** 1946. The effect of penicillin on staphylococci and streptococci commonly associated with bovine mastitis. *J. Food Technol.* **8:**129–133.

**54. Foley, S. L., and A. M. Lynne.** 2008 Food animal-associated *Salmonella* challenges: pathogenicity and antimicrobial resistance. *J. Anim. Sci.* **86**(14 Suppl.)**:**E173–E187.

**55. Frank, U., for the BURDEN Study Group.** 2007. The BURDEN project—assessing the burden of resistance and disease in Europe. *Eurosurveillance* **12:**E070111.

**56. Frye, J. G., and P. J. Fedorka-Cray.** 2007. Prevalence, distribution and characterisation of ceftiofur resistance in *Salmonella enterica* isolated from animals in the USA from 1999 to 2003. *Int. J. Antimicrob. Agents* **30:**134–142.

**57. Ghafir, Y., B. China, K. Dierick, L. De Zutter, and G. Daube.** 2007. A seven-year survey of *Campylobacter* contamination in meat at different production stages in Belgium. *Int. J. Food Microbiol.* **116:**111–120.

**58. Giles, W. P., A. K. Benson, M. E. Olson, R. W. Hutkins, J. M. Whichard, P. L. Winokur, and P. D. Fey.** 2004. DNA sequence analysis of regions surrounding *blaCMY-2* from multiple *Salmonella* plasmid backbones. *Antimicrob. Agents Chemother.* **48:**2845–2852.

**59. Girlich, D., L. Poirel, A. Carattoli, I. Kempf, M.-F. Lartigue, A. Bertini, and P. Nordmann.** 2007. Extended-spectrum β-lactamase CTX-M-1 in *Escherichia coli* isolates from healthy poultry in France. *Appl. Environ. Microbiol.* **73:**4681–4685.

**60. Glynn, M. K., C. Bopp, W. Dewitt, P. Dabney, M. Mokhtar, and F. J. Angulo.** 1998. Emergence of multidrug-resistant *Salmonella enterica* serotype Typhimurium DT104 infections in the United States. *N. Engl. J. Med.* **338:**1333–1338.

**61. Grant, R. B., and L. Di Mambro.** 1977. Antimicrobial resistance and resistance plasmids in *Salmonella* from Ontario, Canada. *Can. J. Microbiol.* **23:**1266–1273.

**62. Gray, J. T., L. L. Hungerford, P. J. Fedorka-Cray, and M. L. Headrick.** 2004. Extended-spectrum-cephalosporin resistance in *Salmonella enterica* isolates of animal origin. *Antimicrob. Agents Chemother.* **48:**3179–3181.

**63. Guardabassi, L., S. Schwarz, and D. H. Lloyd.** 2004. Pet animals as reservoirs of antimicrobial-resistant bacteria. *J. Antimicrob. Chemother.* **54:**321–332.

**64. Gupta, A., J. Fontana, C. Crowe, B. Bolstorff, A. Stout, S. Van Duyne, M. P. Hoekstra, J. M. Whichard, T. J. Barrett, and F. J. Angulo.** 2003. Emergence of multidrug-resistant *Salmonella enterica* serotype Newport infections resistant to expanded-spectrum cephalosporins in the United States. *J. Infect. Dis.* **188:**1707–1716.

**65. Hansen, L. H., E. Johannesen, M. Burmølle, A. H. Sørensen, and S. J. Sørensen.** 2004. Plasmid-encoded multidrug efflux pump conferring resistance to olaquindox in *Escherichia coli. Antimicrob. Agents Chemother.* **48:**3332–3337.

**66. Hansen, L. H., S. J. Sorensen, H. S. Jorgensen, and L. B. Jensen.** 2005. The prevalence of the OqxAB multidrug efflux pump amongst olaquindox-resistant *Escherichia coli* in pigs. *Microb. Drug Resist.* **11:**378–382.

**67. Harakeh, S., H. Yassine, M. Gharios, E. Barbour, S. Hajjar, M. El Fadel, I. Toufeili, and R. Tannous.** 2005. Isolation, molecular characterization and antimicrobial resistance patterns of *Salmonella* and *Escherichia coli* isolates from meat-based fast food in Lebanon. *Sci. Total Environ.* **341:**33–44.

**68. Harbottle, H., S. Thakur, S. Zhao, and D. G. White.** 2006. Genetics of antimicrobial resistance. *Anim. Biotechnol.* **17:**111–124.

**69. Hasman, H., I. Kempf, B. Chidaine, R. Cariolet, A. K. Ersboll, H. Houe, H. C. Bruun Hansen, and F. M. Aarestrup.** 2006. Copper resistance in *Enterococcus faecium,* mediated by the *tcrB* gene, is selected by supplementation of pig feed with copper sulfate. *Appl. Environ. Microbiol.* **72:**5784–5789.

**70. Hayes, J. R., L. L. English, P. J. Carter, T. Proescholdt, K. Y. Lee, D. D. Wagner, and D. G. White.** 2003. Prevalence and antimicrobial resistance of *Enterococcus* species isolated from retail meats. *Appl. Environ. Microbiol.* **69:**7153–7160.

**71. Helms, M., J. Simonsen, and K. Molbak.** 2004. Quinolone resistance is associated with increased risk of invasive illness or death during infection with *Salmonella* serotype Typhimurium. *J. Infect. Dis.* **190:**1652–1654.

**72. Helms, M., J. Simonsen, and K. Molbak.** 2006. Foodborne bacterial infection and hospitalization: a registry-based study. *Clin. Infect. Dis.* **42:**498–506.

73. **Helms, M., J. Simonsen, K. E. Olsen, and K. Molbak.** 2005. Adverse health events associated with antimicrobial drug resistance in *Campylobacter* species: a registry-based cohort study. *J. Infect. Dis.* **191:**1050–1055.
74. **Helms, M., P. Vastrup, P. Gerner-Smidt, and K. Molbak.** 2002. Excess mortality associated with antimicrobial drug-resistant *Salmonella typhimurium*. *Emerg. Infect. Dis.* **8:**490–495.
75. **Hernandez, T., C. Rodriguez-Alvarez, M. P. Arevalo, A. Torres, A. Sierra, and A. Arias.** 2002. Antimicrobial-resistant *Salmonella enterica* serovars isolated from chickens in Spain. *J. Chemother.* **14:**346–350.
76. **Holmberg, S. D., J. G. Wells, and M. L. Cohen.** 1984. Animal-to-man transmission of antimicrobial-resistant *Salmonella:* investigations of U.S. outbreaks, 1971–1983. *Science* **225:**833–835.
77. **Hong, J., J. M. Kim, W. K. Jung, S. H. Kim, W. Bae, H. C. Koo, J. Gil, M. Kim, J. Ser, and Y. H. Park.** 2007. Prevalence and antibiotic resistance of *Campylobacter* spp. isolated from chicken meat, pork, and beef in Korea, from 2001 to 2006. *J. Food Prot.* **70:**860–866.
78. **Huijsdens, X. W., B. J. van Dijke, E. Spalburg, M. G. van Santen-Verheuvel, M. E. Heck, G. N. Pluister, A. Voss, W. J. Wannet, and A. J. de Neeling.** 2006. Community-acquired MRSA and pig-farming. *Ann. Clin. Microbiol. Antimicrob.* **5:**26.
79. **Hummel, R., H. Tschape, and W. Witte.** 1986. Spread of plasmid-mediated nourseothricin resistance due to antibiotic use in animal husbandry. *J. Basic Microbiol.* **26:**461–466.
80. **Hurd, H. S., S. Doores, D. Hayes, A. Mathew, J. Maurer, P. Silley, R. S. Singer, and R. N. Jones.** 2004. Public health consequences of macrolide use in food animals: a deterministic risk assessment. *J. Food Prot.* **67:**980–992.
81. **Hutchison, M. L., L. D. Walters, S. M. Avery, B. A. Synge, and A. Moore.** 2004. Levels of zoonotic agents in British livestock manures. *Lett. Appl. Microbiol.* **39:**207–214.
82. **Johnson, J. R., M. A. Kuskowski, K. Smith, T. T. O'Bryan, and S. Tatini.** 2005. Antimicrobial-resistant and extraintestinal pathogenic *Escherichia coli* in retail foods. *J. Infect. Dis.* **191:**1040–1049.
83. **Johnson, J. R., A. C. Murray, A. Gajewski, M. Sullivan, P. Snippes, M. A. Kuskowski, and K. E. Smith.** 2003. Isolation and molecular characterization of nalidixic acid-resistant extraintestinal pathogenic *Escherichia coli* from retail chicken products. *Antimicrob. Agents Chemother.* **47:**2161–2168.
84. **Johnson, J. R., and T. A. Russo.** 2002. Extraintestinal pathogenic *Escherichia coli:* "the other bad E coli." *J. Lab. Clin. Med.* **139:**155–162.
85. **Johnson, J. R., M. R. Sannes, C. Croy, B. Johnston, C. Clabots, M. A. Kuskowski, J. Bender, K. E. Smith, P. L. Winokur, and E. A. Belongia.** 2007. Antimicrobial drug-resistant *Escherichia coli* from humans and poultry products, Minnesota and Wisconsin, 2002–2004. *Emerg. Infect. Dis.* **13:**838–846.
86. **Johnston, A. M.** 1998. Use of antimicrobial drugs in veterinary practice. *BMJ* **317:**665–667.
87. **Jouini, A., L. Vinué, K. B. Slama, Y. Sáenz, N. Klibi, S. Hammami, A. Boudabous, and C. Torres.** 2007. Characterization of CTX-M and SHV extended-spectrum β-lactamases and associated resistance genes in *Escherichia coli* strains of food samples in Tunisia. *J. Antimicrob. Chemother.* **60:**1137–1141.
88. **Jukes, T. H.** 1977. The history of the "antibiotic growth effect." *Fed. Proc.* **36:**2514–2518.
89. **Kaper, J. B.** 2005. Pathogenic *Escherichia coli*. *Int. J. Med. Microbiol.* **295:**355–356.
90. **Kaper, J. B., J. P. Nataro, and H. L. Mobley.** 2004. Pathogenic *Escherichia coli*. *Nat. Rev. Microbiol.* **2:**123–140.
91. **Kelly, L., D. L. Smith, E. L. Snary, J. A. Johnson, A. D. Harris, M. Wooldridge, and J. G. Morris, Jr.** 2004. Animal growth promoters: to ban or not to ban? A risk assessment approach. *Int. J. Antimicrob. Agents* **24:**205–212.
92. **Khachatryan, A. R., T. E. Besser, D. D. Hancock, and D. R. Call.** 2006. Use of a nonmedicated dietary supplement correlates with increased prevalence of streptomycin-sulfa-tetracycline-resistant *Escherichia coli* on a dairy farm. *Appl. Environ. Microbiol.* **72:**4583–4588.
93. **Khachatryan, A. R., D. D. Hancock, T. E. Besser, and D. R. Call.** 2006. Antimicrobial drug resistance genes do not convey a secondary fitness advantage to calf-adapted *Escherichia coli*. *Appl. Environ. Microbiol.* **72:**443–448.
94. **Kiessling, C. R., M. Jackson, K. A. Watts, M. H. Loftis, W. M. Kiessling, M. B. Buen, E. W. Laster, and J. N. Sofos.** 2007. Antimicrobial susceptibility of *Salmonella* isolated from various products, from 1999 to 2003. *J. Food Prot.* **70:**1334–1338.
95. **Kuhn, I., A. Iversen, L. G. Burman, B. Olsson-Liljequist, A. Franklin, M. Finn, F. Aarestrup, A. M. Seyfarth, A. R. Blanch, X. Vilanova, H. Taylor, J. Caplin, M. A. Moreno, L. Dominguez, I. A. Herrero, and R. Mollby.** 2003. Comparison of enterococcal populations in animals, humans, and the environment—a European study. *Int. J. Food Microbiol.* **88:**133–145.
96. **Lawson, A. J., M. Desai, S. J. O'Brien, R. H. Davies, L. R. Ward, and E. J. Threlfall.** 2004. Molecular characterisation of an outbreak strain of multiresistant *Salmonella enterica*

serovar Typhimurium DT104 in the UK. *Clin. Microbiol. Infect.* **10:**143–147.

97. **LeJeune, J. T., and N. P. Christie.** 2004. Microbiological quality of ground beef from conventionally-reared cattle and "raised without antibiotics" label claims. *J. Food Prot.* **67:**1433–1437.

98. **Levy, S. B., G. B. FitzGerald, and A. B. Macone.** 1976. Changes in intestinal flora of farm personnel after introduction of a tetracycline-supplemented feed on a farm. *N. Engl. J. Med.* **295:**583–588.

99. **Linton, A. H., K. Howe, C. L. Hartley, H. M. Clements, M. H. Richmond, and A. D. Osborne.** 1977. Antibiotic resistance among *Escherichia coli* O-serotypes from the gut and carcasses of commercially slaughtered broiler chickens: a potential public health hazard. *J. Appl. Bacteriol.* **42:**365–378.

100. **Lipsitch, M., R. S. Singer, and B. R. Levin.** 2002. Antibiotics in agriculture: when is it time to close the barn door? *Proc. Natl. Acad. Sci. USA* **99:**5752–5754.

101. **Lisle, J. T., S. C. Broadaway, A. M. Prescott, B. H. Pyle, C. Fricker, and G. A. McFeters.** 1998. Effects of starvation on physiological activity and chlorine disinfection resistance in *Escherichia coli* O157:H7. *Appl. Environ. Microbiol.* **64:**4658–4662.

102. **Luangtongkum, T., T. Y. Morishita, A. J. Ison, S. Huang, P. F. McDermott, and Q. Zhang.** 2006. Effect of conventional and organic production practices on the prevalence and antimicrobial resistance of *Campylobacter* spp. in poultry. *Appl. Environ. Microbiol.* **72:** 3600–3607.

103. **Luo, N., O. Sahin, J. Lin, L. O. Michel, and Q. Zhang.** 2003. In vivo selection of *Campylobacter* isolates with high levels of fluoroquinolone resistance associated with *gyrA* mutations and the function of the CmeABC efflux pump. *Antimicrob. Agents Chemother.* **47:**390–394.

104. **Lynch, M., J. Painter, R. Woodruff, and C. Braden.** 2006. Surveillance for foodborne-disease outbreaks—United States, 1998–2002. *MMWR Surveill. Summ.* **55:**1–42.

105. **Mammina, C., L. Cannova, S. Massa, E. Goffredo, and A. Nastasi.** 2002. Drug resistances in *Salmonella* isolates from animal foods, Italy 1998–2000. *Epidemiol. Infect.* **129:**155–161.

106. **Manges, A. R., J. R. Johnson, B. Foxman, T. T. O'Bryan, K. E. Fullerton, and L. W. Riley.** 2001. Widespread distribution of urinary tract infections caused by a multidrug-resistant *Escherichia coli* clonal group. *N. Engl. J. Med.* **345:**1007–1013.

107. **Martin, L. J., M. Fyfe, K. Doré, J. A. Buxton, F. Pollari, B. Henry, D. Middleton, R. Ahmed, F. Jamieson, B. Ciebin, S. A. McEwen, J. B. Wilson, and the Multi-Provincial *Salmonella* Typhimurium Case-Control Study Steering Committee.** 2004. Increased burden of illness associated with antimicrobial-resistant *Salmonella enterica* serotype Typhimurium infections. *J. Infect. Dis.* **189:** 377–384.

108. **Mayrhofer, S., P. Paulsen, F. J. Smulders, and F. Hilbert.** 2004. Antimicrobial resistance profile of five major food-borne pathogens isolated from beef, pork and poultry. *Int. J. Food Microbiol.* **97:**23–29.

109. **McDermott, P. F., S. M. Bodeis, L. L. English, D. G. White, R. D. Walker, S. Zhao, S. Simjee, and D. D. Wagner.** 2002. Ciprofloxacin resistance in *Campylobacter jejuni* evolves rapidly in chickens treated with fluoroquinolones. *J. Infect. Dis.* **185:**837–840.

110. **McEwen, S. A., and P. J. Fedorka-Cray.** 2002. Antimicrobial use and resistance in animals. *Clin. Infect. Dis.* **34**(Suppl. 3)**:**S93–S106.

111. **McEwen, S. A., and R. S. Singer.** 2006. Stakeholder position paper: the need for antimicrobial use data for risk assessment. *Prev. Vet. Med.* **73:**169–176.

112. **McGowan, L. L., C. R. Jackson, J. B. Barrett, L. M. Hiott, and P. J. Fedorka-Cray.** 2006. Prevalence and antimicrobial resistance of enterococci isolated from retail fruits, vegetables, and meats. *J. Food Prot.* **69:**2976–2982.

113. **McManus, P. S., V. O. Stockwell, G. W. Sundin, and A. L. Jones.** 2002. Antibiotic use in plant agriculture. *Annu. Rev. Phytopathol.* **40:** 443–465.

114. **Mead, P. S., L. Slutsker, V. Dietz, L. F. McCaig, J. S. Bresee, C. Shapiro, P. M. Griffin, and R. V. Tauxe.** 1999. Food-related illness and death in the United States. *Emerg. Infect. Dis.* **5:**607–625.

115. **Meng, J., and M. P. Doyle.** 1997. Emerging issues in microbiological food safety. *Annu. Rev. Nutr.* **17:**255–275.

116. **Meng, J., S. Zhao, M. P. Doyle, and S. W. Joseph.** 1998. Antibiotic resistance of *Escherichia coli* O157:H7 and O157:NM isolated from animals, food, and humans. *J. Food Prot.* **61:**1511–1514.

117. **Mesa, R. J., V. Blanc, A. R. Blanch, P. Cortes, J. J. Gonzalez, S. Lavilla, E. Miro, M. Muniesa, M. Saco, M. T. Tortola, B. Mirelis, P. Coll, M. Llagostera, G. Prats, and F. Navarro.** 2006. Extended-spectrum beta-lactamase-producing Enterobacteriaceae in different environments (humans, food, animal farms and sewage). *J. Antimicrob. Chemother.* **58:** 211–215.

118. **Meunier, D., E. Jouy, C. Lazizzera, M. Kobisch, and J. Y. Madec.** 2006. CTX-M-1- and CTX-M-15-type beta-lactamases in clinical *Escherichia coli* isolates recovered from food-producing animals in France. *Int. J. Antimicrob. Agents* **28:**402–407.
119. **Miriagou, V., P. T. Tassios, N. J. Legakis, and L. S. Tzouvelekis.** 2004. Expanded-spectrum cephalosporin resistance in non-typhoid *Salmonella. Int. J. Antimicrob. Agents* **23:** 547–555.
120. **Miro, E., B. Mirelis, F. Navarro, A. Rivera, R. J. Mesa, M. C. Roig, L. Gomez, and P. Coll.** 2005. Surveillance of extended-spectrum beta-lactamases from clinical samples and faecal carriers in Barcelona, Spain. *J. Antimicrob. Chemother.* **56:**1152–1155.
121. **Molbak, K.** 2005. Human health consequences of antimicrobial drug-resistant *Salmonella* and other foodborne pathogens. *Clin. Infect. Dis.* **41:** 1613–1620.
122. **Mora, A., J. E. Blanco, M. Blanco, M. P. Alonso, G. Dhabi, A. Echeita, E. A. Gonzalez, M. I. Bernardez, and J. Blanco.** 2005. Antimicrobial resistance of Shiga toxin (verotoxin)-producing *Escherichia coli* O157:H7 and non-O157 strains isolated from humans, cattle, sheep and food in Spain. *Res. Microbiol.* **156:** 793–806.
123. **Moulin-Schouleur, M., M. Reperant, S. Laurent, A. Bree, S. Mignon-Grasteau, P. Germon, D. Rasschaert, and C. Schouler.** 2007. Extraintestinal pathogenic *Escherichia coli* strains of avian and human origin: link between phylogenetic relationships and common virulence patterns. *J. Clin. Microbiol.* **45:**3366–3376.
124. **Musgrove, M. T., D. R. Jones, J. K. Northcutt, N. A. Cox, M. A. Harrison, P. J. Fedorka-Cray, and S. R. Ladely.** 2006. Antimicrobial resistance in *Salmonella* and *Escherichia coli* isolated from commercial shell eggs. *Poult. Sci.* **85:**1665–1669.
125. **Nataro, J. P., and J. B. Kaper.** 1998. Diarrheagenic *Escherichia coli. Clin. Microbiol. Rev.* **11:**142–201.
126. **National Research Council.** 1999. *The Use of Drugs in Food Animals: Benefits and Risks.* National Academy Press, Washington, DC.
127. **Nde, C. W., J. S. Sherwood, C. Doetkott, and C. M. Logue.** 2006. Prevalence and molecular profiles of *Salmonella* collected at a commercial turkey processing plant. *J. Food Prot.* **69:** 1794–1801.
128. **Nelson, J. M., K. E. Smith, D. J. Vugia, T. Rabatsky-Ehr, S. D. Segler, H. D. Kassenborg, S. M. Zansky, K. Joyce, N. Marano, R. M. Hoekstra, and F. J. Angulo.** 2004. Prolonged diarrhea due to ciprofloxacin-resistant *Campylobacter* infection. *J. Infect. Dis.* **190:**1150–1157.
129. **Neu, H. C., C. E. Cherubin, E. D. Longo, B. Flouton, and J. Winter.** 1975. Antimicrobial resistance and R-factor transfer among isolates of *Salmonella* in the northeastern United States: a comparison of human and animal isolates. *J. Infect. Dis.* **132:**617–622.
130. **Neu, H. C., E. B. Winshell, J. Winter, and C. E. Cherubin.** 1971. Antibiotic resistance of *Salmonella* in northeastern United States 1968–1969. *N.Y. State J. Med.* **71:**1196–1200.
131. **Norwood, D. E., and A. Gilmour.** 2000. The growth and resistance to sodium hypochlorite of *Listeria monocytogenes* in a steady-state multispecies biofilm. *J. Appl. Microbiol.* **88:**512–520.
132. **Ogeer-Gyles, J., K. Mathews, J. S. Weese, J. F. Prescott, and P. Boerlin.** 2006. Evaluation of catheter-associated urinary tract infections and multi-drug-resistant *Escherichia coli* isolates from the urine of dogs with indwelling urinary catheters. *J. Am. Vet. Med. Assoc.* **229:** 1584–1590.
133. **Ogeer-Gyles, J., K. A. Mathews, W. Sears, J. F. Prescott, J. S. Weese, and P. Boerlin.** 2006. Development of antimicrobial drug resistance in rectal *Escherichia coli* isolates from dogs hospitalized in an intensive care unit. *J. Am. Vet. Med. Assoc.* **229:**694–699.
134. **Orskov, F., and I. Orskov.** 1990. The serology of capsular antigens. *Curr. Top. Microbiol. Immunol.* **150:**43–63.
135. **Orskov, F., and I. Orskov.** 1992. *Escherichia coli* serotyping and disease in man and animals. *Can. J. Microbiol.* **38:**699–704.
136. **Orskov, I., and F. Orskov.** 1985. *Escherichia coli* in extra-intestinal infections. *J. Hyg.* **95:**551–575.
137. **Peirano, G., Y. Agerso, F. M. Aarestrup, and R. D. dos Prazeres.** 2005. Occurrence of integrons and resistance genes among sulphonamide-resistant *Shigella* spp. from Brazil. *J. Antimicrob. Chemother.* **55:**301–305.
138. **Peirano, G., Y. Agerso, F. M. Aarestrup, E. M. dos Reis, and R. D. dos Prazeres.** 2006. Occurrence of integrons and antimicrobial resistance genes among *Salmonella enterica* from Brazil. *J. Antimicrob. Chemother.* **58:**305–309.
139. **Phillips, I.** 2007. Withdrawal of growth-promoting antibiotics in Europe and its effects in relation to human health. *Int. J. Antimicrob. Agents* **30:**101–107.
140. **Phillips, I., M. Casewell, T. Cox, B. De Groot, C. Friis, R. Jones, C. Nightingale,**

**R. Preston, and J. Waddell.** 2004. Does the use of antibiotics in food animals pose a risk to human health? A critical review of published data. *J. Antimicrob. Chemother.* **53:**28–52.

141. **Piddock, L. J.** 1996. Does the use of antimicrobial agents in veterinary medicine and animal husbandry select antibiotic-resistant bacteria that infect man and compromise antimicrobial chemotherapy? *J. Antimicrob. Chemother.* **38:**1–3.

142. **Poppe, C., M. Ayroud, G. Ollis, M. Chirino-Trejo, N. Smart, S. Quessy, and P. Michel.** 2001. Trends in antimicrobial resistance of *Salmonella* isolated from animals, foods of animal origin, and the environment of animal production in Canada, 1994–1997. *Microb. Drug Resist.* **7:**197–212.

143. **Poppe, C., N. Smart, R. Khakhria, W. Johnson, J. Spika, and J. Prescott.** 1998. *Salmonella typhimurium* DT104: a virulent and drug-resistant pathogen. *Can. Vet. J.* **39:**559–565.

144. **Prats, G., B. Mirelis, T. Llovet, C. Muñoz, E. Miró, and F. Navarro.** 2000. Antibiotic resistance trends in enteropathogenic bacteria isolated in 1985–1987 and 1995–1998 in Barcelona. *Antimicrob. Agents Chemother.* **44:**1140–1145.

145. **Prescott, J. F., W. J. Hanna, R. Reid-Smith, and K. Drost.** 2002. Antimicrobial drug use and resistance in dogs. *Can. Vet. J.* **43:** 107–116.

146. **Rabatsky-Ehr, T., J. Whichard, S. Rossiter, B. Holland, K. Stamey, M. L. Headrick, T. J. Barrett, and F. J. Angulo.** 2004. Multidrug-resistant strains of *Salmonella enterica* Typhimurium, United States, 1997–1998. *Emerg. Infect. Dis.* **10:**795–801.

147. **Radu, S., S. A. Mutalib, G. Rusul, Z. Ahmad, T. Morigaki, N. Asai, Y. B. Kim, J. Okuda, and M. Nishibuchi.** 1998. Detection of *Escherichia coli* O157:H7 in the beef marketed in Malaysia. *Appl. Environ. Microbiol.* **64:**1153–1156.

148. **Ramchandani, M., A. R. Manges, C. DebRoy, S. P. Smith, J. R. Johnson, and L. W. Riley.** 2005. Possible animal origin of human-associated, multidrug-resistant, uropathogenic *Escherichia coli*. *Clin. Infect. Dis.* **40:** 251–257.

149. **Ribot, E. M., R. K. Wierzba, F. J. Angulo, and T. J. Barrett.** 2002. *Salmonella enterica* serotype Typhimurium DT104 isolated from humans, United States, 1985, 1990, and 1995. *Emerg. Infect. Dis.* **8:**387–391.

150. **Robins-Browne, R. M., and E. L. Hartland.** 2002. *Escherichia coli* as a cause of diarrhea. *J. Gastroenterol. Hepatol.* **17:**467–475.

151. **Russell, J. B., and A. J. Houlihan.** 2003. Ionophore resistance of ruminal bacteria and its potential impact on human health. *FEMS Microbiol. Rev.* **27:**65–74.

152. **Ryder, R. W., P. A. Blake, A. C. Murlin, G. P. Carter, R. A. Pollard, M. H. Merson, S. D. Allen, and D. J. Brenner.** 1980. Increase in antibiotic resistance among isolates of *Salmonella* in the United States, 1967–1975. *J. Infect. Dis.* **142:**485–491.

153. **Saenz, Y., M. Zarazaga, L. Brinas, M. Lantero, F. Ruiz-Larrea, and C. Torres.** 2001. Antibiotic resistance in *Escherichia coli* isolates obtained from animals, foods and humans in Spain. *Int. J. Antimicrob. Agents* **18:**353–358.

154. **Sanchez, M. X., W. M. Fluckey, M. M. Brashears, and S. R. McKee.** 2002. Microbial profile and antibiotic susceptibility of *Campylobacter* spp. and *Salmonella* spp. in broilers processed in air-chilled and immersion-chilled environments. *J. Food Prot.* **65:**948–956.

155. **Sanchez, S., M. A. McCrackin Stevenson, C. R. Hudson, M. Maier, T. Buffington, Q. Dam, and J. J. Maurer.** 2002. Characterization of multidrug-resistant *Escherichia coli* isolates associated with nosocomial infections in dogs. *J. Clin. Microbiol.* **40:**3586–3595.

156. **Sandvang, D., and F. M. Aarestrup.** 2000. Characterization of aminoglycoside resistance genes and class 1 integrons in porcine and bovine gentamicin-resistant *Escherichia coli*. *Microb. Drug Resist.* **6:**19–27.

157. **Sapkota, A. R., L. B. Price, E. K. Silbergeld, and K. J. Schwab.** 2006. Arsenic resistance in *Campylobacter* spp. isolated from retail poultry products. *Appl. Environ. Microbiol.* **72:** 3069–3071.

158. **Schroeder, C. M., J. Meng, S. Zhao, C. DebRoy, J. Torcolini, C. Zhao, P. F. McDermott, D. D. Wagner, R. D. Walker, and D. G. White.** 2002. Antimicrobial resistance of *Escherichia coli* O26, O103, O111, O128, and O145 from animals and humans. *Emerg. Infect. Dis.* **8:**1409–1414.

159. **Schroeder, C. M., D. G. White, B. Ge, Y. Zhang, P. F. McDermott, S. Ayers, S. Zhao, and J. Meng.** 2003. Isolation of antimicrobial-resistant *Escherichia coli* from retail meats purchased in greater Washington, DC, USA. *Int. J. Food Microbiol.* **85:**197–202.

160. **Schroeder, C. M., C. Zhao, C. DebRoy, J. Torcolini, S. Zhao, D. G. White, D. D. Wagner, P. F. McDermott, R. D. Walker, and J. Meng.** 2002. Antimicrobial resistance of *Escherichia coli* O157 isolated from humans, cattle, swine, and food. *Appl. Environ. Microbiol.* **68:**576–581.

161. **Singer, R. S., L. A. Cox, Jr., J. S. Dickson, H. S. Hurd, I. Phillips, and G. Y. Miller.** 2007. Modeling the relationship between food animal health and human foodborne illness. *Prev. Vet. Med.* **79:**186–203.

162. **Singer, R. S., R. Finch, H. C. Wegener, R. Bywater, J. Walters, and M. Lipsitch.** 2003. Antibiotic resistance—the interplay between antibiotic use in animals and human beings. *Lancet Infect. Dis.* **3:**47–51.

163. **Singer, R. S., and C. L. Hofacre.** 2006. Potential impacts of antibiotic use in poultry production. *Avian Dis.* **50:**161–172.

164. **Singh, R., C. M. Schroeder, J. Meng, D. G. White, P. F. McDermott, D. D. Wagner, H. Yang, S. Simjee, C. DebRoy, R. D. Walker, and S. Zhao.** 2005. Identification of antimicrobial resistance and class 1 integrons in Shiga toxin-producing *Escherichia coli* recovered from humans and food animals. *J. Antimicrob. Chemother.* **56:**216–219.

165. **Smith, D. L., A. D. Harris, J. A. Johnson, E. K. Silbergeld, and J. G. Morris, Jr.** 2002. Animal antibiotic use has an early but important impact on the emergence of antibiotic resistance in human commensal bacteria. *Proc. Natl. Acad. Sci. USA* **99:**6434–6439.

166. **Smith, D. L., J. A. Johnson, A. D. Harris, J. P. Furuno, E. N. Perencevich, and J. G. Morris, Jr.** 2003. Assessing risks for a pre-emergent pathogen: virginiamycin use and the emergence of streptogramin resistance in *Enterococcus faecium*. *Lancet Infect. Dis.* **3:**241–249.

167. **Smith, K. E., J. M. Besser, C. W. Hedberg, F. T. Leano, J. B. Bender, J. H. Wicklund, B. P. Johnson, K. A. Moore, and M. T. Osterholm for The Investigation Team.** 1999. Quinolone-resistant *Campylobacter jejuni* infections in Minnesota, 1992–1998. *N. Engl. J. Med.* **340:**1525–1532.

168. **Snary, E. L., L. A. Kelly, H. C. Davison, C. J. Teale, and M. Wooldridge.** 2004. Antimicrobial resistance: a microbial risk assessment perspective. *J. Antimicrob. Chemother.* **53:**906–917.

169. **Stopforth, J. D., R. O'Connor, M. Lopes, B. Kottapalli, W. E. Hill, and M. Samadpour.** 2007. Validation of individual and multiple-sequential interventions for reduction of microbial populations during processing of poultry carcasses and parts. *J. Food Prot.* **70:** 1393–1401.

170. **Su, L. H., H. L. Chen, J. H. Chia, S. Y. Liu, C. Chu, T. L. Wu, and C. H. Chiu.** 2006. Distribution of a transposon-like element carrying *bla(CMY-2)* among *Salmonella* and other Enterobacteriaceae. *J. Antimicrob. Chemother.* **57:**424–429.

171. **Summers, A. O.** 2006. Genetic linkage and horizontal gene transfer, the roots of the antibiotic multi-resistance problem. *Anim. Biotechnol.* **17:**125–135.

172. **Sunde, M., and M. Norstrom.** 2006. The prevalence of, associations between and conjugal transfer of antibiotic resistance genes in *Escherichia coli* isolated from Norwegian meat and meat products. *J. Antimicrob. Chemother.* **58:**741–747.

173. **Threlfall, E. J.** 2000. Epidemic *Salmonella typhimurium* DT 104—a truly international multiresistant clone. *J. Antimicrob. Chemother.* **46:**7–10.

174. **Threlfall, E. J., M. Day, E. de Pinna, A. Charlett, and K. L. Goodyear.** 2006. Assessment of factors contributing to changes in the incidence of antimicrobial drug resistance in *Salmonella enterica* serotypes Enteritidis and Typhimurium from humans in England and Wales in 2000, 2002 and 2004. *Int. J. Antimicrob. Agents* **28:**389–395.

175. **Threlfall, E. J., J. A. Frost, L. R. Ward, and B. Rowe.** 1996. Increasing spectrum of resistance in multiresistant *Salmonella typhimurium*. *Lancet* **347:**1053–1054.

176. **Threlfall, E. J., B. Rowe, and L. R. Ward.** 1993. A comparison of multiple drug resistance in salmonellas from humans and food animals in England and Wales, 1981 and 1990. *Epidemiol. Infect.* **111:**189–197.

177. **Turnidge, J.** 2004. Antibiotic use in animals—prejudices, perceptions and realities. *J. Antimicrob. Chemother.* **53:**26–27.

177a. **U.S. Department of Health and Human Services, Food and Drug Administration, Center for Veterinary Medicine.** 2003. *Evaulating the Safety of Antimicrobial New Animal Drugs with Regard to Their Microbiological Effects on Bacteria of Human Health Concern.* Guidance 152. Food and Drug Administration, Rockville, MD.

178. **van den Bogaard, A. E., and E. E. Stobberingh.** 2000. Epidemiology of resistance to antibiotics. Links between animals and humans. *Int. J. Antimicrob. Agents* **14:**327–335.

179. **van Loo, I. H. M., B. M. W. Diederen, P. H. M. Savelkoul, J. H. C. Woudenberg, R. Roosendaal, A. van Belkum, N. Lemmens-den Toom, C. Verhulst, P. H. J. van Kenlen, and J. A. J. W. Kluytmans.** 2007. Methicillin-resistant *Staphylococcus aureus* in meat products, the Netherlands. *Emerg. Infect. Dis.* **13:**1753–1755.

180. **Varma, J. K., K. D. Greene, J. Ovitt, T. J. Barrett, F. Medalla, and F. J. Angulo.** 2005. Hospitalization and antimicrobial resistance in *Salmonella* outbreaks, 1984–2002. *Emerg. Infect. Dis.* **11:**943–946.

181. **Varma, J. K., R. Marcus, S. A. Stenzel, S. S. Hanna, S. Gettner, B. J. Anderson, T. Hayes, B. Shiferaw, T. L. Crume, K. Joyce, K. E. Fullerton, A. C. Voetsch, and F. J. Angulo.** 2006. Highly resistant *Salmonella* Newport-MDRAmpC transmitted through the domestic US food supply: a FoodNet case-control study of sporadic *Salmonella* Newport infections, 2002–2003. *J. Infect. Dis.* **194:**222–230.
182. **Varma, J. K., K. Molbak, T. J. Barrett, J. L. Beebe, T. F. Jones, T. Rabatsky-Ehr, K. E. Smith, D. J. Vugia, H. G. Chang, and F. J. Angulo.** 2005. Antimicrobial-resistant nontyphoidal *Salmonella* is associated with excess bloodstream infections and hospitalizations. *J. Infect. Dis.* **191:**554–561.
183. **Voogd, C. E., W. J. van Leeuwen, P. A. Guinee, A. Manten, and J. J. Valkenburg.** 1977. Incidence of resistance to ampicillin, chloramphenicol, kanamycin and tetracycline among *Salmonella* species isolated in the Netherlands in 1972, 1973 and 1974. *Antonie van Leeuwenhoek* **43:**269–281.
184. **Walsh, C., G. Duffy, R. O'Mahony, S. Fanning, I. S. Blair, and D. A. McDowell.** 2006. Antimicrobial resistance in Irish isolates of verocytotoxigenic *Escherichia coli (E. coli)*—VTEC. *Int. J. Food Microbiol.* **109:**173–178.
185. **Wassenaar, T. M., M. Kist, and A. de Jong.** 2007. Re-analysis of the risks attributed to ciprofloxacin-resistant *Campylobacter jejuni* infections. *Int. J. Antimicrob. Agents* **30:**195–201.
186. **Wasteson, Y.** 2001. Zoonotic *Escherichia coli. Acta Vet. Scand. Suppl.* **95:**79–84.
187. **Wegener, H. C.** 2003. Antibiotics in animal feed and their role in resistance development. *Curr. Opin. Microbiol.* **6:**439–445.
188. **Weill, F. X., R. Lailler, K. Praud, A. Kerouanton, L. Fabre, A. Brisabois, P. A. Grimont, and A. Cloeckaert.** 2004. Emergence of extended-spectrum-beta-lactamase (CTX-M-9)-producing multiresistant strains of *Salmonella enterica* serotype Virchow in poultry and humans in France. *J. Clin. Microbiol.* **42:**5767–5773.
189. **Welch, T. J., W. F. Fricke, P. F. McDermott, D. G. White, M. L. Rosso, D. A. Rasko, M. K. Mammel, M. Eppinger, M. J. Rosovitz, D. Wagner, L. Rahalison, J. E. Leclerc, J. M. Hinshaw, L. E. Lindler, T. A. Cebula, E. Carniel, and J. Ravel.** 2007. Multiple antimicrobial resistance in plague: an emerging public health risk. *PLoS ONE* **2:**e309.
190. **Whichard, J. M., K. Joyce, P. D. Fey, J. M. Nelson, F. J. Angulo, and T. J. Barrett.** 2005. β-lactam resistance and *Enterobacteriaceae,* United States. *Emerg. Infect. Dis.* **11:**1464–1466.
191. **White, D. G., A. Datta, P. McDermott, S. Friedman, S. Qaiyumi, S. Ayers, L. English, S. McDermott, D. D. Wagner, and S. Zhao.** 2003. Antimicrobial susceptibility and genetic relatedness of *Salmonella* serovars isolated from animal-derived dog treats in the USA. *J. Antimicrob. Chemother.* **52:**860–863.
192. **White, D. G., S. Zhao, R. Singh, and P. F. McDermott.** 2004. Antimicrobial resistance among gram-negative foodborne bacterial pathogens associated with foods of animal origin. *Foodborne Pathog. Dis.* **1:**137–152.
193. **White, D. G., S. Zhao, R. Sudler, S. Ayers, S. Friedman, S. Chen, P. F. McDermott, S. McDermott, D. D. Wagner, and J. Meng.** 2001. The isolation of antibiotic-resistant *Salmonella* from retail ground meats. *N. Engl. J. Med.* **345:**1147–1154.
194. **Wilson, I. G.** 2004. Antimicrobial resistance of *Salmonella* in raw retail chickens, imported chicken portions, and human clinical specimens. *J. Food Prot.* **67:**1220–1225.
195. **Winokur, P. L., A. Brueggemann, D. L. DeSalvo, L. Hoffmann, M. D. Apley, E. K. Uhlenhopp, M. A. Pfaller, and G. V. Doern.** 2000. Animal and human multidrug-resistant, cephalosporin-resistant *Salmonella* isolates expressing a plasmid-mediated CMY-2 AmpC beta-lactamase. *Antimicrob. Agents Chemother.* **44:**2777–2783.
196. **Winokur, P. L., D. L. Vonstein, L. J. Hoffman, E. K. Uhlenhopp, and G. V. Doern.** 2001. Evidence for transfer of CMY-2 AmpC beta-lactamase plasmids between *Escherichia coli* and *Salmonella* isolates from food animals and humans. *Antimicrob. Agents Chemother.* **45:**2716–2722.
197. **Witte, W.** 2000. Selective pressure by antibiotic use in livestock. *Int. J. Antimicrob. Agents* **16**(Suppl. 1):S19–S24.
197a. **Witte, W.** 2000. Ecological impact of antibiotic use in animals on different complex microflora: environment. *Int. J. Antimicrob. Agents* **14:**321–325.
198. **Witte, W., B. Strommenger, C. Stanek, and C. Cuny.** 2007. Methicillin-resistant *Staphylococcus aureus* ST398 in humans and animals, Central Europe. *Emerg. Infect. Dis.* **13:** 255–258.
199. **Wong, C. S., S. Jelacic, R. L. Habeeb, S. L. Watkins, and P. I. Tarr.** 2000. The risk of the hemolytic-uremic syndrome after antibiotic treatment of *Escherichia coli* O157:H7 infections. *N. Engl. J. Med.* **342:**1930–1936.
200. **Wong, T. L., L. Hollis, A. Cornelius, C. Nicol, R. Cook, and J. A. Hudson.** 2007. Prevalence, numbers, and subtypes of *Campy-*

*lobacter jejuni* and *Campylobacter coli* in uncooked retail meat samples. *J. Food Prot.* **70:**566–573.

**201. World Health Organization.** 1997. *The Medical Impact of the Use of Antimicrobials in Food Animals. Report of a WHO Meeting, Berlin, Germany, 13–17 October 1997.* World Health Organization, Geneva, Switzerland.

**202. World Health Organization.** 1998. *Use of Quinolones in Food Animals and Potential Impact on Human Health. Report of a WHO Meeting, Geneva, Switzerland, 2–5 June 1998.* World Health Organization, Geneva, Switzerland.

**203. World Health Organization.** 2001. *WHO Global Principles for the Containment of Antimicrobial Resistance in Animals Intended for Food. Report of a WHO Consultation with the Participation of the Food and Agriculture Organization of the United Nations and the Office International des Epizooties, Geneva, Switzerland, 5–9 June 2000.* World Health Organization, Geneva, Switzerland.

**204. World Health Organization.** 2004. *Joint FAO/OIE/WHO Expert Workshop on Non-Human Antimicrobial Usage and Antimicrobial Resistance: Scientific Assessment. Geneva, Switzerland, 1–5 December, 2003.* World Health Organization, Geneva, Switzerland.

**205. World Health Organization.** 2005. *Critically Important Antibacterial Agents for Human Medicine for Risk Management Strategies of Non-Human Use. Report of a WHO Working Group Consultation, Canberra, Australia, 15–18 June 2005.* World Health Organization, Geneva, Switzerland.

**206. World Health Organization.** 2007. *Critically Important Antibacterial Agents for Human Medicine: Categorization for the Development of Risk Management Strategies to Contain Antimicrobial Resistance due to Non-Human Use. Report of the Second WHO Expert Meeting, Copenhagen, Denmark, 29–31 May 2007.* World Health Organization, Geneva, Switzerland.

**207. World Health Organization.** 2007. *Joint FAO/WHO/OIE Expert Meeting on Critically Important Antimicrobials. Report of the FAO/WHO/OIE Expert Meeting, FAO, Rome, Italy, 26–30 November 2007.* World Health Organization, Geneva, Switzerland.

**208. Wu, T. L., J. H. Chia, L. H. Su, C. H. Chiu, A. J. Kuo, L. Ma, and L. K. Siu.** 2007. CMY-2 beta-lactamase-carrying community-acquired urinary tract *Escherichia coli:* genetic correlation with *Salmonella enterica* serotypes Choleraesuis and Typhimurium. *Int. J. Antimicrob. Agents* **29:**410–416.

**209. Yan, J. J., C. Y. Hong, W. C. Ko, Y. J. Chen, S. H. Tsai, C. L. Chuang, and J. J. Wu.** 2004. Dissemination of *blaCMY-2* among *Escherichia coli* isolates from food animals, retail ground meats, and humans in southern Taiwan. *Antimicrob. Agents Chemother.* **48:**1353–1356.

**210. Zaidi, M. B., V. Leon, C. Canche, C. Perez, S. Zhao, S. K. Hubert, J. Abbott, K. Blickenstaff, and P. F. McDermott.** 2007. Rapid and widespread dissemination of multidrug-resistant *bla*CMY-2 *Salmonella* Typhimurium in Mexico. *J. Antimicrob. Chemother.* **60:**398–401.

**211. Zaidi, M. B., P. F. McDermott, P. Fedorka-Cray, V. Leon, C. Canche, S. K. Hubert, J. Abbott, M. Leon, S. Zhao, M. Headrick, and L. Tollefson.** 2006. Nontyphoidal *Salmonella* from human clinical cases, asymptomatic children, and raw retail meats in Yucatan, Mexico. *Clin. Infect. Dis.* **42:**21–28.

**212. Zansky, S., B. Wallace, D. Schoonmaker-Bopp, P. Smith, F. Ramsey, J. Painter, A. Gupta, P. Kalluri, and S. Noviello.** 2002. From the Centers for Disease Control and Prevention. Outbreak of multi-drug resistant *Salmonella* Newport—United States, January–April 2002. *JAMA* **288:**951–953.

**213. Zhao, C., B. Ge, J. De Villena, R. Sudler, E. Yeh, S. Zhao, D. G. White, D. Wagner, and J. Meng.** 2001. Prevalence of *Campylobacter* spp., *Escherichia coli,* and *Salmonella* serovars in retail chicken, turkey, pork, and beef from the Greater Washington, D.C., area. *Appl. Environ. Microbiol.* **67:**5431–5436.

**214. Zhao, S., A. R. Datta, S. Ayers, S. Friedman, R. D. Walker, and D. G. White.** 2003. Antimicrobial-resistant *Salmonella* serovars isolated from imported foods. *Int. J. Food Microbiol.* **84:**87–92.

**215. Zhao, S., P. J. Fedorka-Cray, S. Friedman, P. F. McDermott, R. D. Walker, S. Qaiyumi, S. L. Foley, S. K. Hubert, S. Ayers, L. English, D. A. Dargatz, B. Salamone, and D. G. White.** 2005. Characterization of *Salmonella* Typhimurium of animal origin obtained from the National Antimicrobial Resistance Monitoring System. *Foodborne Pathog. Dis.* **2:**169–181.

**216. Zhao, S., P. F. McDermott, S. Friedman, J. Abbott, S. Ayers, A. Glenn, E. Hall-Robinson, S. K. Hubert, H. Harbottle, R. D. Walker, T. M. Chiller, and D. G. White.** 2006. Antimicrobial resistance and genetic relatedness among *Salmonella* from retail foods of animal origin: NARMS retail meat surveillance. *Foodborne Pathog. Dis.* **3:**106–117.

**217. Zhao, S., P. F. McDermott, S. Friedman, S. Qaiyumi, J. Abbott, C. Kiessling, S. Ayers, R. Singh, S. Hubert, J. Sofos,**

**and D. G. White.** 2006. Characterization of antimicrobial-resistant *Salmonella* isolated from imported foods. *J. Food Prot.* **69:**500–507.

**218. Zhao, S., P. F. McDermott, D. G. White, S. Qaiyumi, S. L. Friedman, J. W. Abbott, A. Glenn, S. L. Ayers, K. W. Post, W. H. Fales, R. B. Wilson, C. Reggiardo, and R. D. Walker.** 2007. Characterization of multidrug resistant Salmonella recovered from diseased animals. *Vet. Microbiol.* **123:**122–132.

**219. Zhao, S., S. Qaiyumi, S. Friedman, R. Singh, S. L. Foley, D. G. White, P. F. McDermott, T. Donkar, C. Bolin, S. Munro, E. J. Baron, and R. D. Walker.** 2003. Characterization of *Salmonella enterica* serotype Newport isolated from humans and food animals. *J. Clin. Microbiol.* **41:**5366–5371.

**220. Zhao, S., D. G. White, P. F. McDermott, S. Friedman, L. English, S. Ayers, J. Meng, J. J. Maurer, R. Holland, and R. D. Walker.** 2001. Identification and expression of cephamycinase $bla_{CMY}$ genes in *Escherichia coli* and *Salmonella* isolates from food animals and ground meat. *Antimicrob. Agents Chemother.* **45:** 3647–3650.

**221. Zhao, T., M. P. Doyle, P. J. Fedorka-Cray, P. Zhao, and S. Ladely.** 2002. Occurrence of *Salmonella enterica* serotype Typhimurium DT104A in retail ground beef. *J. Food Prot.* **65:** 403–407.

# COMMENSAL BACTERIA, MICROBIAL ECOSYSTEMS, AND HORIZONTAL GENE TRANSMISSION: ADJUSTING OUR FOCUS FOR STRATEGIC BREAKTHROUGHS AGAINST ANTIBIOTIC RESISTANCE

*Hua H. Wang*

# 14

On the basis of their varied impacts on human and animal health, microbes are often classified as pathogens, opportunistic pathogens, commensal bacteria, and beneficial bacteria. The last group includes microbes of industrial importance (such as fermentation starter cultures) and those believed to help maintain healthy microflora in the host (probiotics). Up to now, pathogens and microbes of direct industrial significance have received the most attention, and they are extensively studied by researchers. However, this by no means reflects the insignificance of commensal bacteria in nature. In fact, pathogens account for only a very small percentage in the microbial population. Unless selectively enriched, many pathogens can hardly be detected directly from the natural environment. There is no doubt that mission-driven investigations on certain microbes have greatly benefited humankind by resulting in the successful control of many formerly deadly diseases and by improving the productivities of various industries. However, the consequent dominance in the literature of a small group of microbes, which has unintentionally resulted in neglect of the importance of the majority of the commensal bacteria present in complex ecosystems, has hindered us from effectively solving problems attributable to complicated interactions among microbes.

By way of example, the rapid emergence of antibiotic-resistant (ART) pathogens is becoming a major public health concern. For a long time it was believed that the uncontrolled usage of antibiotics in clinics and hospitals facilitated the dissemination of antibiotic resistance (AR) in pathogens (42). Thus, limiting the indiscriminate prescription of antibiotics (such as for treatment of likely viral infections) became the general guideline for medical practitioners in the United States, despite the risk of secondary bacterial infections and, in rare cases, death due to occasional misdiagnosis. The isolation of ART pathogens from food animals, plants, the farm environment, and retail meat, poultry, and seafood products has raised a significant concern regarding antibiotic usage in agriculture and aquaculture, and this further suggests a possible route of AR transmission to humans through the food chain (11, 14, 41, 88). Nevertheless, foodborne pathogens, especially ART pathogens, account for only a very small percentage of the microbial population associated with

*Hua H. Wang,* Department of Food Science and Technology and Department of Microbiology, The Ohio State University, Columbus, OH 43210.

*Food-Borne Microbes: Shaping the Host Ecosystem,* Edited by L.-A. Jaykus, H. H. Wang, and L. S. Schlesinger, © 2009 ASM Press, Washington, DC

foods, and they do not appear to serve as a significant avenue in transmitting AR to other organisms. In fact, recent data from both human and food animal studies have shown that limiting the use of antibiotics in food animal production resulted in a modest reduction in the prevalence of resistance in certain bacteria but did not completely eliminate ART microbes. This suggests that the current control strategy is not sufficient to stop AR (for a review, see reference 45). So what have we missed?

Recently, a large AR gene pool was found in dairy, meat, produce, and seafood products, including many ready-to-eat, "healthy" food items. Some items contained up to $10^8$ copies of resistance genes per gram of food (6, 22, 51, 86). Various commensal bacteria, including some beneficial bacteria, were identified to be carriers of AR genes. The resistance genes from the food isolates were transmissible to human residential and pathogenic bacteria in laboratory settings and have been demonstrated to lead to increased resistance in these organisms (86). Certain commensal bacteria with inherited mechanisms can significantly facilitate the dissemination of AR genes in laboratory settings (47). Notably, several studies conducted in Europe revealed that almost 100% of healthy adult and child subjects carried resistant bacteria in their oral cavity, and these individuals did not have recent exposure to antibiotics (39, 64, 85). Fecal samples from humans, including healthy children, also contained a large pool of AR genes (28). These microbial ecosystem-based studies are instrumental because they revealed the actual size of the AR gene pool and demonstrated the importance of community-based pathways for the development of AR, independent of clinical exposure. These studies further identified major partners involved in horizontal gene transmission (HGT) events, as either the AR gene recipients or potential donors. Information such as this justifies the need for enhanced efforts to study bacteria such as *Enterococcus* spp., *Staphylococcus* spp., and certain lactic acid bacteria, not just because of their status as pathogens, opportunistic pathogens, or beneficial bacteria but also as model organisms with inherited HGT mechanisms critical to their success in microbial ecosystems of foods, the environment, and the host. Moreover, the recent findings deliver an urgent message that many microbes, including the bacteria previously believed to be beneficial, are susceptible to AR transmission. It is anticipated that future scientific findings regarding microbial interactions in ecosystems will help shape our views on AR evolution and dissemination, thus enabling the development of more effective control strategies.

This chapter focuses on describing the molecular mechanisms and major routes in the dissemination and persistence of foreign genes in microbial ecosystems. Emphasis is placed on commensal bacteria with respect to the emergence, circulation, and enrichment of AR in the natural environment, the food chain, and the animal or human host.

## INTERACTIONS IN MICROBIAL ECOSYSTEMS

### Microbial Ecosystem Development

A biofilm can be defined as a matrix-enclosed bacterial population in which the cells are adherent to one other and/or to surfaces or interfaces; biofilms include microbial aggregates and floccules and also adherent populations within the pore spaces of porous media (16). Microbial biofilms found in natural environments, industrial settings, and hosts (animals or plants) often consist of heterogeneous microbes, as a result of the interactions among the microbes, which are impacted by a variety of environmental factors. Therefore, they are also referred to as microbial ecosystems. Biofilm development is not a random but a dynamic, selective, and competitive process. In a particular environment, certain microbes may act as primary colonizers attaching to the surface, and others may act as later biofilm partners through establishing interactions with the existing organisms. The early colonizers play a critical role in the selective recruitment of new members into the microbial community, often through ligand-receptor interactions. Compatible microbial cells able to develop such interactions can also form coaggregates when mixed in solutions at a proper ratio (36, 37, 69). Once they become part of the microbial consortium, the new members can attract additional bacteria from the environment by similar means, enabling the dynamic expansion of the microbial community. In response to various environmental challenges, microbes use various mechanisms to maintain the integrity and efficiency of the microbial ecosystem. For example, they can produce various antagonistic compounds to inhibit competitors, promote the synergistic utilization of environmental nutrients among organisms with complementary metabolic machinery, and/or share beneficial traits within the community through HGT.

Microbial cell surface components such as proteins, polysaccharides, and other appendices are the key elements involved in forming the ligand-receptor interactions, and they thus directly coordinate biofilm development. Although it is well known that quorum sensing signals are involved in biofilm development (17), they likely influence biofilm formation by modulating the expression of bacterial cell surface biofilm-forming determinants. Microbial biofilms, especially those dominated by a particular type of organism, can exhibit special architectural features such as the "mushroom" by *Pseudomonas* (17), the "honeycomb" by *Listeria monocytogenes* (52), and the "sponge" and "flat" by *Streptococcus mutans* (44) and *Enterococcus* spp. (G. Dunny, University of Minnesota, personal communication), respectively. These structures often result from the interactions among microbial cells through cell surface-associated components such as polysaccharides and proteins, or extracellular compounds produced by bacteria.

## HGT and Microbial Ecosystem Development

It is anticipated that microbial ecosystems such as biofilms would be suitable for natural HGT events such as conjugation and transformation due to the close proximity of cells, and possibly altered gene expression in the biofilm environment. However, data from recent studies suggest that HGT in fact plays a critical role in microbial ecosystem development. This two-way relationship was demonstrated in both gram-positive and gram-negative bacteria via conjugation and transformation. In the gram-negative bacterium *Escherichia coli,* it was found that conjugal transfer of a surface component-encoding plasmid induced biofilm development which was independent of quorum sensing. In addition, conjugative plasmid transfer drove biofilm expansion in *E. coli* (9, 26, 56, 65, 66). The role of conjugation in biofilm formation was further demonstrated in gram-positive bacteria using a cell clumping-associated high-frequency conjugation *Lactococcus lactis* model system, previously known for its effective transmission of fermentation-related traits promoting lactose and casein utilization. Luo et al. (47) found that conjugation triggered the enhanced expression of the clumping protein CluA, which was a main biofilm attribute in lactococci. Clumping transconjugants further transmitted the biofilm-forming elements among the lactococcal population at a much higher frequency than did the parental nonclumping donor, likely through a yet-to-be-characterized mechanism activated by the initial conjugation, enabling effective sharing of the plasmid in the lactococcal community and rapid expansion of the biofilm network. It is noteworthy that besides *L. lactis,* clumping-associated high-frequency conjugation has also been reported for *Enterococcus, Lactobacillus,* and *Bacillus* (4, 21, 32, 67), suggesting that this mechanism may apply to other bacteria as well. Lactococcal CluA shares significant sequence homology with the reported biofilm-forming adhesins found in many oral streptococci, but most of these are chromosomally encoded. The plasmid-associated element in lactococci offers an evolutionary advantage by effective dissemination within the ecosystem through conjugation to achieve rapid expansion of the biofilm network and the sharing of beneficial traits within the community. In *S. mutans,* it was found that a quorum sensing signal not only facilitated biofilm formation but also significantly increased the competence of the bacteria (44), enabling rapid dissemination of beneficial genes in the ecosystem, including those involved in biofilm development and expansion. Furthermore, the competent cells could readily take up DNA as a nutrient from the surrounding environment, which was facilitated by signal-controlled production of inhibitory compounds such as bacteriocins, which killed competitors under nutrient-limiting conditions. Such mechanisms increased the chances of success of the carrier organism in microbial competition in an adverse environment (38, 54). Therefore, HGT as a natural mechanism benefits microbes greatly and in various ways for their ultimate success in evolution. For the microbial consortium, AR is nothing more than another trait offering a survival advantage in adverse environments. Therefore, improved understanding of the basic molecular mechanisms and critical control steps involved in HGT in microbial ecosystems will shed light on the development of strategies for control of AR dissemination.

## Roles of Commensal Bacteria in HGT

As mentioned above, pathogens account for only a very small percentage of the microbial population found in natural environments. It is known that the frequency of HGT is much higher between organisms with compatible genetic backgrounds than those with divergent ones. Not only do commensal bacteria outnumber pathogens but also they have much more diversified genetic backgrounds. Therefore, the chance of a commensal bacterium or pathogen finding a genetically compatible commensal partner and accomplishing a successful round of AR transmission is much higher than is the chance of a similar event occurring between two pathogens. However, in hospitals, clinics, and patients, the relative density of pathogens is higher than that in the natural environment; thus, the chance of HGT between pathogens can potentially increase. Commensal bacteria can serve as (i) donors transmitting AR genes to commensal or pathogenic recipient strains, (ii) end recipients for AR genes, or (iii) intermediates receiving an AR gene and further transmitting the AR gene to another commensal or pathogenic strain (3). Because the size of the AR gene pool in the microbial population, mostly derived from commensal bacteria, is correlated to the frequency of subsequent HGT events, including AR transmission to pathogens, it can potentially serve as an indicator to predict the likelihood of the emergence of ART pathogens in the corresponding microbial ecosystem.

It is worth noting that in a recent study, Luo at al. (47) showed that clumping-associated *L. lactis,* which had an inherited mechanism for high-frequency conjugation, could also facilitate the transmission of the broad-host-range drug resistance plasmid pAMβ1 to the recipient at a frequency more than 10,000 times greater than that of the donor strain without the high-frequency-conjugation mechanism. The data suggest that commensal bacteria may serve not only as AR gene pools but also as intermediates significantly facilitating AR gene transmission. Because similar clumping-associated high-frequency-conjugation systems have also been reported for *Enterococcus* spp., *Bacillus* spp., and *Lactobacillus* spp., such facilitating mechanisms may have a broad implication in HGT.

In the past, due to technical limitations in population-based genetic studies, ecological investigations on AR in commensal bacteria and AR gene pools in microbial ecosystems were difficult to accomplish. Now with the assistance of modern microbial detection, identification, and quantification tools such as 16S rRNA gene sequence analysis, PCR-denaturing gradient gel electrophoresis, and real-time quantitative PCR, as well as high-throughput genome sequence technologies, comprehensive ecological system studies have become feasible. By examining the microbial composition and the identities of the AR gene carriers, one can reveal the involvement of key and/or unique microbes in HGT events, enabling targeted investigation of the HGT mechanisms associated with these ART bacteria. By using culture-independent molecular methods or by eliminating the enrichment procedure in culture-dependent analysis of ART bacteria, one can assess the size of the AR gene pool or the magnitude of culturable ART bacteria associated with the microbial ecosystem under investigation.

## EMERGENCE, PERSISTENCE, AND DISSEMINATION OF AR

### Origin of Resistance

Various types of resistance genes offer a number of protective mechanisms against antagonistic materials, such as antibiotics and heavy metals. AR transmitted by HGT often involves enzymes modifying the antibiotic target site (or antibiotic itself) or efflux systems which rapidly flush the detrimental materials out of the cells (70). The acquired AR determinants not only can be passed to progenies through vertical transmission (proliferation) and stabilized in the recipient lineages, but also they may be transmitted to other bacteria within or across species or genera through HGT events. Thus, HGT has a great impact on AR dissemination in microbial ecosystems.

One potential source for resistance genes is the organisms that produce antimicrobial compounds (5, 84). Most microbes which produce such substances normally contain mechanisms to shield themselves from the action of the antimicrobial compound. These immunity-encoding genes are also subject to HGT and may provide similar protection to recipient cells, thus becoming the source of resistance. For instance, certain *L. lactis* strains can produce the peptide bacteriocin nisin, which is a broad-spectrum antimicrobial compound. Nisin is used as a functional food-grade additive to inhibit gram-positive food-borne pathogens such as *Listeria monocytogenes* and *Staphylococcus aureus*. It has also been considered as a potential replacement for antibiotic treatment of some infectious diseases (7). However, while the practice of using nisin in the form of pasteurized cheese spread or powder made with nisin-producing *L. lactis* has been quite safe, the use of live bacteriocin-producing cultures to inhibit food-borne pathogens such as *Listeria monocytogenes* (68, 71) can be potentially problematic. Of particular concern is the potential transmission of the bacteriocin immunity genes into the microbial community of the food, leading to acquired resistance. In fact, nisin-producing lactococci contain two immunity modules within a nisin gene cluster, *nis*I and *nis*FEG, for a lipoprotein and transporter proteins, respectively, on a conjugative transposon; HGT of the transposon has been demonstrated within lactococci (62, 63). The cloned immunity genes provided acquired resistance against nisin in both *Bacillus* (82) and *Enterococcus* (8), demonstrating its functionality in other gram-positive bacteria and further supporting concerns about HGT in pathogens such as *L. monocytogenes*. As the identification of new antimicrobial compounds is an active area of research, and bacteriocin-producing organisms frequently have the status of generally recognized as safe (GRAS), clear recognition of the potential risks and implications of HGT events must be continuously stressed.

### Gene Stabilization Mechanisms

While it is the general belief that antibiotic usage is responsible for the selection and maintenance of resistant bacteria, data from a number of laboratories have shown that many resistance genes are actually quite stable in ART bacteria isolated from clinical settings, healthy human subjects, and food items, at least in the absence of the corresponding selective pressure (34, 43, 45, 81). While the selective pressure from antibiotics obviously played a role in the development and amplification of ART bacteria, complete elimination of the use of antibiotics in both humans and animals is impractical. In addition, data already show that banning the use of certain antibiotics did not result in complete elimination of the corresponding resistance in the microbial population (34, 45, 81). Therefore, it is of particular importance to reveal hidden mechanisms that contribute to the maintenance of the resistance in the microbial ecosystem so that additional targeted control strategies can be developed.

## CHROMOSOMAL INTEGRATION

Plasmid-borne genes are generally unstable and can be easily lost if they serve no real beneficial function. Since most AR genes disseminated through HGT are plasmid borne, it is anticipated that the corresponding selective pressure, that being an antibiotic(s), plays an important role in maintaining such genes in the microbial ecosystem. However, recent microbial genome sequencing analyses and other related studies reveal that a number of AR genes, such as those encoding penicillin and tetracycline resistance, have already integrated into the chromosomes of lactic acid bacteria as well as many other organisms (1, 2, 27, 49, 50, 55, 74). The findings suggest that these AR genes will be maintained quite stably in the microbial consortia in the absence of the selective pressure.

## CROSS-SELECTION

It is recognized that certain compounds may cross-select AR traits. For instance, a number of bacterial strains were found to be resistant to heavy metals, such as silver and mercury, and various antibiotics. In many cases, the resistance traits were encoded by the same plasmid (15, 18). Therefore, such compounds can indirectly select for the ART subpopulation in the absence of the corresponding antibiotics. In addition, dual resistance to sanitation agents and antibiotics may result in the same phenomenon (72).

## FITNESS SELECTION INDEPENDENT OF DRUG USAGE

Although it is a general perception that resistance should be a burden, increasing evidence suggests that this may not always be true. Recent reports show that instead of incurring a fitness cost, certain resistance genes may offer an additional survival advantage independent of antibiotic usage. For instance, the *gyrA* mutation in *Campylobacter* not only conferred fluoroquinolone resistance but also offered a fitness advantage over the wild type which led to the rapid dominance of the resistant population in the gut microbiota (48, 87).

In other instances, even if a newly developed resistant mutant initially grows slower than the original parental strain, the fitness cost can often be reduced by acquisition of secondary mutations. Furthermore, many clinical ART clones have spread to a wide range of geographical regions and hospital environments, suggesting that resistant isolates are ecologically fit (for a detailed review, see reference 45).

It has been reported that some plasmids, including certain large ones found in lactic acid bacteria and pseudomonads, are in fact very stable. Many of these plasmids contain genetic elements encoding metabolic traits essential for survival in a particular environment. It therefore appears that the loss of fitness due to the burden of extra replication can be compensated for in some cases (45). For example, in lactococci, the plasmid-encoded lactose fermentation ($Lac^+$) and proteinase ($Prt^+$) activities, essential for efficient growth in milk, are usually unstable and can be lost during consecutive transfer in bacterial media. However, lactose utilization encoded by the plasmid pSK11L is quite stable in *Lactococcus lactis* subsp. *cremoris* strain SK11 (25). The Opp gene cluster, essential for SK11 to utilize peptides and milk proteins as nitrogen sources, is also encoded by this plasmid. Although there is another copy of the Opp homologue in its chromosome with an unidentified functionality, the two Opp homologues share 82% nucleotide sequence identity and may vary in substrate specificity (78). The stability of pSK11L in *L. lactis* subsp. *cremoris* is related to a 3.1-kb PvuII-XbaI fragment on the plasmid (30), which likely corresponds to the 2.6-kb fragment located within the Opp region from the published sequence of pSK11L. Therefore, the presence of genetic elements encoding essential bacterial metabolic activities such as the Opp system might also play an important role in plasmid maintenance. From the point view of evolution, such a mechanism enables the host strain to stably retain a significant amount of extra DNA, which may not have immediate use but be potentially beneficial in response to certain environmental challenges.

#### THE TOXIN-ANTITOXIN PLASMID "ADDICTIVE" MAINTENANCE

Several programmed cell death mechanisms involving a potent toxin and a labile antidote, with the antidote being either the proteic antitoxin (the proteic killer gene system) or antisense RNA, have been reported for various bacteria (for detailed reviews, see references 24 and 33). The proteic toxin-antidote (TA)-encoding genes were initially identified on several low-copy-number *E. coli* plasmids (and some other extrachromosomal elements), involved in plasmid maintenance through post-segregational killing, also known as the plasmid addictive effect. The stable toxins are lethal to the cells by targeting various molecules with essential cellular functions, but the antitoxins bind to the toxins, neutralizing their lethal effects. In progeny cells that have lost the TA-encoding plasmid, the labile antitoxin is degraded by cellular proteinase, leaving the plasmid-free cells susceptible to the stable toxin. Only the progenies carrying the TA-encoding gene will propagate. The TA mechanism plays an important role in plasmid maintenance and retention in these cases. Similar TA systems have also been identified on the chromosomes of a number of bacteria, potentially involved in programmed cell death under various stress conditions. By way of example, the *ccd* system of plasmid F, the *parD* system of plasmid R1, *pem* of plasmid R100, *parDE* of plasmid RP4/RK2, and the ω-ε-ζ system in pSM19035 and pRE25 are all among the well-described model TA systems.

Recently the TA systems, including *mazEF* (a homologue of *parD* and *pem*), and, to a lesser extent, *axe-txe, relBE,* and ω-ε-ζ, were found to be ubiquitous and plasmid encoded in vancomycin-resistant enterococci (57). Particularly, the functional *mazEF* system was identified on most of the *van*A-carrying plasmids, providing a likely explanation of the persistence of the vancomycin-resistant enterococci in microbial ecosystems in the absence of the antibiotic. The ω-ε-ζ system was also found in ART enterococcal isolates from fermented food products, and the resistance trait was stably maintained in the fermentation environment without the corresponding antibiotics (76). The TA system has also been found to be involved in maintaining the erythromycin and vancomycin resistance plasmids in *Streptococcus* spp., *Bacillus subtilis,* and *Enterococcus* spp. in environments where the bacteria were not further exposed to the antibiotics (10, 34, 81, 89). These results suggest that such plasmid retention mechanisms likely play an important role in the commonly observed persistence of AR in microbial ecosystems, even in the absence of the corresponding antibiotic selective pressure.

### Key Factors in AR Dissemination

While the role of antibiotic selective pressure in AR dissemination has been extensively reviewed in the literature, the following factors likely have also played important roles in the spread of AR.

#### TYPE OF ORGANISM

Increasing evidence from recent ecological analyses indicates that certain bacteria might play more important roles than others in HGT in microbial ecosystems. For instance, ART *Enterococcus, Staphylococcus, Carnobacterium,* and *Lactococcus* spp. are prevalent in various food samples, and enterococci are also among the major culturable bacteria in ART populations from human stool samples (43, 73, 77, 86). The AR plasmids from *Enterococcus, Staphylococcus, Carnobacterium,* and *Lactococcus* spp. were successfully transmitted to the human residential bacteria *Streptococcus mutans* and *Enterococcus faecalis* by natural transformation, and electroporation in laboratory settings led to increased resistance to the corresponding antibiotics in the progenies, illustrating the functionality and mobility of the AR-encoding plasmids (43, 86). Most of these gram-positive bacteria are known for their susceptibility to HGT and ability to effectively form biofilms via various mechanisms. Therefore, it is not surprising that such inherited mechanisms serve as a major driving force in microbial ecosystem evolution, making these bacteria "super survivors" in various environments.

It is worth noting that other organisms, such as ART *Pseudomonas,* have also been found in various foods (86). However, my laboratory has failed to demonstrate HGT of the AR genes from *Pseudomonas* spp. to *E. coli* and *S. mutans* by natural transformation, electroporation, or conjugation, despite many trials (unpublished data). The data suggest that the likelihood of AR transmission by such organisms is much less than that by other ART bacteria such as *Enterococcus* and *Staphylococcus* spp. Thus, although the size of the AR gene pool is correlated to the frequency of gene dissemination, the characteristics of the AR carriers likely also have an important impact on the subsequent dissemination of AR genes through HGT events. An improved understanding of molecular mechanisms involved in HGT in key organisms such as *Enterococcus* spp., *Staphylococcus* spp., and certain lactic acid bacteria is critical for understanding associated disease risk and for targeted control strategy development.

## PROBIOTICS

Probiotics as a business has grown rapidly in the last couple of decades based on the belief that consumption of certain lactic acid bacteria and bifidobacteria is beneficial for the maintenance of a healthy gut microflora. This is supported by two main observations: (i) a relatively higher population of bifidobacteria and lactobacilli is found in infants and some long-lived adults than in the general population, and (ii) the dominance of these bacteria may competitively inhibit other bacteria, particularly pathogens. In fact, there are a number of health claims associated with probiotic consumption, such as assisting recovery from diarrhea and stimulating the host immune system. A fundamental question yet to be answered is whether the observed colonization of these bacteria is the cause of a healthy gut microflora or, rather, the indication of a healthy gut environment, which may largely reflect the interplay of the host factors and the microbial population. In fact, many probiotic strains are not retained well by the gut unless consumed regularly. Clearly, many of the beneficial claims of probiotics still need stringent scientific proof, and an improved understanding of the interplay between probiotic strains and the commensal gut microflora is needed in order to confirm such claims.

For a long time, the general perception has been that probiotic bacteria have nothing to do with transmission of "bad" traits such as AR. However, results from microbial genome sequencing studies illustrate that a number of AR genes are already integrated into the chromosomes of several lactic acid bacteria (2, 49). Data from my laboratory further demonstrated that *Enterococcus* spp., *Lactococcus* spp., *Leuconostoc* spp., *Streptococcus thermophilus,* and *Carnobacterium* spp. are among the dominant AR gene carriers found in various food products and are capable of transmitting AR genes (43, 86). As mentioned above, the clumping-associated high-frequency conjugation system is present in several lactic acid bacteria and other gram-positive bacteria, and this system likely facilitates the transmission of AR genes (47). These results suggest that certain strains of lactic acid bacteria and perhaps other probiotic strains, like many other bacteria, are susceptible to HGT events and can become the donor, recipient, or intermediate for AR gene transmission. A number of AR genes have been identified in several potential probiotic strains from Europe and Africa, and some of them are transferable to other bacteria by HGT (59). It was documented that a commercial probiotic *Lactobacillus acidophilus* strain was also able to acquire a vancomycin resistance gene from enterococci during digestive transit in mice, and this strain was persistent in the gut without the constant selective pressure of the antibiotic (53). These observations suggest that caution should be used when introducing large numbers of such organisms into humans, as an unintended consequence may be an increased size of the AR gene pool. Overall, although the intention of probiotic consumption is to improve gut health, our knowledge about the behavior of these strains in complex microbial ecosystems is in fact quite limited. In addition, certain enterococcal strains are considered favorable growth-promoting alternatives to antibiotics in food animal production (75), and some *Carnobacterium* strains are candidates for probiotic applications in aquaculture (40). Without careful evaluation of the safety of these strains, the practice may present an unwanted risk for AR dissemination. Since the propensity for HGT varies greatly among different bacterial species and strains, the establishment of stringent safety screening standards for probiotic strains should be considered, and preferably only those with a low tendency for HGT should be considered for routine use in animals and humans.

### HOSTS

AR became a concern mainly due to its impact on human and animal health. Although bacterial and environmental selective factors are important for AR development and persistence, hosts likely have also played a significant role in the enrichment and circulation of the AR genes. Regardless of pathogenic or commensal status, bacteria that have already adapted to the host environment have a better chance of being selectively enriched in host ecosystems. The ART bacteria already colonizing the host digestive system can further serve as an "immobilized" source of AR genes for potential HGT events, regardless of the presence of antibiotics. Jacobsen et al. (31) recently reported that after inoculation of a tetracycline-resistant *Lactobacillus plantarum* strain and a nonresistant *E. faecalis* strain into the digestive tract of germ-free rats, HGT occurred even in the absence of selective pressure, and the administration of tetracycline only caused a slight increase in resistance transmission. Evidence also suggests AR gene transmission among *Bacteroides* and other genera in the human colon, indicating that the colon likely provides a suitable environment for HGT events (77). In fact, animal and human wastes have been found to be rich in ART bacteria (58, 73; D. Kinkelarr and H. H. Wang, unpublished data). Once released to the environment, these ART bacteria can be spread to other human and animal hosts through contact with water, food, and various environments (12, 35, 60, 80, 83). These organisms likely have a better chance of colonizing human and animal hosts again than do their endogenous environmental counterparts. If not treated properly, human and animal manure and other wastes may be another avenue for circulating such enriched hazardous genes and microbes.

## Major Pathways of AR Dissemination to Humans

### CLINICAL ENVIRONMENT

It is well established that the clinical environment has a high concentration of ART pathogens, giving rise to an increased chance of infection in hospital and clinic patients. It is anticipated that in such environments, the ART commensal bacterial concentration would be even higher than that of the ART pathogens, and certainly above the levels found in the community environment. Therefore, hospital and clinical contacts likely serve as an important avenue for AR dissemination.

### FOOD CHAIN

It is now believed that the food chain may well play a very important role in disseminating AR genes and ART bacteria to the general public (61, 86). The abundance of ART bacteria in the food supply can serve as a constant supply of AR genes to the human digestive microflora (22, 86). The colonization by ART bacteria and associated HGT events inevitably contribute to the increasing resistance to antibiotics as seen in humans (86).

ANIMAL CONTACTS

As mentioned above, ART bacteria adapted to humans or animals (particularly mammals) may have a better chance of colonizing the hosts again than do endogenous environmental isolates. Companion animals are in close contact with humans, providing ample opportunities for transmission of ART bacteria and AR genes. For instance, Simjee at al. (79) reported that a vancomycin-resistant *Enterococcus faecium* strain isolated from a companion dog likely resulted from HGT between human and animal enterococci. Data also showed that the colonization of methicillin-resistant *S. aureus* was significantly higher in veterinary personnel who had frequent contacts with animals, particularly with large animals, than in those without a history of animal contact (29, 46). Animals, like humans, are susceptible hosts to ART bacteria and are exposed in much the same way, through contact with the environment or consumption of food and water, as well as by receiving antibiotic treatments. Whether the ART bacteria going through this additional step of adaptation in companion or farm animals become more "infectious" to humans than the endogenous environmental isolates is worthy of further investigation.

## POTENTIAL BREAKTHROUGHS IN AR CONTROL STRATEGIES

It is obvious now that multiple pathways and complicated mechanisms are involved in the development and maintenance of AR in microbial ecosystems; therefore, simply limiting the use of antibiotics is not sufficient to control the spread of AR. Meanwhile, antibiotics are still essential therapeutic agents, and it is impossible to completely ban their use. Therefore, a systematic evaluation of the contributions of key factors in AR development, dissemination, and maintenance is necessary in order to develop control strategies to reduce problems associated with AR. The following approaches may lead to potential breakthroughs in combating AR.

### Proper Monitoring System

The development of AR takes time, and a proper monitoring system is essential, not only for sounding the alarm but also for the application of different control strategies targeting various stages in AR development. As mentioned above, the size of the AR gene pool in commensal bacteria is much larger than that in pathogens and is correlated with the levels of subsequent HGT events. Therefore, the magnitude of ART commensal bacteria may serve as an indicator for AR status in any given microbial ecosystem. Only pathogens and *Enterococcus* are currently under qualitative monitoring by the National Antimicrobial Resistance Monitoring System (NARMS). Expanded coverage, particularly the quantitative assessment of the AR gene pools in commensal bacteria, could provide a more precise picture of the AR status in the environmental, food, animal, and human microbial ecosystems, enabling prediction of the forthcoming risks associated with AR in targeted pathogens.

### Interrupting the Spread of AR through Main Dissemination Pathways

As discussed above, human and animal manure and other wastes, particularly those from patients being treated with antibiotics, likely contain more "contagious" ART bacteria than do other environmental sources. Therefore, proper treatment of these wastes is critical to interrupt the dissemination of AR genes and ART bacteria. In fact, data from a recent study (13) suggested that methods of manure storage and treatment likely have a substantial impact on the persistence and decline of the AR originating from food animals. In particular, proper treatment of hospital wastes is critical.

AR contamination of water, soil, and the environment in general is not likely to disappear in the short term. Studies have already shown that much of our raw food is contaminated with ART bacteria (86). International travel and globalization of the food supply add an additional layer of complexity to the situation. Certainly, we need to practice proper processing and sanitation measures in an effort to significantly reduce the load of ART bacteria that are directly introduced into the food supply (86). This includes adequate heat processing and utilization of other bactericidal processes, along with proper sanitation during food handling, all the way from food manufacturers, retailers, and restaurants to consumers. In fact, promoting proper personal hygiene practices is probably the most direct means for minimizing human exposure to ART bacteria, along with other pathogens as well.

## Destabilization of Resistance Traits

Many AR traits are quite stable in both the environment and the host, even in the absence of antibiotic selective pressure (34, 81), mostly due to the presence of various plasmid stabilization mechanisms. Thus, development of effective means to destabilize or resensitize resistance traits could be an important control measure. For example, certain small molecules may be used to disrupt plasmid replication and resensitize bacteria to antibiotics (19, 20). Practical application of such molecules to enhance the efficiency of antibiotic treatment can have a very attractive future, but there is much to be learned before the approach becomes a practical solution.

## Novel Antibiotic Delivery Strategies

Currently, systemic administration of antibiotics is most commonly used for the treatment of bacterial infections, including localized and/or superficial infections. Development of novel antibiotic delivery systems which allow selective tissue or pathogen targeting, similar to approaches used in modern cancer therapy, could reduce the potential for global development and dissemination of AR. The goal for such novel delivery systems would be to achieve a local dosage that is much higher than the MICs for the most resistant strains, eliminating even the most resistant pathogens. At the same time, targeted delivery would significantly reduce the exposure of the normal flora to antibiotics, minimizing the unintended consequence of the development of ART strains. Such delivery systems may be possible in the near future, with advancement in nanotechnology, modern drug delivery systems, and improved understanding of various organism- and tissue-specific markers as potential recognition targets (23). This would be a logical approach to the increasingly complex issue of developing new antibiotics, especially since the microbes always seem to eventually win the battle for survival.

## REFERENCES


1. **Aires, J., F. Doucet-Populaire, and M. J. Butel.** 2007. Tetracycline resistance mediated by *tet*(W), *tet*(M), and *tet*(O) genes of *Bifidobacterium* isolates from humans. *Appl. Environ. Microbiol.* **73:**2751–2754.
2. **Ammor, M. S., M. Gueimonde, M. Danielsen, M. Zagorec, A. H. van Hoek, C. G. de Los Reyes-Gavilán, B. Mayo, and A. Margolles.** 2008. Two different tetracycline resistance mechanisms, plasmid-carried *tet*(L) and chromosomally located transposon-associated *tet*(M), coexist in *Lactobacillus sakei* Rits 9. *Appl. Environ. Microbiol.* **74:**1394–1401.
3. **Andremont, A.** 2003. Commensal flora may play key role in spreading antibiotic resistance. *ASM News* **69:**601–607.
4. **Andrup, L., J. Damgaard, and K. Wassermann.** 1993. Mobilization of small plasmids in *Bacillus thuringiensis* subsp. *israelensis* is accompanied by specific aggregation. *J. Bacteriol.* **175:** 6530–6536.
5. **Benveniste, R., and J. Davies.** 1973. Aminoglycoside antibiotic-inactivating enzymes in actinomycetes similar to those present in clinical isolates of antibiotic-resistant bacteria. *Proc. Natl. Acad. Sci. USA* **70:**2276–2280.
6. **Boehme, S., G. Werner, I. Klare, R. Reissbrodt, and W. Witte.** 2004. Occurrence of antibiotic-resistant enterobacteria in agricultural foodstuffs. *Mol. Nutr. Food Res.* **48:**522–531.
7. **Breukink, E., I. Wiedemann, C. van Kraaij, O. P. Kuipers, H. Sahl, and B. de Kruijff.** 1999. Use of the cell wall precursor lipid II by a pore-forming peptide antibiotic. *Science* **286:** 2361–2364.

8. **Bryan, E. M., T. Bae, M. Kleerebezem, and G. M. Dunny.** 2000. Improved vectors for nisin-controlled expression in gram-positive bacteria. *Plasmid* **44:**183–190.
9. **Burmølle, M., M. I. Bahl, L. B. Jensen, S. J. Sørensen, and L. H. Hansen.** 2008. Type 3 fimbriae, encoded by the conjugative plasmid pOLA52, enhance biofilm formation and transfer frequencies in *Enterobacteriaceae* strains. *Microbiology* **154:**187–195.
10. **Ceglowski, P., A. Boitsov, S. Chai, and J. C. Alonso.** 1993. Analysis of the stabilization system of pSM19035-derived plasmid pBT233 in *Bacillus subtilis Gene* **136:**1–12.
11. **Charpentier, E., and P. Courvalin.** 1999. Antibiotic resistance in *Listeria* spp. *Antimicrob. Agents Chemother.* **43:**2103–2108.
12. **Chee-Sanford J. C., R. I. Aminov, I. J. Krapac, N. Garrigures-Jeanjean, and R. I. Mackie.** 2001. Occurrence and diversity of tetracycline resistance gene in lagoons and groundwater underlying two swine production facilities. *Appl. Environ. Microbiol.* **67:**22–32.
13. **Chen, J., Z. Yu, F. C. Michel, Jr., T. Wittum, and M. Morrison.** 2007. Development and application of real-time PCR assays for quantification of *erm* genes conferring resistance to macrolides-lincosamides-streptogramin B in livestock manure and manure management systems. *Appl. Environ. Microbiol.* **73:**4407–4416.
14. **Chen, S., S. Zhao, D. G. White, C. M. Schroeder, R. Lu, H. Yang, P. F. McDermott, S. Ayers, and J. Meng.** 2004. Characterization of multiple-antimicrobial-resistant *Salmonella* serovars isolated from retail meats. *Appl. Environ. Microbiol.* **70:**1–7.
15. **Coleman, D. C., H. Pomeroy, J. K. Estridge, C. T. Keane, M. T. Cafferkey, R. Hone, and T. J. Foster.** 1985. Susceptibility to antimicrobial agents and analysis of plasmids in gentamicin- and methicillin-resistant *Staphylococcus aureus* from Dublin hospitals. *J. Med. Microbiol.* **20:**157–167.
16. **Costerton, J. W., Z. Lewandowski, D. E. Caldwell, D. R. Korber, and H. M. Lappin-Scott.** 1995. Microbial biofilms. *Annu. Rev. Microbiol.* **49:**711–745.
17. **Davies, D. G., M. R. Parsek, J. P. Pearson, B. H. Iglewski, J. W. Costerton, and E. P. Greenberg.** 1998. The involvement of cell-to-cell signals in the development of a bacterial biofilm. *Science* **280:**295–298.
18. **Davis, I. J., A. P. Roberts, D. Ready, H. Richards, M. Wilson, and P. Mullany.** 2005. Linkage of a novel mercury resistance operon with streptomycin resistance on a conjugative plasmid in *Enterococcus faecium*. *Plasmid* **54:**26–38.
19. **DeNap, J. C., and P. J. Hergenrother.** 2005. Bacterial death comes full circle: targeting plasmid replication in drug-resistant bacteria. *Org. Biomol. Chem.* **3:**959–966.
20. **Denap, J. C., J. R. Thomas, D. J. Musk, and P. J. Hergenrother.** 2004. Combating drug-resistant bacteria: small molecule mimics of plasmid incompatibility as antiplasmid compounds. *J. Am. Chem. Soc.* **126:**15402.
21. **Dunny, G. M., B. L. Brown, and D. B. Clewell.** 1978. Induced cell aggregation and mating in *Streptococcus faecalis:* evidence for a bacterial sex pheromone. *Proc. Natl. Acad. Sci. USA* **75:**3479–3483.
22. **Duran, G. M., and D. L. Marshall.** 2005. Ready-to-eat shrimp as an international vehicle of antibiotic-resistant bacteria. *J. Food Prot.* **68:** 2395–2401.
23. **Eckert, R., J. He, D. K. Yarbrough, F. Qi, M. H. Anderson, and W. Shi.** 2006. Targeted killing of *Streptococcus mutans* by a pheromone-guided "smart" antimicrobial peptide. *Antimicrob. Agents Chemother.* **50:**3651–3657.
24. **Engelberg-Kulka, H., B. Sat, M. Reches, S. Amitai, and R. Hazan.** 2004. Bacterial programmed cell death systems as targets for antibiotics. *Trends Microbiol.* **12:**66–71.
25. **Feirtag, J. M., J. P. Petzel, E. Pasalodos, K. A. Baldwin, and L. L. McKay.** 1991. Thermosensitive plasmid replication, temperature-sensitive host growth, and chromosomal plasmid integration conferred by *Lactococcus lactis* subsp. *cremoris* lactose plasmids in *Lactococcus lactis* subsp. *lactis*. *Appl. Environ. Microbiol.* **57:**539–548.
26. **Ghigo, J. M.** 2001. Natural conjugative plasmids induce bacterial biofilm development. *Nature* **412:**442–445.
27. **Girlich, D., R. Leclercq, T. Naas, and P. Nordmann.** 2007. Molecular and biochemical characterization of the chromosome-encoded class A beta-lactamase BCL-1 from *Bacillus clausii*. *Antimicrob. Agents Chemother.* **51:**4009–4014.
28. **Gueimonde, M., S. Salminen, and E. Isolauri.** 2006. Presence of specific antibiotic (tet) resistance genes in infant faecal microbiota. *FEMS Immunol. Med. Microbiol.* **48:**21–25.
29. **Hanselman, B. A., S. A. Kruth, J. Rousseau, D. E. Low, B. M. Willey, A. McGeer, and J. S. Weese.** 2006. Methicillin-resistant *Staphylococcus aureus* colonization in veterinary personnel. *Emerg. Infect. Dis.* **12:**1933–1938.
30. **Horng, J. S., K. M. Polzin, and L. L. McKay.** 1991. Replication and temperature-sensitive maintenance functions of lactose plasmid pSK11L from *Lactococcus lactis* subsp. *cremoris*. *J. Bacteriol.* **173:**7573–7581.

31. **Jacobsen, L., A. Wilcks, K. Hammer, G. Huys, D. Gevers, and S. Andersen.** 2007. Horizontal transfer of *tet*(M) and *erm*(B) resistance plasmids from food strains of *Lactobacillus plantarum* to *Enterococcus faecalis* JH2-2 in the gastrointestinal tract of gnotobiotic rats. *FEMS Microbiol. Ecol.* **59:**158–166.
32. **Jensen, G. B., A. Wilcks, S. S. Petersen, J. Damgaad, J. S. Baum, and L. Andrup.** 1995. The genetic basis of the aggregation system in *Bacillus thuringiensis* subsp. *israelensis* is located on the large conjugative plasmid pXO16. *J. Bacteriol.* **177:**2914–2917.
33. **Jensen, R. B., and K. Gerdes.** 1995. Programmed cell death in bacteria: proteic plasmid stabilization systems. *Mol. Microbiol.* **17:**205–210.
34. **Johnsen, P. J., J. I. Østerhus, H. Sletvold, M. Sørum, H. Kruse, K. Nielsen, G. S. Simonsen, and A. Sundsfjord.** 2005. Persistence of animal and human glycopeptide-resistant enterococci on two Norwegian poultry farms formerly exposed to avoparcin is associated with a widespread plasmid-mediated *van*A element within a polyclonal *Enterococcus faecium* population. *Appl. Environ. Microbiol.* **71:**159–168.
35. **Koike, S., I. G. Krapac, H. D. Oliver, A. C. Yannarell, J. C. Chee-Sanford, R. I. Aminov, and R. I. Mackie.** 2007. Monitoring and source tracking of tetracycline resistance genes in lagoons and groundwater adjacent to swine production facilities over a 3-year period. *Appl. Environ. Microbiol.* **73:**4813–4823.
36. **Kolenbrander, P. E.** 1988. Intergeneric coaggregation among human oral bacteria and ecology of dental plaque. *Annu. Rev. Microbiol.* **42:** 627–656.
37. **Kolenbrander, P. E.** 2000. Oral microbial communities: biofilms, interactions, and genetic systems. *Annu. Rev. Microbiol.* **54:**413–437.
38. **Kreth, J., J. Merritt, W. Shi, and F. Qi.** 2005. Co-ordinated bacteriocin production and competence development: a possible mechanism for taking up DNA from neighbouring species. *Mol. Microbiol.* **57:**392–404.
39. **Lancaster, H., D. Ready, P. Mullany, D. Spratt, R. Bedi, and M. Wilson.** 2003. Prevalence and identification of tetracycline-resistant oral bacteria in children not receiving antibiotic therapy. *FEMS Microbiol. Lett.* **228:**99–104.
40. **Leisner, J. J., G. G. Laursen, H. Prévost, D. Drider, and P. Dalgaard.** 2007. *Carnobacterium:* positive and negative effects in the environment and in foods. *FEMS Microbiol. Rev.* **31:**592–613.
41. **Levy, S. B., G. B. FitzGerald, and A. B. Macone.** 1976. Changes in intestinal flora of farm personnel after introduction of a tetracycline-supplemented feed on a farm. *N. Engl. J. Med.* **295:**583–588.
42. **Levy, S. B., and B. Marshall.** 2004. Antibacterial resistance worldwide: causes, challenges and responses. *Nat. Med.* **10:**S122–S129.
43. **Li, X., and H. H. Wang.** 2008. *Abstr. 108th Gen. Meet. Am. Soc. Microbiol.,* abstr. P-136, p. 513.
44. **Li, Y.-H., N. Tang, M. B. Aspiras, P. C. Y. Lau, J. H. Lee, R. P. Ellen, and D. G. Cvitkovitch.** 2002. A quorum-sensing signaling system essential for genetic competence in *Streptococcus mutans* is involved in biofilm formation. *J. Bacteriol.* **184:**2699–2708.
45. **Livermore, D.** 2004. Can better prescribing turn the tide of resistance? *Nat. Rev. Microbiol.* **2:** 73–78.
46. **Loeffler, A., A. K. Boag, J. Sung, J. A. Lindsay, L. Guardabassi, A. Dalsgaard, H. Smith, K. B. Stevens, and D. H. Lloyd.** 2005. Prevalence of methicillin-resistant *Staphylococcus aureus* among staff and pets in a small animal referral hospital in the UK. *J. Antimicrob. Chemother.* **56:**692–697.
47. **Luo, H., K. Wan, and H. H. Wang.** 2005. High-frequency conjugation system facilitates biofilm formation and pAMβ1 transmission by *Lactococcus lactis*. *Appl. Environ. Microbiol.* **71:** 2970–2978.
48. **Luo, N., S. Pereira, O. Sahin, J. Lin, S. Huang, L. Michel, and Q. Zhang.** 2005. Enhanced in vivo fitness of fluoroquinolone-resistant *Campylobacter jejuni* in the absence of antibiotic selection pressure. *Proc. Natl. Acad. Sci. USA* **102:**541–546.
49. **Makarova, K., A. Slesarev, Y. Wolf, A. Sorokin, B. Mirkin, E. Koonin, A. Pavlov, N. Pavlova, V. Karamychev, N. Polouchine, V. Shakhova, I. Grigoriev, Y. Lou, D. Rohksar, S. Lucas, K. Huang, D. M. Goodstein, T. Hawkins, V. Plengvidhya, D. Welker, J. Hughes, Y. Goh, A. Benson, K. Baldwin, J. H. Lee, I. Díaz-Muñiz, B. Dosti, V. Smeianov, W. Wechter, R. Barabote, G. Lorca, E. Altermann, R. Barrangou, B. Ganesan, Y. Xie, H. Rawsthorne, D. Tamir, C. Parker, F. Breidt, J. Broadbent, R. Hutkins, D. O'Sullivan, J. Steele, G. Unlu, M. Saier, T. Klaenhammer, P. Richardson, S. Kozyavkin, B. Weimer, and D. Mills.** 2006. Comparative genomics of the lactic acid bacteria. *Proc. Natl. Acad. Sci. USA* **103:**15611–15616.
50. **Mammeri, H., F. Eb, A. Berkani, and P. Nordmann.** 2008. Molecular characterization of AmpC-producing *Escherichia coli* clinical isolates

recovered in a French hospital. *J. Antimicrob. Chemother.* **61:**498–503.

51. **Manuzon, M. Y., S. E. Hanna, H. Luo, Z. Yu, W. J. Harper, and H. H. Wang.** 2007. Quantitative assessment of the tetracycline resistance gene pool in cheese samples by real-time TaqMan PCR. *Appl. Environ. Microbiol.* **73:** 1676–1677.
52. **Marsh, E. J., H. Luo, and H. Wang.** 2003. A three-tiered approach to differentiate *Listeria monocytogenes* biofilm-forming abilities. *FEMS Microbiol. Lett.* **228:**203–210.
53. **Mater, D. D., P. Langella, G. Corthier, and M. J. Flores.** 2008. A probiotic *Lactobacillus* strain can acquire vancomycin resistance during digestive transit in mice. *J. Mol. Microbiol. Biotechnol.* **14:**123–127.
54. **Merritt, J., J. Kreth, W. Shi, and F. Qi.** 2005. LuxS controls bacteriocin production in *Streptococcus mutans* through a novel regulatory component. *Mol. Microbiol.* **57:**960–969.
55. **Meziane-Cherif, D., D. Decré, E. A. Høiby, P. Courvalin, and B. Périchon.** 2008. Genetic and biochemical characterization of CAD-1, a chromosomally encoded new class A penicillinase from *Carnobacterium divergens. Antimicrob. Agents Chemother.* **52:**551–556.
56. **Molin, S., and T. Tolker-Nielsen.** 2003. Gene transfer occurs with enhanced efficiency in biofilms and induces enhanced stabilisation of the biofilm structure. *Curr. Opin. Biotechnol.* **14:**255–261.
57. **Moritz, E. M., and P. J. Hergenrother.** 2007. Toxin-antitoxin systems are ubiquitous and plasmid-encoded in vancomycin-resistant enterococci. *Proc. Natl. Acad. Sci. USA* **104:**311–316.
58. **Nandi, S., J. J. Maurer, C. Hofacre, and A. O. Summers.** 2004. Gram-positive bacteria are a major reservoir of class 1 antibiotic resistance integrons in poultry litter. *Proc. Natl. Acad. Sci. USA* **101:**7118–7122.
59. **Ouoba, L. I., V. Lei, and L. B. Jensen.** 2008. Resistance of potential probiotic lactic acid bacteria and bifidobacteria of African and European origin to antimicrobials: determination and transferability of the resistance genes to other bacteria. *Int. J. Food Microbiol.* **12:**217–224.
60. **Peak, N., C. W. Knapp, R. K. Yang, M. M. Hanfelt, M. S. Smith, D. S. Aga, and D. W. Graham.** 2007. Abundance of six tetracycline resistance genes in wastewater lagoons at cattle feedlots with different antibiotic use strategies. *Environ. Microbiol.* **9:**143–151.
61. **Perreten, V., F. Schwarz, L. Cresta, M. Boeglin, G. Dasen, and M. Teuber.** 1997. Antibiotic resistance spread in food. *Nature* **389:** 801–802.
62. **Ra, R., M. M. Beerthuyzen, W. M. de Vos, P. E. Saris, and O. P. Kuipers.** 1999. Effects of gene disruptions in the nisin gene cluster of *Lactococcus lactis* on nisin production and producer immunity. *Microbiology* **145:**1227–1233.
63. **Rauch, P. J. G., and W. M. de Vos.** 1994. Identification and characterization of genes involved in excision of the *Lactococcus lactis* conjugative transposon Tn*5276*. *J. Bacteriol.* **176:**2165–2171.
64. **Ready, D., R. Bedi, D. A. Spratt, P. Mullany, and M. Wilson.** 2003. Prevalence, proportions, and identities of antibiotic-resistant bacteria in the oral microflora of healthy children. *Microb. Drug Resist.* **9:**367–372.
65. **Reisner, A., B. M. Höller, S. Molin, and E. L. Zechner.** 2006. Synergistic effects in mixed *Escherichia coli* biofilms: conjugative plasmid transfer drives biofilm expansion. *J. Bacteriol.* **188:**3582–3588.
66. **Reisner, A., J. A. Haagensen, M. A. Schembri, E. L. Zechner, and S. Molin.** 2003. Development and maturation of *Escherichia coli* K-12 biofilms. *Mol. Microbiol.* **48:**933–946.
67. **Reniero, R., P. Cocconcell, V. Bottazzi, and L. Morelli.** 1992. High frequency of conjugation in *Lactobacillus* mediated by an aggregation-promoting factor. *J. Gen. Microbiol.* **138:**763–768.
68. **Reviriego, C., L. Fernández, and J. M. Rodríguez.** 2007. A food-grade system for production of pediocin PA-1 in nisin-producing and non-nisin-producing *Lactococcus lactis* strains: application to inhibit *Listeria* growth in a cheese model system. *J. Food Prot.* **70:**2512–2517.
69. **Rickard, A. H., P. Gilbert, N. J. High, P. E. Kolenbrander, and P. S. Handley.** 2003. Bacterial coaggregation: an integral process in the development of multi-species biofilms. *Trends Microbiol.* **11:**94–100.
70. **Roberts, M.** 1996. Tetracyline resistance determinants: mechanisms of action, regulation of expression, genetic mobility and distribution. *FEMS Microbiol. Rev.* **19:**1–24.
71. **Rodríguez, E., J. Tomillo, M. Nuñez, and M. Medina.** 1997. Combined effect of bacteriocin-producing lactic acid bacteria and lactoperoxidase system activation on *Listeria monocytogenes* in refrigerated raw milk. *J. Appl. Microbiol.* **83:**389–395.
72. **Russell, A. D.** 1997. Plasmids and bacterial resistance to biocides. *J. Appl. Microbiol.* **83:**155–165.
73. **Salyers, A. A., A. Gupta, and Y. Wang.** 2004. Human intestinal bacteria as reservoir for antibiotic resistance. *Trends Microbiol.* **12:**412–416.

74. **Sawant, A. A., N. V. Hegde, B. A. Straley, S. C. Donaldson, B. C. Love, S. J. Knabel, and B. M. Jayarao.** 2007. Antimicrobial-resistant enteric bacteria from dairy cattle. *Appl. Environ. Microbiol.* **73:**156–163.
75. **Scharek, L., J. Guth, M. Filter, and M. F. Schmidt.** 2007. Impact of the probiotic bacteria *Enterococcus faecium* NCIMB 10415 (SF68) and *Bacillus cereus* var. toyoi NCIMB 40112 on the development of serum IgG and faecal IgA of sows and their piglets. *Arch. Anim. Nutr.* **61:**223–234.
76. **Schwarz, F. V., V. Perreten, and M. Teuber.** 2001. Sequence of the 50-kb conjugative multiresistance plasmid pRE25 from *Enterococcus faecalis* RE25. *Plasmid* **46:**170–187.
77. **Shoemaker, N. B., K. Vlamakis, K. Hayes. and A. A. Salyers.** 2001. Evidence for extensive resistance gene transfer among *Bacteroides* spp. and among *Bacteroides* and other genera in the human colon. *Appl. Environ. Microbiol.* **67:**561–568.
78. **Siezen, R. J., B. Renckens, I. van Swam, S. Peters, R. van Kranenburg, M. Kleerebezem, and W. M. de Vos.** 2005. Complete sequences of four plasmids of *Lactococcus lactis* subsp. *cremoris* SK11 reveal extensive adaptation to the dairy environment. *Appl. Environ. Microbiol.* **71:** 8371–8382.
79. **Simjee, S., D. G. White, P. F. McDermott, D. D. Wagner, M. J. Zervos, S. M. Donabedian, L. L. English, J. R. Hayes, and R. D. Walker.** 2002. Characterization of Tn*1546* in vancomycin-resistant *Enterococcus faecium* isolated from canine urinary tract infections: evidence of gene exchange between human and animal enterococci. *J. Clin. Microbiol.* **40:**4659–4665.
80. **Smith, M. S., R. K. Yang, C. W. Knapp, Y. Niu, N. Peak, M. M. Hanfelt, J. C. Galland, and D. W. Graham.** 2004. Quantification of tetracycline resistance genes in feedlot lagoons by real-time PCR. *Appl. Environ. Microbiol.* **70:** 7372–7377.
81. **Sørum, M., P. J. Johnsen, B. Aasnes, T. Rosvoll, H. Kruse, A. Sundsfjord, and G. S. Simonsen.** 2006. Prevalence, persistence, and molecular characterization of glycopeptide-resistant enterococci in Norwegian poultry and poultry farmers 3 to 8 years after the ban on avoparcin. *Appl. Environ. Microbiol.* **72:**516–521.
82. **Stein, T., S. Heinzmann, I. Solovieva, and K. D. Entian.** 2003. Function of *Lactococcus lactis* nisin immunity genes *nis*I and *nis*FEG after coordinated expression in the surrogate host *Bacillus subtilis*. *J. Biol. Chem.* **278:**89–94.
83. **Stine, O. C., J. A. Johnson, A. Keefer-Norris, K. L. Perry, J. Tigno, S. Qaiyumi, M. S. Stine, and J. G. Morris, Jr.** 2007. Widespread distribution of tetracycline resistance genes in a confined animal feeding facility. *Int. J. Antimicrob. Agents* **29:**348–352.
84. **Thompson, C. J., and G. S. Gray.** 1983. Nucleotide sequence of a streptomycete aminoglycoside phosphotransferase gene and its relationship to phosphotransferases encoded by resistance plasmids. *Proc. Natl. Acad. Sci. USA* **80:**5190–5194.
85. **Villedieu, A., M. L. Diaz-Torres, A. P. Roberts, N. Hunt, R. McNab, D. A. Spratt, M. Wilson, and P. Mullany.** 2004. Genetic basis of erythromycin resistance in oral bacteria. *Antimicrob. Agents Chemother.* **48:**2298–2301.
86. **Wang, H. H., M. Manuzon, M. Lehman, K. Wan, H. Luo, T. E. Wittum, A. Yousef, and L. O. Bakaletz.** 2006. Food commensal microbes as a potentially important avenue in transmitting antibiotic resistance genes. *FEMS Microbiol. Lett.* **254:**226–231. (Erratum, **255:**328.)
87. **Zhang, Q., O. Sahin, P. F. McDermott, and S. Payot.** 2006. Fitness of antimicrobial-resistant *Campylobacter* and *Salmonella*. *Microbes Infect.* **8:** 1972–1978.
88. **Zhao, C., B. Ge, J. De Villena, R. Sudler, E. Yeh, S. Zhao, D. G. White, D. Wagner, and J. Meng.** 2001. Prevalence of *Campylobacter* spp., *Escherichia coli,* and *Salmonella* serovars in retail chicken, turkey, pork, and beef from the Greater Washington, D.C., area. *Appl. Environ. Microbiol.* **67:**5431–5436.
89. **Zielenkiewicz, U., and P. Ceglowski.** 2005. The toxin-antitoxin system of the streptococcal plasmid pSM19035. *J. Bacteriol.* **187:**6094–6105.

# MODEL ORGANISMS

# ANTIBIOTIC RESISTANCE AND FITNESS OF ENTERIC PATHOGENS

*Qijing Zhang and Dan I. Andersson*

# 15

For the last 60 years, antimicrobials have been the key weapons used by humans for treating bacterial infections in humans and animals. Unfortunately, bacterial pathogens have become increasingly resistant to various antibiotics, compromising the effectiveness of clinical treatments and threatening public health (41). The growing concern about antibiotic resistance has led to the call for prudent and restricted use of antimicrobials in both human medicine and animal production (40, 76). Acquisition of antibiotic resistance not only confers on bacteria the ability to survive antibiotic treatment but also affects bacterial physiology and consequently their fitness in various environments (8, 39). Understanding the impact of antimicrobial resistance on bacterial fitness and the associated mechanisms will help us predict and control the transmission and spread of antimicrobial pathogens.

Food-borne diseases cause significant morbidity and mortality each year, and hundreds of millions of people worldwide suffer from these illnesses (45, 51). Among bacterial pathogens, *Campylobacter* and *Salmonella* are the leading causes of human bacterial food-borne illnesses in the United States and other developed nations (45, 51, 65). Other important bacterial causes include *Clostridium perfringens, Shigella, Yersinia, Listeria monocytogenes, Vibrio,* and Shiga toxin-producing *Escherichia coli.* Food-borne microbes are primarily intestinal organisms and are commonly present in livestock and poultry as well as wild animals and birds. Since antibiotics are widely used for modern animal production, food-borne microbes are often exposed to antibiotic selection pressure (76). A direct consequence of the exposure is the selection of antibiotic-resistant pathogens and commensal organisms that can potentially be transmitted to humans via the food chain (41, 74, 77). Thus, there is an urgent need for developing strategies to control the emergence and persistence of antibiotic-resistant microbes in animal reservoirs. Toward that end, we need to understand and predict the evolutionary consequences of antimicrobial resistance. In this chapter, we discuss how antimicrobial resistance impacts bacterial fitness and how bacteria adapt to restore fitness in the absence of antibiotic selection pressure. This topic has been studied for many different bacterial pathogens (3), but here we

*Qijing Zhang,* Department of Veterinary Microbiology and Preventive Medicine, Iowa State University, Ames, IA 50011. *Dan I. Andersson,* Department of Medical Biochemistry and Microbiology, Uppsala University, Box 582, S-751 23 Uppsala, Sweden.

*Food-Borne Microbes: Shaping the Host Ecosystem,* Edited by L.-A. Jaykus, H. H. Wang, and L. S. Schlesinger, © 2009 ASM Press, Washington, DC

review information only on food-borne bacteria, with a particular emphasis on *E. coli, Salmonella,* and *Campylobacter.*

## FITNESS CHANGES ASSOCIATED WITH ANTIMICROBIAL RESISTANCE

Bacteria acquire resistance to antimicrobials by various mechanisms. In general, resistance occurs via either mutations in chromosomal loci or acquisition of horizontally transferred mobile genetic elements such as plasmids, phages, transposons, and integrons (58). Development of resistance often has deleterious effects on the vital physiological processes in microorganisms, imposing an initial fitness cost in bacteria in the absence of antibiotic selection pressure (2–4, 33, 39, 58). "Fitness cost" is defined as reduced growth and persistence of resistant bacteria in a given environment or reduced transmission between hosts (2, 4). But in some situations, resistance imposes no apparent burden or even enhances fitness (9, 22, 24, 28, 31, 35, 43, 55, 68). Thus, the impact of resistance on bacterial fitness is diverse and varies with the resistance mechanisms, the particular pathogens carrying the resistance, and the environments in which bacterial fitness is measured.

### Resistance Conferred by Mobile Elements

Many studies have documented the effects of resistance-carrying plasmids on bacterial fitness (15, 18, 19, 37, 56). In general, acquisition of plasmids imposes an initial fitness cost on the host in the absence of a specific selection pressure. For example, constitutive expression of tetracycline resistance carried on pBR322 (37) produced a clear adverse effect on the fitness of *E. coli* in culture media. In another example, Nguyen et al. (56) examined the effect of a plasmid-carried and Tn*10*-encoded tetracycline resistance operon on *E. coli* fitness. When induced or constitutively expressed, the resistance operon caused a significant reduction in the growth rate of *E. coli* K-12, while possession of the tightly repressed resistance operon itself did not have a negative effect on the fitness of the host. In another example, experimental cloning of *ampC* (encoding β-lactamase) from *Enterobacter cloacae* into *Salmonella enterica* serotype Typhimurium reduced its growth rate in culture media and its invasiveness to MDCK cells (54). However, when *ampC* is cotransferred with *ampR,* which encodes a repressor that inhibits the expression of *ampC,* the fitness cost associated with *ampC* production was eliminated, suggesting that overexpression of *ampC* was deleterious to *Salmonella* (54). In a later study by Hossain et al. (29) using a clinical *Salmonella* isolate harboring pCMY-7, a natural plasmid carrying $bla_{CMY-7}$ (encoding AmpC β-lactamase), it was found that high-level expression of the AmpC β-lactamase from pCMY-7 in the clinical isolate did not affect its growth rate or ability to invade mammalian cells. When pCMY-7 was conjugated to a naïve strain of *Salmonella* (LT2), it did not impose a burden on the growth rate or invasiveness of LT2; however, when $bla_{CMY-7}$ was cloned from pCMY-7 to a cloning vector and then transferred to LT2, it caused a significant reduction in growth rate and invasiveness, suggesting that other functions encoded by pCMY-7 amended the cost imposed by $bla_{CMY-7}$ (29). These findings indicate that overexpressing AmpC β-lactamase, without the inhibitory effect of AmpR or the compensatory functions of the carrying plasmid, imposes a fitness burden on *Salmonella.* Recently, it was reported that the plasmid-borne SME-1 carbapenemase enzyme conferred a fitness cost when expressed in *E. coli* (49). Interestingly, the fitness cost appeared to be associated with the signal sequence of the β-lactamase via a still-unidentified mechanism.

There are also reports indicating that carrying resistance-conferring plasmids or transposons may actually enhance the fitness of the hosts. For example, Enne et al. (20) investigated four *sul-2* (encoding sulfamethoxazole resistance)-containing plasmids from clinical *E. coli* isolates. Three of them imposed a fitness cost on a naïve *E. coli* host (JM109), while one (p9123) of them enhanced the fitness of both JM109 and its original host. Using natural plasmids isolated from commensal *E. coli* of calves, Yates et al. (81) showed that two of the three conjugative plasmids carrying apramycin resistance, when transferred to a naïve *E. coli* host (MG1655), conferred on the new host a fitness advantage. Another example is with Tn*5*-carried bleomycin resistance. Although bleomycin has never been used for treating bacterial infections, bacterial resistance to bleomycin has been reported, and it is found that the Tn*5*-carried bleomycin resistance gene *(ble)* conferred a growth and survival advantage on *E. coli* in the absence of bleomycin (12, 14). The resistance and fitness benefit conferred by *ble* required the cooperation of host-encoded factors, such as AidC (13). Subsequent studies further demonstrated that *ble* encodes a binding protein that binds to bleomycin and blocks its antibacterial and DNA-cleaving activities (1, 36). Calf-adapted *E. coli* stains carrying resistance to streptomycin, sulfadiazine, and tetracycline were shown to be more fit than antibiotic-susceptible strains in neonatal calves, but a subsequent study demonstrated that the enhanced fitness was not linked to the resistance genes (32, 33).

## Resistance Conferred by Chromosomal Mutations

Effects of chromosomally encoded resistance on bacterial fitness have been observed in a number of studies (2, 3), many of which were performed with *Salmonella* (9–11, 48, 57). A general finding from these studies is that resistance-conferring mutations impose an initial fitness cost on bacteria. For example, Bjorkman et al. (9) showed that *Salmonella* mutants that were resistant to streptomycin, fusidic acid, nalidixic acid, and rifampin lost virulence in the mouse model. Acquiring resistance to actinonin (a deformylase inhibitor) conferred a fitness cost on *Salmonella* in LB broth and in mice (57). *Salmonella* mutants that were highly resistant to fluoroquinolones (FQs) exhibited a reduced growth rate on solid media as reflected by smaller colony sizes (25). Subsequent experiments demonstrated that FQ-resistant *Salmonella* mutants selected in vitro or in vivo (chicken) showed different growth characteristics in culture medium and in the chicken host (26). The *Salmonella* mutants selected by in vitro plating were highly resistant to FQs but grew significantly slower in culture medium, and they failed to colonize chickens. On the contrary, the in vivo-selected resistant isolates exhibited intermediate susceptibility to FQs, had normal growth in liquid medium, and were able to colonize chickens at an extent comparable to, or lower than, that of the wild-type strains (26). In the case with FQ-resistant *E. coli,* single mutations in DNA gyrase or topoisomerase IV confer low-level resistance to FQs, while accumulation of multiple mutations in the enzymes results in high-level resistance (34). Interestingly, the single mutations were not associated with fitness reduction, but the mutants with multiple mutations showed a significant fitness disadvantage when assayed in competition experiments in vitro and in a mouse model (34).

Although the majority of chromosomal mutations that mediate antibiotic resistance impose a fitness loss, some of them do not impose an apparent growth disadvantage, or they even enhance bacterial fitness. One example is FQ resistance in *Campylobacter,* which is mediated by single point mutations in DNA gyrase (61, 82). Using isogenic strains, Luo et al. (43) evaluated the impact of FQ resistance on *Campylobacter* fitness in chicken, a natural host for this food-borne pathogen. The FQ-resistant *Campylobacter* was able to colonize and persist in chickens as efficiently as the isogenic FQ-susceptible strains in the absence of FQ antimicrobials. When examined by pairwise competitions, the FQ-resistant *Campylobacter* rapidly outcompeted the FQ-susceptible strains and became the dominant population in the chicken host. By using various mutant constructs, the fitness change was directly linked to the C257T mutation in *gyrA,* which confers high-level resistance to FQ antimicrobials in *Campylobacter* (43). Similarly, a streptomycin-resistant *Salmonella* mutant carrying the K42T change in the ribosomal protein S12 (encoded by the *rpsL* gene) did not show a fitness disadvantage and was fully virulent in mice (9).

The examples discussed above clearly illustrate that acquisition of antimicrobial resistance has diverse effects on bacterial fitness, and the degree of impact varies greatly with different types of mutations and pathogens. It is also important to emphasize that resistant mutants showing a fitness disadvantage in one environment may not demonstrate a fitness reduction in another environment. For example, the K43R mutation in ribosomal protein S12 imposed a fitness cost on *E. coli* in culture media, but the mutant was as fit as the wild type when inoculated into pigs (21). Likewise, *E. coli* carrying plasmid R46 showed a fitness reduction in vitro but displayed no fitness disadvantage or even showed enhanced fitness in the pig model (21). Thus, the findings derived from in vitro experiments do not necessarily predict the fitness of bacteria in vivo, emphasizing the importance of in vivo data for relevant predictions of the expected effects of resistance on bacterial growth and spread.

## MOLECULAR MECHANISMS ASSOCIATED WITH FITNESS CHANGES

The physiological pathways (e.g., ribosome, DNA gyrase, RNA polymerase, and cell wall biosynthesis) targeted by antimicrobials are essential for bacterial growth. To be able to resist the deleterious effects of antimicrobials, bacteria have to acquire mechanisms that confer resistance. However, acquisition of antibiotic resistance alters the normal metabolic state and often leads to a reduced fitness in the absence of antibiotic selection pressure (2–4, 33, 39, 58). For plasmid-encoded resistance, the resistance itself or other functions encoded by a plasmid may affect bacterial growth in antibiotic-free environments. Mutation-mediated antibiotic resistance often occurs in genes encoding products that are involved in vital cellular processes (e.g., DNA gyrase and 23S rRNA), and the resistance-conferring mutations often affect the normal physiological functions of the products, leading to reduced growth rates. For efflux-mediated resistance, production of efflux pumps on bacterial cell membranes is energy expensive and may impose a physiological burden on bacteria in the absence of antibiotics. Some known mechanisms of resistance-associated fitness changes are discussed in the following paragraphs.

### Reduced Protein Synthesis

Certain resistance mechanisms affect the rates of protein synthesis. For example, fusidic acid resistance conferred by mutations in the *fusA* gene (encoding elongation factor G [EF-G]) and streptomycin resistance mediated by mutations in the *rpsL* gene (encoding ribosomal protein S12) decrease the rate of protein synthesis in *Salmonella* and consequently result in reduced growth in culture medium and mice (10, 11). Actinonin inhibits the activity of peptide deformylase (encoded by *def*), which removes the formyl group from the N-terminal methionine of nascent peptides and is required for protein maturation. Bacterial resistance to actinonin is usually mediated by mutations in the *fmt* gene, encoding methionyl-tRNA formyltransferase, or the *folD* gene, encoding an enzyme involved in the production of 10-formyl-$H_4$-folate (57). Mutations in both genes result in the loss of function of their encoded products and force bacteria to synthesize proteins with unformylated methionyl initiator tRNA, which slows down protein translation and bacterial growth (57).

### Changes in Global Regulation

Mutations in the *fusA* gene of fusidic acid-resistant *Salmonella* mutants not only reduce the rate of protein synthesis but also modulate the level of global regulators, including ppGpp and RpoS. ppGpp, a global regulator for gene expression, was reduced in fusidic acid-resistant *Salmonella* mutants (47). In addition, the mutants also showed a reduced ability to produce RpoS in the stationary phase (46). RpoS is a global regulator for the general stress response and an important virulence factor in *Salmonella.* The reduced induction of RpoS in the fusidic acid-resistant mutants contributed to their low fitness in vivo (46). These examples illustrate the pleiotropic effects of *fusA* mutations on the cellular and molecular processes in *Salmonella,* which result in a low growth rate, abnormal cell size, poor survival in vivo, and sensitivity to oxidative stress.

### Alteration of DNA Supercoiling

Development of FQ resistance often involves mutations in DNA gyrases and topoisomerases. These resistance-conferring mutations may alter the functions of the enzymes. For example, a study conducted with *E. coli* indicated that resistance-conferring mutations in GyrA, GyrB, and ParC had varied effects on the supercoiling activity of the enzymes (6, 7). Another study by Kugelberg et al. (35) demonstrated that FQ-resistant mutants of *Pseudomonas aeruginosa* had decreased DNA supercoiling and reduced fitness. Alteration of DNA supercoiling by mutations in TopA and GyrB affected the relative abundance of more than 80 proteins in *E. coli* (73). In *Campylobacter,* the Thr-86→Ile mutation in GyrA confers high-level resistance to FQs (61) and significantly reduces the supercoiling function of the mutant gyrase (28a). These examples illustrate that point mutations in gyrases and topoisomerases not only confer resistance to quinolone antimicrobials but also modulate DNA supercoiling in the mutant organisms. Since DNA supercoiling affects DNA replication and gene transcription in bacterial organisms and plays an important role in regulating bacterial adaptive responses to environmental challenges such as stress (16, 52, 53, 59, 75), it is conceivable that alteration of DNA supercoiling would consequently influence bacterial fitness. However, direct evidence that links resistance-associated supercoiling change with fitness alteration has yet not been reported.

In a few cases, development of antibiotic resistance actually enhances bacterial fitness. One example is Tn*5*-carried bleomycin resistance in *E. coli* (1). How *ble* confers a fitness advantage in *E. coli* is unclear, but it is speculated that enhanced DNA repair or protection against DNA cleavage is involved in the fitness increase (1). Interestingly, the resistance and the fitness benefit conferred by *ble* are separable, because a mutant *ble* did not confer resistance to bleomycin but retained the beneficial effect on the *E. coli* host (36). In addition, the C257T *gyrA* mutation conferring FQ resistance enhanced the fitness of *Campylobacter* in chickens (43). Although it is known that this C257T mutation reduced the supercoiling activity of the mutant gyrase (28a), it is still unclear how this change impacts *Campylobacter* fitness. Notably, the same C257T mutation imposed a fitness cost in some isolates, suggesting that this mutation interacts with different genetic backgrounds, which can either enhance or reduce the fitness of FQ-resistant *Campylobacter.*

## EVOLUTION TO RESTORE FITNESS

As discussed above, development of antibiotic resistance in bacteria often imposes an initial fitness cost, but bacteria can rapidly adapt to restore fitness by in vitro or in vivo evolution (2, 39). The evolutionary process often involves accumulation of compensatory mutations that offset the deleterious effect of resistance-conferring elements or mutations on bacterial fitness. As a result of compensatory evolution, antimicrobial-resistant bacteria are able to stabilize and persist in environments even in the absence of antibiotic usage (3, 8, 39, 58, 68).

### Plasmid-Host Coevolution

In an early work conducted by Bouma and Lenski (15), a nonconjugative plasmid (p-ACYC184, conferring resistance to chloramphenicol and tetracycline) imposed an initial fitness cost on a naïve *E. coli* host in antibiotic-free medium, but after 500 generations of evolution in culture medium in the presence of chloramphenicol, the host-plasmid association enhanced the fitness of the evolved host even in the absence of the antibiotic. This adaptation was due to changes in the host genome but not in the plasmid (15). A subsequent work (38) further revealed that the chloramphenicol resistance carried by pACYC184 imposed a fitness cost on both the naïve and evolved *E. coli* hosts, while the tetracycline resistance (mediated by an efflux mechanism) on the plasmid had a neutral effect on the naïve host but a beneficial effect on the evolved host. These findings demonstrated that the same plasmid imposed a fitness cost on the naïve *E. coli* but conferred a fitness advantage on the evolved host, suggesting that the interaction of a plasmid with different host genetic backgrounds can have different fitness outcomes.

In addition to evolution of the host, plasmids themselves can adapt for better fitness. For example, Dahlberg and Chao (18) reported that conjugative plasmids R1 and RP4 imposed an initial fitness burden in *E. coli,* but after 1,100 generations of evolution in culture medium, both the plasmids and the host mutated to reduce the fitness cost. Thus, the evolved plasmids, when transferred to the initial *E. coli* host, no longer caused a fitness burden, and the evolved host was more fit than the ancestral host. Similarly, plasmid R1 showed an initial fitness cost in *E. coli,* but after 420 generations of evolution in culture medium, both the evolved plasmid and the *E. coli* host adapted with increased fitness compared to the ancestral plasmid and the original *E. coli* cells (19). Notably, some lineages of the evolved R1 plasmid were even able to confer a significant fitness advantage on new *E. coli* hosts as well as on *Salmonella* (19). These findings reveal the dynamic nature of plasmid-host coevolution and suggest that natural plasmids in clinical isolates may have adapted well in their natural hosts and thus may not impose a fitness burden on their hosts (20, 81).

## Compensatory Mutations

For chromosomally encoded resistance, intragenic or extragenic mutations often occur to compensate the fitness loss caused by the resistance-conferring mutations. One example is fusidic acid resistance in *Salmonella* mediated by mutations in *fusA,* encoding EF-G. The resistance-conferring mutations in EF-G decrease the rate of nucleotide exchange (GDP to GTP) and reduce the rate of protein synthesis (47), but intragenic compensatory mutations can fully or partly restore the growth rate, probably by promoting the transition of EF-G from a GDP-binding state to a GTP-binding state (30). For streptomycin-resistant *rpsL* mutants of *Salmonella,* different compensatory mutations identified in S4, S5, S12, and L19 partly restored the fitness (48). These changes presumably decrease translation fidelity and offset the overincreased rate of proofreading occurring in streptomycin-resistant mutants, thereby restoring the normal rate of protein synthesis (48). Similar to the situation in *Salmonella,* streptomycin resistance mediated by *rpsL* mutations imposed a significant fitness cost on *E. coli* because the resistance-associated mutations impaired the process of peptide elongation (70). Adaptation in culture medium rapidly restored the rate of protein synthesis and consequently the fitness of *E. coli* (70), and it was suggested that this adaptation was due to unidentified compensatory mutations that were not present in *rpsL* (70). Likewise, mutations in RNA polymerase (RpoB) confer rifampin resistance and result in a fitness cost in *E. coli,* which is due to an impaired transcription rate in the host (67). Evolution to restore fitness in the absence of antibiotic selection did not result in the loss of the resistance-associated mutation. Instead, secondary compensatory mutations occurred in *rpoB* and other genes, which restored transcription and fitness (67). These examples illustrate that compensatory mutations evolve rapidly in antibiotic-resistant mutants and potentially prevent the reversion of the mutants to susceptibility.

The extent to which fitness was restored by compensatory mutations varied greatly, and multiple compensatory mutations may be required for fitness restoration. For *Salmonella* resistance to streptomycin and nalidixic acid, most of the compensatory mutations restored fitness in mice and in laboratory medium to wild-type or near-wild-type levels, and some of the evolved mutants even exhibited fitness qualities superior to those of the wild-type strains (9, 10). On the other hand, the relative fitness of the majority of fusidic acid-resistant *Salmonella* isolates in mice (but not in the culture medium) was still lower than that of the wild-type strains even after compensatory evolution took place (10).

### Gene Amplification

Gene amplification (increased copy number) is also involved in the evolutionary process of fitness restoration. As reported by Nilsson et al. (57), mechanisms utilized by *Salmonella* to reduce the fitness cost associated with actinonin resistance are via acquisition of intragenic compensatory mutations in the *fmt* or *folD* gene or amplification (duplication) of the *metZW* genes (encoding unformylated initiator tRNA). In another example, mupirocin resistance mediated by chromosomal mutations in *ileS* imposed a fitness cost and *Salmonella* restored fitness by acquiring compensatory mutation in *ileS* or amplification of the *ileS* gene (60). These examples illustrate that both compensatory mutations and gene amplification can be involved in compensatory evolution.

### Environmental Influence

The evolution of compensatory mutations that restore fitness can be influenced by the environment in which the compensatory mutations evolve. For example, Bjorkman et al. (9, 10) showed that all streptomycin-resistant *Salmonella* isolates that evolved in vitro accumulated extragenic compensatory mutations in the *rpsD* and *rpsE* genes, while the mutants which evolved in mice all contained intragenic suppressor mutations in the *rpsL* gene. On the other hand, 26 of the 28 fusidic acid-resistant *Salmonella* mutants which evolved in culture media had intragenic suppressor mutations in the *fusA* gene, while the majority (14 of 25) of the mutants which evolved in mice were true revertants and the rest of them (11 of 25) harbored intragenic suppressor mutations (10). Both the true revertants and mutants compensated for fusidic acid resistance showed a fully restored fitness in culture media, but only the true revertants were fully fit in mice. These examples demonstrate that different environments may favor the development and selection of different types of mutations and that compensated mutants which evolve in one environment may not be fully fit in another milieu (10).

### Coexistence of Compensatory Mutations and Resistance-Conferring Mutations

Compensatory mutations must coexist with resistance-associated mutations for fitness restoration, as they often are deleterious to the evolved organisms when they are separated from the resistance-conferring mutations (2). For example, most extragenic compensatory mutations of streptomycin-resistant *Salmonella* mutants, when separated from the resistance-conferring mutations in the compensated mutants or when introduced into an antibiotic-susceptible wild-type background, impose a fitness cost on *Salmonella* (9–11). Similarly, with streptomycin-resistant *E. coli,* separation of compensatory mutations from the resistance-associated mutations in *rpsL* made the *E. coli* host less fit (70, 71). These examples indicate that once a resistance-conferring mutation is acquired by bacteria, it may be difficult to revert it if compensatory mutations have already occurred, because the reversion to susceptibility without concomitant reversion of compensatory mutations will make the pathogen unfit (8).

## WILL WITHDRAWAL OF ANTIBIOTICS REDUCE ANTIMICROBIAL RESISTANCE IN FOOD-BORNE BACTERIA?

Food-borne bacteria are exposed to antibiotics that are used for animal production. Since some of the antibiotics used on farms are also used for human therapy, there has been a major concern about the development and persistence of antibiotic-resistant food-borne microbes, which can be transmitted to humans via the food chain (5, 74, 76, 80). This public health concern has led the European Union to ban the use of growth-promoting antibiotics that are in the same classes of drugs as used for human medicine (78). Since the ban, a significant reduction has been observed in the resistance of enterococci of animal origin to glycopeptides, macrolides, and other antibiotics (78). It was also observed in Denmark that ending the use of tylosin as a growth promoter significantly reduced the prevalence of erythromycin-resistant *Campylobacter coli* in pigs (79). These observations provide compelling evidence that decreased use of antibiotics may reduce the prevalence of resistance.

There are also examples that suggest that reduction or withdrawal of antibiotic usage will not simply make antimicrobial resistance disappear, at least for the short term (23, 64, 72). For example, Enne et al. (23) investigated the persistence of sulfonamide resistance in clinical *E. coli* isolates in the United Kingdom. Despite the drastic decrease in the prescription of sulfonamide between 1991 and 1999, a decline in sulfonamide resistance was not observed and the prevalence of resistance remained high during that period. The authors attributed the persistence of sulfonamide resistance to genetic linkage of the resistance to other resistance elements. In the United States, enrofloxacin (an FQ antimicrobial) was used in poultry as a therapeutic agent to treat certain bacterial infections until 2005. Although enrofloxacin was not intended for treating *Campylobacter* infection in poultry, its use selects for and promotes the prevalence of FQ-resistant *Campylobacter,* which was shown in multiple studies (27, 44, 50). FQ resistance in *Campylobacter* is conferred by single-step mutations in *gyrA* (61, 82). The well-known effect of FQ treatment on the prevalence of FQ-resistant *Campylobacter* in poultry and the potential transfer of FQ-resistant *Campylobacter* via the food-borne route to humans led the Food and Drug Administration to withdraw the approval of FQ antimicrobials for poultry production in the United States (http://www.fda.gov/oc/antimicrobial/baytril.html).

Despite the ban of FQ antimicrobials for poultry production in the United States, the question of whether FQ-resistant *Campylobacter* will disappear from poultry and consequently from humans remains. Some surveys suggest that FQ-resistant *Campylobacter* will continue to persist in poultry even in the absence of FQ antimicrobials. For example, Price et al. (63, 64) conducted two consecutive studies in the United States to monitor the rates of FQ resistance in *Campylobacter* from poultry, and their work revealed that FQ-resistant *Campylobacter* continued to be prevalent in chickens from farms that had stopped using FQ antimicrobials for 4 years. Similarly, the work conducted by Petersen and Wedderkopp (62) in Denmark found that FQ-resistant *Campylobacter* still existed on farms that had ceased using FQs for 4 years. Analysis of the isolates by pulsed-field gel electrophoresis revealed that some of the FQ-resistant *Campylobacter* clones were able to persist for several rotations on the farms in the absence of FQ usage (62), suggesting that these resistant clones were ecologically competent in the farm environment. These observations are consistent with the laboratory finding by Luo et al. (43) that FQ-resistant *Campylobacter* shows an enhanced fitness and is able to persist in chickens in the absence of selection pressure. Together, these findings suggest that FQ-resistant *Campylobacter,* once evolved, may continue to persist in chicken flocks regardless of the use of FQ antimicrobials.

Another example of persistence of antibiotic resistance in the absence of selection pressure is tetracycline resistance in food-borne bacteria. There are many studies showing that tetracycline-resistant bacteria, such as *E. coli, Salmonella,* and *Campylobacter,* are widely distributed in food-producing animals regardless of their exposure to this class of antibiotics (17, 42, 66, 69, 72). For example, work conducted by Luangtongkum et al. (42) revealed the high prevalence and dynamic change of tetracycline-resistant *Campylobacter* strains on organic poultry farms, where no antibiotics were used for production. It was found that tetracycline-resistant *Campylobacter* carrying the *tet*(O) gene coexisted with tetracycline-susceptible strains in chicken flocks in the absence of antibiotic usage (42). Cui et al. (17) also reported the high prevalence of tetracycline-resistant *Campylobacter* in organic chickens from retail stores. In addition, tetracycline-resistant *Campylobacter* and *Salmonella* were also frequently isolated from organic dairy farms (66, 69). Together, these examples suggest that tetracycline-resistant food-borne pathogens have evolutionarily adapted to the animal production environment and may continue to persist on farms regardless of the use of tetracycline antibiotics.

## CONCLUSIONS

Antimicrobial resistance has profound effects on the biological fitness of bacterial organisms. Most resistance-conferring mechanisms impose a fitness cost, while others have low or no cost or even enhance the fitness of their hosts. This diverse impact is further complicated by the dynamic and constant evolutionary process in bacteria for better fitness and the fact that nonresistance genes, environmental conditions, and host factors also influence the fitness of microbes. Thus, once enriched and selected, antimicrobial resistance may or may not decrease even after antibiotic selection pressure is removed. This situation will have a significant impact on our effort to control the persistence and spread of antimicrobial-resistant pathogens that are detrimental to animal and human health.

Food-borne bacteria (commensals and pathogens) are often exposed to antibiotic selection in agricultural settings. The potential transfer of antibiotic-resistant organisms or genes to humans represents a major public health concern, which has led to the call for prudent and judiciary use of antimicrobials in agricultural production. However, restricted use of antimicrobials alone may not solve the problem, because some antibiotic-resistant organisms and resistance gene pools already exist at a high level. In order to effectively control the persistence and transmission of antimicrobial resistance in food-borne bacteria, we must have a better understanding of if and how antibiotic resistance affects bacterial adaptation and evolution and, in particular, how antibiotic-resistant bacteria interact with their environments and animal hosts in the absence of selection pressure. Information such as this will ultimately facilitate the development of strategies to predict and prevent transmission of antibiotic-resistant food-borne pathogens.

## ACKNOWLEDGMENTS

Work in Qijing Zhang's laboratory was supported by Public Health Service grant DK063008 from the National Institutes of Health and by National Research Initiative Competitive Grants 2005-51110-03273 and 2007-35201-18278 from the USDA Cooperative State Research, Education, and Extension Service. Work in Dan I. Andersson's laboratory was supported by the Swedish Research Council, European Union Framework Programme 6, and Uppsala University.

## REFERENCES

1. **Adam, E., M. R. Volkert, and M. Blot.** 1998. Cytochrome c biogenesis is involved in the transposon Tn5-mediated bleomycin resistance and the associated fitness effect in *Escherichia coli*. *Mol. Microbiol.* **28:**15–24.
2. **Andersson, D. I.** 2003. Persistence of antibiotic resistant bacteria. *Curr. Opin. Microbiol.* **6:**452–456.
3. **Andersson, D. I.** 2006. The biological cost of mutational antibiotic resistance: any practical conclusions? *Curr. Opin. Microbiol.* **9:**461–465.
4. **Andersson, D. I., and B. R. Levin.** 1999. The biological cost of antibiotic resistance. *Curr. Opin. Microbiol.* **2:**489–493.
5. **Angulo, F. J., V. N. Nargund, and T. C. Chiller.** 2004. Evidence of an association between use of anti-microbial agents in food animals and anti-microbial resistance among bacteria isolated from humans and the human health consequences of such resistance. *J. Vet. Med. Ser. B* **51:**374–379.
6. **Bagel, S., V. Hullen, B. Wiedemann, and P. Heisig.** 1999. Impact of *gyrA* and *parC* mutations on quinolone resistance, doubling time, and supercoiling degree of *Escherichia coli*. *Antimicrob. Agents Chemother.* **43:**868–875.
7. **Barnard, F. M., and A. Maxwell.** 2001. Interaction between DNA gyrase and quinolones: effects of alanine mutations at GyrA subunit res-

idues Ser[83] and Asp[87]. *Antimicrob. Agents Chemother.* **45:**1994–2000.

8. **Bjorkman, J., and D. I. Andersson.** 2000. The cost of antibiotic resistance from a bacterial perspective. *Drug Resist. Updates* **3:**237–245.
9. **Bjorkman, J., D. Hughes, and D. I. Andersson.** 1998. Virulence of antibiotic-resistant *Salmonella typhimurium*. *Proc. Natl. Acad. Sci. USA* **95:**3949–3953.
10. **Bjorkman, J., I. Nagaev, O. G. Berg, D. Hughes, and D. I. Andersson.** 2000. Effects of environment on compensatory mutations to ameliorate costs of antibiotic resistance. *Science* **287:**1479–1482.
11. **Bjorkman, J., P. Samuelsson, D. I. Andersson, and D. Hughes.** 1999. Novel ribosomal mutations affecting translational accuracy, antibiotic resistance and virulence of *Salmonella typhimurium*. *Mol. Microbiol.* **31:**53–58.
12. **Blot, M., B. Hauer, and G. Monnet.** 1994. The Tn5 bleomycin resistance gene confers improved survival and growth advantage on *Escherichia coli*. *Mol. Gen. Genet.* **242:**595–601.
13. **Blot, M., J. Heitman, and W. Arber.** 1993. Tn5-mediated bleomycin resistance in *Escherichia coli* requires the expression of host genes. *Mol. Microbiol.* **8:**1017–1024.
14. **Blot, M., J. Meyer, and W. Arber.** 1991. Bleomycin-resistance gene derived from the transposon Tn5 confers selective advantage to *Escherichia coli* K-12. *Proc. Natl. Acad. Sci. USA* **88:** 9112–9116.
15. **Bouma, J. E., and R. E. Lenski.** 1988. Evolution of a bacteria/plasmid association. *Nature* **335:**351–352.
16. **Champoux, J. J.** 2001. DNA topoisomerases: structure, function, and mechanism. *Annu. Rev. Biochem.* **70:**369–413.
17. **Cui, S., B. Ge, J. Zheng, and J. Meng.** 2005. Prevalence and antimicrobial resistance of *Campylobacter* spp. and *Salmonella* serovars in organic chickens from Maryland retail stores. *Appl. Environ. Microbiol.* **71:**4108–4111.
18. **Dahlberg, C., and L. Chao.** 2003. Amelioration of the cost of conjugative plasmid carriage in *Eschericha coli* K12. *Genetics* **165:**1641–1649.
19. **Dionisio, F., I. C. Conceicao, A. C. Marques, L. Fernandes, and I. Gordo.** 2005. The evolution of a conjugative plasmid and its ability to increase bacterial fitness. *Biol. Lett.* **1:**250–252.
20. **Enne, V. I., P. M. Bennett, D. M. Livermore, and L. M. Hall.** 2004. Enhancement of host fitness by the sul2-coding plasmid p9123 in the absence of selective pressure. *J. Antimicrob. Chemother.* **53:**958–963.
21. **Enne, V. I., A. A. Delsol, G. R. Davis, S. L. Hayward, J. M. Roe, and P. M. Bennett.** 2005. Assessment of the fitness impacts on *Escherichia coli* of acquisition of antibiotic resistance genes encoded by different types of genetic element. *J. Antimicrob. Chemother.* **56:**544–551.
22. **Enne, V. I., A. A. Delsol, J. M. Roe, and P. M. Bennett.** 2004. Rifampicin resistance and its fitness cost in *Enterococcus faecium*. *J. Antimicrob. Chemother.* **53:**203–207.
23. **Enne, V. I., D. M. Livermore, P. Stephens, and L. M. Hall.** 2001. Persistence of sulphonamide resistance in *Escherichia coli* in the UK despite national prescribing restriction. *Lancet* **357:** 1325–1328.
24. **Gillespie, S. H., L. L. Voelker, and A. Dickens.** 2002. Evolutionary barriers to quinolone resistance in *Streptococcus pneumoniae*. *Microb. Drug Resist.* **8:**79–84.
25. **Giraud, E., A. Brisabois, J. L. Martel, and E. Chaslus-Dancla.** 1999. Comparative studies of mutations in animal isolates and experimental in vitro- and in vivo-selected mutants of *Salmonella* spp. suggest a counterselection of highly fluoroquinolone-resistant strains in the field. *Antimicrob. Agents Chemother.* **43:**2131–2137.
26. **Giraud, E., A. Cloeckaert, S. Baucheron, C. Mouline, and E. Chaslus-Dancla.** 2003. Fitness cost of fluoroquinolone resistance in *Salmonella enterica* serovar Typhimurium. *J. Med. Microbiol.* **52:**697–703.
27. **Griggs, D. J., M. M. Johnson, J. A. Frost, T. Humphrey, F. Jorgensen, and L. J. Piddock.** 2005. Incidence and mechanism of ciprofloxacin resistance in *Campylobacter* spp. isolated from commercial poultry flocks in the United Kingdom before, during, and after fluoroquinolone treatment. *Antimicrob. Agents Chemother.* **49:** 699–707.
28. **Gustafsson, I., O. Cars, and D. I. Andersson.** 2003. Fitness of antibiotic resistant *Staphylococcus epidermidis* assessed by competition on the skin of human volunteers. *J. Antimicrob. Chemother.* **52:**258–263.

28a. **Han, J., Y.-W. Barton, and Q. Zhang.** 2005. Resistance-conferring *gyrA* mutations in fluoroquinolone-resistant *Campylobacter* affect the supercoiling activity of gyrase, p. 92. *Abstr. Conf. Res. Workers Anim. Dis.* Conference of Research Workers in Animal Disease, St. Louis, MO, 3 to 6 December 2005.

29. **Hossain, A., M. D. Reisbig, and N. D. Hanson.** 2004. Plasmid-encoded functions compensate for the biological cost of AmpC overexpression in a clinical isolate of *Salmonella typhimurium*. *J. Antimicrob. Chemother.* **53:**964–970.
30. **Johanson, U., A. Aevarsson, A. Liljas, and D. Hughes.** 1996. The dynamic structure of EF-G studied by fusidic acid resistance and internal revertants. *J. Mol. Biol.* **258:**420–432.

31. **Kanai, K., K. Shibayama, S. Suzuki, J. Wachino, and Y. Arakawa.** 2004. Growth competition of macrolide-resistant and -susceptible *Helicobacter pylori* strains. *Microbiol. Immunol.* **48:**977–980.

32. **Khachatryan, A. R., D. D. Hancock, T. E. Besser, and D. R. Call.** 2004. Role of calf-adapted *Escherichia coli* in maintenance of antimicrobial drug resistance in dairy calves. *Appl. Environ. Microbiol.* **70:**752–757.

33. **Khachatryan, A. R., D. D. Hancock, T. E. Besser, and D. R. Call.** 2006. Antimicrobial drug resistance genes do not convey a secondary fitness advantage to calf-adapted *Escherichia coli. Appl. Environ. Microbiol.* **72:**443–448.

34. **Komp, L. P., L. L. Marcusson, D. Sandvang, N. Frimodt-Moller, and D. Hughes.** 2005. Biological cost of single and multiple norfloxacin resistance mutations in *Escherichia coli* implicated in urinary tract infections. *Antimicrob. Agents Chemother.* **49:**2343–2351.

35. **Kugelberg, E., S. Lofmark, B. Wretlind, and D. I. Andersson.** 2005. Reduction of the fitness burden of quinolone resistance in *Pseudomonas aeruginosa. J. Antimicrob. Chemother.* **55:**22–30.

36. **Kumagai, T., T. Nakano, M. Maruyama, H. Mochizuki, and M. Sugiyama.** 1999. Characterization of the bleomycin resistance determinant encoded on the transposon Tn5. *FEBS Lett.* **442:**34–38.

37. **Lee, S. W., and G. Edlin.** 1985. Expression of tetracycline resistance in pBR322 derivatives reduces the reproductive fitness of plasmid-containing *Escherichia coli. Gene* **39:**173–180.

38. **Lenski, R. E., S. C. Simpson, and T. T. Nguyen.** 1994. Genetic analysis of a plasmid-encoded, host genotype-specific enhancement of bacterial fitness. *J. Bacteriol.* **176:**3140–3147.

39. **Levin, B. R., V. Perrot, and N. Walker.** 2000. Compensatory mutations, antibiotic resistance and the population genetics of adaptive evolution in bacteria. *Genetics* **154:**985–997.

40. **Levy, S. B.** 1997. Antibiotic resistance: an ecological imbalance. *Ciba Found. Symp.* **207:**1–9.

41. **Levy, S. B., and B. Marshall.** 2004. Antibacterial resistance worldwide: causes, challenges and responses. *Nat. Med.* **10:**S122–S129.

42. **Luangtongkum, T., T. Y. Morishita, A. J. Ison, S. Huang, P. F. McDermott, and Q. Zhang.** 2006. Effect of conventional and organic production practices on the prevalence and antimicrobial resistance of *Campylobacter* spp. in poultry. *Appl. Environ. Microbiol.* **72:**3600–3607.

43. **Luo, N., S. Pereira, O. Sahin, J. Lin, S. Huang, L. Michel, and Q. Zhang.** 2005. Enhanced in vivo fitness of fluoroquinolone-resistant *Campylobacter jejuni* in the absence of antibiotic selection pressure. *Proc. Natl. Acad. Sci. USA* **102:**541–546.

44. **Luo, N., O. Sahin, J. Lin, L. O. Michel, and Q. Zhang.** 2003. In vivo selection of *Campylobacter* isolates with high levels of fluoroquinolone resistance associated with *gyrA* mutations and the function of the CmeABC efflux pump. *Antimicrob. Agents Chemother.* **47:**390–394.

45. **MacKenzie, A. A., D. G. Allard, E. Perez, and S. Hathaway.** 2004. Food systems and the changing patterns of foodborne zoonoses. *Rev. Sci. Tech.* **23:**677–684.

46. **Macvanin, M., J. Bjorkman, S. Eriksson, M. Rhen, D. I. Andersson, and D. Hughes.** 2003. Fusidic acid-resistant mutants of *Salmonella enterica* serovar Typhimurium with low fitness in vivo are defective in RpoS induction. *Antimicrob. Agents Chemother.* **47:**3743–3749.

47. **Macvanin, M., U. Johanson, M. Ehrenberg, and D. Hughes.** 2000. Fusidic acid-resistant EF-G perturbs the accumulation of ppGpp. *Mol. Microbiol.* **37:**98–107.

48. **Maisnier-Patin, S., O. G. Berg, L. Liljas, and D. I. Andersson.** 2002. Compensatory adaptation to the deleterious effect of antibiotic resistance in *Salmonella typhimurium. Mol. Microbiol.* **46:**355–366.

49. **Marciano, D. C., O. Y. Karkouti, and T. Palzkill.** 2007. A fitness cost associated with the antibiotic resistance enzyme SME-1 β-lactamase. *Genetics* **176:**2381–2392.

50. **McDermott, P. F., S. M. Bodeis, L. L. English, D. G. White, R. D. Walker, S. Zhao, S. Simjee, and D. D. Wagner.** 2002. Ciprofloxacin resistance in *Campylobacter jejuni* evolves rapidly in chickens treated with fluoroquinolones. *J. Infect. Dis.* **185:**837–840.

51. **Mead, P. S., L. Slutsker, V. Dietz, L. F. McCaig, J. S. Bresee, C. Shapiro, P. M. Griffin, and R. V. Tauxe.** 1999. Food-related illness and death in the United States. *Emerg. Infect. Dis.* **5:**607–625.

52. **Menzel, R., and M. Gellert.** 1983. Regulation of the genes for *E. coli* DNA gyrase: homeostatic control of DNA supercoiling. *Cell* **34:**105–113.

53. **Menzel, R., and M. Gellert.** 1987. Modulation of transcription by DNA supercoiling: a deletion analysis of the *Escherichia coli gyrA* and *gyrB* promoters. *Proc. Natl. Acad. Sci. USA* **84:**4185–4189.

54. **Morosini, M. I., J. A. Ayala, F. Baquero, J. L. Martinez, and J. Blazquez.** 2000. Biological cost of AmpC production for *Salmonella enterica* serotype Typhimurium. *Antimicrob. Agents Chemother.* **44:**3137–3143.

55. **Nagaev, I., J. Bjorkman, D. I. Andersson, and D. Hughes.** 2001. Biological cost and

compensatory evolution in fusidic acid-resistant *Staphylococcus aureus*. *Mol. Microbiol.* **40:**433–439.

56. **Nguyen, T. N., Q. G. Phan, L. P. Duong, K. P. Bertrand, and R. E. Lenski.** 1989. Effects of carriage and expression of the Tn10 tetracycline-resistance operon on the fitness of *Escherichia coli* K12. *Mol. Biol. Evol.* **6:**213–225.
57. **Nilsson, A. I., A. Zorzet, A. Kanth, S. Dahlstrom, O. G. Berg, and D. I. Andersson.** 2006. Reducing the fitness cost of antibiotic resistance by amplification of initiator tRNA genes. *Proc. Natl. Acad. Sci. USA* **103:**6976–6981.
58. **Normark, B. H., and S. Normark.** 2002. Evolution and spread of antibiotic resistance. *J. Intern. Med.* **252:**91–106.
59. **Oh, T. J., and I. G. Kim.** 1999. The expression of *Escherichia coli* SOS genes *recA* and *uvrA* is inducible by polyamines. *Biochem. Biophys. Res. Commun.* **264:**584–589.
60. **Paulander, W., S. Maisnier-Patin, and D. I. Andersson.** 2007. Multiple mechanisms to ameliorate the fitness burden of mupirocin resistance in *Salmonella typhimurium*. *Mol. Microbiol.* **64:** 1038–1048.
61. **Payot, S., J. M. Bolla, D. Corcoran, S. Fanning, F. Megraud, and Q. Zhang.** 2006. Mechanisms of fluoroquinolone and macrolide resistance in *Campylobacter* spp. *Microbes Infect.* **8:** 1967–1971.
62. **Petersen, L., and A. Wedderkopp.** 2001. Evidence that certain clones of *Campylobacter jejuni* persist during successive broiler flock rotations. *Appl. Environ. Microbiol.* **67:**2739–2745.
63. **Price, L. B., E. Johnson, R. Vailes, and E. Silbergeld.** 2005. Fluoroquinolone-resistant *Campylobacter* isolates from conventional and antibiotic-free chicken products. *Environ. Health Perspect.* **113:**557–560.
64. **Price, L. B., L. G. Lackey, R. Vailes, and E. Silbergeld.** 2007. The persistence of fluoroquinolone-resistant *Campylobacter* in poultry production. *Environ. Health Perspect.* **115:**1035–1039.
65. **Rautelin, H., and M. L. Hanninen.** 2000. Campylobacters: the most common bacterial enteropathogens in the Nordic countries. *Ann. Med.* **32:**440–445.
66. **Ray, K. A., L. D. Warnick, R. M. Mitchell, J. B. Kaneene, P. L. Ruegg, S. J. Wells, C. P. Fossler, L. W. Halbert, and K. May.** 2006. Antimicrobial susceptibility of *Salmonella* from organic and conventional dairy farms. *J. Dairy Sci.* **89:**2038–2050.
67. **Reynolds, M. G.** 2000. Compensatory evolution in rifampin-resistant *Escherichia coli*. *Genetics* **156:**1471–1481.
68. **Sander, P., B. Springer, T. Prammananan, A. Sturmfels, M. Kappler, M. Pletschette, and E. C. Bottger.** 2002. Fitness cost of chromosomal drug resistance-conferring mutations. *Antimicrob. Agents Chemother.* **46:**1204–1211.
69. **Sato, K., P. C. Bartlett, J. B. Kaneene, and F. P. Downes.** 2004. Comparison of prevalence and antimicrobial susceptibilities of *Campylobacter* spp. isolates from organic and conventional dairy herds in Wisconsin. *Appl. Environ. Microbiol.* **70:** 1442–1447.
70. **Schrag, S. J., and V. Perrot.** 1996. Reducing antibiotic resistance. *Nature* **381:**120–121.
71. **Schrag, S. J., V. Perrot, and B. R. Levin.** 1997. Adaptation to the fitness costs of antibiotic resistance in Escherichia coli. *Proc. Biol. Sci.* **264:** 1287–1291.
72. **Smith, H. W., and M. A. Lovell.** 1981. *Escherichia coli* resistant to tetracyclines and to other antibiotics in the faeces of U.K. chickens and pigs in 1980. *J. Hyg.* **87:**477–483.
73. **Steck, T. R., R. J. Franco, J. Y. Wang, and K. Drlica.** 1993. Topoisomerase mutations affect the relative abundance of many *Escherichia coli* proteins. *Mol. Microbiol.* **10:**473–481.
74. **Teuber, M.** 1999. Spread of antibiotic resistance with food-borne pathogens. *Cell. Mol. Life Sci.* **56:**755–763.
75. **Urios, A., G. Herrera, V. Aleixandre, and M. Blanco.** 1990. Expression of the *recA* gene is reduced in *Escherichia coli* topoisomerase I mutants. *Mutat. Res.* **243:**267–272.
76. **van den Bogaard, A. E., and E. E. Stobberingh.** 1999. Antibiotic usage in animals: impact on bacterial resistance and public health. *Drugs* **58:**589–607.
77. **Wang, H. H., M. Manuzon, M. Lehman, K. Wan, H. Luo, T. E. Wittum, A. Yousef, and L. O. Bakaletz.** 2006. Food commensal microbes as a potentially important avenue in transmitting antibiotic resistance genes. *FEMS Microbiol. Lett.* **254:**226–231.
78. **Wegener, H. C.** 2003. Antibiotics in animal feed and their role in resistance development. *Curr. Opin. Microbiol.* **6:**439–445.
79. **Wegener, H. C.** 2003. Ending the use of antimicrobial growth promoters is making a difference. *ASM News* **69:**443–448.
80. **Wegener, H. C., F. M. Aarestrup, P. Gerner-Smidt, and F. Bager.** 1999. Transfer of antibiotic resistant bacteria from animals to man. *Acta Vet. Scand. Suppl.* **92:**51–57.
81. **Yates, C. M., D. J. Shaw, A. J. Roe, M. E. Woolhouse, and S. G. Amyes.** 2006. Enhancement of bacterial competitive fitness by apramycin resistance plasmids from non-pathogenic *Escherichia coli*. *Biol. Lett.* **2:**463–465.
82. **Zhang, Q., J. Lin, and S. Pereira.** 2003. Fluoroquinolone-resistant *Campylobacter* in animal reservoirs: dynamics of development, resistance mechanisms and ecological fitness. *Anim. Health Res. Rev.* **4:**63–71.

# *Staphylococcus aureus:* THE "SUPERBUG"

*Michael Otto*

# 16

*Staphylococcus aureus* is one of the most versatile and successful bacterial pathogens. It is known to the food microbiology research community as a common cause of food poisoning, but it has become much more notorious due to its leading role in hospital-associated infections and the surge in frequency of *S. aureus*-induced mild to severe skin and soft tissue infections. In addition, *S. aureus* can cause dramatic and even fatal infections such as endocarditis, necrotizing fasciitis, and toxic shock syndrome (TSS) (54).

Together with its less virulent cousin *Staphylococcus epidermidis, S. aureus* ranks far ahead of any other cause of hospital-associated (nosocomial) infections (2). In addition, while hospitals have traditionally been the most common places to acquire *S. aureus* infections, especially those caused by antibiotic-resistant strains, the recent emergence of community-associated methicillin-resistant *S. aureus* (CA-MRSA) poses a novel and extremely severe threat to the public health system (7). These strains combine antibiotic resistance with hypervirulence and may cause infections in otherwise healthy people outside the hospital setting.

As a permanent or transient colonizer of the human body, *S. aureus* normally has a relatively benign relationship with its host. Colonization by or direct infection with virulent strains and a compromised immune system such as in AIDS patients are risk factors for the development of disease. A mere abrasion of one of the most effective barriers that we have against infection, the human skin, represents a frequent origin of *S. aureus* infection. Once it has penetrated through the human skin into the body, *S. aureus* has a variety of mechanisms to evade the host's immune defenses (21). It is the combination of its extraordinary capacities to colonize body surfaces and evade immune defenses during infection that makes *S. aureus* one of the most dangerous and persistent threats to human health. Here, I review our knowledge of those mechanisms, with a special emphasis on the recent emergence of community-associated antibiotic-resistant strains.

## COLONIZATION

Colonization of the epithelial and mucosal surfaces of animals is a feature shared by all members of the genus *Staphylococcus* (41). While *S. epidermidis* and other coagulase-

*Michael Otto,* National Institute of Allergy and Infectious Diseases, National Institutes of Health, Bethesda, MD 20892.

*Food-Borne Microbes: Shaping the Host Ecosystem,* Edited by L.-A. Jaykus, H. H. Wang, and L. S. Schlesinger, © 2009 ASM Press, Washington, DC

negative staphylococci colonize most parts of the human body (65), *S. aureus* is found mostly in the nares. Whether this is due to a dedicated repertoire of colonization factors specifically developed for those areas and/or inferiority for the other skin areas compared to other staphylococcal species is not well understood. Most likely, physiological adaptations of different species for a certain ecological niche on the human body play a role in this specification together with, possibly, bacterial interference (26, 66).

Two families of molecules have an especially important role in the colonization of human body surfaces: (i) bacterial surface-bound proteins that bind to human matrix proteins for a tight interaction with epithelial tissues and (ii) ion transporters that cope with the high-salt and low-pH environment of the skin.

Surface-bound bacterial proteins that interact with human matrix proteins are called MSCRAMMs, for "microbial surface components recognizing adhesive matrix molecules" (67). The overwhelming number and frequent functional redundancy in *S. aureus* of MSCRAMMs underpin the importance of those molecules for bacterial survival and demonstrate the extreme versatility of *S. aureus* as a pathogen. Staphylococcal surface proteins may be bound covalently and noncovalently to the bacterial surface. The former are linked to the staphylococcal cell wall by an enzyme called sortase, whose sequence and function are conserved in many bacteria (57). Most surface-bound proteins in *S. aureus* are linked by sortase A, the classical member of the sortase enzyme family (56). In the sortase substrate proteins, sortase-catalyzed anchoring can be recognized by a series of conserved features, which comprise the N-terminal secretion signal, a C-terminal membrane-spanning domain, and the amino acid sequence LPXTG situated close to the latter. The threonyl residue within this sequence is linked to the amino acid in position 3 of the peptide stem in peptidoglycan, in the case of *S. aureus*, L-lysine (87). This family of staphylococcal surface-bound proteins contains, for example, proteins that interact with fibrinogen, fibronectin, and elastin (22). Several have been well characterized, such as the clumping factors A and B, the fibronectin-binding proteins A and B, and certain members of the SDR (serine/aspartate repeat) family of surface proteins (36). In addition, genome sequencing has allowed us to assemble a complete list of LPXTG motif-containing proteins in *S. aureus* (24). Some of those await characterization, while others, notably the IsdA (SasE) and IsdB (SasJ) proteins, have recently been investigated in-depth and found to have a crucial function in heme acquisition (58).

The second class of surface-bound proteins is linked to the surface in a noncovalent manner that is less well understood. Most likely, these proteins interact with the bacterial surface by electrostatic attraction between oppositely charged structures (62). This notion is supported, for example, by the influence that D-alanylation of teichoic acids, which considerably changes the overall anionic character of the teichoic acids, has on the interaction with autolysins (18, 19). In addition to the autolysins, which often combine human matrix protein binding function with their primary function in bacterial cell division, noncovalently bound staphylococcal surface binding proteins contain proteins with a task in the interaction with fibrinogen (Fib and Efb), fibronection (Ebh), and elastin (Ebp) (24). Finally, the extracellular adhesin binding protein Eap (or Map) has a broader specificity (35, 43).

*S. aureus* is very well adapted to deal with osmotic stress and changes in pH as encountered on the human skin and mucous surfaces, although its relative *S. epidermidis* might do even better in that regard. *S. aureus* possesses seven sodium ion/proton exchangers and six transport systems for osmoprotection molecules, such as proline and glycine betaine, and potassium ion channels (24). Together, this enormously broad coverage of adaptive mechanisms to survive in the hostile environment of human body surfaces is very likely one of the main reasons for which *S. aureus* is a premier colonizer of the human body. A second, equally important feature is its enormous capacity to resist the human immune defense, which is discussed in the following section.

## INTERACTION WITH THE HOST'S IMMUNE SYSTEM

Evading the human immune defenses is a premier interest of *S. aureus* both during colonization of the human epithelial surfaces, i.e., the noninfectious lifestyle, and after penetrating through the skin and entering the bloodstream, i.e., during infection. Here, I try to distinguish between mechanisms that appear to have been originally invented for survival on the exterior body surfaces, such as biofilm formation or pigmentation, and which may have risen to additional importance during infection, and those that appear to be exclusively aimed to interfere with the systemic human immune defense machinery, such as protein A, complement blockers, or toxins. However, a clear distinction cannot always be made.

### Biofilms

The formation of biofilms, while recognized as the common mode of growth for most microorganisms in their natural environment (9), facilitates the survival of *S. aureus* on body surfaces during colonization and chronic infection. Biofilms provide protection from mechanical and chemical stresses, including attacks by the innate and acquired immune systems and antibiotics (10). This resistance is based mainly on the specific physiology of a biofilm and the extracellular biofilm matrix. In addition to slow growth, which renders ineffective antibiotics that target only growing cells, such as β-lactams, transcriptional profiling studies have shown that staphylococcal biofilms show specific metabolic adaptations and overexpress resistance factors (4, 98).

Biofilm development has often been described as a three-step process with a primary attachment phase, a proliferation phase that produces the typical mature biofilm structure, and a final detachment process (47). I have already described the initial attachment phase, which occurs by surface protein-mediated binding to host tissues. Alternatively or in addition, staphylococci can adhere directly to plastic material as used for indwelling medical devices, a process most likely primarily determined by the hydrophobicity of the bacterial cell surface. The second phase is characterized by the production of the typical biofilm components, mostly exopolymeric substances, and the formation of a three-dimensional biofilm structure with channels in between mushroom-like cellular agglomerations. A characteristic component of the staphylococcal biofilm matrix is the biofilm exopolysaccharide PIA/PNAG (polysaccharide intercellular adhesin or poly-*N*-acetylglucosamine) (11, 55). Other molecules appear to help in the formation of the biofilm matrix, although it is evident that this is not their primary function, such as teichoic acids (29) and extracellular DNA (76)—the latter a remainder of lysed cells. PIA/PNAG in particular has been shown to have a specific function in protection against phagocytosis and antimicrobial peptides (AMPs) (94) in addition to its role as a building substance for the biofilm matrix, which as a whole has a similar role (83). The final detachment process is deemed vital for the dissemination of a biofilm-associated infection to other sites in the human body, but the underlying mechanism is poorly understood. Recent studies in my laboratory suggest an involvement of the detergent-like phenol-soluble modulins (PSMs), especially those of the β subtype, in this process (42).

### Capsule

*S. aureus* produces a series of capsular polysaccharides of different serotypes that protect against phagocytosis (64). However, the role of the capsule in virulence is ambivalent, as masking of MSCRAMMs by encapsulated strains may be counterproductive during adhesive types of *S. aureus* infection, such as infective endocarditis. Capsule production is subject to regulation by environmental factors and quorum sensing, suggesting that its role in virulence varies with different stages of *S. au-*

*reus* infection. It is not clear whether capsule production is aimed only to protect against phagocytosis or whether there is an original role in colonization and resistance to environmental influences, which appears likely.

### Staphyloxanthin

The carotenoid staphyloxanthin is the typical yellow pigment that has given *S. aureus* its name (*aureus,* Latin for golden). Its biosynthesis has recently been elucidated in detail (68). While this pigment most likely has an original role in providing protection from UV radiation and radicals in general—such as most carotenoids, including vitamin A—recent studies have demonstrated that staphyloxanthin has a crucial role in immune evasion (53) and may represent a promising drug target (52).

### Antimicrobial Peptides

Antibody-based, acquired immunity does not play a major role on the body surfaces. Nor do we find professional pathogen-ingesting cells like macrophages or neutrophils (or polymorphonuclear phagocytes) in that place, except for those on the respiratory mucosae involved in bronchoalveolar clearance. Instead, the major part of innate host defense that is present on the human skin and mucosal surfaces is the evolutionarily ancient class of AMPs, small proteins secreted by epithelial cells, among many cell types, to prevent bacterial colonization (30). Additionally, AMPs form part of the antimicrobial activity found inside human neutrophils. Due to the million-year-long interplay between bacterial pathogens and AMPs during evolution, we can find evidence for adaptations on both sides: bacteria have developed resistance mechanisms against AMPs and mechanisms to sense the presence of AMPs, while mammals have developed pathogen-resistant and more efficacious AMPs (72).

There is a multitude of sophisticated defense mechanisms to resist AMPs in *S. aureus* (69). This pathogen has at least one transporter that provides protection from the antimicrobial activity of AMPs, VraFG, most likely by active export of the AMPs (49). Additional mechanisms mainly function by changing the charge of the bacterial cell surface to provide less electrostatic attraction for the commonly cationic AMPs, i.e., by making the cell surface less anionic in its character. This may be achieved, for example, by including the phospholipid lysyl-phosphatidylglycerol with the cationic lysyl group in the cytoplasmic membrane, a reaction catalyzed by the MprF enzyme (70), or by the D-alanylation of teichoic acids, which decreases the negative charge of those molecules and is catalyzed by the proteins encoded in the *dlt* locus (71).

Interestingly, *S. aureus* can sense the presence of AMPs and coordinate the expression of the *dlt, mprF,* and the *vraFG* transporter loci in response to AMPs (49). This mechanism, equivalent but not homologous to the PhoP/PhoQ sensor in gram-negative bacteria, (3), is encoded by the *aps* locus, which codes for a classical two-component system and a third protein of unknown function, all of which are needed for the signal transduction process (50). While the prototype system has been investigated in *S. epidermidis* (50), it needs to be stressed that *aps* has been shown to contribute to the pathogenic success of the exceptionally virulent prototype CA-MRSA strain MW2 (49).

Proteolytic digestion of AMPs is a relatively simple answer of the bacteria to the antimicrobial activity of AMPs, but AMPs with a higher degree of resistance to proteolytic digestion, such as by pronounced cystine cross-linking found, for example, in the defensin class of AMPs, are widespread (31). Interestingly, cationic AMPs and the anionic AMP dermcidin can induce increased production of exoproteases in *S. epidermidis,* most likely by a relatively nonspecific general stress response-like mechanism (45). It is likely that this mechanism also exists in *S. aureus.* The anionic dermcidin, an important AMP in human sweat (82), appears to be exceptionally sensitive to proteolytic digestion while not subject to the resistance mechanisms aimed to counteract cationic AMPs. This is an obvious example of how the human body has invented an AMP not subject to the main bacterial resistance mechanisms, whereas the bacteria have learned to fight back by a simple upregulation of exoprotease production.

## Dedicated Mechanisms To Evade Host Defenses inside the Human Body

Inside the human body, *S. aureus* has to cope with a multitude of defense mechanisms of the innate and acquired immune systems, in addition to the changed environmental conditions. It is fascinating to see that not only has this pathogen developed resistance against virtually every path of immune defense, but also it often even offers several resistance mechanisms for the same kind of immune system attack in a pronounced functional redundancy. It is important to stress that innate host defense is most likely by far the most important part of host defense with the task of eliminating invading *S. aureus*. In contrast, the acquired immune system plays a less crucial role, as can be seen by the considerable amount of antistaphylococcal antibodies circulating in healthy humans (34) without facilitating acquired immunity against staphylococcal infections. Resistance to antibody-based host defenses is in part due to specific defense against antibodies using protein A, but it is also caused by the often very efficacious mechanisms of *S. aureus* to escape elimination by professional phagocytes (90), irrespective of whether this elimination was initiated by the innate or acquired immune system. Finally, *S. aureus* may survive in phagocytes over certain periods without lysing them (28, 44).

## Iron Sequestration

Low availability of free iron in the body fluids forms an extremely effective part of the human immune defense for protection against almost all microorganisms. *S. aureus* has learned to deal with this situation extremely well. It has a series of secreted molecules that tightly complex iron, called siderophores (51). Further, it has mechanisms to import heme and sequester iron from that source (58, 88). Finally, there is a multitude of ferric and ferrous iron ABC import systems, indicating that iron acquisition is a serious priority for the pathogen (24).

## Protein A

Protein A is probably the best-known *S. aureus* protein due to its use in the laboratory for antibody purification, which is based on the interaction of protein A with the Fc part of immunoglobulin G molecules (46). During pathogenesis, this feature enables *S. aureus* to sequester nonspecific antibodies on its surface, which serve as a "camouflage coat" for the bacteria and protect them efficiently from attacks by the innate and acquired immune systems (20). Protein A also has a more specific role in the pathogenesis of airway infections by interacting with the tumor necrosis factor alpha receptor on airway epithelia (27).

## Evasion of Innate Host Defense

*S. aureus* has an astounding repertoire of factors allowing the bacterium to hide from mechanisms of innate host defense, which includes the phagocytic activity of white blood cells, complement, and AMPs, which I have already discussed above due to their role on the body surface. These add to the protective effects of biofilm, the extracellular matrix, and the capsule.

## Inhibition of the Complement Cascade

Several immune evasion molecules appear to be arranged in genomic clusters, which have been called IECs, for "immune evasion clusters" (79). IEC-1 contains the chemotaxis-inhibitory protein CHIPS (12) and the staphylococcal complement inhibitor SCIN (78), while IEC-2 contains Efb (extracellular fibrinogen-binding protein) and Ecb (extracellular complement-binding protein). All these factors inhibit complement and its activation. CHIPS binds specifically to two receptors, C5aR and the formyl-peptide receptor (FPR) and blocks calcium ion mobilization induced by these receptors (12). Another *S. aureus* protein called FLIPr blocks the FPR-like receptor FPLR1 (74). SCIN, Efb, and Ecb block the complement cascade by interfering with the C3 convertase. Staphylokinase and several staphylococcal superantigen-like proteins interact with C5 and C3b, respectively (79, 80). Thus, inhibition of the complement cascade is one of the most intriguing examples of the functional redundancy of *S. aureus* in attacking many steps of a host defense pathway with several dedicated molecules.

## *S. aureus* Toxins

The immune evasion strategies that I have described so far are rather "passive"; i.e., they either provide the pathogen with molecules necessary to survive under the hostile conditions encountered on or inside the human body, are aimed to hide from the immune system in a relatively benign fashion, or finally, sabotage mechanisms of the host that would kill *S. aureus*. However, while this may be the strategy of many among its relatives, such as *S. epidermidis, S. aureus* produces a series of toxins that can cause serious damage to the host and, as a consequence, a variety of extremely acute diseases. Notably, the production of toxins is strain specific, defining to a large extent the acute pathogenic potential of a strain.

Many *S. aureus* toxins form pores in host cells, thereby causing their lysis. Several are aimed at the destruction of leukocytes, such as the leukocidin family (37), alpha-toxin (5), and the recently identified PSMs (95). Several of these toxins, such as the PSMs and alpha-toxin, have relatively low host cell specificity and often also lyse erythrocytes, whereas many leukocidins are specific for leukocytes (37). Leukolytic activity may be limited to leukocyte subclasses: alpha-toxin, for example, does not lyse neutrophils (89). Panton-Valentine leukocidin (PVL), a member of the two-component leukocidin family that also includes gamma-toxin, for example, has recently drawn a lot of interest due to its alleged role in the virulence of CA-MRSA, a role it was hypothesized to have based on a pronounced epidemiological correlation of the PVL-encoding genes *lukS-PV* and *lukF-PV,* with a predominant occurrence in CA-MRSA strains (25). However, molecular studies and animal models indicate that PVL is not a major virulence factor in CA-MRSA disease (91). Finally, beta-toxin is a sphingomyelinase that leads to a species-specific lysis of erythrocytes and lymphocytes (33).

Further, *S. aureus* produces a series of superantigenic toxins, a class of toxins that activate T cells without the need for the presence of an antigen on an antigen-presenting cell, by cross-linking the T-cell receptor with the major histocompatibility complex class II on the surface of the antigen-presenting cell (59). In *S. aureus,* these include the TSS toxin, exfoliative toxins involved in staphylococcal scalded skin syndrome, and the staphylococcal enterotoxins (SEs) (17). TSS is a severe acute disease and may be of menstrual or nonmenstrual origin. Menstrual TSS is caused by *S. aureus* colonization of tampons (6). Nonmenstrual TSS may also be caused by enterotoxins, which, however, have a more notorious role as the cause of *S. aureus* food poisoning.

In fact, *S. aureus* is the second most important cause of food-borne diseases after *Salmonella* spp. (48). It may be present in a variety of foods, such as meat and dairy products. The toxicity of *S. aureus* as a food-borne pathogen is entirely due to the production of the SEs, the genes for at least one of which are present in ~25% of *S. aureus* strains detected in food samples. The SEs are members of the pyrogenic toxin class, which also comprise the exfoliative toxins A and B and TSS toxin of *S. aureus,* in addition to the *Streptococcus* pyrogenic toxins. They are different from the other pyrogenic toxins inasmuch as they contain an additional protein domain that has emetic activity. The mechanism of how the SEs cause emesis and the biological function of this activity in *S. aureus* physiology and colonization of the human body are not understood. Although most recent interest in *S. aureus* as a pathogen has focused on diseases other than food poisoning, the potential use of SE-producing strains as biological weapons of bioterrorism warrants more in-depth investigation of the SEs.

### Degradative Exoenzymes

*S. aureus* secretes a series of degradative exoenzymes with an assumed main task in acquiring nutrients from host tissue during prolonged infection. Similar to toxins, they presumably may cause severe tissue damage, although the role of these enzymes in virulence is not well understood. They include nucleases, lipases, and proteases. The last have an additional function in eliminating antibacterial proteins and peptides (84).

## ANTIMICROBIAL RESISTANCE

*S. aureus* has a relatively high degree of natural resistance to antimicrobials due to nonspecific mechanisms such as the formation of biofilms and broad-range drug exporters. However, it has become better known as a bacterium with pronounced acquired resistance to a variety of antibiotics, most notably, β-lactam antibiotics such as the penicillins and cephalosporins. Resistance to β-lactam antibiotics is due to the enzyme β-lactamase, which is widespread among *S. aureus* strains (77). When the β-lactamase-resistant methicillin was introduced, MRSA was discovered only some months thereafter, exemplifying how fast *S. aureus* strains may arise that are resistant to a specific antibiotic (13). Nowadays, MRSA is the rule rather than the exception in the hospital setting. Resistance to other antibiotics is also frequent and often combined with methicillin resistance. Multidrug-resistant strains not only are emerging in hospitals but also are now found in community-associated infections (15, 73), indicating that the acquisition of multidrug-resistant strains is not paired with a significant cost in bacterial fitness. On the other hand, while strains resistant to vancomycin, often cited as the antibiotic of last resort for staphylococcal infections, have been detected (32), these strains have not spread to a considerable extent. This is possibly caused by a higher fitness cost to strains of carrying the *van* resistance genes, but it may also be due to infection control (85).

## COMMUNITY-ASSOCIATED MRSA

In 1999, infectious disease physicians and researchers working on staphylococcal pathogenesis were equally surprised to see infections with MRSA emerge in the community without any connection to the hospital setting (1). First reported to occur in children in the northwestern United States, CA-MRSA has spread with astounding speed. In the last couple of years, outbreaks have been reported to occur in professional football players (38) and players of other sports, in prison inmates, and among other communities in which hygienic procedures are frequently disregarded and close body-to-body contacts occur, clearly indicating that MRSA has developed infectivity high enough to cause disease in otherwise healthy individuals (60). In the United States, almost all MRSA infections are due to a clonally related background called sequence type USA300, which by now represents the most frequent cause of all infections reported to emergency departments in the United States (61). Further underpinning the danger that CA-MRSA represents predominantly in the United States, strains have been reported more recently that combine resistance to methicillin with resistance to a series of other antibiotics (multidrug-resistant CA-MRSA) (15, 16). So far, these strains appear to have spread more rapidly in the gay communities of several American cities than elsewhere (15). Whether infections with multidrug-resistant CA-MRSA have occurred by sexual contacts and are due to certain sexual behaviors is certainly a valid hypothesis that needs to be investigated, most notably to assess the risks of further distribution. However, sexual transmission of CA-MRSA has only been clearly documented among heterosexual couples (8). Regardless, the fast spread of CA-MRSA possibly using entirely new routes of infection reminds us again of the surprising versatility of *S. aureus,* which astounds us with every new epidemic wave that we face.

The molecular reason for which CA-MRSA and especially USA300 are so much more infective than their hospital-associated counterparts is still puzzling researchers. Initially, PVL was declared as the major factor responsible for the exceptional virulence and success of USA300 and other CA-MRSA strains worldwide. However, as mentioned above, several studies using animal infection models resembling the various types of CA-MRSA disease in humans have demonstrated that PVL is not a major factor affecting CA-MRSA virulence (91, 96). In contrast, more recent studies suggest that the basis of virulence in CA-MRSA is multifactorial and comprises increased expression of known and newly identified virulence factors. The α-type PSMs have been shown to represent a major factor influencing virulence of CA-MRSA in animal models of bacteremia and skin and soft tissue infections (95), which represent the main types of disease caused by CA-MRSA. Of note, the PSMs are expressed, at least on average, to a higher extent in CA-MRSA than in hospital-associated MRSA strains, thus providing CA-MRSA strains with increased virulence. PSMs work by forming pores in human neutrophils, thus damaging the main cell type responsible for eliminating invading bacteria, similar to the leukocidins, but apparently with a more important role during infection. In addition, the well-known highly potent pore former alpha-toxin was shown to be the major driving force behind necrotizing pneumonia, a dramatic complication seen in some CA-MRSA patients (96). Remarkably, antibodies against alpha-toxin reduced experimental pneumonia caused by CA-MRSA (97), indicating that at least against this type of disease, an antibody-based approach against one major virulence factor may lead to the development of a promising therapy (14).

Other researchers focus on a potentially increased capacity of CA-MRSA to colonize humans. The arginine catabolic mobile element is present only in USA300 and is discussed as a factor that gives USA300 a colonization advantage by neutralizing the acid environment on the human skin via the production of ammonia (16). Thus, the first studies suggest that the molecular basis of the exceptional virulence of USA300 and other CA-MRSA strains is multifactorial and combines hypervirulence with increased colonization capacity.

## THE ACCESSORY GENE REGULATOR *agr* AND LIFESTYLE SWITCHES

The initial findings on the molecular basis of virulence in CA-MRSA exemplify why *S. aureus* is such a versatile pathogen and emphasize the importance of gene expression for the establishment of virulence. Both the PSMs and alpha-toxin are under strong regulation by the quorum sensing regulator *agr* (75, 95). In the case of the PSMs, it has been shown that high-level expression of PSMs in CA-MRSA is usually accompanied by high-level expression of RNAIII, the internal regulatory molecule of the *agr* system, indicating overall strong expression of this regulator (95). Thus, high-level expression of *agr,* which has long been recognized as a main virulence regulator in *S. aureus* (63), may contribute to CA-MRSA virulence, a hypothesis that remains to be investigated.

On the other hand, it appears that *S. aureus* adapts to specific environments in the human body, and certain types of infection during pathogenesis, by rendering the *agr* system nonfunctional. In fact, the *agr* locus is known to be subject to a high frequency of mutations in vitro (86). Furthermore, animal infection models and epidemiological studies indicate that mutation in *S. aureus* or *S. epidermidis agr,* rendering this regulatory locus nonfunctional, occurs naturally and at a high frequency (23, 93). It has been demonstrated that a dysfunctional *agr* system facilitates chronic bacteremia (23) and the colonization of indwelling medical devices by biofilm formation (92, 93) and that it leads to increased antibiotic resistance (81), suggesting a lifestyle switch to a less aggressive, chronic mode of infection. Thus, irreversible genetic change aimed to render the *agr* system nonfunctional appears to represent a method of adaptation to a changing environment. We do not know yet whether a part of the population maintains a functional *agr* system in those cases to be able to switch back to a more aggressive lifestyle, but this seems likely. Additional support for the hypothesis that strains with a nonfunctional *agr* system have an advantage in causing chronic infection comes from a genome-wide investigation of a series of USA300 strains that were found to be genetically very closely related (39). In that study, one *agr*-negative strain was isolated from a case of osteomyelitis, a typically chronic form of infection, whereas all others originated from acute types of infection. At least one mechanism by which *agr* is rendered nonfunctional is by genetic insertion elements (92). These elements have also been found to integrate into other crucial regulatory systems (40). Thus, mutation of global regulatory loci appears to represent a general, fast, and easy way to change gene expression in a subset of *S. aureus* populations to adapt to changing environments in a lifestyle switch-like fashion.

## CONCLUDING REMARKS

In conclusion, *S. aureus* combines a series of features that make it an exceptionally successful and difficult-to-combat pathogen. First, as a typical colonizer of humans, it has developed multiple molecules that ensure tight binding to our body surfaces. As a consequence, it is always already "there" and waiting for us to become weak or develop abrasions, wounds, etc., to initiate its infectious lifestyle—either by infection from colonization sites such as the nose or by interaction with other humans that carry infectious strains. Second, it has developed multiple ways to evade our immune defenses in an astounding functional redundancy, covering every possible aspect of especially innate host defense. Third, it is collecting antibiotic resistance genes to a large extent which add to the relatively high degree of nonspecific antibiotic resistance that *S. aureus* has naturally. Finally, swift changes in gene expression appear to facilitate adaptation to different ecological niches in the human body and different types of *S. aureus* infection.

## REFERENCES


1. **Anonymous.** 1999. From the Centers for Disease Control and Prevention. Four pediatric deaths from community-acquired methicillin-resistant *Staphylococcus aureus*—Minnesota and North Dakota, 1997–1999. *JAMA* **282:**1123–1125.
2. **Anonymous.** 2004. National Nosocomial Infections Surveillance (NNIS) System Report, data summary from January 1992 through June 2004, issued October 2004. *Am. J. Infect. Control* **32:** 470–485.
3. **Bader, M. W., S. Sanowar, M. E. Daley, A. R. Schneider, U. Cho, W. Xu, R. E. Klevit, H. Le Moual, and S. I. Miller.** 2005. Recognition of antimicrobial peptides by a bacterial sensor kinase. *Cell* **122:**461–472.
4. **Beenken, K. E., P. M. Dunman, F. McAleese, D. Macapagal, E. Murphy, S. J. Projan, J. S. Blevins, and M. S. Smeltzer.** 2004. Global gene expression in *Staphylococcus aureus* biofilms. *J. Bacteriol.* **186:**4665–4684.
5. **Bhakdi, S., and J. Tranum-Jensen.** 1991. Alpha-toxin of *Staphylococcus aureus. Microbiol. Rev.* **55:**733–751.
6. **Bohach, G. A., D. J. Fast, R. D. Nelson, and P. M. Schlievert.** 1990. Staphylococcal and streptococcal pyrogenic toxins involved in toxic shock syndrome and related illnesses. *Crit. Rev. Microbiol.* **17:**251–272.
7. **Chambers, H. F.** 2005. Community-associated MRSA—resistance and virulence converge. *N. Engl. J. Med.* **352:**1485–1487.
8. **Cook, H. A., E. Y. Furuya, E. Larson, G. Vasquez, and F. D. Lowy.** 2007. Heterosexual transmission of community-associated methicillin-resistant *Staphylococcus aureus. Clin. Infect. Dis.* **44:**410–413.
9. **Costerton, J. W., Z. Lewandowski, D. E. Caldwell, D. R. Korber, and H. M. Lappin-Scott.** 1995. Microbial biofilms. *Annu. Rev. Microbiol.* **49:**711–745.
10. **Costerton, J. W., P. S. Stewart, and E. P. Greenberg.** 1999. Bacterial biofilms: a common cause of persistent infections. *Science* **284:**1318–1322.
11. **Cramton, S. E., C. Gerke, N. F. Schnell, W. W. Nichols, and F. Gotz.** 1999. The intercellular adhesion *(ica)* locus is present in *Staphylococcus aureus* and is required for biofilm formation. *Infect. Immun.* **67:**5427–5433.

12. **de Haas, C. J., K. E. Veldkamp, A. Peschel, F. Weerkamp, W. J. Van Wamel, E. C. Heezius, M. J. Poppelier, K. P. Van Kessel, and J. A. van Strijp.** 2004. Chemotaxis inhibitory protein of *Staphylococcus aureus,* a bacterial antiinflammatory agent. *J. Exp. Med.* **199:**687–695.
13. **de Lencastre, H., D. Oliveira, and A. Tomasz.** 2007. Antibiotic resistant *Staphylococcus aureus:* a paradigm of adaptive power. *Curr. Opin. Microbiol.* **10:**428–435.
14. **Deleo, F. R., and M. Otto.** 2008. An antidote for *Staphylococcus aureus* pneumonia? *J. Exp. Med.* **205:**271–274. (Erratum, **205:**739.)
15. **Diep, B. A., H. F. Chambers, C. J. Graber, J. D. Szumowski, L. G. Miller, L. L. Han, J. H. Chen, F. Lin, J. Lin, T. H. Phan, H. A. Carleton, L. K. McDougal, F. C. Tenover, D. E. Cohen, K. H. Mayer, G. F. Sensabaugh, and F. Perdreau-Remington.** 2008. Emergence of multidrug-resistant, community-associated, methicillin-resistant *Staphylococcus aureus* clone USA300 in men who have sex with men. *Ann. Intern. Med.* **148:**249–257.
16. **Diep, B. A., S. R. Gill, R. F. Chang, T. H. Phan, J. H. Chen, M. G. Davidson, F. Lin, J. Lin, H. A. Carleton, E. F. Mongodin, G. F. Sensabaugh, and F. Perdreau-Remington.** 2006. Complete genome sequence of USA300, an epidemic clone of community-acquired methicillin-resistant *Staphylococcus aureus. Lancet* **367:**731–739.
17. **Dinges, M. M., P. M. Orwin, and P. M. Schlievert.** 2000. Exotoxins of *Staphylococcus aureus. Clin. Microbiol. Rev.* **13:**16–34.
18. **Fedtke, I., D. Mader, T. Kohler, H. Moll, G. Nicholson, R. Biswas, K. Henseler, F. Gotz, U. Zahringer, and A. Peschel.** 2007. A *Staphylococcus aureus ypfP* mutant with strongly reduced lipoteichoic acid (LTA) content: LTA governs bacterial surface properties and autolysin activity. *Mol. Microbiol.* **65:**1078–1091.
19. **Fischer, W., P. Rosel, and H. U. Koch.** 1981. Effect of alanine ester substitution and other structural features of lipoteichoic acids on their inhibitory activity against autolysins of *Staphylococcus aureus. J. Bacteriol.* **146:**467–475.
20. **Forsgren, A., and K. Nordstrom.** 1974. Protein A from *Staphylococcus aureus:* the biological significance of its reaction with IgG. *Ann. N.Y. Acad. Sci.* **236:**252–266.
21. **Foster, T. J.** 2005. Immune evasion by staphylococci. *Nat. Rev. Microbiol.* **3:**948–958.
22. **Foster, T. J., and M. Hook.** 1998. Surface protein adhesins of *Staphylococcus aureus. Trends Microbiol.* **6:**484–488.
23. **Fowler, V. G., Jr., G. Sakoulas, L. M. McIntyre, V. G. Meka, R. D. Arbeit, C. H. Cabell, M. E. Stryjewski, G. M. Eliopoulos, L. B. Reller, G. R. Corey, T. Jones, N. Lucindo, M. R. Yeaman, and A. S. Bayer.** 2004. Persistent bacteremia due to methicillin-resistant *Staphylococcus aureus* infection is associated with *agr* dysfunction and low-level in vitro resistance to thrombin-induced platelet microbicidal protein. *J. Infect. Dis.* **190:**1140–1149.
24. **Gill, S. R., D. E. Fouts, G. L. Archer, E. F. Mongodin, R. T. Deboy, J. Ravel, I. T. Paulsen, J. F. Kolonay, L. Brinkac, M. Beanan, R. J. Dodson, S. C. Daugherty, R. Madupu, S. V. Angiuoli, A. S. Durkin, D. H. Haft, J. Vamathevan, H. Khouri, T. Utterback, C. Lee, G. Dimitrov, L. Jiang, H. Qin, J. Weidman, K. Tran, K. Kang, I. R. Hance, K. E. Nelson, and C. M. Fraser.** 2005. Insights on evolution of virulence and resistance from the complete genome analysis of an early methicillin-resistant *Staphylococcus aureus* strain and a biofilm-producing methicillin-resistant *Staphylococcus epidermidis* strain. *J. Bacteriol.* **187:**2426–2438.
25. **Gillet, Y., B. Issartel, P. Vanhems, J. C. Fournet, G. Lina, M. Bes, F. Vandenesch, Y. Piemont, N. Brousse, D. Floret, and J. Etienne.** 2002. Association between *Staphylococcus aureus* strains carrying gene for Panton-Valentine leukocidin and highly lethal necrotising pneumonia in young immunocompetent patients. *Lancet* **359:**753–759.
26. **Goerke, C., M. Kummel, K. Dietz, and C. Wolz.** 2003. Evaluation of intraspecies interference due to agr polymorphism in *Staphylococcus aureus* during infection and colonization. *J. Infect. Dis.* **188:**250–256.
27. **Gomez, M. I., A. Lee, B. Reddy, A. Muir, G. Soong, A. Pitt, A. Cheung, and A. Prince.** 2004. *Staphylococcus aureus* protein A induces airway epithelial inflammatory responses by activating TNFR1. *Nat. Med.* **10:**842–848.
28. **Gresham, H. D., J. H. Lowrance, T. E. Caver, B. S. Wilson, A. L. Cheung, and F. P. Lindberg.** 2000. Survival of *Staphylococcus aureus* inside neutrophils contributes to infection. *J. Immunol.* **164:**3713–3722.
29. **Gross, M., S. E. Cramton, F. Gotz, and A. Peschel.** 2001. Key role of teichoic acid net charge in *Staphylococcus aureus* colonization of artificial surfaces. *Infect. Immun.* **69:**3423–3426.
30. **Hancock, R. E., and G. Diamond.** 2000. The role of cationic antimicrobial peptides in innate host defences. *Trends Microbiol.* **8:**402–410.
31. **Harder, J., and J. M. Schroder.** 2005. Antimicrobial peptides in human skin. *Chem. Immunol. Allergy* **86:**22–41.
32. **Hiramatsu, K.** 2001. Vancomycin-resistant *Staphylococcus aureus:* a new model of antibiotic resistance. *Lancet Infect. Dis.* **1:**147–155.

33. **Huseby, M., K. Shi, C. K. Brown, J. Digre, F. Mengistu, K. S. Seo, G. A. Bohach, P. M. Schlievert, D. H. Ohlendorf, and C. A. Earhart.** 2007. Structure and biological activities of beta toxin from *Staphylococcus aureus*. *J. Bacteriol.* **189:**8719–8726.
34. **Jensen, K.** 1958. A normally occurring *Staphylococcus* antibody in human serum. *Acta Pathol. Microbiol. Scand.* **44:**421–428.
35. **Jonsson, K., D. McDevitt, M. H. McGavin, J. M. Patti, and M. Hook.** 1995. *Staphylococcus aureus* expresses a major histocompatibility complex class II analog. *J. Biol. Chem.* **270:**21457–21460.
36. **Josefsson, E., K. W. McCrea, D. Ni Eidhin, D. O'Connell, J. Cox, M. Hook, and T. J. Foster.** 1998. Three new members of the serine-aspartate repeat protein multigene family of *Staphylococcus aureus*. *Microbiology* **144**(Pt. 12)**:** 3387–3395.
37. **Kaneko, J., and Y. Kamio.** 2004. Bacterial two-component and hetero-heptameric pore-forming cytolytic toxins: structures, pore-forming mechanism, and organization of the genes. *Biosci. Biotechnol. Biochem.* **68:**981–1003.
38. **Kazakova, S. V., J. C. Hageman, M. Matava, A. Srinivasan, L. Phelan, B. Garfinkel, T. Boo, S. McAllister, J. Anderson, B. Jensen, D. Dodson, D. Lonsway, L. K. McDougal, M. Arduino, V. J. Fraser, G. Killgore, F. C. Tenover, S. Cody, and D. B. Jernigan.** 2005. A clone of methicillin-resistant *Staphylococcus aureus* among professional football players. *N. Engl. J. Med.* **352:**468–475.
39. **Kennedy, A. D., M. Otto, K. R. Braughton, A. R. Whitney, L. Chen, B. Mathema, J. R. Mediavilla, K. A. Byrne, L. D. Parkins, F. C. Tenover, B. N. Kreiswirth, J. M. Musser, and F. R. DeLeo.** 2008. Epidemic community-associated methicillin-resistant *Staphylococcus aureus:* recent clonal expansion and diversification. *Proc. Natl. Acad. Sci. USA* **105:** 1327–1332.
40. **Kiem, S., W. S. Oh, K. R. Peck, N. Y. Lee, J. Y. Lee, J. H. Song, E. S. Hwang, E. C. Kim, C. Y. Cha, and K. W. Choe.** 2004. Phase variation of biofilm formation in *Staphylococcus aureus* by IS 256 insertion and its impact on the capacity adhering to polyurethane surface. *J. Korean Med. Sci.* **19:**779–782.
41. **Kloos, W., and K. H. Schleifer.** 1986. Staphylococcus, p. 1013–1035. *In* P. H. A. Sneath, N. S. Mair, M. E. Sharpe, and J. G. Holt (ed.), *Bergey's Manual of Systematic Bacteriology,* vol. 2. Williams & Wilkins, Baltimore, MD.
42. **Kong, K. F., C. Vuong, and M. Otto.** 2006. *Staphylococcus* quorum sensing in biofilm formation and infection. *Int. J. Med. Microbiol.* **296:** 133–139.
43. **Kreikemeyer, B., D. McDevitt, and A. Podbielski.** 2002. The role of the Map protein in *Staphylococcus aureus* matrix protein and eukaryotic cell adherence. *Int. J. Med. Microbiol.* **292:**283–295.
44. **Kubica, M., K. Guzik, J. Koziel, M. Zarebski, W. Richter, B. Gajkowska, A. Golda, A. Maciag-Gudowska, K. Brix, L. Shaw, T. Foster, and J. Potempa.** 2008. A potential new pathway for *Staphylococcus aureus* dissemination: the silent survival of *S. aureus* phagocytosed by human monocyte-derived macrophages. *PLoS ONE* **3:**e1409.
45. **Lai, Y., A. E. Villaruz, M. Li, D. J. Cha, D. E. Sturdevant, and M. Otto.** 2007. The human anionic antimicrobial peptide dermcidin induces proteolytic defence mechanisms in staphylococci. *Mol. Microbiol.* **63:**497–506.
46. **Langone, J. J.** 1982. Protein A of *Staphylococcus aureus* and related immunoglobulin receptors produced by streptococci and pneumococci. *Adv. Immunol.* **32:**157–252.
47. **Lappin-Scott, H. M., and C. Bass.** 2001. Biofilm formation: attachment, growth, and detachment of microbes from surfaces. *Am. J. Infect. Control* **29:**250–251.
48. **Le Loir, Y., F. Baron, and M. Gautier.** 2003. *Staphylococcus aureus* and food poisoning. *Genet. Mol. Res.* **2:**63–76.
49. **Li, M., D. J. Cha, Y. Lai, A. E. Villaruz, D. E. Sturdevant, and M. Otto.** 2007. The antimicrobial peptide-sensing system *aps* of *Staphylococcus aureus*. *Mol. Microbiol.* **66:**1136–1147.
50. **Li, M., Y. Lai, A. E. Villaruz, D. J. Cha, D. E. Sturdevant, and M. Otto.** 2007. Gram-positive three-component antimicrobial peptide-sensing system. *Proc. Natl. Acad. Sci. USA* **104:** 9469–9474.
51. **Lindsay, J. A., and T. V. Riley.** 1994. Staphylococcal iron requirements, siderophore production, and iron-regulated protein expression. *Infect. Immun.* **62:**2309–2314.
52. **Liu, C. I., G. Y. Liu, Y. Song, F. Yin, M. E. Hensler, W. Y. Jeng, V. Nizet, A. H. Wang, and E. Oldfield.** 2008. A cholesterol biosynthesis inhibitor blocks *Staphylococcus aureus* virulence. *Science* **319:**1391–1394.
53. **Liu, G. Y., A. Essex, J. T. Buchanan, V. Datta, H. M. Hoffman, J. F. Bastian, J. Fierer, and V. Nizet.** 2005. *Staphylococcus aureus* golden pigment impairs neutrophil killing and promotes virulence through its antioxidant activity. *J. Exp. Med.* **202:**209–215.
54. **Lowy, F. D.** 1998. *Staphylococcus aureus* infections. *N. Engl. J. Med.* **339:**520–532.
55. **Mack, D., W. Fischer, A. Krokotsch, K. Leopold, R. Hartmann, H. Egge, and R. Laufs.** 1996. The intercellular adhesin involved

in biofilm accumulation of *Staphylococcus epidermidis* is a linear beta-1,6-linked glucosaminoglycan: purification and structural analysis. *J. Bacteriol.* **178:**175–183.

56. **Marraffini, L. A., A. C. Dedent, and O. Schneewind.** 2006. Sortases and the art of anchoring proteins to the envelopes of gram-positive bacteria. *Microbiol. Mol. Biol. Rev.* **70:** 192–221.
57. **Mazmanian, S. K., G. Liu, H. Ton-That, and O. Schneewind.** 1999. *Staphylococcus aureus* sortase, an enzyme that anchors surface proteins to the cell wall. *Science* **285:**760–763.
58. **Mazmanian, S. K., E. P. Skaar, A. H. Gaspar, M. Humayun, P. Gornicki, J. Jelenska, A. Joachmiak, D. M. Missiakas, and O. Schneewind.** 2003. Passage of heme-iron across the envelope of *Staphylococcus aureus*. *Science* **299:** 906–909.
59. **McCormick, J. K., J. M. Yarwood, and P. M. Schlievert.** 2001. Toxic shock syndrome and bacterial superantigens: an update. *Annu. Rev. Microbiol.* **55:**77–104.
60. **Miller, L. G., and B. A. Diep.** 2008. Clinical practice: colonization, fomites, and virulence: rethinking the pathogenesis of community-associated methicillin-resistant *Staphylococcus aureus* infection. *Clin. Infect. Dis.* **46:**752–760.
61. **Moran, G. J., A. Krishnadasan, R. J. Gorwitz, G. E. Fosheim, L. K. McDougal, R. B. Carey, and D. A. Talan.** 2006. Methicillin-resistant *S. aureus* infections among patients in the emergency department. *N. Engl. J. Med.* **355:** 666–674.
62. **Navarre, W. W., and O. Schneewind.** 1999. Surface proteins of gram-positive bacteria and mechanisms of their targeting to the cell wall envelope. *Microbiol. Mol. Biol. Rev.* **63:**174–229.
63. **Novick, R. P.** 2003. Autoinduction and signal transduction in the regulation of staphylococcal virulence. *Mol. Microbiol.* **48:**1429–1449.
64. **O'Riordan, K., and J. C. Lee.** 2004. *Staphylococcus aureus* capsular polysaccharides. *Clin. Microbiol. Rev.* **17:**218–234.
65. **Otto, M.** 2004. Virulence factors of the coagulase-negative staphylococci. *Front. Biosci.* **9:** 841–863.
66. **Otto, M., H. Echner, W. Voelter, and F. Gotz.** 2001. Pheromone cross-inhibition between *Staphylococcus aureus* and *Staphylococcus epidermidis*. *Infect. Immun.* **69:**1957–1960.
67. **Patti, J. M., B. L. Allen, M. J. McGavin, and M. Hook.** 1994. MSCRAMM-mediated adherence of microorganisms to host tissues. *Annu. Rev. Microbiol.* **48:**585–617.
68. **Pelz, A., K. P. Wieland, K. Putzbach, P. Hentschel, K. Albert, and F. Götz.** 2005. Structure and biosynthesis of staphyloxanthin from *Staphylococcus aureus*. *J. Biol. Chem.* **280:** 32493–32498.
69. **Peschel, A.** 2002. How do bacteria resist human antimicrobial peptides? *Trends Microbiol.* **10:**179–186.
70. **Peschel, A., R. W. Jack, M. Otto, L. V. Collins, P. Staubitz, G. Nicholson, H. Kalbacher, W. F. Nieuwenhuizen, G. Jung, A. Tarkowski, K. P. van Kessel, and J. A. van Strijp.** 2001. *Staphylococcus aureus* resistance to human defensins and evasion of neutrophil killing via the novel virulence factor MprF is based on modification of membrane lipids with L-lysine. *J. Exp. Med.* **193:**1067–1076.
71. **Peschel, A., M. Otto, R. W. Jack, H. Kalbacher, G. Jung, and F. Gotz.** 1999. Inactivation of the *dlt* operon in *Staphylococcus aureus* confers sensitivity to defensins, protegrins, and other antimicrobial peptides. *J. Biol. Chem.* **274:** 8405–8410.
72. **Peschel, A., and H. G. Sahl.** 2006. The co-evolution of host cationic antimicrobial peptides and microbial resistance. *Nat. Rev. Microbiol.* **4:** 529–536.
73. **Pillar, C. M., D. C. Draghi, D. J. Sheehan, and D. F. Sahm.** 2008. Prevalence of multidrug-resistant, methicillin-resistant *Staphylococcus aureus* in the United States: findings of the stratified analysis of the 2004 to 2005 LEADER Surveillance Programs. *Diagn. Microbiol. Infect. Dis.* **60:**221–224.
74. **Prat, C., J. Bestebroer, C. J. de Haas, J. A. van Strijp, and K. P. van Kessel.** 2006. A new staphylococcal anti-inflammatory protein that antagonizes the formyl peptide receptor-like 1. *J. Immunol.* **177:**8017–8026.
75. **Recsei, P., B. Kreiswirth, M. O'Reilly, P. Schlievert, A. Gruss, and R. P. Novick.** 1986. Regulation of exoprotein gene expression in *Staphylococcus aureus* by *agr*. *Mol. Gen. Genet.* **202:**58–61.
76. **Rice, K. C., E. E. Mann, J. L. Endres, E. C. Weiss, J. E. Cassat, M. S. Smeltzer, and K. W. Bayles.** 2007. The CidA murein hydrolase regulator contributes to DNA release and biofilm development in *Staphylococcus aureus*. *Proc. Natl. Acad. Sci. USA* **104:**8113–8118.
77. **Richmond, M. H.** 1979. Beta-lactam antibiotics and beta-lactamases: two sides of a continuing story. *Rev. Infect. Dis.* **1:**30–38.
78. **Rooijakkers, S. H., M. Ruyken, A. Roos, M. R. Daha, J. S. Presanis, R. B. Sim, W. J. van Wamel, K. P. van Kessel, and J. A. van Strijp.** 2005. Immune evasion by a staphylococcal complement inhibitor that acts on C3 convertases. *Nat. Immunol.* **6:**920–927.

79. **Rooijakkers, S. H., K. P. van Kessel, and J. A. van Strijp.** 2005. Staphylococcal innate immune evasion. *Trends Microbiol.* **13:**596–601.
80. **Rooijakkers, S. H., W. J. van Wamel, M. Ruyken, K. P. van Kessel, and J. A. van Strijp.** 2005. Anti-opsonic properties of staphylokinase. *Microbes Infect.* **7:**476–484.
81. **Sakoulas, G., G. M. Eliopoulos, R. C. Moellering, Jr., R. P. Novick, L. Venkataraman, C. Wennersten, P. C. DeGirolami, M. J. Schwaber, and H. S. Gold.** 2003. *Staphylococcus aureus* accessory gene regulator *(agr)* group II: is there a relationship to the development of intermediate-level glycopeptide resistance? *J. Infect. Dis.* **187:**929–938.
82. **Schittek, B., R. Hipfel, B. Sauer, J. Bauer, H. Kalbacher, S. Stevanovic, M. Schirle, K. Schroeder, N. Blin, F. Meier, G. Rassner, and C. Garbe.** 2001. Dermcidin: a novel human antibiotic peptide secreted by sweat glands. *Nat. Immunol.* **2:**1133–1137.
83. **Shiau, A. L., and C. L. Wu.** 1998. The inhibitory effect of *Staphylococcus epidermidis* slime on the phagocytosis of murine peritoneal macrophages is interferon-independent. *Microbiol. Immunol.* **42:**33–40.
84. **Sieprawska-Lupa, M., P. Mydel, K. Krawczyk, K. Wojcik, M. Puklo, B. Lupa, P. Suder, J. Silberring, M. Reed, J. Pohl, W. Shafer, F. McAleese, T. Foster, J. Travis, and J. Potempa.** 2004. Degradation of human antimicrobial peptide LL-37 by *Staphylococcus aureus*-derived proteinases. *Antimicrob. Agents Chemother.* **48:**4673–4679.
85. **Sievert, D. M., J. T. Rudrik, J. B. Patel, L. C. McDonald, M. J. Wilkins, and J. C. Hageman.** 2008. Vancomycin-resistant *Staphylococcus aureus* in the United States, 2002–2006. *Clin. Infect. Dis.* **46:**668–674.
86. **Somerville, G. A., S. B. Beres, J. R. Fitzgerald, F. R. DeLeo, R. L. Cole, J. S. Hoff, and J. M. Musser.** 2002. In vitro serial passage of *Staphylococcus aureus:* changes in physiology, virulence factor production, and *agr* nucleotide sequence. *J. Bacteriol.* **184:**1430–1437.
87. **Ton-That, H., S. K. Mazmanian, K. F. Faull, and O. Schneewind.** 2000. Anchoring of surface proteins to the cell wall of *Staphylococcus aureus.* Sortase catalyzed in vitro transpeptidation reaction using LPXTG peptide and $NH_2$-$Gly_3$ substrates. *J. Biol. Chem.* **275:**9876–9881.
88. **Torres, V. J., G. Pishchany, M. Humayun, O. Schneewind, and E. P. Skaar.** 2006. *Staphylococcus aureus* IsdB is a hemoglobin receptor required for heme iron utilization. *J. Bacteriol.* **188:**8421–8429.
89. **Valeva, A., I. Walev, M. Pinkernell, B. Walker, H. Bayley, M. Palmer, and S. Bhakdi.** 1997. Transmembrane beta-barrel of staphylococcal alpha-toxin forms in sensitive but not in resistant cells. *Proc. Natl. Acad. Sci. USA* **94:**11607–11611.
90. **Voyich, J. M., K. R. Braughton, D. E. Sturdevant, A. R. Whitney, B. Said-Salim, S. F. Porcella, R. D. Long, D. W. Dorward, D. J. Gardner, B. N. Kreiswirth, J. M. Musser, and F. R. DeLeo.** 2005. Insights into mechanisms used by *Staphylococcus aureus* to avoid destruction by human neutrophils. *J. Immunol.* **175:**3907–3919.
91. **Voyich, J. M., M. Otto, B. Mathema, K. R. Braughton, A. R. Whitney, D. Welty, R. D. Long, D. W. Dorward, D. J. Gardner, G. Lina, B. N. Kreiswirth, and F. R. DeLeo.** 2006. Is Panton-Valentine leukocidin the major virulence determinant in community-associated methicillin-resistant *Staphylococcus aureus* disease? *J. Infect. Dis.* **194:**1761–1770.
92. **Vuong, C., S. Kocianova, Y. Yao, A. B. Carmody, and M. Otto.** 2004. Increased colonization of indwelling medical devices by quorum-sensing mutants of *Staphylococcus epidermidis* in vivo. *J. Infect. Dis.* **190:**1498–1505.
93. **Vuong, C., H. L. Saenz, F. Gotz, and M. Otto.** 2000. Impact of the *agr* quorum-sensing system on adherence to polystyrene in *Staphylococcus aureus. J. Infect. Dis.* **182:**1688–1693.
94. **Vuong, C., J. M. Voyich, E. R. Fischer, K. R. Braughton, A. R. Whitney, F. R. DeLeo, and M. Otto.** 2004. Polysaccharide intercellular adhesin (PIA) protects *Staphylococcus epidermidis* against major components of the human innate immune system. *Cell. Microbiol.* **6:**269–275.
95. **Wang, R., K. R. Braughton, D. Kretschmer, T. H. Bach, S. Y. Queck, M. Li, A. D. Kennedy, D. W. Dorward, S. J. Klebanoff, A. Peschel, F. R. Deleo, and M. Otto.** 2007. Identification of novel cytolytic peptides as key virulence determinants for community-associated MRSA. *Nat. Med.* **13:**1510–1514.
96. **Wardenburg, J. B., T. Bae, M. Otto, F. R. Deleo, and O. Schneewind.** 2007. Poring over pores: alpha-hemolysin and Panton-Valentine leukocidin in *Staphylococcus aureus* pneumonia. *Nat. Med.* **13:**1405–1406.
97. **Wardenburg, J. B., and O. Schneewind.** 2008. Vaccine protection against *Staphylococcus aureus* pneumonia. *J. Exp. Med.* **205:**287–294.
98. **Yao, Y., D. E. Sturdevant, and M. Otto.** 2005. Genomewide analysis of gene expression in *Staphylococcus epidermidis* biofilms: insights into the pathophysiology of *S. epidermidis* biofilms and the role of phenol-soluble modulins in formation of biofilms. *J. Infect. Dis.* **191:**289–298.

# *Mycobacterium avium* subsp. *paratuberculosis*: AN UNCONVENTIONAL PATHOGEN?

*Srinand Sreevatsan, Natalia Cernicchiaro, and Radhey Kaushik*

# 17

Mycobacteria in general, and the *Mycobacterium avium* complex in particular, are a closely related group of microorganisms which do not readily lend themselves to simplified identification or differentiation. The *M. avium* complex includes the closely related *Mycobacterium avium* subsp. *avium, Mycobacterium avium* subsp. *paratuberculosis,* and *Mycobacterium intracellulare,* as well as the wood pigeon bacillus. This complex is characterized by over 90% similarity at the nucleotide level, but the members differ widely in terms of their host tropisms, microbiological and disease phenotypes, and pathogenicities. They are associated with animal and human diseases, including infections of the lung, lymph nodes, skin, bones, and gastrointestinal and genitourinary tracts (30, 31). In recent years, *M. avium* complex strains have assumed greater importance in human medicine, largely because of intractable *Mycobacterium avium* complex infections in AIDS patients and also because of the possible association of *M. avium* subsp. *paratuberculosis* with Crohn's disease (30, 33).

*M. avium* subsp. *paratuberculosis* is the causative agent of Johne's disease (or paratuberculosis), a debilitating chronic enteritis in ruminants (27, 33). Crohn's disease is also a chronic inflammation of the distal intestines and exhibits pathology similar to that of Johne's disease in ruminants. Several studies have associated *M. avium* subsp. *paratuberculosis* with a proportion of Crohn's disease cases (3, 56). Although strain sharing has been documented (16), the evidence for a link remains controversial, and a causal role for *M. avium* subsp. *paratuberculosis* in Crohn's disease has not been definitively demonstrated (21, 26). Johne's disease is now recognized to have significant economic and health consequences in domesticated ruminant species (primarily dairy and beef cattle, sheep, and goats) throughout the world (33, 74). Johne's disease has the greatest economic impact in dairy cattle, for which premature culling, reduced carcass value, and decreased weight gain and milk production result in estimated losses up to $250 million annually (33, 46, 69).

*Srinand Sreevatsan,* Department of Veterinary Population Medicine and Department of Veterinary Biomedical Sciences, University of Minnesota, St. Paul, MN 55108. *Natalia Cernicchiaro,* Department of Veterinary Population Medicine, University of Minnesota, St. Paul, MN 55108. *Radhey Kaushik,* Department of Biology and Microbiology and Department of Veterinary Sciences, NPB Rm. 252A, Box 2140D, South Dakota State University, Brookings, SD 57007.

*Food-Borne Microbes: Shaping the Host Ecosystem,* Edited by L.-A. Jaykus, H. H. Wang, and L. S. Schlesinger, © 2009 ASM Press, Washington, DC

## MICROBIOLOGY AND GENETIC DIVERSITY OF *M. AVIUM* SUBSP. *PARATUBERCULOSIS*

*Mycobacterium avium* subsp. *paratuberculosis* is a slow-growing species of the *Mycobacteriaceae*. Among the *Mycobacterium* species of importance in veterinary medicine are two major complexes, *Mycobacterium tuberculosis* and *Mycobacterium avium*. The *Mycobacterium tuberculosis* complex includes the two major pathogens *M. tuberculosis* (the cause of tuberculosis in humans, primates, dogs, and other animals) and *Mycobacterium bovis* (the causative agent of tuberculosis in ruminants, humans, primates, and swine, among others). Their differentiation at the phenotypic level is extremely difficult, but it is now established that *M. bovis* is marginally resistant to isoniazid and resistant to pyrazinamide, two frontline antitubercular drugs. Genotypically, the two organisms may be differentiated on the basis of single nucleotide polymorphisms in several segments across their genomes (13, 23, 24, 64, 65). The *Mycobacterium avium* complex includes *M. avium* subsp. *avium, Mycobacterium avium* subsp. *silvaticum, M. avium* subsp. *paratuberculosis,* and *M. intracellulare*.

*M. avium* subsp. *paratuberculosis* is an obligate parasite that may exist in a (presumed) dormant or metabolically inactive form in the environment. *M. avium* subsp. *paratuberculosis* is an acid-fast, mycobactin-dependent organism. The organism possesses a unique insertion sequence, IS*900,* that is represented in 14 to 18 copies within its genome. As a multispecies pathogen, its transmission has been speculated to be zoonotic (44).

On solid media, *M. avium* subsp. *paratuberculosis* grows at 37°C under aerobic conditions. Colonies are small (1 to 2 mm), unpigmented, and domed shaped, with a margin. Rough and pigmented colonies are rarely seen. In liquid culture media, *M. avium* subsp. *paratuberculosis* also grows at 37°C under aerobic conditions. The organism has a thick waxy cell wall containing about 60% lipid, which confers the properties of acid fastness, hydrophobicity, and increased resistance to chemicals (such as chlorine) and physical processes (such as pasteurization) (53). The lipid wall gives mycobacteria a survival advantage, but as a consequence, the organism grows slowly because the uptake of nutrients through the cell wall is somewhat restricted. In addition, the inability to produce mycobactin, a siderophore responsible for the binding or transport of iron into cells, causes slow growth. Because of the importance of iron in electron transport and as a key component of various metabolic enzymes, the organism requires iron supplementation in order to survive and multiply.

Sequencing of the complete genome of *M. avium* subsp. *paratuberculosis* strain K-10 (GenBank accession no. AE016958) has revealed several unique characteristics. In the single circular genome of 4,829,781 bp with a G+C content of 69.3%, there is a 72.2-kb segment comprised of repetitive DNA-like insertion sequences, multigene families, and duplicate housekeeping genes. Seventeen copies of the insertion sequence IS*900* were identified, as well as seven copies of IS*1311* and three copies of the IS*Mav2*. Sixteen additional insertion sequence elements were identified, totaling 19 different insertion sequences with 58 total copies in the K-10 genome. Several insertion sequences with no identifiable homologues among other mycobacteria were identified, including IS_*Map 02,* present in six copies, and IS_*Map 04,* present in four copies.

The K-10 genome contains 4,350 open reading frames 114 to 19,155 bp in length. A cluster of 10 genes called *mbtA* to *mbtJ* responsible for the production of mycobactin and the transport of iron also have been described for *M. tuberculosis* (51). *M. avium* subsp. *paratuberculosis* possesses a homologous cluster in which *mbtA* is truncated relative to the orthologues identified in *M. tuberculosis* and *M. avium*. This truncation was speculated to be a possible reason for the mycobactin dependence of *M. avium* subsp. *paratuberculosis* (34) in vitro. Two other possible reasons for the slow growth of *M. avium* subsp. *paratuberculosis* are the presence of an insertion sequence (Map0028c/IS*1311*) close to *oriC* (origin of replication) and the presence of a higher substitution rate in the gene Map0638 *(purF),* which encodes an amidophosphoribosyltransferase involved in purine metabolism. The former phenomenon may be detrimental to chromosomal replication, causing an increased generation interval. The presence of more nucleotide substitutions in the gene responsible for regulation of purine synthesis and cell growth may also translate functionally to slower growth (34, 53).

Previous studies have identified 185 mono-, di-, and trinucleotide repeat sequences dispersed in the *M. avium* subsp. *paratuberculosis* genome, with 79 of them being perfect repeats (1). A total of 362 additional sequences 6 to 74 bp in length with a repeat number ranging from 2 to 16 have been identified in the genome of *M. avium* subsp. *paratuberculosis* (34). The presence of these repetitive elements provides additional strain differentiation capabilities through the use of molecular typing approaches such as multilocus short sequence repeat analysis or variable-number tandem-repeat analysis.

Current research based on understanding of the genomic diversity among *M. avium* subsp. *paratuberculosis* strains is now enabling elucidation of mechanisms of survival in the environment, host specificity, and the association of specific genotypes with overt disease versus subclinical states (16, 28, 42–44). In Johne's disease, two host-specific subgroups of *M. avium* subsp. *paratuberculosis* isolates have been recognized recently. In addition to phenotypic data and epidemiological observations, all subtyping techniques, irrespective of their discriminatory power, concur in this respect. Almost without exception and regardless of geographical location, isolates from cattle belong to the C type (type II), as have most isolates from goats and deer. In complete contrast, isolates from sheep have been of the C, S (type I), or I (intermediate) type, with most countries tending to have only one type in their sheep population. The underrepresentation of S isolates in cattle, deer, and goats might reflect the difficulty of laboratory culture of such isolates, but their identification in sheep from a similar range of countries suggests that cultivability alone does not explain the apparent strain segregation within host species (70). This raises the possibility that the *M. avium* subsp. *paratuberculosis* types display a degree of host specificity or at least host preference, although definitive evidence for this is lacking and the true degree of host adaptation of these isolates remains unknown.

The possibility that a particular group of *M. avium* subsp. *paratuberculosis* isolates may be involved in Crohn's disease also exists. Using oligonucleotide primers specific for IS*900* and DNA amplification by PCR, several groups reported a positive correlation between the presence of IS*900* DNA and Crohn's disease. However, the frequency of detection of the IS*900* sequences differed substantially from one study to another. Based on these data, *M. avium* subsp. *paratuberculosis* appears to be present in only a subgroup of Crohn's disease patients. So far, only a limited number of human *M. avium* subsp. *paratuberculosis* isolates are available for typing, a situation which has hampered progress in understanding the relationship between this organism and Crohn's disease. There is, nonetheless, consistency in the observation that *M. avium* subsp. *paratuberculosis* isolates from Crohn's disease patients show limited diversity (4, 15, 16, 47, 50). Clustering of the human and animal isolates is suggestive of inter- and intraspecies transmission (15, 16, 70) and an association between a few animal *M. avium* subsp. *paratuberculosis* isolates and the pathobiology of Crohn's disease.

## PATHOGENESIS OF JOHNE'S DISEASE

Johne's disease progression may be divided into four stages: silent infection and subclinical, clinical, and advanced clinical disease (5, 68). After entry through the intestinal wall, *M. avium* subsp. *paratuberculosis* bacilli are phagocytosed by subepithelial macrophages, in which they reside and multiply in phagosomes or early endosomes. The infected macrophages play a critical role in the pathogenesis of the disease and dictate the disease outcome. We and others have investigated the nature of macrophage-*M. avium* subsp. *paratuberculosis* interactions; this has also been extensively reviewed (5, 9, 10, 63, 66).

Johne's disease lesions begin as discrete granulomas largely restricted to the middle and distal parts of the small intestine, especially the ileum and draining lymph nodes (8, 62, 68). With time, the inflammatory process progresses to granuloma-like or diffuse granulomatous inflammation with a large number of *M. avium* subsp. *paratuberculosis* bacilli in the macrophages, leading to blunted villi and a severely thickened and corrugated intestinal epithelium (5, 9, 63). This finally results in the malabsorption/malnutrition syndrome (weight loss, decreased milk production, and diarrhea) commonly observed in clinical Johne's disease and which may lead to the death of the infected animals.

## A ROLE FOR INTESTINAL EPITHELIAL CELLS IN JOHNE'S DISEASE

*M. avium* subsp. *paratuberculosis* is naturally transmitted to the ruminant hosts through the fecal-oral route. The organism must traverse the intestinal epithelial barrier in order to reach subepithelial macrophages, which are the ultimate target cells for this organism. The small intestinal epithelial cells, including M cells, are considered one of the primary targets for *M. avium* subsp. *paratuberculosis* entry and transport to the subepithelial intestinal tissue; however, the exact anatomical site and cell types used by *M. avium* subsp. *paratuberculosis* to cross the intestinal epithelium are unresolved to date (5, 63). A few studies using ileal loops of cattle, sheep, goats, and mice (41, 59, 60; J. F. Garcia Marin, S. Benazzi, V. Peres, and J. J. Badiola, presented at the Proceedings of the Third International Colloquium on Paratuberculosis, Madison, WI, 1991) suggest that M cells located on the follicle-associated epithelium (11, 32), especially in the distal ileum, play a critical role in *M. avium* subsp. *paratuberculosis* translocation across the intestinal barrier. In animal studies (41, 60; Garcia Marin et al., presented previously), most bacteria have been identified subepithelially in the dome and interfollicular areas. Based on these observations, it was concluded that M cells present in the follicle-associated epithelium of ileal Peyer's patches served as the primary portal of entry for *M. avium* subsp. *paratuberculosis*. In all these studies, some bacteria were also present in the villi of the epithelial lining or in lamina propria adjacent to domes, suggesting a possible role for non-M-cell (enterocyte)-mediated processes in the uptake of *M. avium* subsp. *paratuberculosis*. It has also been established that the fibronectin attachment protein homologue of *M. avium* subsp. *paratuberculosis* (FAP-P) mediates fibronectin (FN) binding to the bacterium, which, in turn, facilitates *M. avium* subsp. *paratuberculosis* attachment to intestinal epithelial cells (57, 58). Further, a mouse model was used to show that nonopsonized *M. avium* subsp. *paratuberculosis* invaded M cells in murine gut loops; however, a small percentage of the villus epithelium also showed evidence of *M. avium* subsp. *paratuberculosis* adhesion and entry. When FN-opsonized *M. avium* subsp. *paratuberculosis* organisms were used, FN selectively increased the ability of the organisms to invade M cells but not villus epithelial cells (59). Taken together, these studies show that although M cells play a major role in the uptake of *M. avium* subsp. *paratuberculosis,* villus epithelial cells also participate, albeit to a lesser extent, in the uptake of this organism.

Recent studies have also demonstrated binding of *M. avium* subsp. *paratuberculosis* to the absorptive epithelium of the intestinal villi of cattle and goat small intestinal tissues (55, 61). In one study (55), five radiolabeled *M. avium* subsp. *paratuberculosis* strains were used on short-term bovine intestinal organ cultures to study attachment to the intestinal epithelium. In this case, *M. avium* subsp. *paratuberculosis* strains were shown to bind to enterocytes in organ cultures. This study also showed significant differences in the abilities of different strains to attach to the villus epithelium but did not show any differences in the attachment to different regions of the intestinal tract. Furthermore, opsonization of bacteria with FN did not affect the overall attachment of *M. avium* subsp. *paratuberculosis* to intestinal epithelial cells. In another study (61), an everted-sleeve model was used to study the attachment of a radiolabeled *M. avium* subsp. *paratuberculosis* strain to intestinal mucosa of jejunum from goats, both in areas with and without Peyer's patches. This study showed that *M. avium* subsp. *paratuberculosis* bacilli entered the subepithelial intestinal tissues through both M cells and enterocytes. The findings from these two studies clearly suggest a role for enterocytes in the translocation of *M. avium* subsp. *paratuberculosis* across the intestinal barrier. This argument is further substantiated by a study using a mouse gut loop and *Mycobacterium avium* subsp. *avium,* a pathogen with >96% nucleotide similarity to *M. avium* subsp. *paratuberculosis*. In this case, *M. avium* subsp. *avium* invaded the intestine preferentially through enterocytes and not M cells (54). Taken together, these studies strongly suggest a significant role for the absorptive epithelium in *M. avium* subsp. *paratuberculosis* uptake and transmission and in the early stages of pathogenesis of paratuberculosis.

It should be clear from the discussion above that host-pathogen interaction studies using various *M. avium* subsp. *paratuberculosis* strains have identified specific cell types involved in the modus operandi of *M. avium* subsp. *paratuberculosis*. Findings such as these will facilitate future in vitro functional studies with this organism. Studies involving the responses of this organism in a host cellular environment are lacking and will be necessary in order to fully understand the functional genome of this organism.

## *M. AVIUM* SUBSP. *PARATUBERCULOSIS* AND THE ENVIRONMENT

A unique feature of *M. avium* subsp. *paratuberculosis* is that it can survive in the environment for several months. Indeed, this characteristic has been a major hurdle in effectively controlling Johne's disease within herds. Further, nonruminant and ruminant wildlife reservoirs, as well as metazoan parasites and protozoa, are thought to contribute to the environmental persistence of *M. avium* subsp. *paratuberculosis* (6, 7).

While it is well established that *M. avium* subsp. *paratuberculosis* is able to survive for up to 11 months in bovine feces, less than 30 days in bovine urine, 9 to 19 months in tap water at pH 5.0 to 8.5, and 113 to 218 days in river water under different ambient conditions (−7 to 18°C), the precise mechanisms by which such environmental persistence is achieved are not fully understood (71–73).

Comparative genomic analysis has revealed several lateral gene transfer events in *M. avium* subsp. *paratuberculosis* that may have occurred at different phases of the organism's divergence, evolution, and specialization into a pathogen (37). Notably, 48 of the 275 genes showing evidence of horizontal transfer events shared significant homology with genes found in the proteobacteria, which are common environmental organisms. In fact, a microbial population structure analysis of dairy runoff wastes (38, 39) as well as anaerobic composted dairy waste (40) performed in California identified several microbial subpopulations, including proteobacteria, cyanobacteria, and actinobacteria, with which *M. avium* subsp. *paratuberculosis* shares in excess of 95 of the 275 lateral gene transfer events. Functional characterization of these lateral gene transfer events from environmental bacteria into *M. avium* subsp. *paratuberculosis* is expected to provide critical missing information on the physiology and mechanisms of its survival in the environment. In silico analysis has identified two severe stress-related elements in *M. avium* subsp. *paratuberculosis:* a *dps*-like genetic element and *relA* (71). While their presence in the genome is evidence that *M. avium* subsp. *paratuberculosis* may use them under specific stress conditions, a formal "stressome" analysis under naturally occurring conditions will be necessary to define the repertoire of genes needed to force *M. avium* subsp. *paratuberculosis* into dormancy.

The physical nature of the environments in which *M. avium* subsp. *paratuberculosis* survives was evaluated by sampling fecal material from soil and grass in pasture plots and boxes at defined intervals for up to 117 weeks. *M. avium* subsp. *paratuberculosis* survived for up to 55 weeks in a dry, fully shaded environment and for up to 24 weeks on grass in shaded boxes. Shorter survival times were found in unshaded locations. Dam water and related sediment analysis has also revealed that *M. avium* subsp. *paratuberculosis* is able to survive for up to 48 weeks in water or sediment in the shade and for 36 weeks in semiexposed locations. Survival in sediment was 12 to 26 weeks longer than in the water column, and only 12 weeks in soil or feces in a ground-shaded environment. Thus, it has been suggested that water may be a significant reservoir of *M. avium* subsp. *paratuberculosis* and that mechanisms to prevent contamination of water should be put into place to prevent transmission of the organism (71, 73). Prospective environmental testing on sheep flocks with a history of Johne's disease before and up to 9 months after removal of the infected stock showed that *M. avium* subsp. *paratuberculosis* could be detected just before destocking, while the likelihood of positive cultures was very low 5 months after animal removal (72). While not conclusive, these studies enable us to make informed decisions on animal restocking and the potential for exposure of wildlife populations in the "open" periods.

## POTENTIAL SOURCES FOR HUMAN EXPOSURE TO *M. AVIUM* SUBSP. *PARATUBERCULOSIS*

Contaminated domestic water (29, 49), pasteurized milk (12, 17–20), contaminated meat or other food, environmental sources, and direct animal exposures have all been suggested as possible mechanisms of human exposure to *M. avium* subsp. *paratuberculosis*.

Contaminated water supplies have been documented to be a potential environmental source of several mycobacterial diseases. Pickup et al. tested for the presence of *M. avium* subsp. *paratuberculosis* in the catchment area and waters of the River Tywi, located in South Wales, United Kingdom (48, 49). They showed that 69% of river water samples were positive by IS*900* PCR. Furthermore, one of 54 domestic cold-water tanks tested positive, as well as sediments from 9 of the 10 lakes that received inflow from these catchments. This study showed that *M. avium* subsp. *paratuberculosis* is able to directly enter and survive in the domestic water supply, providing evidence for its potential cycling through human populations.

One other possible mechanism of water contamination and subsequent environmental survival is via infection of free-living protozoa. Much of the analyses have been performed using *M. avium,* an opportunistic human pathogen, as a model organism. Cirillo et al. showed that *M. avium* enters and replicates in *Acanthamoeba castellanii* (7). Akin to the events that occur in macrophages, *M. avium* inhibits phagolysosomal fusion and replicates in vacuoles that are tightly juxtaposed to the bacterial surfaces within amoebae. These findings have led to the speculation that protozoa present in water environments could serve as maintenance and amplification hosts for pathogenic mycobacteria.

*M. avium* subsp. *paratuberculosis* has also been shown to replicate and persist for long periods within the environmental protist *Acanthamoeba polyphaga* (45, 67). Interactions between *M. avium* subsp. *paratuberculosis* and free-living protozoa in water are likely to occur in nature. Spiked populations of *M. avium* subsp. *paratuberculosis* were ingested by *Acanthamoeba* spp. *(A. castellanii* and *A. polyphaga),* and *M. avium* subsp. *paratuberculosis* bacilli were found within vacuoles of the amoebae. After being exposed to chlorine, ingested bacteria were more resistant to chlorine inactivation than free bacilli, suggesting that the organism may "hitchhike" in the protozoa for nutrition and protection from environmental stressors (67).

Carriage of *M. avium* subsp. *paratuberculosis* in nematodes is also a likely mode of survival in the environment and a source of transmission to animals and/or humans. Since *M. avium* subsp. *paratuberculosis* has been shown to persist in moist, shady areas of pastures, it is likely that the organism comes in contact with nematodes that share similar environments. Lloyd et al. demonstrated that larvae of *Haemonchus contortus, Ostertagia circumcincta,* and *Trichostrongylus colubriformis* present in sheep feces can ingest *M. avium* subsp. *paratuberculosis* (35). Thus, nematode larvae may also represent a viable means of transmission of *M. avium* subsp. *paratuberculosis* (23). Other studies have found that earthworms and insects such as diptera and cockroaches may also contribute to survival and transmission of *M. avium* subsp. *paratuberculosis* in the environment (14).

Numerous microorganisms such as *Escherichia coli* O157:H7, *Salmonella* spp., *Listeria monocytogenes,* and *M. avium* subsp. *paratuberculosis* can be found in livestock manure. Despite such heavy microbial loads in manure, high-temperature composting has proven to reduce the persistence of *M. avium* subsp. *paratuberculosis* after 3 days (22, 25). Other pack storage or lagoon storage conditions have not been found to be as efficient in destroying *M. avium* subsp. *paratuberculosis.* Thus, consideration of manure storage, treatment, and recycling on farms is critical to controlling the recycling of infection within herds. Appropriate on-site treatment of manure would be expected to reduce pathogen loads entering the environment and/or the human food cycle.

In a recent review, Rowe and Grant suggested that even though *M. avium* subsp. *paratuberculosis* has not been isolated from biofilms in water distribution systems, biofilm formation by this organism would not be unexpected given that the closely related *M. avium* subsp. *avium* has a particular niche in hot-water systems (53). Thus, aerosol inhalation could serve as a route of transmission of *M. avium* subsp. *paratuberculosis* to animals and humans, especially since the mycobacterial lipid-rich cell wall might facilitate concentration in rising bubbles in water columns and aerosolization at the air-water interface (53). These speculations will need to be substantiated in both epidemiological and mechanistic studies.

Ruminant and nonruminant domestic animals and wildlife also contribute to environmental contamination and persistence of *M. avium* subsp. *paratuberculosis.* For example, there is an association between infected herds and contaminated environments, and this association emphasizes the critical importance of farm management strategies for the control of Johne's disease transmission (2, 36, 52, 73). The unique ability of *M. avium* subsp. *paratuberculosis* to survive diverse environments allows the organism regular contact with susceptible hosts, providing for a sustained cycle of infection within herds. This special characteristic also suggests that there may be widespread exposure to humans via a variety of sources.

## CONCLUDING REMARKS

*M. avium* subsp. *paratuberculosis* has been shown to survive diverse conditions in either the environment (water, soil, and temperature) or the host. These environments expose the organism to numerous stressors, over which *M. avium* subsp. *paratuberculosis* is able to prevail. Encrypted in its genome of about 4.5 Mb is functional code that helps the organism in its fight for survival in stressful environments. Comprehensive analysis of these survival mechanisms at the transcriptome and proteome levels is expected to improve our understanding of host-pathogen and pathogen-environment interactions of *M. avium* subsp. *paratuberculosis.* Together with improved in vitro culture systems for rapid diagnosis, these research areas should aid in the development of vaccines to prevent and control infection in animal populations and in elucidation of the role (if any) of *M. avium* subsp. *paratuberculosis* in human disease.

## REFERENCES

1. **Amonsin, A., L. L. Li, Q. Zhang, J. P. Bannantine, A. S. Motiwala, S. Sreevatsan, and V. Kapur.** 2004. Multilocus short sequence repeat sequencing approach for differentiating among *Mycobacterium avium* subsp. *paratuberculosis* strains. *J. Clin. Microbiol.* **42:**1694–1702.
2. **Berghaus, R. D., T. B. Farver, R. J. Anderson, C. C. Jaravata, and I. A. Gardner.** 2006. Environmental sampling for detection of Mycobacterium avium ssp. paratuberculosis on large California dairies. *J. Dairy Sci.* **89:**963–970.

3. **Bull, T. J., E. J. McMinn, K. Sidi-Boumedine, A. Skull, D. Durkin, P. Neild, G. Rhodes, R. Pickup, and J. Hermon-Taylor.** 2003. Detection and verification of *Mycobacterium avium* subsp. *paratuberculosis* in fresh ileocolonic mucosal biopsy specimens from individuals with and without Crohn's disease. *J. Clin. Microbiol.* **41:**2915–2923.
4. **Bull, T. J., K. Sidi-Boumedine, E. J. McMinn, K. Stevenson, R. Pickup, and J. Hermon-Taylor.** 2003. Mycobacterial interspersed repetitive units (MIRU) differentiate *Mycobacterium avium* subspecies *paratuberculosis* from other species of the *Mycobacterium avium* complex. *Mol. Cell. Probes* **17:**157–164.
5. **Chacon, O., L. E. Bermudez, and R. G. Barletta.** 2004. Johne's disease, inflammatory bowel disease, and Mycobacterium paratuberculosis. *Annu. Rev. Microbiol.* **58:**329–363.
6. **Cirillo, J. D., S. L. Cirillo, L. Yan, L. E. Bermudez, S. Falkow, and L. S. Tompkins.** 1999. Intracellular growth in *Acanthamoeba castellanii* affects monocyte entry mechanisms and enhances virulence of *Legionella pneumophila*. *Infect. Immun.* **67:**4427–4434.
7. **Cirillo, J. D., S. Falkow, L. S. Tompkins, and L. E. Bermudez.** 1997. Interaction of *Mycobacterium avium* with environmental amoebae enhances virulence. *Infect. Immun.* **65:**3759–3767.
8. **Clarke, C. J.** 1997. The pathology and pathogenesis of paratuberculosis in ruminants and other species. *J. Comp. Pathol.* **116:**217–261.
9. **Coussens, P. M.** 2004. Model for immune responses to *Mycobacterium avium* subspecies *paratuberculosis* in cattle. *Infect. Immun.* **72:**3089–3096.
10. **Coussens, P. M.** 2001. Mycobacterium paratuberculosis and the bovine immune system. *Anim. Health Res. Rev.* **2:**141–161.
11. **Didierlaurent, A., J. C. Sirard, J. P. Kraehenbuhl, and M. R. Neutra.** 2002. How the gut senses its content. *Cell. Microbiol.* **4:**61–72.
12. **Ellingson, J. L., J. L. Anderson, J. J. Koziczkowski, R. P. Radcliff, S. J. Sloan, S. E. Allen, and N. M. Sullivan.** 2005. Detection of viable Mycobacterium avium subsp. paratuberculosis in retail pasteurized whole milk by two culture methods and PCR. *J. Food Prot.* **68:**966–972.
13. **Filliol, I., A. S. Motiwala, M. Cavatore, W. Qi, M. H. Hazbon, M. Bobadilla del Valle, J. Fyfe, L. Garcia-Garcia, N. Rastogi, C. Sola, T. Zozio, M. I. Guerrero, C. I. Leon, J. Crabtree, S. Angiuoli, K. D. Eisenach, R. Durmaz, M. L. Joloba, A. Rendon, J. Sifuentes-Osornio, A. Ponce de Leon, M. D. Cave, R. Fleischmann, T. S. Whittam, and D. Alland.** 2006. Global phylogeny of *Mycobacterium tuberculosis* based on single nucleotide polymorphism (SNP) analysis: insights into tuberculosis evolution, phylogenetic accuracy of other DNA fingerprinting systems, and recommendations for a minimal standard SNP set. *J. Bacteriol.* **188:**759–772.
14. **Fischer, O., L. Matlova, L. Dvorska, P. Svastova, J. Bartl, I. Melicharek, R. T. Weston, and I. Pavlik.** 2001. Diptera as vectors of mycobacterial infections in cattle and pigs. *Med. Vet. Entomol.* **15:**208–211.
15. **Francois, B., R. Krishnamoorthy, and J. Elion.** 1997. Comparative study of *Mycobacterium paratuberculosis* strains isolated from Crohn's disease and Johne's disease using restriction fragment length polymorphism and arbitrarily primed polymerase chain reaction. *Epidemiol. Infect.* **118:** 227–233.
16. **Ghadiali, A. H., M. Strother, S. A. Naser, E. J. Manning, and S. Sreevatsan.** 2004. *Mycobacterium avium* subsp. *paratuberculosis* strains isolated from Crohn's disease patients and animal species exhibit similar polymorphic locus patterns. *J. Clin. Microbiol.* **42:**5345–5348.
17. **Grant, I. R.** 2003. Mycobacterium paratuberculosis and milk. *Acta Vet. Scand.* **44:**261–266.
18. **Grant, I. R.** 2005. Zoonotic potential of Mycobacterium avium ssp. paratuberculosis: the current position. *J. Appl. Microbiol.* **98:**1282–1293.
19. **Grant, I. R., H. J. Ball, S. D. Neill, and M. T. Rowe.** 1996. Inactivation of *Mycobacterium paratuberculosis* in cows' milk at pasteurization temperatures. *Appl. Environ. Microbiol.* **62:**631–636.
20. **Grant, I. R., H. J. Ball, and M. T. Rowe.** 2002. Incidence of *Mycobacterium paratuberculosis* in bulk raw and commercially pasteurized cows' milk from approved dairy processing establishments in the United Kingdom. *Appl. Environ. Microbiol.* **68:**2428–2435.
21. **Greenstein, R. J.** 2003. Is Crohn's disease caused by a mycobacterium? Comparisons with leprosy, tuberculosis, and Johne's disease. *Lancet Infect. Dis.* **3:**507–514.
22. **Grewal, S. K., S. Rajeev, S. Sreevatsan, and F. C. Michel, Jr.** 2006. Persistence of *Mycobacterium avium* subsp. *paratuberculosis* and other zoonotic pathogens during simulated composting, manure packing, and liquid storage of dairy manure. *Appl. Environ. Microbiol.* **72:**565–574.
23. **Gutacker, M. M., B. Mathema, H. Soini, E. Shashkina, B. N. Kreiswirth, E. A. Graviss, and J. M. Musser.** 2006. Single-nucleotide polymorphism-based population genetic analysis of Mycobacterium tuberculosis strains from 4 geographic sites. *J. Infect. Dis.* **193:**121–128.
24. **Gutacker, M. M., J. C. Smoot, C. A. Migliaccio, S. M. Ricklefs, S. Hua, D. V. Cous-**

**ins, E. A. Graviss, E. Shashkina, B. N. Kreiswirth, and J. M. Musser.** 2002. Genome-wide analysis of synonymous single nucleotide polymorphisms in Mycobacterium tuberculosis complex organisms: resolution of genetic relationships among closely related microbial strains. *Genetics* **162:**1533–1543.

25. **Gwóźdź, J. M.** 2006. Comparative evaluation of two decontamination methods for the isolation of Mycobacterium avium subspecies paratuberculosis from faecal slurry and sewage. *Vet. Microbiol.* **115:**358–363.
26. **Harris, J. E., and A. M. Lammerding.** 2001. Crohn's disease and *Mycobacterium avium* subsp. *paratuberculosis*: current issues. *J. Food Prot.* **64:** 2103–2110.
27. **Harris, N. B., and R. G. Barletta.** 2001. *Mycobacterium avium* subsp. *paratuberculosis* in veterinary medicine. *Clin. Microbiol. Rev.* **14:**489–512.
28. **Harris, N. B., J. B. Payeur, V. Kapur, and S. Sreevatsan.** 2006. Short-sequence-repeat analysis of *Mycobacterium avium* subsp. *paratuberculosis* and *Mycobacterium avium* subsp. *avium* isolates collected from animals throughout the United States reveals both stability of loci and extensive diversity. *J. Clin. Microbiol.* **44:**2970–2973.
29. **Hermon-Taylor, J., T. J. Bull, J. M. Sheridan, J. Cheng, M. L. Stellakis, and N. Sumar.** 2000. Causation of Crohn's disease by Mycobacterium avium subspecies paratuberculosis. *Can. J. Gastroenterol.* **14:**521–539.
30. **Karakousis, P. C., R. D. Moore, and R. E. Chaisson.** 2004. Mycobacterium avium complex in patients with HIV infection in the era of highly active antiretroviral therapy. *Lancet Infect. Dis.* **4:**557–565.
31. **Katoch, V. M.** 2004. Infections due to nontuberculous mycobacteria (NTM). *Indian J. Med. Res.* **120:**290–304.
32. **Kraehenbuhl, J. P., and M. R. Neutra.** 2000. Epithelial M cells: differentiation and function. *Annu. Rev. Cell Dev. Biol.* **16:**301–332.
33. **Lauzi, S., D. Pasotto, M. Amadori, I. L. Archetti, G. Poli, and L. Bonizzi.** 2000. Evaluation of the specificity of the gamma-interferon test in Italian bovine tuberculosis-free herds. *Vet. J.* **160:**17–24.
34. **Li, L., J. P. Bannantine, Q. Zhang, A. Amonsin, B. J. May, D. Alt, N. Banerji, S. Kanjilal, and V. Kapur.** 2005. The complete genome sequence of Mycobacterium avium subspecies paratuberculosis. *Proc. Natl. Acad. Sci. USA* **102:**12344–12349.
35. **Lloyd, J. B., R. J. Whittington, C. Fitzgibbon, and R. Dobson.** 2001. Presence of Mycobacterium avium subspecies paratuberculosis in suspensions of ovine trichostrongylid larvae produced in faecal cultures artificially contaminated with the bacterium. *Vet. Rec.* **148:**261–263.
36. **Lombard, J. E., B. A. Wagner, R. L. Smith, B. J. McCluskey, B. N. Harris, J. B. Payeur, F. B. Garry, and M. D. Salman.** 2006. Evaluation of environmental sampling and culture to determine Mycobacterium avium subspecies paratuberculosis distribution and herd infection status on US dairy operations. *J. Dairy Sci.* **89:** 4163–4171.
37. **Marri, P. R., J. P. Bannantine, M. L. Paustian, and G. B. Golding.** 2006. Lateral gene transfer in Mycobacterium avium subspecies paratuberculosis. *Can. J. Microbiol.* **52:**560–569.
38. **McGarvey, J. A., W. G. Miller, S. Sanchez, C. J. Silva, and L. C. Whitehand.** 2005. Comparison of bacterial populations and chemical composition of dairy wastewater held in circulated and stagnant lagoons. *J. Appl. Microbiol.* **99:** 867–877.
39. **McGarvey, J. A., W. G. Miller, S. Sanchez, and L. Stanker.** 2004. Identification of bacterial populations in dairy wastewaters by use of 16S rRNA gene sequences and other genetic markers. *Appl. Environ. Microbiol.* **70:**4267–4275.
40. **McGarvey, J. A., W. G. Miller, R. Zhang, Y. Ma, and F. Mitloehner.** 2007. Bacterial population dynamics in dairy waste during aerobic and anaerobic treatment and subsequent storage. *Appl. Environ. Microbiol.* **73:**193–202.
41. **Momotani, E., D. L. Whipple, A. B. Thiermann, and N. F. Cheville.** 1988. Role of M cells and macrophages in the entrance of Mycobacterium paratuberculosis into domes of ileal Peyer's patches in calves. *Vet. Pathol.* **25:**131–137.
42. **Motiwala, A. S., H. K. Janagama, M. L. Paustian, X. Zhu, J. P. Bannantine, V. Kapur, and S. Sreevatsan.** 2006. Comparative transcriptional analysis of human macrophages exposed to animal and human isolates of *Mycobacterium avium* subspecies *paratuberculosis* with diverse genotypes. *Infect. Immun.* **74:**6046–6056.
43. **Motiwala, A. S., L. Li, V. Kapur, and S. Sreevatsan.** 2006. Current understanding of the genetic diversity of Mycobacterium avium subsp. paratuberculosis. *Microbes Infect.* **8:**1406–1418.
44. **Motiwala, A. S., M. Strother, A. Amonsin, B. Byrum, S. A. Naser, J. R. Stabel, W. P. Shulaw, J. P. Bannantine, V. Kapur, and S. Sreevatsan.** 2003. Molecular epidemiology of *Mycobacterium avium* subsp. *paratuberculosis:* evidence for limited strain diversity, strain sharing, and identification of unique targets for diagnosis. *J. Clin. Microbiol.* **41:**2015–2026.
45. **Mura, M., T. J. Bull, H. Evans, K. Sidi-Boumedine, L. McMinn, G. Rhodes, R.**

**Pickup, and J. Hermon-Taylor.** 2006. Replication and long-term persistence of bovine and human strains of *Mycobacterium avium* subsp. *paratuberculosis* within *Acanthamoeba polyphaga. Appl. Environ. Microbiol.* **72:**854–859.

46. **Ott, S. L., S. J. Wells, and B. A. Wagner.** 1999. Herd-level economic losses associated with Johne's disease on US dairy operations. *Prev. Vet. Med.* **40:**179–192.

47. **Overduin, P., L. Schouls, P. Roholl, A. van der Zanden, N. Mahmmod, A. Herrewegh, and D. van Soolingen.** 2004. Use of multilocus variable-number tandem-repeat analysis for typing *Mycobacterium avium* subsp. *paratuberculosis. J. Clin. Microbiol.* **42:**5022–5028.

48. **Pickup, R. W., G. Rhodes, S. Arnott, K. Sidi-Boumedine, T. J. Bull, A. Weightman, M. Hurley, and J. Hermon-Taylor.** 2005. *Mycobacterium avium* subsp. *paratuberculosis* in the catchment area and water of the river Taff in South Wales, United Kingdom, and its potential relationship to clustering of Crohn's disease cases in the city of Cardiff. *Appl. Environ. Microbiol.* **71:** 2130–2139.

49. **Pickup, R. W., G. Rhodes, T. J. Bull, S. Arnott, K. Sidi-Boumedine, M. Hurley, and J. Hermon-Taylor.** 2006. *Mycobacterium avium* subsp. *paratuberculosis* in lake catchments, in river water abstracted for domestic use, and in effluent from domestic sewage treatment works: diverse opportunities for environmental cycling and human exposure. *Appl. Environ. Microbiol.* **72:**4067–4077.

50. **Pillai, S. R., B. M. Jayarao, J. D. Gummo, E. C. Hue, D. Tiwari, J. R. Stabel, and R. H. Whitlock.** 2001. Identification and subtyping of *Mycobacterium avium* subsp. *paratuberculosis* and *Mycobacterium avium* subsp. *avium* by randomly amplified polymorphic DNA. *Vet. Microbiol.* **79:**275–284.

51. **Quadri, L. E., J. Sello, T. A. Keating, P. H. Weinreb, and C. T. Walsh.** 1998. Identification of a Mycobacterium tuberculosis gene cluster encoding the biosynthetic enzymes for assembly of the virulence-conferring siderophore mycobactin. *Chem. Biol.* **5:**631–645.

52. **Raizman, E. A., S. J. Wells, S. M. Godden, R. F. Bey, M. J. Oakes, D. C. Bentley, and K. E. Olsen.** 2004. The distribution of Mycobacterium avium ssp. paratuberculosis in the environment surrounding Minnesota dairy farms. *J. Dairy Sci.* **87:**2959–2966.

53. **Rowe, M. T., and I. R. Grant.** 2006. Mycobacterium avium ssp. paratuberculosis and its potential survival tactics. *Lett. Appl. Microbiol.* **42:** 305–311.

54. **Sangari, F. J., J. Goodman, M. Petrofsky, P. Kolonoski, and L. E. Bermudez.** 2001. *Mycobacterium avium* invades the intestinal mucosa primarily by interacting with enterocytes. *Infect. Immun.* **69:**1515–1520.

55. **Schleig, P. M., C. D. Buergelt, J. K. Davis, E. Williams, G. R. Monif, and M. K. Davidson.** 2005. Attachment of Mycobacterium avium subspecies paratuberculosis to bovine intestinal organ cultures: method development and strain differences. *Vet. Microbiol.* **108:**271–279.

56. **Schwartz, D., I. Shafran, C. Romero, C. Piromalli, J. Biggerstaff, N. Naser, W. Chamberlin, and S. A. Naser.** 2000. Use of short-term culture for identification of Mycobacterium avium subsp. paratuberculosis in tissue from Crohn's disease patients. *Clin. Microbiol. Infect.* **6:** 303–307.

57. **Secott, T. E., T. L. Lin, and C. C. Wu.** 2001. Fibronectin attachment protein homologue mediates fibronectin binding by *Mycobacterium avium* subsp. *paratuberculosis. Infect. Immun.* **69:**2075–2082.

58. **Secott, T. E., T. L. Lin, and C. C. Wu.** 2002. Fibronectin attachment protein is necessary for efficient attachment and invasion of epithelial cells by *Mycobacterium avium* subsp. *paratuberculosis. Infect. Immun.* **70:**2670–2675.

59. **Secott, T. E., T. L. Lin, and C. C. Wu.** 2004. *Mycobacterium avium* subsp. *paratuberculosis* fibronectin attachment protein facilitates M-cell targeting and invasion through a fibronectin bridge with host integrins. *Infect. Immun.* **72:**3724–3732.

60. **Sigurðardóttir, Ó. G., C. M. Press, and Ø. Evensen.** 2001. Uptake of Mycobacterium avium subsp. paratuberculosis through the distal small intestinal mucosa in goats: an ultrastructural study. *Vet. Pathol.* **38:**184–189.

61. **Sigurðardóttir, Ó. G., A. M. Bakke-McKellep, B. Djønne, and Ø. Evensen.** 2005. Mycobacterium avium subsp. paratuberculosis enters the small intestinal mucosa of goat kids in areas with and without Peyer's patches as demonstrated with the everted sleeve method. *Comp. Immunol. Microbiol. Infect. Dis.* **28:**223–230.

62. **Sigurðardóttir, Ó. G., C. M. Press, F. Saxegaard, and Ø. Evensen.** 1999. Bacterial isolation, immunological response, and histopathological lesions during the early subclinical phase of experimental infection of goat kids with Mycobacterium avium subsp. paratuberculosis. *Vet. Pathol.* **36:**542–550.

63. **Sigurðardóttir, Ó. G., M. Valheim, and C. M. Press.** 2004. Establishment of Mycobacterium avium subsp. paratuberculosis infection in the intestine of ruminants. *Adv. Drug Deliv. Rev.* **56:**819–834.

64. **Sreevatsan, S., P. Escalante, X. Pan, D. A. Gillies II, S. Siddiqui, C. N. Khalaf, B. N.**

**Kreiswirth, P. Bifani, L. G. Adams, T. Ficht, V. S. Perumaalla, M. D. Cave, J. D. A. van Embden, and J. M. Musser.** 1996. Identification of a polymorphic nucleotide in *oxyR* specific for *Mycobacterium bovis*. *J. Clin. Microbiol.* **34:**2007–2010.

65. **Sreevatsan, S., X. Pan, Y. Zhang, B. N. Kreiswirth, and J. M. Musser.** 1997. Mutations associated with pyrazinamide resistance in *pncA* of *Mycobacterium tuberculosis* complex organisms. *Antimicrob. Agents Chemother.* **41:**636–640.
66. **Valentin-Weigand, P., and R. Goethe.** 1999. Pathogenesis of Mycobacterium avium subspecies paratuberculosis infections in ruminants: still more questions than answers. *Microbes Infect.* **1:** 1121–1127.
67. **Whan, L., I. R. Grant, and M. T. Rowe.** 2006. Interaction between Mycobacterium avium subsp. paratuberculosis and environmental protozoa. *BMC Microbiol.* **6:**63.
68. **Whitlock, R. H., and C. Buergelt.** 1996. Preclinical and clinical manifestations of paratuberculosis (including pathology). *Vet. Clin. N. Am. Food Anim. Pract.* **12:**345–356.
69. **Whitlock, R. H., S. J. Wells, R. W. Sweeney, and J. Van Tiem.** 2000. ELISA and fecal culture for paratuberculosis (Johne's disease): sensitivity and specificity of each method. *Vet. Microbiol.* **77:**387–398.
70. **Whittington, R. J., A. F. Hope, D. J. Marshall, C. A. Taragel, and I. Marsh.** 2000. Molecular epidemiology of *Mycobacterium avium* subsp. *paratuberculosis:* IS*900* restriction fragment length polymorphism and IS*1311* polymorphism analyses of isolates from animals and a human in Australia. *J. Clin. Microbiol.* **38:**3240–3248.
71. **Whittington, R. J., I. B. Marsh, and L. A. Reddacliff.** 2005. Survival of *Mycobacterium avium* subsp. *paratuberculosis* in dam water and sediment. *Appl. Environ. Microbiol.* **71:**5304–5308.
72. **Whittington, R. J., I. B. Marsh, P. J. Taylor, D. J. Marshall, C. Taragel, and L. A. Reddacliff.** 2003. Isolation of Mycobacterium avium subsp. paratuberculosis from environmental samples collected from farms before and after destocking sheep with paratuberculosis. *Aust. Vet. J.* **81:**559–563.
73. **Whittington, R. J., D. J. Marshall, P. J. Nicholls, I. B. Marsh, and L. A. Reddacliff.** 2004. Survival and dormancy of *Mycobacterium avium* subsp. *paratuberculosis* in the environment. *Appl. Environ. Microbiol.* **70:**2989–3004.
74. **Whittington, R. J., and E. S. Sergeant.** 2001. Progress towards understanding the spread, detection and control of *Mycobacterium avium* subsp. *paratuberculosis* in animal populations. *Aust. Vet. J.* **79:**267–278.

# EMERGING TOOLS AND ISSUES

# MOLECULAR METHODS TO STUDY COMPLEX MICROBIAL COMMUNITIES

*Dionysios A. Antonopoulos, Jennifer M. Brulc, Anthony Yannarell, and Bryan A. White*

# 18

The use of modern molecular ecology techniques based on nucleic acid sequence comparisons and genomics-based approaches provide molecular characterization of host-microbe interactions that can inform predictions of virulence and identify epidemiological patterns of transmission. The ecology of microorganisms impacts food safety in several capacities, including, but not limited to, the distribution and spread of pathogens, pathogen-host interactions, pathogen-commensal microbial interactions, and the effects of environment on these interactions (e.g., antibiotics and animal housing strategies). Moreover, the use of molecular methods for monitoring populations of microbes and important gene sets (e.g., antibiotic resistance genes) allows scientists to design testable hypotheses and intervention strategies. The importance of understanding the complexity of food safety host-microbe interactions has been eloquently described by Gilmore and Ferretti (46) as "the thin line between gut commensal and pathogen." Furthermore, they pose seminal questions related to the acquisition of virulence factors, antibiotic resistance genes, and traits that allow host colonization, as well as the potential role of commensal bacteria as a reservoir for these traits.

The gastrointestinal tract (GIT) of production animals is a major reservoir of food-borne microbes that have the potential to cause disease. It is estimated that the complex intestinal microbiome of mammals contains as many as 500 to 1,000 different species and reaches population densities as high as $10^{12}$ microorganisms, depending on the site being sampled (54, 122, 134, 140, 158). Furthermore, each microbial species can be present at vastly different concentrations. For example, microbes such as the *Bacteroides* spp., which are strict anaerobes, are present in much higher concentrations ($\sim 10^{10}$ to $10^{12}$ per g of feces) than other microbes such as *Escherichia coli* ($\sim 10^4$ to $10^5$ per g of feces). Most GIT microbes have not been grown in culture, and their identities remain unknown. Cultivation-based approaches have been hampered by numerous factors, including unknown growth require-

*Dionysios A. Antonopoulos,* Institute for Genomics and Systems Biology, Argonne National Laboratory, Argonne, IL 60439. *Jennifer M. Brulc,* Department of Animal Sciences, The Institute for Genomic Biology, University of Illinois at Urbana-Champaign, Urbana, IL 61801. *Anthony Yannarell,* The Institute for Genomic Biology, University of Illinois at Urbana-Champaign, Urbana, IL 61801. *Bryan A. White,* Departments of Animal Sciences and Pathobiology, Division of Nutritional Sciences, and Institute for Genomic Biology, University of Illinois at Urbana-Champaign, Urbana, IL 61801.

*Food-Borne Microbes: Shaping the Host Ecosystem,* Edited by L.-A. Jaykus, H. H. Wang, and L. S. Schlesinger, © 2009 ASM Press, Washington, DC

ments, the use of selective or habitat-simulating media, the stress of the cultivation/isolation protocol and the need for strictly anaerobic or microaerophilic conditions, and a lack of knowledge about environmental requirements, which are mediated by interactions with other microbes or the host. This is in spite of the relatively high culturability of GIT microbes compared to those from other microbiomes (3, 150, 165). Therefore, over the past 20 years, molecular technologies that are culture independent have been implemented in studies of the microbial ecology of the GIT.

These molecular methods have allowed a more detailed and defined assessment of the abundance and diversity of the microbes present in the GIT, as well as their metabolic potential and the competitive and synergistic relationships between the host animal and the resident microbial consortium. Although the extent of known microbial diversity is expansive (the number of major taxonomic divisions [phyla] within the bacterial domain has grown from 12 in the mid-1980s [39] to more than 80 today [40, 59]), the tools based on this vantage point are somewhat limited by broad-range PCR primer bias and the specificities of hybridization probes used for organism detection. Despite these limitations, molecular approaches to studying microbial communities provide a wealth of data that can be analyzed, using classical concepts from macroecology, to measure microbial diversity. Diversity measures describing both the composition ("richness") and structure ("evenness") of communities can be used to understand the role of the organisms beyond simple identification. This chapter provides a survey of established molecular approaches as well as emerging ones for studying microbial communities in gastrointestinal systems.

## ISOLATE FINGERPRINTING

While molecular studies suggest an abundance of not-yet-cultured microorganisms in natural communities, the use of molecular typing methods for single isolates from cultivation studies remains an important epidemiological tool for studying the relationship between pathogens, commensals, and the host. Comparison of complete genome nucleotide sequences is considered the ideal approach for determining precise estimates of genetic relatedness (147). However, for large genomes, or when a large number of isolates of a pathogen are to be compared, the time and cost expenditures of nucleotide sequencing can be prohibitive. Nonetheless, comparative genomics has been applied to numerous food-borne microbes, including *Salmonella enterica* (28) and *Campylobacter jejuni* (38).

Whole-genome typing methods are cost-effective alternatives for assessing genetic relatedness. These molecular techniques provide more specificity than phenotypic identification of culturable microbes and include pulsed-field gel electrophoresis (PFGE), rep-PCR (see below), multilocus sequence typing (MLST), and octamer-based genome scanning (OGBS). In order for a genotyping method to be acceptable for making inferences about the source and transmission of a food-borne microbe, it should be able to accurately represent the broad spectrum of genetic diversity within populations (14, 43, 77, 105, 106, 111). It should also be able to differentiate closely related isolates so that transmission links can be identified accurately.

PFGE, which separates DNA fragments produced by digestion with restriction enzymes targeting specific nucleotide sequences in accordance with their size, has been a standard whole-genome typing method for food-borne microbes (48, 105, 142). Another genotyping method used in recent years for *Enterobacteriaceae* has been a PCR-based technique using primers targeting repetitive sequences (rep-PCR), which also produces numerous DNA fragments that are separable by electrophoresis (25, 48, 60, 85, 133, 154). The rep-PCR approach collectively refers to different repeating sequences found in *Enterobacteriaceae* that are used as targets, including the 35- to 40-bp repetitive extragenic palindromic (REP) sequence for REP-PCR, the 124- to

127-bp enterobacterial repetitive intergenic consensus (ERIC) sequence for ERIC-PCR, and the 154-bp BOX element for BOX-PCR. For all three variations, a genetic profile based on whole-genome fragmentation patterns can serve as the basis for estimating genetic relatedness and characterizing isolates of a pathogen (120).

MLST is another PCR-enabled genotyping method that has found widespread use for subtyping a variety of pathogens. The technique is an extension of multilocus enzyme electrophoresis and was first described in 1998 using *Neisseria meningitidis* as an example (90). PCR primers specific for internal regions of several conserved housekeeping genes are used to generate multiple sequences for comparing closely related organisms. These regions are short enough to be covered by a single sequencing reaction. Since the multiple gene targets being analyzed are conserved, a high level of discrimination is possible, facilitating binning of closely related organisms. The number of online MLST databases has grown quickly in an effort to distribute information easily without requiring the exchange of potentially harmful organisms themselves (89).

OBGS, a powerful method enabled by genome sequence information, was developed as a result of determining the complete genome sequence of *Escherichia coli* (64). In the case of *E. coli* K-12, there are over 150 different overrepresented oligomers whose occurrence is skewed to one strand of the genome (121). Of these, 23 are octamers with frequencies of occurrence ranging from 515 to 867 times within the genome. To perform OBGS, Kim et al. (64) prepared fluorescent PCR primers specific to octamers biased to the leading strand and unlabeled primers for octamers biased to the lagging strand (i.e., complements of the leading-strand octamers). Following PCR, amplified products are resolved on denaturing polyacrylamide gels, permitting size resolution at the single-nucleotide level. The OBGS method revealed a bifurcation within *E. coli*, with most isolates obtained from animals and farm environments falling within one "lineage" and most isolates from infected persons falling into a second lineage. Some of these polymorphisms have been recovered, cloned, and sequenced (65). Moreover, this approach has been recently used to design a lineage-specific polymorphism assay (six markers) that can distinguish different subpopulations (162). For these reasons, OBGS offers a powerful way not only to examine population structure but also to identify genome alterations (deletions, rearrangements, or insertions) in a large number of bacteria.

## COMMUNITY FINGERPRINTING

Analyzing the microbial composition from an environmental sample has traditionally relied on culture-based techniques. However, culture-based techniques not only fail to adequately characterize the consortia of complex organisms inhabiting a particular environment but also can be arduous and time-consuming. Researchers using molecular community fingerprinting methods can rapidly infer the genetic composition or community structure following the analysis of a region of amplified DNA. These methods can be partitioned into schemes that resolve individual organisms by fragment size and number (terminal restriction fragment length polymorphism [T-RFLP], amplified rDNA restriction analysis [ARDRA], ribosomal intergenic spacer analysis [RISA] and automated RISA [ARISA], amplified fragment length polymorphism [AFLP], and random amplified polymorphic DNA [RAPD]) or by nucleotide sequence composition (denaturing gradient gel electrophoresis [DGGE] and temperature gradient gel electrophoresis [TGGE]). Since most of the indigenous microbes in the GIT of mammals have not been cultivated in the laboratory, studies to measure community composition must depend on techniques that do not require growth of the microbes. Of the many genetic tools that have been developed in the last decade or so to measure phylogeny, analysis of the gene encoding the small-subunit (SSU, or the 16S) rRNA has emerged as the leading tool. The advantage of using a detection protocol based on the SSU rRNA gene is that this gene is present in all organisms. The SSU rRNA gene is composed of ~1,500 bp and contains conserved sequences that flank more unique variable regions (V1 to V9 [9]). These variable regions can be utilized as a means to distinguish between different microorganisms in complex mixtures.

All community fingerprinting methods begin with extraction of total genomic DNA from the sample. Direct DNA extraction followed by PCR amplification and subsequent restriction digestion is the basis for T-RFLP. T-RFLP was first implemented to monitor changes in aquatic community structure (82). Since its development, there have been many applications of this technique, including exploring microbial communities in soils (36, 104), ocean water and sediments (20, 149), wastewater sludge (119), and gastrointestinal environments (32, 58). In T-RFLP, a region of DNA is PCR amplified with fluorescently labeled primers that result in end-labeled amplicons. (The gene encoding the SSU rRNA is most commonly used as the template, but studies have alternatively used a functional gene of interest [18]). These PCR products are then digested with a restriction enzyme and separated by size (along with a comigrating internal size standard) using a capillary DNA sequencer. The resulting pattern of T-RFLP fragments corresponding to individual phylotypes can then be correlated to species richness (number of peaks) and abundance (the area bound by each peak) within the environmental sample.

T-RFLP fragments can also be compared to databases of known SSU rRNA gene terminal fragments to infer organism identification. Although direct sequence information cannot be determined from the results alone, virtual digests of database sequences can be performed in silico and matched with the recovered T-RFLP pattern. Care must be taken with interpreting T-RFLP patterns with in silico (virtual) digests because a single peak may correspond to more than one individual microorganism, many isolates of the same species can have different banding patterns, and one organism may have multiple bands due to multiple copies of the rRNA operon (55). Virtual digests are potentially unreliable, since organisms that are quite similar in sequence may yield fragments of the same size, even when using two restriction enzymes for greater resolution. Additionally, T-RFLP is influenced by the same biases as PCR and electrophoresis in general. The absence of a specific peak does not verify the absence of an organism, and similarly, the presence of a peak does not necessarily correspond to the presence of a species (8). Therefore, primer selection, restriction enzyme selection, and the number of amplification cycles are crucial parameters to take into consideration when performing this method.

Similar to T-RFLP, ARDRA is another PCR-based technique that begins with amplification of the SSU rRNA gene. However, all of the resulting size fragments, not solely the terminal fragments, are analyzed on a gel following restriction digestion of the PCR products. Shifts in population structure can be determined via analysis of the fragment pattern that is specific to a particular community. Further organism identification can be determined via hybridization with genus-specific probes. ARDRA, however, is severely limited because there are only a few restriction enzymes that can be used to obtain a unique banding pattern for each community and also generate bands large enough to be easily resolved using electrophoresis (113). The technique has been successfully applied in creating a fingerprint for communities found in groundwater and soils (95, 98).

RISA and ARISA (16, 37) use the intergenic spacer region between the genes encoding the SSU and large-subunit rRNAs. A forward primer and reverse primer that bind to conserved regions in the two genes are used to amplify the intervening region. PCR products are then resolved on a gel (RISA) or by capillary gel electrophoresis (ARISA), from which the sizes of whole amplicons are analyzed. The amplified region of DNA can vary anywhere from 50 bp to 1.5 kb for each species, providing enhanced resolution compared to SSU rRNA gene amplification alone (113). Amplifying a larger amount of the adjacent conserved SSU rRNA gene sequence can provide an added level of taxonomic identification, and bands can even be subsequently excised, cloned, and sequenced (44). One of the most significant limitations in this method is the lack of intergenic spacer sequences present in the database compared with an overwhelming number of available SSU rRNA gene sequences (compare 9,462 intergenic spacer sequences in GenBank with 440,891 SSU rRNA gene sequences in Ribosomal Database Project release 9.55, 11 October 2007).

ARISA has been applied to community profiling of forest soil and freshwater lakes and for observing shifts in bacterial communities due to increasing concentrations of a fungicide (16, 37). The technique, in concert with DGGE, has also been used to identify differences in rumen microbial populations. Specifically, comparisons were made between liquid fractions, weakly fiber-associated fractions, or fiber-adherent fractions in sheep that were fed a fiber diet or a fiber-grain mixed diet (72). Although in this study DGGE yielded significantly more bands than ARISA, both techniques demonstrated that microbial populations (represented by their respective banding patterns) were unique to each individual animal yet could be clustered according to diet. ARISA PCR products were cloned, yielding 359 unique phylotypes from 514 clones, and rarefaction analysis indicated that approximately 50% of the diversity had been achieved within these ARISA-based libraries. Animals consuming a fiber diet yielded a greater number of phylotypes than the animals on a fiber-grain mixed diet. Additionally, diversity measurements (indices) indicated that fiber-adherent fractions possessed more complex and distinct populations than did those found in the liquid fractions.

AFLP is a technique that begins with genomic DNA digestion followed by PCR amplification of selected sets of restriction fragments. After genomic DNA has been digested with two different enzymes (most commonly EcoRI [6-bp cutter] and MseI [4-bp cutter]), adaptors with sticky ends complementary to the restriction sites are ligated to the ends of the fragments. The fragments can then be amplified with primers that not only are specific to the adaptor sequences but also include an extension of a single nucleotide base at the end of each primer. Amplicons are subjected to another round of amplification with fluorescently end-labeled primers, enabling the fragments to be read on a capillary DNA sequencer, and the extension of two more nucleotide bases at the ends of the secondary amplicons further increases specificity (97). Fragment profiles can then be interpreted with software used for community fingerprint identification (152).

The use of fluorescently labeled primers increases the throughput of this method because a capillary DNA sequencer can be used for detection of amplicons produced using up to four different primers at one time. Sets of restriction fragments can also be analyzed without any prior sequence knowledge. However, this technique cannot be further expanded because of instrument restrictions, and because different primer combinations can greatly affect the quality of the AFLP profiles (97). Often bands of the same length can represent more than one stretch of DNA, and there can be pattern differences between electrophoresis runs (13).

AFLP has been successfully applied to analyzing intestinal bacterial strains in pure culture and in isolates from pooled bovine fecal samples with high reproducibility (19). The technique has also been used for taxonomic discrimination in other bovine intestinal studies and to evaluate isolates in clinical studies, and it has been commonly used for molecular typing studies in crop science (63, 123).

PCR amplification of genomic DNA using short random primers (~10 bp) combined with a low annealing temperature generates different-sized products using the RAPD approach (155, 157). These PCR products are then resolved on a gel, and the pattern is analyzed by computer software. Advantages to RAPD include the absence of a need for a priori sequence knowledge as well as the ability to distinguish polymorphisms between closely related strains. However, band intensity, quantity of DNA loaded, and gel-to-gel variation can contribute to poor reproducibility of RAPD patterns when analyzing electrophoresis-based results using a binary (presence/absence) scoring system (102). Also, RAPD fails to provide any taxonomic or sequence information about the microbial communities, and the number of RAPD fragments from a mixed microbial community does not necessarily correlate to microbial richness or diversity (113). However, RAPD is attractive for many microbial ecology studies since the requirement for high-quality input DNA is not as stringent as for other techniques. The RAPD approach has been used to observe differences in anaerobic and aerobic zones in aquatic communities (41).

DGGE and TGGE are two of the most common community fingerprinting techniques utilized in microbial ecology (99). In these methods, total genomic DNA is extracted and SSU rRNA gene fragments are amplified by PCR, after which DGGE or TGGE is applied to separate amplicons of the same size in a polyacrylamide gel. By increasing the amounts of denaturing agents as a gradient in the gel, DNA strands progressively dissociate as they move through the gel, thereby decreasing their mobility (sequences with a higher GC content will have more hydrogen bonding and therefore will not dissociate as easily). The location of the immobilized fragments is ultimately dependent on their sequence composition. The gels can be silver stained to detect bands in low abundance, or oligonucleotide probes can be hybridized to recovered fragments to detect the presence of specific organisms of interest (100).

With the addition of known standards and the inclusion of a GC clamp at one end of the DNA fragment, one can analyze and compare numerous samples simultaneously on a gel and allow for better separation among bands (68). One advantage of this method relative to other fingerprinting approaches is the ability to excise and sequence bands of interest, enabling construction of a clone library specific to a particular community. However, DGGE will only separate sequences 200 to 700 bp in length, somewhat limiting the taxonomic information that can be recovered using this technique. Choice of stain is also important due to differences between methodologies (the use of silver staining eliminates the possibility of subsequent hybridization) and the ability to resolve less dominant members of the community (100). Additionally, bands with different sequences have been shown to migrate to the same area, and the reproducibility of DGGE is low.

Despite these drawbacks, DGGE has been used successfully to observe changes in bacterial populations in gastrointestinal environments and manure-treated or fertilizer-treated sediment samples, as well as continuous population changes in food animal species (127, 128, 135, 165). Using the example of the porcine gastrointestinal microbial community, SSU rRNA gene techniques were used to observe shifts in the community after introduction of an exogenous *Lactobacillus reuteri* strain (127). Initial DGGE profiles indicated that each individual animal maintained a stable, unique gastrointestinal microbial community. Primers specific to a SSU rRNA gene variable region (V1) were then used to restrict the DGGE analysis to a selected group of lactobacilli to monitor the effects of dosing. *L. reuteri* was distinguishable within V1 banding patterns based upon its relative band migration distance and sequence determination. The V1 DGGE amplicon pattern observed for fecal lactobacilli was much simpler than whole-community DGGE banding patterns. These studies showed that DGGE can be applied effectively to the porcine gastrointestinal microbial community to monitor changes in bacterial populations and that the use of primers specific for a selected group of organisms (*Lac-*

*tobacillus* in this case) permitted a detailed look into interspecies relationships and abundance as influenced by various environmental factors. Even though this approach has been applied to the porcine food aminal model, there are limited studies using samples from either cattle or poultry. One study analyzed the rumen microbial community of cattle maintained on two different diets, corn and hay, and found that bacterial populations clustered according to the diets (67), whereas a study using poultry cecal samples showed both qualitative and quantitative differences in the cecal microbial community in birds reared under similar conditions (164).

## SSU rRNA GENE SEQUENCING

Although fingerprinting methods can provide important information regarding microbial community structure, they lack the resolution to make specific identifications of community members in a high-throughput manner. The most common approach for identifying community members is by the construction of SSU rRNA gene libraries from a given sample (although non-SSU rRNA gene targets have also been used for characterizing populations of microorganisms in complex communities [30]). As with the community fingerprinting approaches described above, extracted genomic DNA samples serve as the template for PCR, which is then followed by cloning and sequence determination. The resulting library of sequences thereby represents the organisms present in the original sample. In most cases, "full-length" SSU rRNA gene amplicons (~1,500 bases) are cloned, but libraries can also be generated using individual variable regions (V1 to V9) as sequence tags, or a combination of them (V7 to V9). In each of these cases, library sequences can then be compared with previously characterized sequences available in several online databases (The Ribosomal Database Project [22; http://rdp.cme.msu.edu/], Greengenes [31; http://greengenes.lbl.gov/cgi-bin/nph-index.cgi], and the European Ribosomal RNA database [http://www.psb.ugent.be/rRNA/index.html]). These databases contain sequences both from pure cultures and from other culture-independent SSU rRNA gene libraries.

Although the chance of finding exact matches with sequences from well-described, cultured organisms is minimized when working from complex communities, analytical tools have been developed that can be used to at least make inferences about higher-level taxonomic assignments. Many tools used to identify species and estimate diversity are currently based on the definition of an operational taxonomic unit (OTU) or phylotype, which is not standardized. Indeed, a range of sequence differences for defining a species (anywhere from 1 to 5%) have been used to make comparisons of SSU rRNA gene library sequences. This can have dramatic effects on statistical comparisons of data sets from different experiments and environments if different OTU definitions have been used.

Cloning and sequencing of SSU rRNA genes to describe microbial communities have been widely applied to the intestinal tracts of cattle (137, 138, 156), swine (75), and poultry (87, 88, 164). SSU rRNA gene sequence inventories from the rumen of cattle have demonstrated that there is a large microbial component that has yet to be cultured (67, 79–81, 101, 132, 136, 138, 139, 156) and that a high proportion of the fibrolytic population has not been cultured or described (70, 72, 101, 132). A meta-analysis of these libraries predicted a total of 341 OTUs when using 95% sequence similarity as the identity criterion for the definition of the OTU (33). Similar OTU estimates were derived from an analysis of microbial populations in the porcine GIT as part of the Pig Intestinal Molecular Biology Project that led to the most extensive SSU rRNA analysis of the porcine GIT to date (75). Leser et al. (75) described the complete SSU rRNA sequences of 4,270 clones from 52 samples isolated from various portions of 24 different porcine GITs. Using a 97% similarity criterion, they were able to assign the sequences to 375 OTUs. These estimates are not available for poultry, even though this approach has been used with broilers and laying hens (87, 88, 164).

However, one of the most serious limitations to the use of SSU rRNA gene libraries is the labor and cost involved in generating clones and for sequencing. Therefore, the upper limit of these libraries is about 5,000 sequences, with many libraries consisting of less than 1,000 sequences. As a consequence, not all of the predicted OTUs are recovered in these studies. For example, only 177 of the predicted 341 OTUs were recovered in the meta-analysis of the rumen SSU rRNA gene libraries (33). This becomes even more troubling when one considers that other calculators of microbial diversity would estimate 16,000 phylotypes (at a 95% sequence identity) in the rumen (24, 33).

Approaches for increasing sequencing depth (throughput) are now being applied to microbial community structure projects to increase coverage of the diversity present. One such approach is serial analysis of V1 ribosomal sequence tags (SARST-V1 [103, 104]) or V6 ribosomal sequence tags (SARST-V6 [71]). In both of these approaches, concatemers of ribosomal sequence "tags" are generated and cloned. Thus, one can generate larger sequence data sets from a fewer number of clones, increasing the sequence depth (the average number of ribosomal sequence tags per clone has been six). For example, this approach was used to characterize the rumen microbiome. Yu et al. analyzed 190 clones which contained a total of 1,055 ribosomal sequence tags and identified 236 of the predicted 353 OTUs present in the rumen microbiome under study (163). This approach represents a cost-effective method to increase sequence coverage using Sanger-based sequencing.

Recent advances using the massively parallel DNA sequencing capacity of 454 Life Sciences pyrosequencing-based technology are now being applied to SSU rRNA gene analysis (93). Sogin et al. (131) recently used this technology to develop a global, in-depth description of the diversity of deep-sea microbes by increasing the number of sampled PCR amplicons in environmental surveys by orders of magnitude (71, 131). In this case, ribosomal sequence tags from the V6 regions of the SSU rRNA-encoding gene are generated by PCR and the amplicon mixture is then directly sequenced (thereby circumventing cloning). The analysis of 118,000 sequence tags resulted in the discovery of a large number of diverse, low-abundance OTUs that populate what Sogin calls a largely unexplored "rare biosphere." More recently, this approach has also been applied to soil microbiomes by Roesch et al. (117). Using pyrosequencing of ribosomal sequence tag rRNA gene libraries, Roesch et al. estimated that the maximal diversity of bacterial phylotypes in soil ranged from 26,140 to 53,533.

While there is concern that the error rate and short sequence reads that result from pyrosequencing may hinder the usefulness of this next-generation sequence technology, recent reports suggest otherwise. First, Huse et al. (61) calculated a 99.5% accuracy rate in unassembled sequences from the Roche Genome Sequencer GS20 and identified factors that can be used to remove a small percentage of low-quality reads, improving the accuracy to 99.75% or better. Therefore, the use of pyrosequencing in molecular ecological applications can surpass the accuracy of traditional capillary methods. Second, Liu et al. (83) have shown that SSU rRNA gene surveys using reads of 100 to 200 nucleotides (pyrosequencing) yield the same clustering of full-length sequences and therefore allow substantial resolution of biologically meaningful similarities and differences in microbiome samples. Indeed, they recommend that sequencing efforts be directed towards gathering more short sequences rather than fewer long sequences (83), and a subsequent study using multiplex pyrosequencing generated 437,544 SSU rRNA gene tags, nearly as many as have already been generated by Sanger sequencing (50). This approach is sure to be applied to food animal species and will be an important tool for detecting minor species that are below the sequencing depth of current studies.

## HYBRIDIZATION-BASED TECHNOLOGIES

There are several approaches that can be used to determine microbiome community structure that use nucleic acid hybridization as the defining principle. Perhaps the most commonly used method is dot blot hybridization. A phylogenetically specific DNA probe is hybridized to a sample preparation of SSU rRNA genes or rRNA genes bound to a nylon membrane. This method results in a quantitative measure of a target (SSU rRNA or rRNA genes) present in a microbiome relative to the total concentration of SSU rRNA or rRNA genes. While quantitation by this method is accurate, and the procedure has been validated and widely used (114), there are several issues that need to be taken into account when using this approach. First, SSU rRNA gene sequence information for organisms that are being specifically quantified is necessary for probe development. Second, the quantitative power of the method is somewhat compromised, as the measurement is relative to total SSU rRNA or rRNA genes; e.g., the concentration of SSU rRNA (number of ribosomes) will change according to growth rate and stage of growth. Additionally, the numbers of copies of rRNA genes on the genome also vary among microbes (66). These factors are important considerations in the interpretation of dot blot-based data. More recently, the widespread use of real-time PCR for monitoring specific populations has increased as the technology has become more affordable and available.

An additional approach for identifying differences between microbial communities is suppressive subtractive hybridization (SSH). SSH enables the identification of unique microbial DNA sequences present in one microbiome sample relative to another by subtracting the ones that are common to both and amplifying the unique sequences. SSH is a PCR-based technique that was originally developed to amplify rare mRNA transcripts for full representation in mRNA libraries from eukaryotic tissues (15). An amended version of SSH utilizing specially designed PCR adaptors to minimize the amplification of small DNA fragments and enrich for larger, differential ones was demonstrated by Akopyants et al. (1) in comparing the genomes of two strains of *Helicobacter pylori*. Mau and Timmis (96) have used SSH to design habitat-based specific 16S rRNA gene probes. This approach resulted in probes for minor and abundant microorganisms present in the habitat of choice. Furthermore, these probes targeted phylogenetically isolated taxa that are poorly characterized, showing that this approach has the potential of identifying novel organisms and physiologies from as-yet-uncultured microbes in the microbiome.

One advance that has expanded the power of hybridization-based strategies is the use of rRNA-targeted oligonucleotide microarray slides called PhyloChips (21, 35, 69, 110, 130, 148). These DNA-based microarrays allow for large sample throughputs and, depending on the number of sequences on the chip, a highly parallel detection of much of the diversity present in complex microbial communities. Multiple probes can be designed to detect target organisms at the same or different taxonomic levels. The advantages of this approach are that the resolution of organism identification is high and that many microorganisms of interest are detected in a single assay by multiple probes with various levels of specificity. However, as with dot blot hybridization, accurate quantification from these multiple hybridization signals can be problematic. Nonetheless, the recent development of high-density PhyloChips containing nearly 9,000 taxa highlights the potential of this approach (17).

Fluorescence in situ hybridization (FISH) is another technique that has been applied to microbial community analysis (165). This culture-independent method uses SSU rRNA gene probes in combination with microscopy (epifluorescent light or confocal laser) or flow cytometry for direct quantitation of a phylogroup. Although quantitative, this method has several limitations. First, the lower limit of detection is approximately $10^6$ cells per g of intestinal contents (165). Second, FISH is limited to a few probes per experiment and detects only known phylotypes (i.e., based on the specificity of the probe being used). Furthermore, there are issues with the accessibility of the probe target due to cell permeability, as well as rRNA genome and ribosome copy numbers.

## NUMERICAL ANALYSIS OF MOLECULAR DATA

The goal of studying microbial communities is to determine which microbes are present in different habitats and environments, or under different experimental conditions, and to use these data to make inferences about the functions, interactions, and basic ecology of the organisms present in these systems. Most of the approaches described earlier can be broadly viewed as belonging to two major analytical classes: OTU-based approaches and community-based approaches. Each of these classes focuses on a different level of biological organization, each has its own inherent assumptions and biases, and each lends itself to a different type of numerical analysis.

OTU-based approaches are those that focus primarily on distinguishing one organism or population of organisms from others, classifying each into a unique strain or OTU. Examples include all of the molecular genotyping methods, as well as SSU rRNA gene clone libraries, SARST, sequence tag-based pyrosequencing, and many of the hybridization techniques (dot blot, microarray, and FISH). Because these techniques generate DNA sequence data or fingerprint data derived directly from the genomes of isolates, they are excellent for establishing phylogenetic relationships. Obtaining DNA sequences allows access to a variety of phylogenetic tree-building methods, including neighbor joining, maximum likelihood, parsimony, and Bayesian inference (57). Some of the genotyping methods also allow for the construction of phylogenetic trees, as long as one can assume that the differences in the fingerprinting patterns translate directly into evolutionary differences between the strains. In addition to uncovering the evolutionary relationships of the organisms under study, phylogenies are used in microbial ecology to identify and differentiate environmental isolates or uncultivated organisms present in samples.

From a community-level perspective, the principal drawback of many OTU-based approaches is the potential for undersampling of the true taxonomic diversity of the sample. One can never be sure that all of the species in a sample have been accounted for, and therefore caution must be used when making inferences about community composition based on these types of data. Numerical techniques such as rarefaction can help determine the adequacy of the sampling effort, and there are a variety of statistical and nonparametric procedures that can estimate the true OTU richness of a sample based on a subsampling of populations (124). This problem of undersampling may be alleviated in highly simplified, taxon-poor communities or by techniques, such as sequence tag-based pyrosequencing or PhyloChips, that generate large amounts of sequence data.

Community-based approaches are those primarily focused on distinguishing between assemblages of microorganisms. This is primarily the province of the community fingerprinting techniques (e.g., DGGE, T-RFLP, and ARISA), although certain applications of hybridization-based studies (e.g., PhyloChip and GeoChip) and large-clone library studies can produce community-level data as well. The advantage of community-based studies, especially the various fingerprinting techniques, is the speed and low cost with which a large number of samples can be analyzed. This high sample throughput makes these techniques ideal for the generation of time-series data about microbial community dynamics, for comparison of many samples from different environments, or for studies of large spatial extent or high sample replication. The immense sample processing power comes with the trade-off that the information gained about any particular organism in the samples will be less detailed than that provided by the OTU-based approaches.

For community comparison with fingerprinting, it must be assumed that each variation in the fingerprint pattern corresponds directly with variation in the composition of the microbial communities under consideration. Each band or peak in the profile is taken to represent a distinct microbial phylotype, and so the data from fingerprinting studies are analogous to the species lists produced in plant and animal ecology. Thus, the data are typically analyzed in a traditional multivariate framework using principal-component analysis, correspondence analysis, or principal-coordinate analysis (62, 74). These analyses seek to summarize the variation between different communities and the correlation structure between the occurrences of different taxa. They do this mathematically by representing sites and species as points on a graph of low dimensionality (ideally two or three axes), where sites of similar species composition, and species with correlated distributions, cluster together. The causes of the resultant patterns must be inferred, either indirectly by correlating explanatory data with the major axes of variation or directly by incorporating such data into the analysis by using the so-called "canonical" forms, i.e., redundancy analysis or canonical correspondence analysis (62, 74). For molecular fingerprinting studies, the most widely used multivariate analysis is nonmetric multidimensional scaling because of the flexibility and relative insensitivity of this method to a variety of underlying data distributions. Perhaps the greatest drawback of the community-based approaches is the use of molecular proxies (the fingerprints) for actual species abundance data. One can never be entirely certain that each band or peak corresponds to a single species across all samples, and thus, some changes in community composition may go entirely unnoticed by these studies.

As mentioned above, certain molecular approaches do not fit strictly into one or the other of these categories, and this is especially true of large clone libraries, metagenomic surveys, sequence tag-based pyrosequencing, and microarray hybridization. These approaches seek to produce phylogenetically informative data about an entire microbial community, and thus they could be considered a combination of both OTU- and community-based analyses. There is a growing numerical tool kit designed to analyze phylogenetic data in a community-centered context. In addition, neither phylogenetic trees nor multivariate analyses are implicitly hypothesis-testing techniques, and many of the new numerical approaches seek to include hypothesis testing as a primary focus. Many of these approaches utilize phylogenetic relatedness as a means to weight differences in community composition, under the assumption that two communities with closely related organisms ought to be considered more similar than two others with distantly related organisms, even if both sets of communities have the exact same number of species in common. The approach taken by LIBSHUFF (129) and its integral form, ∫-LIBSHUFF (126), examines the mutual "coverage" of clone libraries from different samples and determines the statistical likelihood that the resultant clone libraries would have resulted from random sampling if both samples had the same underlying community composition.

Other approaches seek to establish the significance of the correlation between the OTU phylogeny and sample distribution, as described by Martin (94) and implemented by various programs such as Treeclimber (125) and UNIFRAC (86). The interest in merging phylogenetic data with ecological interpretation has also led to techniques for determining whether the observed composition of different communities was likely to have resulted from competition or from environmental filtering (52, 153).

## EMERGING TECHNIQUES FOR LINKING MICROBIAL COMMUNITY STRUCTURE TO FUNCTION

The greatest challenge to microbial community analysis is linking the community structure at the phylogenetic level with its metabolic potential and subsequent functionality in the environment being studied. An extensive suite of molecular approaches developed over the past decade has enabled the metagenomic study of uncultured microorganisms. In several cases, these have been the direct outcome of the use of SSU rRNA gene targets utilized to study microbial diversity (109, 143, 145). One example has been the application of FISH in combination with microautoradiography, which permits both the detection of microbial populations and the measurement of substrate utilization within a microbial community (47, 103). The metabolic potential of microbiomes can also be assessed by the use of stable isotope labeling in which metabolically active species are preferentially labeled (92). Coupling this enrichment with the approaches described above can link the phylogenetic content to the metabolic potential (107). Indeed, the combined use of stable isotope labeling with in situ reverse transcription-PCR (2–7, 23, 56) might well be a key technique for linking community structure with metabolic activity.

Construction of metagenomic libraries using bacterial artificial chromosome vectors (12, 118) is another route for gaining insight into the functional nature of mixed microbial samples. Beja et al. uncovered large genomic fragments (~80 kb) from a diverse population of largely uncultivated microorganisms and assigned functional properties to many of the genes on a 60-kb sequence from an archaeal clone (12). Specifically, an association of three genes possibly involved in a novel membrane-associated proteolytic system was observed (12). Direct cloning approaches have also been applied to construct soil DNA libraries to facilitate screening for specific enzyme activity (53, 91) and to bioprospect for novel antibiotics (45). Most recently, Riesenfeld et al. used the metagenomic approach to identify new antibiotic resistance genes from uncultured soil bacteria (116). The sequencing and assembly of these large gene insert libraries have also been hypothesized to lead to reconstruction of a nearly complete microbial genome (26, 27).

There are limitations to applying the bacterial artificial chromosome vector approach to metagenomes that include issues of representative cloning and quantitative lysis (53). Furthermore, the number of clones needed to represent the entire metagenome is staggering (51). Nonetheless, this approach does allow one to begin to harvest the remarkable and vast diversity present in a given metagenome. To reduce the complexity of metagenomic sequences, SSH has been used to compare metagenomes of microbial communities from food animals (42). Based on simple quality control assessments of the SSH libraries constructed, the technique is capable of discerning and identifying DNA fragments unique to one complex microbial community relative to another.

An additional challenge with libraries constructed from metagenomes is to link the functional and metabolic potential within a specific clone to the organism from which it came. The link can be made by screening metagenome clones for SSU rRNA genes. SSU rRNA gene screening via PCR with primers specific for individual species, genera, or higher taxa has been applied (11, 12, 78, 84, 112, 151). This phylogenetic screening allows one to determine the genotypes of the DNA flanking the SSU rRNA genes for functions that could yield important information on the physiology or ecology of the organism from which the DNA originated (11, 12, 78, 84, 112). For example, Beja et al. (10) discovered the presence of bacteriorhodopsin in an uncultured marine bacterium, which suggests that this organism gains energy from sunlight (10).

DNA microarray technology has also been applied to determine patterns of metabolic gene expression within mixed microbial systems (29, 141). Microarrays have tremendous potential for characterizing microbial communities and how they function in a given environment (49). The application of microarrays to mixed microbial communities include gene expression studies (29) and the detection of functional genes using array-based genome technologies (115, 141, 144, 159–161). Although these methods provide a wealth of information about microbial community structure, they are limited to the set of sequences spotted on the array.

The advances of the massively parallel sequencing capacity of 454 Life Sciences pyrosequencing technology (93) are now being applied to the sequencing and analysis of metagenomes. Edwards et al. used this technology to analyze the metabolic potential of a given microbiome by analyzing both SSU rRNA gene and functional subsystems in the metagenome of the Soudan Mine (34). A similar approach was used as part of an analysis of the gastrointestinal microbiomes of C57BL/6J mice with or without a mutation in the leptin gene (146). The results from this study were used to make links between microbial community metabolic potential and the physiological condition of the host (obesity). Both of these surveys are based on global views of the genomic content of the samples, in contrast to the approach of Roesch and others (117) and Sogin et al. (131), which focuses on a limited portion of the SSU rRNA gene molecule. In this manner, the capacity of the sequencing throughput is focused on a specific target to increase the number of sampled SSU rRNA gene PCR amplicons by several orders of magnitude.

## CONCLUSION

Over the past decade, an extensive set of molecular approaches has been developed that enable the study of both cultured and uncultured microorganisms in a variety of environments, including the GITs of production animals. Many of these approaches are based on the use of the gene encoding the SSU rRNA as the target molecule. The approaches have been widely utilized to study microbial community structure and richness and the interactions of the constituent microbes within these communities (108, 143, 145). The application of new numerical techniques, as well as traditional phylogenetic and multivariate approaches, is helping to unlock the ecological potential of molecular studies in microbial ecology. Additionally, the slowly emerging view that species of low abundance are having major metabolic effects will undoubtedly change the way in which we understand most microbial communities in their natural habitat.

## REFERENCES

1. **Akopyants, N. S., A. Fradkov, L. Diatchenko, J. E. Hill, P. D. Siebert, S. A. Lukyanov, E. D. Sverdlov, and D. E. Berg.** 1998. PCR-based subtractive hybridization and differences in gene content among strains of Helicobacter pylori. *Proc. Natl. Acad. Sci. USA* **95:** 13108–13113.
2. **Amann, R., J. Snaidr, M. Wagner, W. Ludwig, and K. H. Schleifer.** 1996. In situ visualization of high genetic diversity in a natural microbial community. *J. Bacteriol.* **178:**3496–3500.
3. **Amann, R. I.** 1995. Fluorescently labelled, rRNA-targeted oligonucleotide probes in the study of microbial ecology. *Mol. Ecol.* **4:**543–554.
4. **Amann, R. I., L. Krumholz, and D. A. Stahl.** 1990. Fluorescent-oligonucleotide probing of whole cells for determinative, phylogenetic, and environmental studies in microbiology. *J. Bacteriol.* **172:**762–770.
5. **Amann, R. I., C. Lin, R. Key, L. Montgomery, and D. A. Stahl.** 1992. Diversity among Fibrobacter isolates: towards a phylogenetic classification. *Syst. Appl. Microbiol.* **15:**23–31.
6. **Amann, R. I., W. Ludwig, and K. H. Schleifer.** 1994. Identification of uncultured bacteria: a challenging task for molecular taxonomists. *ASM News* **60:**360–365.
7. **Amann, R. I., W. Ludwig, and K. H. Schleifer.** 1995. Phylogenetic identification and in situ detection of individual microbial cells without cultivation. *Microbiol. Rev.* **59:**143–169.
8. **Anderson, I. C., and J. W. Cairney.** 2004. Diversity and ecology of soil fungal communities:

increased understanding through the application of molecular techniques. *Environ. Microbiol.* **6:** 769–779.

9. **Ashelford, K. E., N. A. Chuzhanova, J. C. Fry, A. J. Jones, and A. J. Weightman.** 2005. At least 1 in 20 16S rRNA sequence records currently held in public repositories is estimated to contain substantial anomalies. *Appl. Environ. Microbiol.* **71:**7724–7736.
10. **Beja, O., L. Aravind, E. V. Koonin, M. T. Suzuki, A. Hadd, L. P. Nguyen, S. B. Jovanovich, C. M. Gates, R. A. Feldman, J. L. Spudich, E. N. Spudich, and E. F. DeLong.** 2000. Bacterial rhodopsin: evidence for a new type of phototrophy in the sea. *Science* **289:**1902–1906.
11. **Beja, O., E. V. Koonin, L. Aravind, L. T. Taylor, H. Seitz, J. L. Stein, D. C. Bensen, R. A. Feldman, R. V. Swanson, and E. F. DeLong.** 2002. Comparative genomic analysis of archaeal genotypic variants in a single population and in two different oceanic provinces. *Appl. Environ. Microbiol.* **68:**335–345.
12. **Beja, O., M. T. Suzuki, E. V. Koonin, L. Aravind, A. Hadd, L. P. Nguyen, R. Villacorta, M. Amjadi, C. Garrigues, S. B. Jovanovich, R. A. Feldman, and E. F. DeLong.** 2000. Construction and analysis of bacterial artificial chromosome libraries from a marine microbial assemblage. *Environ. Microbiol.* **2:**516–529.
13. **Bensch, S., and M. Akesson.** 2005. Ten years of AFLP in ecology and evolution: why so few animals? *Mol. Ecol.* **14:**2899–2914.
14. **Beyer, J., D. Fichtner, H. Schirrmeier, H. Granzow, U. Polster, E. Weiland, A. Berndt, and H. Wege.** 1998. Arterivirus PRRSV. Experimental studies on the pathogenesis of respiratory disease. *Adv. Exp. Med. Biol.* **440:**593–599.
15. **Bonaldo, M. F., G. Lennon, and M. B. Soares.** 1996. Normalization and subtraction: two approaches to facilitate gene discovery. *Genome Res.* **6:**791–806.
16. **Borneman, J., and E. W. Triplett.** 1997. Molecular microbial diversity in soils from eastern Amazonia: evidence for unusual microorganisms and microbial population shifts associated with deforestation. *Appl. Environ. Microbiol.* **63:**2647–2653.
17. **Brodie, E. L., T. Z. Desantis, D. C. Joyner, S. M. Baek, J. T. Larsen, G. L. Andersen, T. C. Hazen, P. M. Richardson, D. J. Herman, T. K. Tokunaga, J. M. Wan, and M. K. Firestone.** 2006. Application of a high-density oligonucleotide microarray approach to study bacterial population dynamics during uranium reduction and reoxidation. *Appl. Environ. Microbiol.* **72:**6288–6298.
18. **Bruce, K. D.** 1997. Analysis of *mer* gene subclasses within bacterial communities in soils and sediments resolved by fluorescent-PCR–restriction fragment length polymorphism profiling. *Appl. Environ. Microbiol.* **63:**4914–4919.
19. **Burtscher, M. M., K. E. Kollner, R. Sommer, K. Keiblinger, A. H. Farnleitner, and R. L. Mach.** 2006. Development of a novel amplified fragment length polymorphism (AFLP) typing method for enterococci isolates from cattle faeces and evaluation of the single versus pooled faecal sampling approach. *J. Microbiol. Methods* **67:** 281–293.
20. **Castro-Gonzalez, M., G. Braker, L. Farias, and O. Ulloa.** 2005. Communities of nirS-type denitrifiers in the water column of the oxygen minimum zone in the eastern South Pacific. *Environ. Microbiol.* **7:**1298–1306.
21. **Chandler, D. P., G. J. Newton, J. A. Small, and D. S. Daly.** 2003. Sequence versus structure for the direct detection of 16S rRNA on planar oligonucleotide microarrays. *Appl. Environ. Microbiol.* **69:**2950–2958.
22. **Cole, J. R., B. Chai, R. J. Farris, Q. Wang, S. A. Kulam, D. M. McGarrell, G. M. Garrity, and J. M. Tiedje.** 2005. The Ribosomal Database Project (RDP-II): sequences and tools for high-throughput rRNA analysis. *Nucleic Acids Res.* **33:**D294–D296.
23. **Colombi, M., L. Moro, N. Zoppi, and S. Barlati.** 1993. Quantitative evaluation of m-RNAs by in situ hybridization and image analysis: principles and applications. *DNA Cell Biol.* **12:** 629–636.
24. **Curtis, T. P., W. T. Sloan, and J. W. Scannell.** 2002. Estimating prokaryotic diversity and its limits. *Proc. Natl. Acad. Sci. USA* **99:**10494–10499.
25. **de Bruijn, F. J.** 1992. Use of repetitive (repetitive extragenic palindromic and enterobacterial repetitive intergeneric consensus) sequences and the polymerase chain reaction to fingerprint the genomes of *Rhizobium meliloti* isolates and other soil bacteria. *Appl. Environ. Microbiol.* **58:**2180–2187.
26. **DeLong, E. F.** 2002. Microbial population genomics and ecology. *Curr. Opin. Microbiol.* **5:** 520–524.
27. **DeLong, E. F.** 2004. Microbial population genomics and ecology: the road ahead. *Environ. Microbiol.* **6:**875–878.
28. **Deng, W., S. R. Liou, G. Plunkett III, G. F. Mayhew, D. J. Rose, V. Burland, V. Kodoyianni, D. C. Schwartz, and F. R. Blattner.** 2003. Comparative genomics of *Salmonella*

*enterica* serovar Typhi strains Ty2 and CT18. *J. Bacteriol.* **185:**2330–2337.

29. **Dennis, P., E. A. Edwards, S. N. Liss, and R. Fulthorpe.** 2003. Monitoring gene expression in mixed microbial communities by using DNA microarrays. *Appl. Environ. Microbiol.* **69:** 769–778.

30. **Deplancke, B., K. R. Hristova, H. A. Oakley, V. J. McCracken, R. Aminov, R. I. Mackie, and H. R. Gaskins.** 2000. Molecular ecological analysis of the succession and diversity of sulfate-reducing bacteria in the mouse gastrointestinal tract. *Appl. Environ. Microbiol.* **66:**2166–2174.

31. **DeSantis, T. Z., P. Hugenholtz, N. Larsen, M. Rojas, E. L. Brodie, K. Keller, T. Huber, D. Dalevi, P. Hu, and G. L. Andersen.** 2006. Greengenes, a chimera-checked 16S rRNA gene database and workbench compatible with ARB. *Appl. Environ. Microbiol.* **72:**5069–5072.

32. **Dicksved, J., H. Floistrup, A. Bergstrom, M. Rosenquist, G. Pershagen, A. Scheynius, S. Roos, J. S. Alm, L. Engstrand, C. Braun-Fahrlander, E. von Mutius, and J. K. Jansson.** 2007. Molecular fingerprinting of the fecal microbiota of children raised according to different lifestyles. *Appl. Environ. Microbiol.* **73:**2284–2289.

33. **Edwards, J. E., N. R. McEwan, A. J. Travis, and R. J. Wallace.** 2004. 16S rDNA library-based analysis of ruminal bacterial diversity. *Antonie van Leeuwenhoek* **86:**263–281.

34. **Edwards, R. A., B. Rodriguez-Brito, L. Wegley, M. Haynes, M. Breitbart, D. M. Peterson, M. O. Saar, S. Alexander, E. C. Alexander, and F. Rohwer.** 2006. Using pyrosequencing to shed light on deep mine microbial ecology. *BMC Genomics* **7:**57–69.

35. **El Fantroussi, S., H. Urakawa, A. E. Bernhard, J. J. Kelly, P. A. Noble, H. Smidt, G. M. Yershov, and D. A. Stahl.** 2003. Direct profiling of environmental microbial populations by thermal dissociation analysis of native rRNAs hybridized to oligonucleotide microarrays. *Appl. Environ. Microbiol.* **69:**2377–2382.

36. **Fierer, N., and R. B. Jackson.** 2006. The diversity and biogeography of soil bacterial communities. *Proc. Natl. Acad. Sci. USA* **103:**626–631.

37. **Fisher, M. M., and E. W. Triplett.** 1999. Automated approach for ribosomal intergenic spacer analysis of microbial diversity and its application to freshwater bacterial communities. *Appl. Environ. Microbiol.* **65:**4630–4636.

38. **Fouts, D. E., E. F. Mongodin, R. E. Mandrell, W. G. Miller, D. A. Rasko, J. Ravel, L. M. Brinkac, R. T. DeBoy, C. T. Parker, S. C. Daugherty, R. J. Dodson, A. S. Durkin, R. Madupu, S. A. Sullivan, J. U. Shetty, M. A. Ayodeji, A. Shvartsbeyn, M. C. Schatz, J. H. Badger, C. M. Fraser, and K. E. Nelson.** 2005. Major structural differences and novel potential virulence mechanisms from the genomes of multiple campylobacter species. *PLoS Biol.* **3:**e15.

39. **Fox, G. E., E. Stackebrandt, R. B. Hespell, J. Gibson, J. Maniloff, T. A. Dyer, R. S. Wolfe, W. E. Balch, R. S. Tanner, L. J. Magrum, L. B. Zablen, R. Blakemore, R. Gupta, L. Bonen, B. J. Lewis, D. A. Stahl, K. R. Luehrsen, K. N. Chen, and C. R. Woese.** 1980. The phylogeny of prokaryotes. *Science* **209:**457–463.

40. **Fox, J. L.** 2005. Ribosomal gene milestone met, already left in dust. *ASM News* **71:**6–7.

41. **Franklin, R. B., D. R. Taylor, and A. L. Mills.** 1999. Characterization of microbial communities using randomly amplified polymorphic DNA (RAPD). *J. Microbiol. Methods* **35:**225–235.

42. **Galbraith, E. A., D. A. Antonopoulos, and B. A. White.** 2004. Suppressive subtractive hybridization as a tool for identifying genetic diversity in an environmental metagenome: the rumen as a model. *Environ. Microbiol.* **6:**928–937.

43. **Garaizar, J., N. Lopez-Molina, I. Laconcha, D. Lau Baggesen, A. Rementeria, A. Vivanco, A. Audicana, and I. Perales.** 2000. Suitability of PCR fingerprinting, infrequent-restriction-site PCR, and pulsed-field gel electrophoresis, combined with computerized gel analysis, in library typing of *Salmonella enterica* serovar Enteritidis. *Appl. Environ. Microbiol.* **66:**5273–5281.

44. **Garcia-Martinez, J., S. G. Acinas, A. I. Anton, and F. Rodriguez-Valera.** 1999. Use of the 16S–23S ribosomal genes spacer region in studies of prokaryotic diversity. *J. Microbiol. Methods* **36:**55–64.

45. **Gillespie, D. E., S. F. Brady, A. D. Bettermann, N. P. Cianciotto, M. R. Liles, M. R. Rondon, J. Clardy, R. M. Goodman, and J. Handelsman.** 2002. Isolation of antibiotics turbomycin A and B from a metagenomic library of soil microbial DNA. *Appl. Environ. Microbiol.* **68:** 4301–4306.

46. **Gilmore, M. S., and J. J. Ferretti.** 2003. Microbiology. The thin line between gut commensal and pathogen. *Science* **299:**1999–2002.

47. **Gray, N. D., R. Howarth, R. W. Pickup, J. G. Jones, and I. M. Head.** 2000. Use of combined microautoradiography and fluorescence in situ hybridization to determine carbon metabolism in mixed natural communities of uncultured bacteria from the genus *Achromatium*. *Appl. Environ. Microbiol.* **66:**4518–4522.

48. **Gurtler, V., and B. C. Mayall.** 2001. Genomic approaches to typing, taxonomy and evolution of bacterial isolates. *Int. J. Syst. Evol. Microbiol.* **51:** 3–16.
49. **Guschin, D. Y., B. K. Mobarry, D. Proudnikov, D. A. Stahl, B. E. Rittmann, and A. D. Mirzabekov.** 1997. Oligonucleotide microchips as genosensors for determinative and environmental studies in microbiology. *Appl. Environ. Microbiol.* **63:**2397–2402.
50. **Hamady, M., J. J. Walker, J. K. Harris, N. J. Gold, and R. Knight.** 10 February 2008, posting date. Error-correcting barcoded primers for pyrosequencing hundreds of samples in multiplex. *Nat. Methods* doi:10.1038/nmeth.1182.
51. **Handelsman, J., M. R. Rondon, S. F. Brady, J. Clardy, and R. M. Goodman.** 1998. Molecular biological access to the chemistry of unknown soil microbes: a new frontier for natural products. *Chem. Biol.* **5:**R245–R249.
52. **Helmus, M. R., T. J. Bland, C. K. Williams, and A. R. Ives.** 2007. Phylogenetic measures of biodiversity. *Am. Nat.* **169:**E68–E83.
53. **Henne, A., R. Daniel, R. A. Schmitz, and G. Gottschalk.** 1999. Construction of environmental DNA libraries in *Escherichia coli* and screening for the presence of genes conferring utilization of 4-hydroxybutyrate. *Appl. Environ. Microbiol.* **65:**3901–3907.
54. **Hespell, R. B., D. E. Akin, and B. A. Dehority.** 1997. Bacteria, fungi, and protozoa of the rumen, p. 59–141. *In* R. I. Mackie, B. A. White, and R. E. Isaacson (ed.), *Gastrointestinal Microbes and Host Interactions,* vol. 2. Chapman and Hall, New York, NY.
55. **Hiraishi, A., M. Iwasaki, and H. Shinjo.** 2000. Terminal restriction pattern analysis of 16S rRNA genes for the characterization of bacterial communities of activated sludge. *J. Biosci. Bioeng.* **90:**148–156.
56. **Hodson, R. E., W. A. Dustman, R. P. Garg, and M. A. Moran.** 1995. In situ PCR for visualization of microscale distribution of specific genes and gene products in prokaryotic communities. *Appl. Environ. Microbiol.* **61:**4074–4082.
57. **Holder, M., and P. O. Lewis.** 2003. Phylogeny estimation: traditional and Bayesian approaches. *Nat. Rev. Genet.* **4:**275–284.
58. **Hongoh, Y., P. Deevong, T. Inoue, S. Moriya, S. Trakulnaleamsai, M. Ohkuma, C. Vongkaluang, N. Noparatnaraporn, and T. Kudo.** 2005. Intra- and interspecific comparisons of bacterial diversity and community structure support coevolution of gut microbiota and termite host. *Appl. Environ. Microbiol.* **71:**6590–6599.
59. **Hugenholtz, P., B. M. Goebel, and N. R. Pace.** 1998. Impact of culture-independent studies on the emerging phylogenetic view of bacterial diversity. *J. Bacteriol.* **180:**4765–4774.
60. **Hulton, C. S., C. F. Higgins, and P. M. Sharp.** 1991. ERIC sequences: a novel family of repetitive elements in the genomes of Escherichia coli, Salmonella typhimurium and other enterobacteria. *Mol. Microbiol.* **5:**825–834.
61. **Huse, S. M., J. A. Huber, H. G. Morrison, M. L. Sogin, and D. M. Welch.** 2007. Accuracy and quality of massively parallel DNA pyrosequencing. *Genome Biol.* **8:**R143.
62. **Jongman, R. H. G., C. J. F. ter Braak, and O. F. R. van Tongeren (ed.).** 1995. *Data Analysis in Community and Landscape Ecology.* Cambridge University Press, New York, NY.
63. **Kiehnbaum, L. A., A. Amonsin, S. J. Wells, and V. Kapur.** 2005. Amplified fragment length polymorphism to detect clonal diversity and distribution of Mycobacterium avium subspecies paratuberculosis in selected Minnesota dairy cattle. *J. Vet. Diagn. Investig.* **17:**311–315.
64. **Kim, J., J. Nietfeldt, and A. K. Benson.** 1999. Octamer-based genome scanning distinguishes a unique subpopulation of Escherichia coli O157:H7 strains in cattle. *Proc. Natl. Acad. Sci. USA* **96:**13288–13293.
65. **Kim, J., J. Nietfeldt, J. Ju, J. Wise, N. Fegan, P. Desmarchelier, and A. K. Benson.** 2001. Ancestral divergence, genome diversification, and phylogeographic variation in subpopulations of sorbitol-negative, beta-glucuronidase-negative enterohemorrhagic *Escherichia coli* O157. *J. Bacteriol.* **183:**6885–6897.
66. **Klappenbach, J. A., P. R. Saxman, J. R. Cole, and T. M. Schmidt.** 2001. rrndb: the ribosomal RNA operon copy number database. *Nucleic Acids Res.* **29:**181–184.
67. **Kocherginskaya, S. A., R. I. Aminov, and B. A. White.** 2001. Analysis of the rumen bacterial diversity under two different diet conditions using denaturing gradient gel electrophoresis, random sequencing, and statistical ecology approaches. *Anaerobe* **7:**119–134.
68. **Kocherginskaya, S. A., I. K. O. Cann, and R. I. Mackie.** 2005. Denaturing gradient gel electrophoresis, p. 119–128. *In* H. P. S. Makkar and C. S. McSweeney (ed.), *Methods in Gut Microbial Ecology for Ruminants.* Springer, Dordrecht, The Netherlands.
69. **Koizumi, Y., J. J. Kelly, T. Nakagawa, H. Urakawa, S. El-Fantroussi, S. Al-Muzaini, M. Fukui, Y. Urushigawa, and D. A. Stahl.** 2002. Parallel characterization of anaerobic toluene- and ethylbenzene-degrading microbial consortia by PCR-denaturing gradient gel electrophoresis, RNA-DNA membrane hybridiza-

tion, and DNA microarray technology. *Appl. Environ. Microbiol.* **68:**3215–3225.

70. **Krause, D. O., B. P. Dalrymple, W. J. Smith, R. I. Mackie, and C. S. McSweeney.** 1999. 16S rDNA sequencing of Ruminococcus albus and Ruminococcus flavefaciens: design of a signature probe and its application in adult sheep. *Microbiology* **145**(Pt 7):1797–1807.
71. **Kysela, D. T., C. Palacios, and M. L. Sogin.** 2005. Serial analysis of V6 ribosomal sequence tags (SARST-V6): a method for efficient, high-throughput analysis of microbial community composition. *Environ. Microbiol.* **7:**356–364.
72. **Larue, R., Z. Yu, V. A. Parisi, A. R. Egan, and M. Morrison.** 2005. Novel microbial diversity adherent to plant biomass in the herbivore gastrointestinal tract, as revealed by ribosomal intergenic spacer analysis and rrs gene sequencing. *Environ. Microbiol.* **7:**530–543.
73. Reference deleted.
74. **Legendre, P., and L. Legendre.** 1998. *Numerical Ecology,* 2nd English ed. Elsevier Science BV, Amsterdam, The Netherlands.
75. **Leser, T. D., J. Z. Amenuvor, T. K. Jensen, R. H. Lindecrona, M. Boye, and K. Moller.** 2002. Culture-independent analysis of gut bacteria: the pig gastrointestinal tract microbiota revisited. *Appl. Environ. Microbiol.* **68:**673–690.
76. Reference deleted.
77. **Liebana, E.** 2002. Molecular tools for epidemiological investigations of S. enterica subspecies enterica infections. *Res. Vet. Sci.* **72:**169–175.
78. **Liles, M. R., B. F. Manske, S. B. Bintrim, J. Handelsman, and R. M. Goodman.** 2003. A census of rRNA genes and linked genomic sequences within a soil metagenomic library. *Appl. Environ. Microbiol.* **69:**2684–2691.
79. **Lin, C., B. Flesher, W. C. Capman, R. I. Amann, and D. Stahl.** 1994. Taxon specific hybridization probes for fiber-digesting bacteria suggest novel gut-associated Fibrobacter. *Syst. Appl. Microbiol.* **17:**418–424.
80. **Lin, C., and D. A. Stahl.** 1995. Taxon-specific probes for the cellulolytic genus *Fibrobacter* reveal abundant and novel equine-associated populations. *Appl. Environ. Microbiol.* **61:**1348–1351.
81. **Lin, C. Z., L. Raskin, and D. A. Stahl.** 1997. Microbial community structure in gastrointestinal tracts of domestic animals: comparative analyses using rRNA-targeted oligonucleotide probes. *FEMS Microbiol. Ecol.* **22:**281–294.
82. **Liu, W. T., T. L. Marsh, H. Cheng, and L. J. Forney.** 1997. Characterization of microbial diversity by determining terminal restriction fragment length polymorphisms of genes encoding 16S rRNA. *Appl. Environ. Microbiol.* **63:**4516–4522.
83. **Liu, Z. Z., C. Lozupone, M. Hamady, F. D. Bushman, and R. Knight.** 2007. Short pyrosequencing reads suffice for accurate microbial community analysis. *Nucleic Acids Res.* **35:**1–10.
84. **Lopez-Garcia, P., C. Brochier, D. Moreira, and F. Rodriguez-Valera.** 2004. Comparative analysis of a genome fragment of an uncultivated mesopelagic crenarchaeote reveals multiple horizontal gene transfers. *Environ. Microbiol.* **6:**19–34.
85. **Louws, F. J., D. W. Fulbright, C. T. Stephens, and F. J. de Bruijn.** 1994. Specific genomic fingerprints of phytopathogenic *Xanthomonas* and *Pseudomonas* pathovars and strains generated with repetitive sequences and PCR. *Appl. Environ. Microbiol.* **60:**2286–2295.
86. **Lozupone, C., and R. Knight.** 2005. UniFrac: a new phylogenetic method for comparing microbial communities. *Appl. Environ. Microbiol.* **71:** 8228–8235.
87. **Lu, J., U. Idris, B. Harmon, C. Hofacre, J. J. Maurer, and M. D. Lee.** 2003. Diversity and succession of the intestinal bacterial community of the maturing broiler chicken. *Appl. Environ. Microbiol.* **69:**6816–6824.
88. **Lu, J., S. Sanchez, C. Hofacre, J. J. Maurer, B. G. Harmon, and M. D. Lee.** 2003. Evaluation of broiler litter with reference to the microbial composition as assessed by using 16S rRNA and functional gene markers. *Appl. Environ. Microbiol.* **69:**901–908.
89. **Maiden, M. C.** 2006. Multilocus sequence typing of bacteria. *Annu. Rev. Microbiol.* **60:**561–588.
90. **Maiden, M. C., J. A. Bygraves, E. Feil, G. Morelli, J. E. Russell, R. Urwin, Q. Zhang, J. Zhou, K. Zurth, D. A. Caugant, I. M. Feavers, M. Achtman, and B. G. Spratt.** 1998. Multilocus sequence typing: a portable approach to the identification of clones within populations of pathogenic microorganisms. *Proc. Natl. Acad. Sci. USA* **95:**3140–3145.
91. **Majernik, A., G. Gottschalk, and R. Daniel.** 2001. Screening of environmental DNA libraries for the presence of genes conferring $Na^+(Li^+)/H^+$ antiporter activity on *Escherichia coli:* characterization of the recovered genes and the corresponding gene products. *J. Bacteriol.* **183:**6645–6653.
92. **Manefield, M., A. S. Whiteley, R. I. Griffiths, and M. J. Bailey.** 2002. RNA stable isotope probing, a novel means of linking microbial community function to phylogeny. *Appl. Environ. Microbiol.* **68:**5367–5373.
93. **Margulies, M., M. Egholm, W. E. Altman, S. Attiya, J. S. Bader, L. A. Bemben, J. Berka, M. S. Braverman, Y. J. Chen, Z. Chen, S. B. Dewell, L. Du, J. M. Fierro, X. V. Gomes, B. C. Godwin, W. He, S.

**Helgesen, C. H. Ho, G. P. Irzyk, S. C. Jando, M. L. Alenquer, T. P. Jarvie, K. B. Jirage, J. B. Kim, J. R. Knight, J. R. Lanza, J. H. Leamon, S. M. Lefkowitz, M. Lei, J. Li, K. L. Lohman, H. Lu, V. B. Makhijani, K. E. McDade, M. P. McKenna, E. W. Myers, E. Nickerson, J. R. Nobile, R. Plant, B. P. Puc, M. T. Ronan, G. T. Roth, G. J. Sarkis, J. F. Simons, J. W. Simpson, M. Srinivasan, K. R. Tartaro, A. Tomasz, K. A. Vogt, G. A. Volkmer, S. H. Wang, Y. Wang, M. P. Weiner, P. Yu, R. F. Begley, and J. M. Rothberg.** 2005. Genome sequencing in microfabricated high-density picolitre reactors. *Nature* **437:**376–380.

94. **Martin, A. P.** 2002. Phylogenetic approaches for describing and comparing the diversity of microbial communities. *Appl. Environ. Microbiol.* **68:**3673–3682.

95. **Massol-Deya, A., R. Weller, L. Rios-Hernandez, J. Z. Zhou, R. F. Hickey, and J. M. Tiedje.** 1997. Succession and convergence of biofilm communities in fixed-film reactors treating aromatic hydrocarbons in groundwater. *Appl. Environ. Microbiol.* **63:**270–276.

96. **Mau, M., and K. N. Timmis.** 1998. Use of subtractive hybridization to design habitat-based oligonucleotide probes for investigation of natural bacterial communities. *Appl. Environ. Microbiol.* **64:**185–191.

97. **Meudt, H. M., and A. C. Clarke.** 2007. Almost forgotten or latest practice? AFLP applications, analyses and advances. *Trends Plant Sci.* **12:**106–117.

98. **Mohamed, M. A. N., L. Ranjard, C. Catroux, G. Catroux, and A. Hartmann.** 2005. Effect of natamycin on the enumeration, genetic structure and composition of bacterial community isolated from soils and soybean rhizosphere. *J. Microbiol. Methods* **60:**31–40.

99. **Muyzer, G., E. C. de Waal, and A. G. Uitterlinden.** 1993. Profiling of complex microbial populations by denaturing gradient gel electrophoresis analysis of polymerase chain reaction-amplified genes coding for 16S rRNA. *Appl. Environ. Microbiol.* **59:**695–700.

100. **Muyzer, G., and K. Smalla.** 1998. Application of denaturing gradient gel electrophoresis (DGGE) and temperature gradient gel electrophoresis (TGGE) in microbial ecology. *Antonie van Leeuwenhoek* **73:**127–141.

101. **Nelson, K. E., S. H. Zinder, I. Hance, P. Burr, D. Odongo, D. Wasawo, A. Odenyo, and R. Bishop.** 2003. Phylogenetic analysis of the microbial populations in the wild herbivore gastrointestinal tract: insights into an unexplored niche. *Environ. Microbiol.* **5:**1212–1220.

102. **Neppelenbroek, K. H., N. H. Campanha, D. M. P. Spolidorio, L. C. Spolidorio, R. S. Seó, and A. C. Pavarina.** 2006. Molecular fingerprinting methods for the discrimination between C. albicans and C. dubliniensis. *Oral Dis.* **12:**242–253.

103. **Nielsen, J. L., D. Christensen, M. Kloppenborg, and P. H. Nielsen.** 2003. Quantification of cell-specific substrate uptake by probe-defined bacteria under in situ conditions by microautoradiography and fluorescence in situ hybridization. *Environ. Microbiol.* **5:**202–211.

104. **Noll, M., D. Matthies, P. Frenzel, M. Derakshani, and W. Liesack.** 2005. Succession of bacterial community structure and diversity in a paddy soil oxygen gradient. *Environ. Microbiol.* **7:**382–395.

105. **Olsen, J. E., M. N. Skov, E. J. Threlfall, and D. J. Brown.** 1994. Clonal lines of Salmonella enterica serotype Enteritidis documented by IS200-, ribo-, pulsed-field gel electrophoresis and RFLP typing. *J. Med. Microbiol.* **40:**15–22.

106. **On, S. L., and D. L. Baggesen.** 1997. Determination of clonal relationships of Salmonella typhimurium by numerical analysis of macrorestriction profiles. *J. Appl. Microbiol.* **83:**699–706.

107. **Ouverney, C. C., and J. A. Fuhrman.** 1999. Combined microautoradiography-16S rRNA probe technique for determination of radioisotope uptake by specific microbial cell types in situ. *Appl. Environ. Microbiol.* **65:**1746–1752.

108. **Pace, N. R.** 1997. A molecular view of microbial diversity and the biosphere. *Science* **276:** 734–740.

109. Reference deleted.

110. **Peplies, J., F. O. Glockner, and R. Amann.** 2003. Optimization strategies for DNA microarray-based detection of bacteria with 16S rRNA-targeting oligonucleotide probes. *Appl. Environ. Microbiol.* **69:**1397–1407.

111. **Punia, P., M. D. Hampton, A. M. Ridley, L. R. Ward, B. Rowe, and E. J. Threlfall.** 1998. Pulsed-field electrophoretic fingerprinting of Salmonella indiana and its epidemiological applicability. *J. Appl. Microbiol.* **84:**103–107.

112. **Quaiser, A., T. Ochsenreiter, H. P. Klenk, A. Kletzin, A. H. Treusch, G. Meurer, J. Eck, C. W. Sensen, and C. Schleper.** 2002. First insight into the genome of an uncultivated crenarchaeote from soil. *Environ. Microbiol.* **4:** 603–611.

113. **Ranjard, L., F. Poly, and S. Nazaret.** 2000. Monitoring complex bacterial communities using culture-independent molecular techniques: application to soil environment. *Res. Microbiol.* **151:**167–177.

114. **Raskin, L., W. C. Capman, R. Sharp, L. K. Poulsen, and D. A. Stahl.** 1997. Molecular ecology of gastrointestinal ecosystems, p. 243–298. *In* R. I. Mackie, B. A. White, and R. E. Isaacson (ed.), *Gastrointestinal Microbes and Host Interactions,* vol. 2. Chapman and Hall, New York, NY.

115. **Rhee, S. K., X. Liu, L. Wu, S. C. Chong, X. Wan, and J. Zhou.** 2004. Detection of genes involved in biodegradation and biotransformation in microbial communities by using 50-mer oligonucleotide microarrays. *Appl. Environ. Microbiol.* **70:**4303–4317.

116. **Riesenfeld, C. S., R. M. Goodman, and J. Handelsman.** 2004. Uncultured soil bacteria are a reservoir of new antibiotic resistance genes. *Environ. Microbiol.* **6:**981–989.

117. **Roesch, L. F. W., R. R. Fulthorpe, A. Riva, G. Casella, A. K. M. Hadwin, A. D. Kent, S. H. Daroub, F. A. O. Camargo, W. G. Farmerie, and E. W. Triplett.** 2007. Pyrosequencing enumerates and contrasts soil microbial diversity. *ISME J.* **1:**283–290.

118. **Rondon, M. R., P. R. August, A. D. Bettermann, S. F. Brady, T. H. Grossman, M. R. Liles, K. A. Loiacono, B. A. Lynch, I. A. MacNeil, C. Minor, C. L. Tiong, M. Gilman, M. S. Osburne, J. Clardy, J. Handelsman, and R. M. Goodman.** 2000. Cloning the soil metagenome: a strategy for accessing the genetic and functional diversity of uncultured microorganisms. *Appl. Environ. Microbiol.* **66:**2541–2547.

119. **Saikaly, P. E., P. G. Stroot, and D. B. Oerther.** 2005. Use of 16S rRNA gene terminal restriction fragment analysis to assess the impact of solids retention time on the bacterial diversity of activated sludge. *Appl. Environ. Microbiol.* **71:**5814–5822.

120. **Salamon, H., M. A. Behr, J. T. Rhee, and P. M. Small.** 2000. Genetic distances for the study of infectious disease epidemiology. *Am. J. Epidemiol.* **151:**324–334.

121. **Salzberg, S. L., A. J. Salzberg, A. R. Kerlavage, and J. F. Tomb.** 1998. Skewed oligomers and origins of replication. *Gene* **217:**57–67.

122. **Savage, D. C.** 1977. Microbial ecology of the gastrointestinal tract. *Annu. Rev. Microbiol.* **31:** 107–133.

123. **Savelkoul, P. H., H. J. Aarts, J. de Haas, L. Dijkshoorn, B. Duim, M. Otsen, J. L. Rademaker, L. Schouls, and J. A. Lenstra.** 1999. Amplified-fragment length polymorphism analysis: the state of an art. *J. Clin. Microbiol.* **37:** 3083–3091.

124. **Schloss, P. D., and J. Handelsman.** 2005. Introducing DOTUR, a computer program for defining operational taxonomic units and estimating species richness. *Appl. Environ. Microbiol.* **71:**1501–1506.

125. **Schloss, P. D., and J. Handelsman.** 2006. Introducing TreeClimber, a test to compare microbial community structures. *Appl. Environ. Microbiol.* **72:**2379–2384.

126. **Schloss, P. D., B. R. Larget, and J. Handelsman.** 2004. Integration of microbial ecology and statistics: a test to compare gene libraries. *Appl. Environ. Microbiol.* **70:**5485–5492.

127. **Simpson, J. M., V. J. McCracken, H. R. Gaskins, and R. I. Mackie.** 2000. Denaturing gradient gel electrophoresis analysis of 16S ribosomal DNA amplicons to monitor changes in fecal bacterial populations of weaning pigs after introduction of *Lactobacillus reuteri* strain MM53. *Appl. Environ. Microbiol.* **66:**4705–4714.

128. **Simpson, J. M., V. J. McCracken, B. A. White, H. R. Gaskins, and R. I. Mackie.** 1999. Application of denaturant gradient gel electrophoresis for the analysis of the porcine gastrointestinal microbiota. *J. Microbiol. Methods* **36:**167–179.

129. **Singleton, D. R., M. A. Furlong, S. L. Rathbun, and W. B. Whitman.** 2001. Quantitative comparisons of 16S rRNA gene sequence libraries from environmental samples. *Appl. Environ. Microbiol.* **67:**4374–4376.

130. **Small, J., D. R. Call, F. J. Brockman, T. M. Straub, and D. P. Chandler.** 2001. Direct detection of 16S rRNA in soil extracts by using oligonucleotide microarrays. *Appl. Environ. Microbiol.* **67:**4708–4716.

131. **Sogin, M. L., H. G. Morrison, J. A. Huber, D. M. Welch, S. M. Huse, P. R. Neal, J. M. Arrieta, and G. J. Herndl.** 2006. Microbial diversity in the deep sea and the underexplored "rare biosphere." *Proc. Natl. Acad. Sci. USA* **103:**12115–12120.

132. **Stahl, D. A., B. Flesher, H. R. Mansfield, and L. Montgomery.** 1988. Use of phylogenetically based hybridization probes for studies of ruminal microbial ecology. *Appl. Environ. Microbiol.* **54:**1079–1084.

133. **Stern, M. J., E. Prossnitz, and G. F. Ames.** 1988. Role of the intercistronic region in post-transcriptional control of gene expression in the histidine transport operon of Salmonella typhimurium: involvement of REP sequences. *Mol. Microbiol.* **2:**141–152.

134. **Stewart, C. S.** 1997. Microorganisms in hindgut fermentors, p. 142–186. *In* R. I. Mackie, B. A. White, and R. E. Isaacson (ed.), *Gastrointestinal Microbes and Host Interactions,* vol. 2. Chapman and Hall, New York, NY.

135. **Sun, H. Y., S. P. Deng, and W. R. Raun.** 2004. Bacterial community structure and diver-

sity in a century-old manure-treated agroecosystem. *Appl. Environ. Microbiol.* **70:**5868–5864.

136. **Tajima, K., R. I. Aminov, T. Nagamine, H. Matsui, M. Nakamura, and Y. Benno.** 2001. Diet-dependent shifts in the bacterial population of the rumen revealed with real-time PCR. *Appl. Environ. Microbiol.* **67:**2766–2774.

137. **Tajima, K., R. I. Aminov, T. Nagamine, K. Ogata, M. Nakamura, H. Matsui, and Y. Benno.** 1999. Rumen bacterial diversity as determined by sequence analysis of 16S rDNA libraries. *FEMS Microbiol. Ecol.* **29:**159–169.

138. **Tajima, K., S. Arai, K. Ogata, T. Nagamine, H. Matsui, M. Nakamura, R. I. Aminov, and Y. Benno.** 2000. Rumen bacterial community transition during adaptation to high-grain diet. *Anaerobe* **6:**273–284.

139. **Tajima, K., T. Nagamine, H. Matsui, M. Nakamura, and R. Aminov.** 2001. Phylogenetic analysis of archaeal 16S rRNA libraries from the rumen suggests the existence of a novel group of archaea not associated with known methanogens. *FEMS Microbiol. Ecol.* **200:**67–72.

140. **Tannock, G. W.** 1997. Normal microbiota of the gastrointestinal tract of rodents, p. 187–215. *In* R. I. Mackie, B. A. White, and R. E. Isaacson (ed.), *Gastrointestinal Microbes and Host Interactions,* vol. 2. Chapman and Hall, New York, NY.

141. **Taroncher-Oldenburg, G., E. M. Griner, C. A. Francis, and B. B. Ward.** 2003. Oligonucleotide microarray for the study of functional gene diversity in the nitrogen cycle in the environment. *Appl. Environ. Microbiol.* **69:**1159–1171.

142. **Tenover, F. C., R. D. Arbeit, R. V. Goering, P. A. Mickelsen, B. E. Murray, D. H. Persing, and B. Swaminathan.** 1995. Interpreting chromosomal DNA restriction patterns produced by pulsed-field gel electrophoresis: criteria for bacterial strain typing. *J. Clin. Microbiol.* **33:**2233–2239.

143. **Theron, J., and T. E. Cloete.** 2000. Molecular techniques for determining microbial diversity and community structure in natural environments. *Crit. Rev. Microbiol.* **26:**37–57.

144. **Tiquia, S. M., L. Wu, S. C. Chong, S. Passovets, D. Xu, Y. Xu, and J. Zhou.** 2004. Evaluation of 50-mer oligonucleotide arrays for detecting microbial populations in environmental samples. *Biotechniques* **36:**664–670, 672, 674–675.

145. **Torsvik, V., and L. Ovreas.** 2002. Microbial diversity and function in soil: from genes to ecosystems. *Curr. Opin. Microbiol.* **5:**240–245.

146. **Turnbaugh, P. J., R. E. Ley, M. A. Mahowald, V. Magrini, E. R. Mardis, and J. I. Gordon.** 2006. An obesity-associated gut microbiome with increased capacity for energy harvest. *Nature* **444:**1027–1031.

147. **Upholt, W. B.** 1977. Estimation of DNA sequence divergence from comparison of restriction endonuclease digests. *Nucleic Acids Res.* **4:** 1257–1265.

148. **Urakawa, H., S. El Fantroussi, H. Smidt, J. C. Smoot, E. H. Tribou, J. J. Kelly, P. A. Noble, and D. A. Stahl.** 2003. Optimization of single-base-pair mismatch discrimination in oligonucleotide microarrays. *Appl. Environ. Microbiol.* **69:**2848–2856.

149. **Urakawa, H., T. Yoshida, M. Nishimura, and K. Ohwada.** 2000. Characterization of depth-related population variation in microbial communities of a coastal marine sediment using 16S rDNA-based approaches and quinone profiling. *Environ. Microbiol.* **2:**542–554.

150. **Vaughan, E. E., F. Schut, H. G. Heilig, E. G. Zoetendal, W. M. de Vos, and A. D. Akkermans.** 2000. A molecular view of the intestinal ecosystem. *Curr. Issues Intest. Microbiol.* **1:**1–12.

151. **Vergin, K. L., E. Urbach, J. L. Stein, E. F. DeLong, B. D. Lanoil, and S. J. Giovannoni.** 1998. Screening of a fosmid library of marine environmental genomic DNA fragments reveals four clones related to members of the order *Planctomycetales. Appl. Environ. Microbiol.* **64:**3075–3078.

152. **Vos, P., R. Hogers, M. Bleeker, M. Reijans, T. van de Lee, M. Hornes, A. Fritjers, J. Pot, J. Paleman, M. Kuiper, and M. Zabeau.** 1995. AFLP: a new technique for DNA fingerprinting. *Nucleic Acids Res.* **23:**4407–4414.

153. **Webb, C. O., D. D. Ackerly, M. A. McPeek, and M. J. Donoghue.** 2002. Phylogenies and community ecology. *Annu. Rev. Ecol. Syst.* **33:**475–505.

154. **Weigel, R. M., B. Qiao, B. Teferedegne, D. K. Suh, D. A. Barber, R. E. Isaacson, and B. A. White.** 2004. Comparison of pulsed field gel electrophoresis and repetitive sequence polymerase chain reaction as genotyping methods for detection of genetic diversity and inferring transmission of Salmonella. *Vet. Microbiol.* **100:**205–217.

155. **Welsh, J., and M. McClelland.** 1990. Fingerprinting genomes using PCR with arbitrary primers. *Nucleic Acids Res.* **18:**7213–7218.

156. **Whitford, M. F., R. J. Forster, C. E. Beard, J. Gong, and R. M. Teather.** 1998. Phylogenetic analysis of rumen bacteria by comparative sequence analysis of cloned 16S rRNA genes. *Anaerobe* **4:**153–163.

157. **Williams, J. G., A. R. Kubelik, K. J. Livak, J. A. Rafalski, and S. V. Tingey.** 1990. DNA

polymorphisms amplified by arbitrary primers are useful as genetic markers. *Nucleic Acids Res.* **18:**6531–6535.

158. **Wilson, K. H.** 1997. Biota of the human gastrointestinal tract, p. 39–58. *In* R. I. Mackie, B. A. White, and R. E. Isaacson (ed.), *Gastrointestinal Microbes and Host Interactions,* vol. 2. Chapman and Hall, New York, NY.

159. **Wu, L., X. Liu, C. W. Schadt, and J. Zhou.** 2006. Microarray-based analysis of subnanogram quantities of microbial community DNAs by using whole-community genome amplification. *Appl. Environ. Microbiol.* **72:**4931–4941.

160. **Wu, L., D. K. Thompson, G. Li, R. A. Hurt, J. M. Tiedje, and J. Zhou.** 2001. Development and evaluation of functional gene arrays for detection of selected genes in the environment. *Appl. Environ. Microbiol.* **67:**5780–5790.

161. **Wu, L., D. K. Thompson, X. Liu, M. W. Fields, C. E. Bagwell, J. M. Tiedje, and J. Zhou.** 2004. Development and evaluation of microarray-based whole-genome hybridization for detection of microorganisms within the context of environmental applications. *Environ. Sci. Technol.* **38:**6775–6782.

162. **Yang, Z., J. Kovar, J. Kim, J. Nietfeldt, D. R. Smith, R. A. Moxley, M. E. Olson, P. D. Fey, and A. K. Benson.** 2004. Identification of common subpopulations of non-sorbitol-fermenting, beta-glucuronidase-negative *Escherichia coli* O157:H7 from bovine production environments and human clinical samples. *Appl. Environ. Microbiol.* **70:**6846–6854.

163. **Yu, Z., M. Yu, and M. Morrison.** 2006. Improved serial analysis of V1 ribosomal sequence tags (SARST-V1) provides a rapid, comprehensive, sequence-based characterization of bacterial diversity and community composition. *Environ. Microbiol.* **8:**603–611.

164. **Zhu, X. Y., T. Zhong, Y. Pandya, and R. D. Joerger.** 2002. 16S rRNA-based analysis of microbiota from the cecum of broiler chickens. *Appl. Environ. Microbiol.* **68:**124–137.

165. **Zoetendal, E. G., C. T. Collier, S. Koike, R. I. Mackie, and H. R. Gaskins.** 2004. Molecular ecological analysis of the gastrointestinal microbiota: a review. *J. Nutr.* **134:**465–472.

# MATHEMATICAL MODELING OF MICROBIAL ECOLOGY: SPATIAL DYNAMICS OF INTERACTIONS IN BIOFILMS AND GUTS

*Jan-Ulrich Kreft*

# 19

This chapter tries to present interesting examples of the use of mathematical models in microbial ecology, since comprehensive coverage is not possible. The examples were also chosen to introduce various types of mathematical models to give the reader some guidance on the choice of suitable types of models to address a particular problem. Furthermore, the selection of examples is somewhat biased towards mechanistic rather than descriptive models because, in my opinion, the mechanistic approach leads to a better understanding of the system.

The common thread running through the variety of examples and models presented is how spatial structure can change the interactions of microbes with each other and the host. To this end, I begin with a model of a well-mixed system, the chemostat, as a reference point and then move to spatially structured systems, including chemostats with wall growth, plug flow reactors (PFRs), colonies on agar plates, and finally biofilms. In so doing, I investigate the effect of spatial structure on competition, on cooperation within species and between species (mutualism), and on communication. If this chapter manages to convince the reader that models of microbial interactions should be spatially explicit, it will have achieved its aim.

## THE CHEMOSTAT AS A NONSPATIAL "CONTROL"

The chemostat was invented by Monod as an experimental approach (Fig. 1) which allows the investigator to control the specific growth rate of the cell and to study, under steady-state conditions, how the physiology of microbes changes with their specific growth rate (67). In contrast to this view of the chemostat as a system achieving constant chemistry and therefore constant biology, Novick and Szilard (75) simultaneously invented the chemostat independently from Monod to study evolution, which is dealt with below. Although we will later see that chemostat models and experiments fail to predict what will happen in most natural systems which are spatially structured, it is nevertheless good to understand the chemostat as the simplest system, not so much in its own right, but as a "control" for comparisons with studies of spatially structured systems.

*Jan-Ulrich Kreft,* Centre for Systems Biology, School of Biosciences, University of Birmingham, Edgbaston, Birmingham B15 2TT, United Kingdom.

*Food-Borne Microbes: Shaping the Host Ecosystem,* Edited by L.-A. Jaykus, H. H. Wang, and L. S. Schlesinger, © 2009 ASM Press, Washington, DC

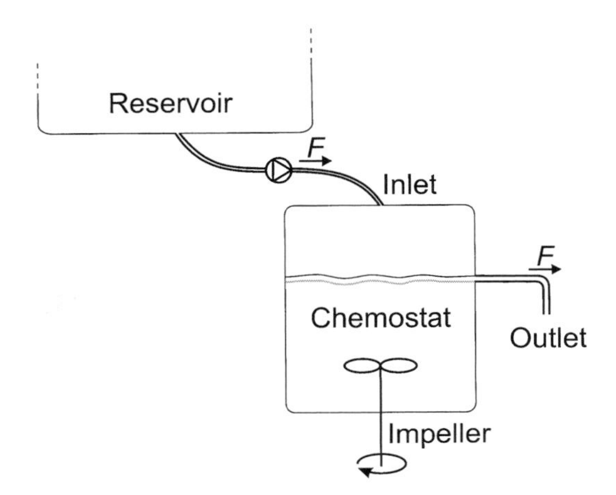


**FIGURE 1** Schematic drawing of the essentials of a chemostat. The (idealized) chemostat is characterized by continuous flowthrough of material and perfect mixing of material within the reaction vessel. A vessel of volume, $V$, is fed at a constant flow rate, $F$, from a reservoir with substrate concentration, $S_0$, replenishing the substrate concentration in the chemostat, $S$. The reactor volume remains constant because the bulk liquid containing bacteria, substrate, and metabolites is removed from the reactor at the same $F$. This leads to dilution of all contents of the chemostat bulk liquid (bacteria, substrate, and metabolites) with dilution rate, $D$, determined as $F/V$. When the state of this system fluctuates or oscillates, the term chemostat, suggesting static conditions, is better replaced by the more technical term "continuous-flow stirred-tank reactor" (CSTR).

## The Basic Chemostat Model

To derive the basic mathematical model of a chemostat (Fig. 1) comprising one planktonic population, $P$, and one growth-limiting substrate, $S$, one must consider how the physical processes of inflow of substrate-containing medium and outflow of bulk liquid which contains the remaining substrate and microbes, and the biological processes of substrate uptake and metabolism coupled to growth, change the biomass density and substrate concentration in the chemostat. Thus, the differential equations for these changes consist of the following terms: a term for biomass growth, where $\mu(S)$ models the dependence of the specific growth rate, $\mu$, on substrate concentration, $S$; a term for substrate consumption related to the growth term by the biomass yield, $Y$ (biomass formed per substrate consumed); a term for the continuous inflow of medium with substrate concentration, $S_0$, at rate $F$; and a term for the outflow of bulk liquid, including biomass and substrate, with the same rate, $F$, which guarantees constant volume, $V$. The inflow of medium results in the dilution of the chemostat content with dilution rate, determined as $D = F/V$; i.e., dilution increases with flow rate and decreases with volume. I phrase the equation in terms of dilution rate rather than flow rate because dilution rate is independent of volume:

| Change of plankton | Growth | Inflow | Outflow |
|---|---|---|---|
| $\frac{dP}{dt} =$ | $+\ \mu(S)P$ | | $-\ DP$ |

**(1a)**

| Change of substrate | Substrate consumption | Inflow | Outflow |
|---|---|---|---|
| $\frac{dS}{dt} =$ | $-\ Y^{-1}\mu(S)P$ | $+\ DS_0$ | $-\ DS$ |

**(1b)**

The key assumption on which chemostat models are based is that of perfect mixing, so that the system is uniform in space; this rules out formation of microbial aggregates. Our basic model of chemostat growth makes a lot of further *implicit* simplifying assumptions, many of which are replaced while the chapter progresses towards more and more realistic and complex models. Some of these assumptions include the following: (i) growth on the chemostat wall does not occur (included below in the Freter model); (ii) microbes do not adapt physiologically—i.e., they have the same kinetic parameters regardless of specific growth rate, substrate concentration, etc.; (iii) evolution of the genotype does not occur; (iv) maintenance can be neglected so that the yield is independent of the specific growth rate; (v) cells remain perfectly viable forever; (vi) the substrate is completely metabolized after uptake rather than stored (storage compounds can be included in a model with structured biomass); (vii) metabolites and products have no inhibitory or other effects (can be included with appropriate kinetic equations); (viii) growth depends on a single growth-limiting substrate (mixed substrate growth is discussed later); and (ix) all individuals in a population are identical (not so in individual-based models).

### Dependence of Growth on Substrate Concentration

The growth and substrate consumption terms deserve more attention, since this is where all the biology is located. Moreover, the same terms are also used in more complex models, e.g., for biofilms, because after all, the organism's genome remains the same. So I discuss the dependence of growth on a single growth-limiting substrate and then mixed-substrate growth here in the chemostat section—noting that the same terms will also be found in biofilm models.

Although a number of alternative kinetics have been put forward over the years, such as exponential and logarithmic functions of substrate concentration, the most widely used model remains that of Monod (65) (Fig. 2a). Experimental support for the Monod model comes from the studies of Egli and coworkers (60, 86), who measured the dependence of the specific growth rate of *Escherichia coli* on the concentration of various sugars in the chemostat, finding that the Monod equation was either the best or at least as good as a variety of alternative kinetic equations. Substrate and product inhibition (68) are also commonly modeled by Monod-type kinetics.

Substrate affinity is an important component of the competitiveness of microbes in environments with (temporarily) scarce resources or limited mass transfer into microbial aggregates. Typically, substrate affinity is expressed in terms of the parameter $K_S$ in the Monod equation, where a higher $K_S$ means lower affinity (Fig. 2a). However, Button et al. (12, 13) argued that $K_S$ is not a good measure of substrate affinity, because it depends on the maximum growth rate ($\mu_{max}$) (Fig. 2a). Instead, they defined specific affinity as the initial slope of the Monod curve, $a = \mu_{max}/K_S$, which is a more fundamental property of the kinetics, as it is independent of $\mu_{max}$ and can be measured on its own (12, 13). For example, organisms with a higher concentration of substrate transporters have a competitive advantage when nutrients are sparse. Their higher affinity results from a higher frequency of successful nutrient-transporter collisions. But the increased flux might not result in a decreased $K_S$, because the maximal rate might increase as well (12, 13). Thus, we can rewrite the Monod function, $\mu(S) = \mu_{max}S/(K_S + S)$, in terms of the two primary parameters, i.e., maximal specific growth rate and specific affinity in the sense of Button, as $\mu(S) = \mu_{max}aS/(\mu_{max} + aS)$ and use this as the more straightforward way of comparing the kinetics of competing species.

### Mixed-Substrate Growth

In contrast to many laboratory studies of microbial growth, heterotrophic microbes in natural environments encounter a large variety of carbon and energy sources, often at very low concentrations. Any one of these scarce substrates alone would not support the growth rates necessary to compensate losses due to phages, predators, etc., but simultaneous use of many carbon sources allows them to grow fast enough to survive (23, 59). It has become increasingly clear that diauxie, the classic finding of Monod (66) that bacteria utilize substrates sequentially, first consuming the "preferred" substrate and then, in a second growth phase, the other substrate, only occurs when an excess of a preferred substrate is available, as is typical for batch cultures (23). In fact, if *E. coli* is grown under single-sugar limitation, numerous alternative catabolic functions and high-affinity periplasmic binding proteins are up-regulated, in spite of the absence of the respective inducers. If grown in carbon-limited complex medium chemostat cultures, *E. coli* is able to oxidize 43 substrates. For further detail, see reference 28 and references therein.

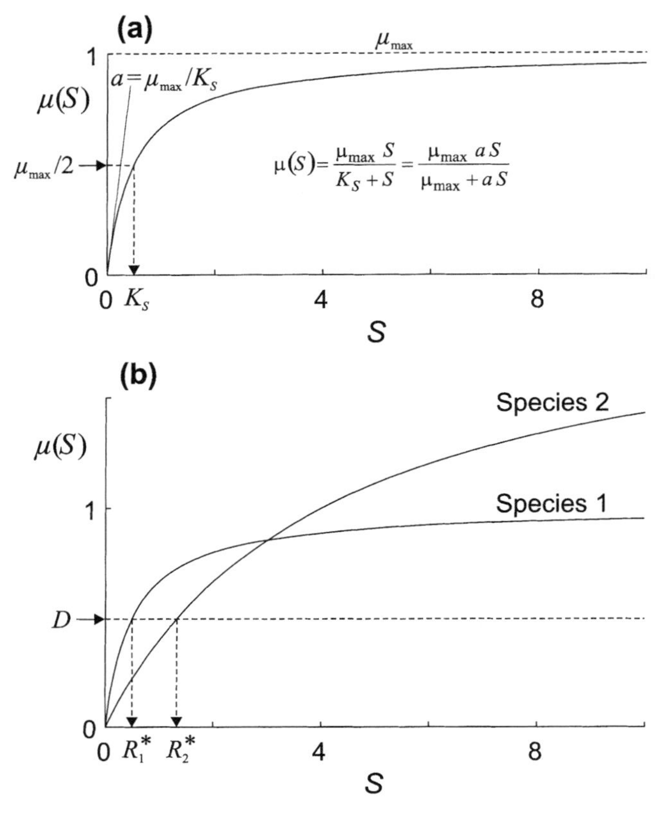


**FIGURE 2** (a) Graph of the Monod function, the standard model describing the dependence of specific growth rate, $\mu(S)$, on the substrate concentration, $S$. $K_S$ is the substrate concentration at which half the maximal specific growth rate, $\mu_{max}$, is reached; it therefore depends on $\mu_{max}$. The specific affinity sensu Button, $a = \mu_{max}/K_S$, is the initial slope of the curve and independent of $\mu_{max}$. (b) Graphs of the Monod functions of two competitors for the most interesting case where the curves cross at a particular substrate concentration. This requires that the specific affinity of species 1 is higher than that of species 2, while the maximal specific growth rate of species 1 is lower than that of species 2. In the steady state of the chemostat, the specific growth rate becomes equal to the dilution rate set by the experimenter, indicated by the horizontal line. The competitor that can achieve this specific growth rate at a lower steady-state substrate concentration, $R^\star$, will win the competition and invade a chemostat system dominated by the other species, because it has a positive net rate of increase if $S$ is above its own $R^\star$, which includes the $R^\star$ of the weaker competitor. In short, the competitor with the lower $R^\star$ will win, which is known as the $R^\star$ rule (91). While microbiologists use the term substrate, ecologists use the more general term resource, and that is where the $R^\star$ rule originates. In this case, competitor 1 has the lower $R^\star$ if the dilution rate is below the crossover point, but above, competitor 2 has the lower $R^\star$.

But why is a certain substrate preferred under conditions of excess, and why are substrates used simultaneously under limiting conditions? Narang et al. (69–73) developed a structured yet simple model of the intracellular dynamics that can explain all the possible outcomes of mixed-substrate growth, namely, (i) diauxic growth, (ii) simultaneous utilization of substrates, and (iii) bistable growth, where the cells use only one of the substrates, but which one depends on preculture conditions. (In this context, "structured model" refers to structured biomass, as opposed to the ecological term structured model, which often refers to structure in the population, such as age, physiological status, or size.)

The intracellular components in the minimal version of the model (71) are the inducer and inducible enzyme for each substrate and the remainder of the biomass (Fig. 3). The key to understanding the dynamics is that the (lumped) enzymes for uptake and peripheral metabolism of substrates are induced by their own intermediates; e.g., allolactose induces the *lac* operon. So the enzyme becomes more and more induced and therefore takes up or produces more and more inducer, in effect rendering enzyme synthesis autocatalytic (72). Further, the increased uptake and metabolism of the substrate increase the rate of growth, i.e., the rate of production of cellular volume. This in effect dilutes all cellular contents, including the inducible enzyme itself but also the inducible enzyme for uptake and peripheral metabolism of the other substrate (72). Mutual inhibition of the enzymes by regulatory interactions is not required for diauxie; in fact, mutual inhibition would hamper simultaneous use at lower growth rates (71). Now it is clear that the preferred substrate is simply that substrate supporting the higher growth rate, equivalent to the higher rate of dilution of intracellular compounds, so the competing enzyme is diluted out, and its substrate can no longer be used by the cell. The better substrate in the sense of enabling a higher growth rate of the cell, and therefore higher fitness, becomes automatically the preferred substrate (73). This shows that there is no need for evolving regulatory circuits to choose substrates according to their fitness rank, as evaluated by natural selection.

Interestingly, by mathematical comparison of structured models for mixed-substrate growth with the classical Lotka-Volterra model for competition of species, Narang found that the two processes show the same dynamic behavior (69, 71). Comparing these mechanistically unrelated interactions can, of course, only be achieved by phrasing the dynamics in the same language, i.e., by comparing the mathematical models representing the

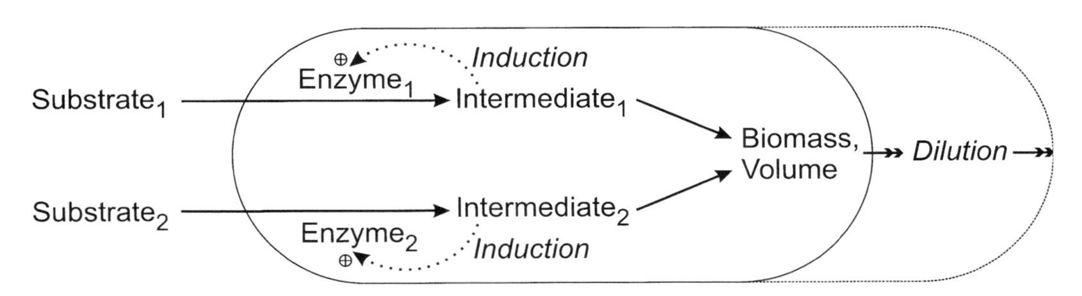


**FIGURE 3** Scheme of the minimal structured model that can produce all observed phenomena of mixed substrate growth: diauxie, simultaneous use, and bistability. The enzyme dynamics of autocatalytic synthesis of the inducible enzymes and inhibition by dilution of cellular contents are analogous to the population dynamics of autocatalytic growth and resource competition. The model specifies the kinetics of all the lumped enzymatic reactions shown. Components are in roman type, and processes are in italics. Adapted from reference 71.

dynamics. Autocatalytic growth of species corresponds to autocatalytic synthesis of inducible enzymes, and competition of species for resources corresponds to mutual "inhibition" of the enzymes by dilution. Diauxic growth corresponds to "extinction" of one of the enzymes if the intracellular dilution rate is above a critical transition dilution rate, similar to washout of a species in a chemostat that occurs if the chemostat dilution rate is above critical. Simultaneous use of substrates corresponds to "coexistence" of the enzymes. Bistable growth corresponds to a "priority effect."

## Chemostat Dynamics

The discussion of chemostat dynamics focuses on the steady state, although more interesting dynamics such as damped or sustained oscillations can occur even in single-substrate, single-species chemostats, e.g., due to the slow induction of a transporter causing a delayed response (87).

Apart from such cases, the key characteristic of the chemostat is that it reaches either one of two steady states, where the system's state (the set of all variables characterizing the system) is constant, in other words, when all rates of change are zero. In our case,

$$\frac{dP}{dt} = 0 = \mu(S)P - DP = [\mu(S) - D]P \quad \text{(2a)}$$

$$\frac{dS}{dt} = 0 = -Y^{-1}\mu(S)P + DS_0 - DS = D(S_0 - S) - Y^{-1}\mu(S)P \quad \text{(2b)}$$

If the specific growth rate equals the dilution rate, equation 2a (planktonic population does not change) is satisfied. Equation 2b (substrate does not change) is satisfied if substrate influx is balanced by substrate outflow and substrate consumption by growth. This steady state will be referred to as the colonization steady state because the chemostat is colonized by a microbial population:

$$\mu(S^\star) = D \text{ and } P^\star = Y(S_0 - S^\star) \quad \text{(3)}$$

The asterisk is often used to denote steady-state variables. Assuming Monod kinetics, we can write equation 3 as $\mu_{max}S^\star/(K_S + S^\star) = D$, which can be solved for $S^\star$: $S^\star = DK_S/(\mu_{max} - D)$. The trivial or washout steady state, $P^\star = 0$ and $S^\star = S_0$, also satisfies equation 2. See Fig. 4 for graphs of the steady-state variables as a function of dilution rate.

Using linear stability analysis, it can be shown that the colonization steady state is stable if and only if it exists. If it does not exist because the dilution rate is above the critical dilution rate, $D_c = \mu_{max}S_0/(K_S + S_0)$ (which is the maximal rate at which the microbes can grow in the chemostat, since the substrate concentration in the chemostat cannot exceed the substrate concentration in the feed), the washout steady state is the only steady state and therefore automatically stable. For an accessible introduction to linear stability analysis, which also uses the chemostat equations as an

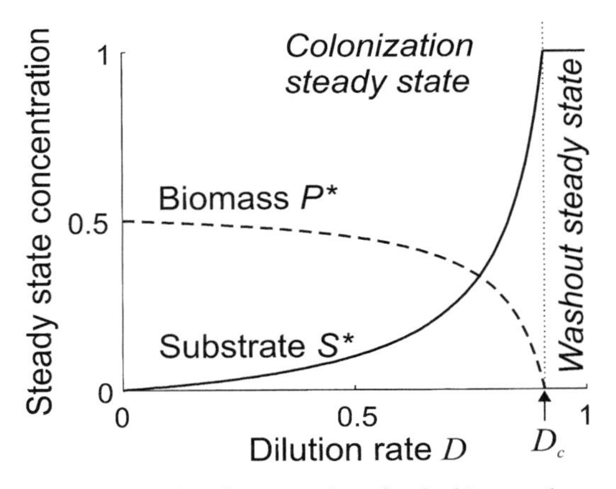


**FIGURE 4** Steady-state values for the biomass density of the planktonic population, $P^\star$, and substrate concentration, $S^\star$, in the (idealized) chemostat as a function of dilution rate, $D$. The colonization steady state, $P^\star = Y(S_0 - S^\star)$ and $S^\star = DK_S/(\mu_{max} - D)$, is stable below the critical dilution rate, $D_c = \mu_{max}S_0/(K_S + S_0)$. Above, the washout steady state, $P^\star = 0$ and $S^\star = S_0$, is the only existing steady state; it is therefore stable. Parameters used in this example were $\mu_{max} = 1$, $K_S = 0.1$, $S_0 = 1$, and $Y = 0.5$.

example, see reference 22. In summary, the chemostat reaches a stable steady state where the specific growth rate is equal to the dilution rate, which the experimenter can control at will, provided the critical dilution rate is not exceeded.

## CHEMOSTAT COMPETITION: THE COMPETITIVE EXCLUSION PRINCIPLE

The competitive exclusion principle has been studied experimentally and analyzed mathematically based on the chemostat (3, 91) because the chemostat is the simplest appropriate system for such studies (Fig. 1). The characteristics of the chemostat relevant for competition are that mixing leads to (i) uniform substrate concentration and biomass densities, (ii) global competition for substrates and other interactions such as interference by toxins or metabolites, and (iii) random short-term encounters of cells rather than long-term stable associations. Early work (3, 91) led to a rather simple competitive exclusion principle, which states that only the best competitor can survive if species compete for a single resource. Which organism is the best competitor can be decided by the $R^\star$ rule (91), explained in Fig. 2b. When species compete for several resources, the number of coexisting species cannot exceed the number of resources (or "niches" or "limiting factors"). However, this simple principle is in striking contrast to the amazing diversity of species encountered in many environments, including the human oral cavity and gut, which each harbor up to 1,000 bacterial species (50, 62). So one might ask, what has gone wrong with the competitive exclusion principle?

One key assumption behind the competitive exclusion principle is that competing species approach some steady state, and this is indeed the case in competition for one or two limiting resources in the chemostat (4). However, with three or more resources, the competitors can chase each other in time, as was strikingly demonstrated by Huisman and Weissing (40) with a standard chemostat model of phytoplankton competition (Fig. 5). The equations for this model are simply our base model (equation 1) extended for $n$ species and $k$ resources,

| Change of plankton | | Growth | Inflow-outflow | $n$ Species |
|---|---|---|---|---|
| $\frac{dP_i}{dt}$ | $=$ | $\mu_i(S_1,\ldots,S_k)P_i$ | $-\ DP_i$ | $i = 1,\ldots,n$ |

**(4a)**

| Change of substrate | | Substrate consumption | Inflow-outflow | $k$ Substrates |
|---|---|---|---|---|
| $\frac{dS_j}{dt}$ | $=$ | $-\sum_{i=1}^{n} c_{ji}\ \mu_i(S_1,\ldots,S_k)P_i$ | $D(S_j^0 - S_j)$ | $j = 1,\ldots,k$ |

**(4b)**

For example, when considering algal growth, all of following abiotic resources are required simultaneously, including carbon, nitrogen, phosphorus, iron, and silicon, according to the content of resource $j$ in species $i$, $c_{ji}$. Growth on such essential resources can be described by Liebig's law of the minimum,

$$\mu_i(S_1,\ldots,S_k) = \min\left(\frac{\mu_{\max}S_1}{K_{1i} + S_1},\ldots,\frac{\mu_{\max}S_k}{K_{ki} + S_k}\right) \quad \textbf{(4c)}$$

In other words, at any one time, growth is limited by a single resource, namely, that resource with the smallest Monod term. The maximal growth rates of all the species were assumed to be equal.

To understand how this cyclic chasing in time (Fig. 5) is generated, consider three competitors, A, B, and C. If B is better than A, and C is better than B, one is tempted to conclude that C is better than A. But A may well be better than C, if competitiveness is cyclic rather than transitive. In the plankton competition model, A is the best competitor on resource a but becomes limited by resource b, on which competitor B is best, replacing A. Now competitor B becomes limited by re-

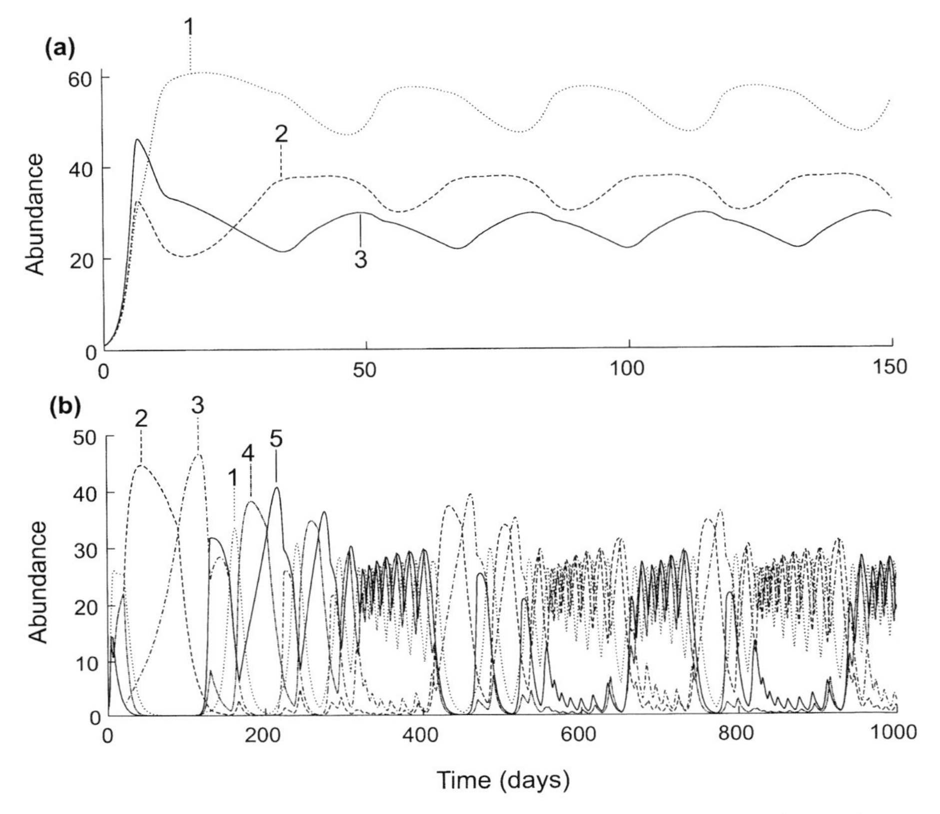


**FIGURE 5** The standard model of chemostat competition can generate oscillations if competitiveness is not transitive; there is no need for external fluctuations to maintain biodiversity. The example is based on the growth of algae, which is limited by a handful of abiotic resources. (a) Competition of three species on three resources can sustain periodic oscillations, which allow coexistence of more species than resources. There is a cyclic chasing of competitors in time. Parameters for this simulation were taken from the first three species of Fig. 1c of reference 40. (b) Chaotic oscillations can result for a range of parameter values so that each species is an intermediate competitor for the resources that most limit its growth rate. Here, five species compete for three resources. Parameters for this simulation correspond to Fig. 1 of reference 41.

source c, on which competitor C is best, replacing B. Then competitor C becomes limited by resource a, on which competitor A is best, replacing C, and so on. These oscillations are driven by the competitive interactions themselves, rather than being forced by external fluctuations. Depending on the parameters, stable oscillations (limit cycles) or even deterministic chaos can result from the interactions. Such chaos, produced by entirely deterministic equations, is defined as bounded fluctuations which are not periodic and where the states of two systems with minute differences in the initial conditions diverge exponentially with time.

In other cases without a clear competitive hierarchy, the competitors can also chase each other in space. A well-studied example uses colicin-producing (C), colicin-resistant (R), and colicin-sensitive (S) strains of *E. coli* as the three competitors (16, 47). Like above, there is a cyclic structure of competitiveness: the toxic C strain will kill the S strain and then be outgrown by the R strain, which does not pay the cost of producing the toxin. The R strain will in turn be outgrown by the S strain, which pays the cost for neither the toxin nor the resistance but will then be killed by the C strain, and so on (16, 47). This system has been studied using game theory, as an example of the game of rock-paper-scissors, which has the right payoffs (38, 64). Since the colicin will kill all sensitive cells it reaches, a well-mixed liquid culture such as a chemostat does not enable coexistence in this case; rather, spatial structure is required (16, 38, 47), which is not hard to find in natural systems. We come back to this spatial chasing example for coexistence in the section on cellular automata (below) as a modeling approach for spatially structured systems.

Another neglected aspect in early formulations of the competitive exclusion principle is the need to distinguish abiotic from biotic resources because biotic resources regenerate. If the consumption of the biotic resource is described by a nonlinear function, the variance of the resource abundance behaves like an additional resource because the average of the values of a nonlinear function over time is not the same as the function value of the average (61). This only works if the system varies in time, a further reason to criticize the steady-state assumption on which early formulations of the competitive exclusion principle were founded. An example with biotic resources is predator-prey interactions, which at intermediate dilution rates can generate periodic oscillations where the predator lags behind by a quarter period of the cycle in the chemostat (32). Such intrinsically generated fluctuations can allow the coexistence of two predators on a single prey (49). A predator can also mediate coexistence of two prey populations in a chemostat, if the predator prefers the better resource competitor (7). Such a system can generate not only stable limit cycles but also chaotic fluctuations at intermediate dilution rates (7), as in the plankton dynamics (40).

In conclusion, key factors which allow the high diversity of species in nature are (i) intrinsic or extrinsic temporal fluctuations, (ii) spatial heterogeneity, (iii) predation, and (iv) mutualism. Trade-offs are an essential ingredient of the mechanisms behind the dynamics in cases ii and iii, since they constrain evolution of competitiveness so that no overall best competitor can arise.

## CHEMOSTATS AND EVOLUTION OF CROSS-FEEDING

In our basic chemostat model, we have assumed that the growth parameters of the microbes do not change, but the chemostat cannot prevent change by evolution. In fact, the selective pressure in the chemostat is intense, and that was the reason Novick and Szilard (75) developed the use of the chemostat independently of Monod to study evolution. Ever since, the chemostat has generated a lot of insight into evolutionary processes (1). For example, Adams and coworkers have found that in long-term glucose-limited continuous cultures, a clonal population of *E. coli* repeatedly evolves into a polymorphic population consisting of at least two strains (1, 36). The entirely metabolic polymorphism is based on the parental strain evolving to excrete acetate as a metabolic intermediate of glucose catabolism, followed by the rise of an acetate-scavenging strain (1). Note that the two strains do not compete for the same resource, so this is in line with the competitive exclusion principle. Acetate excretion evolves because it allows faster growth under a range of glucose-limiting conditions in the chemostat. Since acetate excretion means truncation of the complete pathway of substrate degradation, the shorter pathway apparently allows faster growth. Why is this so?

Pfeiffer and Bonhoeffer (80) have shown that the kinetic theory of optimal pathway length developed by Heinrich and coworkers (34, 35) can explain the occurrence of shorter pathways in rich habitats. In addition to the above example of acetate excretion, another well-known example is baker's yeast, which starts to ferment glucose to ethanol under aerobic conditions if the dilution rate in the chemostat rises above a certain threshold, because the shorter pathway of fermentation allows faster growth. For further examples, see reference 14.

The kinetic theory of optimal pathway length makes three general assumptions (34, 35): (i) evolution maximizes the rate of ATP production by a catabolic pathway because this is equivalent to maximizing the growth rate (provided the amount of biomass formed per mole of ATP [$Y_{ATP}$] is approximately constant [90]), (ii) the total concentration of enzymes in a catabolic pathway is minimized because enzyme synthesis is costly or because of competition for the limiting enzyme synthetic capacity of the cell, and (iii) the total concentration of intermediates is also minimized because intermediates are costly (for example, intermediates can be toxic or lost by leakage or decay, group carriers need to be synthesized, etc.). Let us now consider what happens if a given pathway would be extended by an additional step (14). Since the total enzyme concentration is constrained (assumption ii), synthesis of an additional enzyme has to be compensated by reducing the concentration of the other enzymes. Further, since the total intermediate concentration is also constrained (assumption iii), the additional intermediate means that the concentrations of the other intermediates have to be reduced. After extension of the pathway, the distribution of enzyme concentrations over the pathway must be optimized again in order to maximize the flux of substrate through the pathway (assumption i). Therefore, a longer pathway leads to reduced concentrations of enzymes and intermediates, and both reduce the flux. In the simplest case of linear kinetics, the rate of an enzyme reaction is proportional to the product of the enzyme and intermediate concentrations, and that is why the substrate flux decreases with the square of the pathway length (Fig. 6). Since the ATP yield increases with pathway length, the rate of ATP production typically has an optimum at a certain pathway length; the higher the substrate concentration and the toxicity or costs of the intermediates, the shorter the optimal pathway length (Fig. 6).

Costa et al. (14) have argued that this kinetic theory in fact explains why nitrification, the oxidation of ammonia to nitrate, proceeds in two steps carried out by different groups of prokaryotes, a fact discovered by Winogradsky in 1890. The first group, the ammonia oxidizers, presumably evolved the shorter pathway of incomplete oxidation of ammonia to nitrite because this increased their growth rate. The excreted nitrite then sustains the second group of nitrifiers, which oxidize nitrite to nitrate, and this is analogous to acetate cross-feeding.

However, there is a downside to pathway truncation because the shortened pathway, or incomplete conversion of the substrate, will have reduced ATP production and therefore growth yield (Fig. 6), explaining the observed trade-off between growth rate and yield (44, 52, 74, 81). We have seen that the yield is not relevant for fitness in chemostats, where fitness is solely determined by specific growth rate in accordance with assumption i of the kinetic theory, but we will see in the biofilm section how spatial structure of the biomass combined with growth limitation by diffusional flux into the biomass can, in the extreme, lead to the opposite result, namely, that only yield determines fitness (52). For that reason, Costa et al. (14) proposed that under such conditions, complete degradation by long pathways can evolve, and therefore that a one-step nitrifying organism amalgamating the metabolism of ammonia and nitrite-oxidizing prokaryotes can be expected to exist.

Considering this kinetic theory combined with the importance of yield in biofilms, one would expect the evolution of a metabolic division of labor giving rise to a community of specialists at high substrate concentrations, or high turnover rates, in reactors where mixing of individuals and substrates leads to global competition (perhaps as is the case in the rumen). On the other hand, under conditions of low substrate concentrations, or low turnover rates, and incomplete mixing of individuals

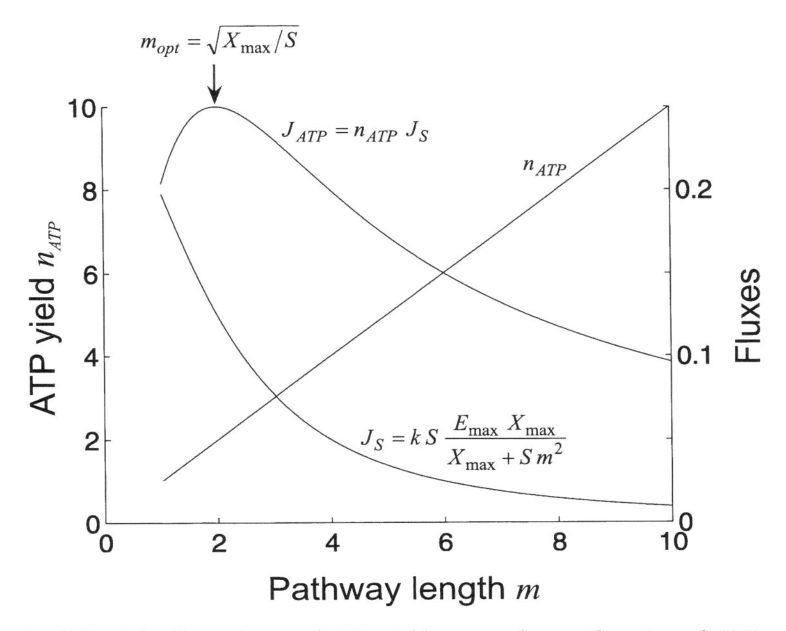


**FIGURE 6** Dependence of ATP yield, $n_{ATP}$, substrate flux, $J_S$, and ATP flux, $J_{ATP}$, on pathway length, $m$, according to the kinetic theory of optimal pathway length (35, 80). For the sake of simplicity, one assumes a linear pathway with linear, irreversible, and identical kinetics for all enzymes. Further, ATP yield is assumed to be proportional to pathway length. In this simplest case, analytical expressions for the fluxes can be derived. $k$ is the rate constant for all enzymes, $S$ is the substrate concentration, the total concentration of all enzymes is restricted to $E_{max}$, and the total concentration of all intermediates is restricted to $X_{max}$. For this example, all parameters were set to unity, apart from $S = 1/4$. Since ATP flux is the product of substrate flux and ATP yield, an optimal pathway length exists if the substrate flux decreases more steeply than the ATP yield increases. The optimal length of a pathway is shorter at higher substrate concentration and at higher metabolic costs of intermediates (restricting $X_{max}$ to lower values). Qualitatively the same results are obtained for reversible and/or Michaelis-Menten kinetics (14). Adapted from reference 14.

and substrate gradients, the food chain would be much shorter, consisting of fewer generalists with longer pathways, perhaps as is the case towards the end of a PFR such as a gut.

## COLONIZATION RESISTANCE AND WALL GROWTH: THE FRETER MODEL

The indigenous flora of the intestinal tract is quite diverse and stable, implying that many species have been able to colonize the gut and coexist for a long time, yet the resident flora is resistant to colonization by new invaders. Given the large daily intake of a plethora of food-borne microbes, this colonization resistance is quite surprising (29). Even if the invader is a wild-type strain of an indigenous species (e.g., an *E. coli* strain recently isolated from the feces), its ingestion will rarely result in colonization of the host. Remarkably, the same *E. coli* strain will usually colonize quite well when it is introduced into a germfree animal first, and the indigenous microflora is allowed to colonize only afterwards (29), demonstrating a priority effect that is also seen in plant and animal ecology.

Since bacterial species not indigenous to the intestinal flora are often able to colonize germfree animals, even when the hosts are immunized, but not conventional animals with a resident flora, the mechanism(s) of this colonization resistance must be based on interactions between invaders and the resident flora (29). One candidate interaction is competition for limiting nutrients. Since the competitor with the better Monod kinetics will be able to invade a system dominated by another strain (as explained above in the chemostat section), this mechanism fails to explain the priority effect that is characteristic of colonization resistance.

In order to understand which mechanism(s) could explain this priority effect, Freter (29) developed a simple mathematical model of chemostats with wall growth. This was done in an effort to determine whether competition for nutrients and/or attachment sites on the gut wall could be responsible for colonization resistance. Thus, he extended the classical chemostat model by mass action kinetic terms to include growth of microbes on the wall, as well as exchanges between the planktonic and wall-bound biofilm population via attachment and detachment, and also overspill of excess wall growth. Crucially, the biofilm population is not diluted by the flow through the reactor.

The assumptions of the Freter model include those discussed in the chemostat section and the following: (i) resident and invader strains compete for the same limiting nutrient and for the same adhesion sites (i.e., the invader has been isolated from the indigenous microbiota and reintroduced so it is the same strain as the resident); (ii) planktonic bacteria attach to adhesion sites in proportion both to their density and to the fraction of free attachment sites; (iii) biofilm bacteria detach from adhesion sites proportional to their density; (iv) a fraction $[G(W)]$ of the offspring of biofilm cells remain attached, depending on the fraction of free attachment sites available, while the complement $[1 - G(W)]$ joins the planktonic population (Fig. 7); (v) mortality and maintenance can be neglected; (vi) wall

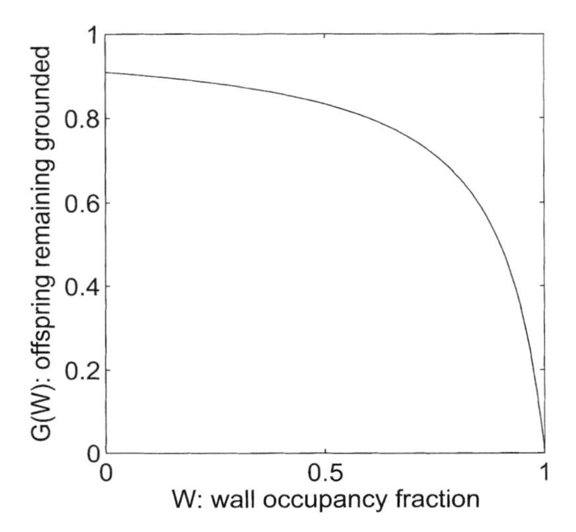


**FIGURE 7** Graph of the function $G(W) = (1 - W)/(1.1 - W)$, which models the fraction of biofilm offspring that remains grounded or bound to the wall because a free attachment site in the vicinity was available, which is therefore a monotone decreasing function of the fraction of the wall occupied, $W = \Sigma_i B_i / W_{max}$, where $W_{max}$ is the maximum number of attachment sites on the wall.

and lumen populations are described by identical growth functions because the biomass on the wall is thought to be infinitesimally thin so that substrate or physiological gradients cannot develop; and (vii) the substrate concentration represents that portion of limiting nutrient not utilized by the background flora. These assumptions are expressed in the differential equations of the Freter model:

$$\underbrace{\frac{dP_i}{dt}}_{\text{Change of plankton}} = \underbrace{+\mu_i(S)P_i}_{\text{Growth}} \underbrace{- DP_i}_{\text{Dilution}} \underbrace{- k_{at,i}(1 - W)P_i}_{\text{Attachment}} \underbrace{+ \{\mu_i(S)[1 - G_i(W)] + k_{de,i}\}\delta B_i}_{\text{Loss from biofilm: growth overspill + detachment}} \tag{5a}$$

$$\underbrace{\frac{dB_i}{dt}}_{\text{Change of biofilm}} = \underbrace{+\mu_i(S)B_iG_i(W)}_{\text{Growth remaining on wall}} \underbrace{+ k_{at,i}(1 - W)\delta^{-1}P_i}_{\text{Attachment}} \underbrace{- k_{de,i}B_i}_{\text{Detachment}} \tag{5b}$$

| Change of substrate | | Inflow − outflow | Substrate consumption summed over all planktonic and biofilm populations |
|---|---|---|---|
| $\frac{dS}{dt}$ | = | $D(S_0 - S)$ | $- \Sigma_i Y_i^{-1} \mu_i(S)(P_i + \delta B_i)$ |

**(5c)**

where terms are defined as follows: $P$, planktonic cell density; $B$, biofilm cell density; $i$, index for species (also indicating species-specific parameters or functions); $S$, substrate concentration; $S_0$, substrate concentration in the feed; $D$, dilution rate; $k_{at}$ or $k_{de}$, first-order rate constants for attachment or detachment, respectively; $\delta$, area of the reactor wall divided by the volume of the reactor, a factor which converts the areal density of the biofilm population into the volume-based density of the planktonic population; $G(W)$, the fraction of offspring remaining grounded on the wall as a function of $W$ (Fig. 7); $W$, wall occupancy fraction; $\mu_i(S)$, specific growth rate as a function of substrate concentration, typically a Monod function; and $Y_i$, yield of biomass per substrate consumed.

As a result, the Freter model predicts colonization resistance only if attachment, detachment, and growth on the gut wall are included (Fig. 8). To understand how competition for adhesion sites brings about colonization resistance, consider the dynamics occurring after ingestion of food-borne microbes. First, we consider invasion by one of the resident strains, so invader and resident are equally good competitors for the same resources and attachment sites. The invader must find a free attachment site, which is a rare event, so most cells of the invader population are diluted from the system before they have found an attachment site. The resident population, having colonized the mucosa, continuously sheds cells into the lumen, maintaining a high abundance of residents in the lumen despite dilution, in contrast to the invader, which does not have this source for replenishment of its population.

Based on the Freter model, it should be possible to prevent gastrointestinal infections by improving colonization resistance. Now we consider introduction of commensals that are more fit under in situ conditions, both in nutrient competition and in competition for wall attachment sites. This can overcome colonization resistance, especially if they are introduced repeatedly, thus replenishing the overwhelming fraction of colonizers that are washed out before happening upon an attachment site (6). After the more fit commensals have been established, it will be less likely that a pathogenic invader will exceed the fitness of the resident microflora. From the above considerations, it should be clear that our understanding of microbial competition based on chemostat experiments and models would be

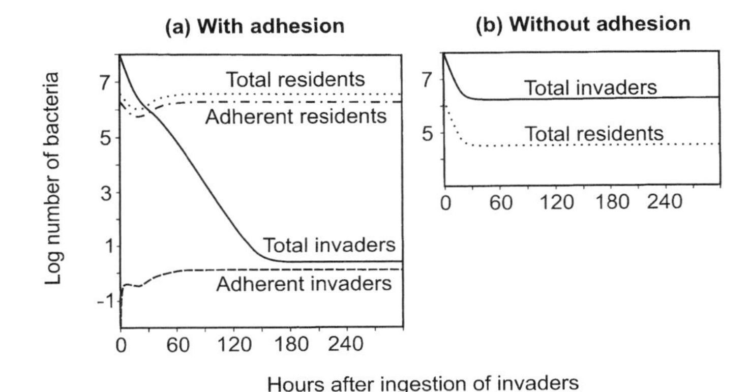


**FIGURE 8** Competition between resident and invader for the same limiting nutrients and adhesion sites in a chemostat. The resident and invader are the same strain, so they are equally fit, i.e., have equal parameters. (a) With adhesion to the gut wall; (b) without adhesion to the gut wall. Redrawn and adapted from reference 29.

misleading if applied to systems with wall growth, such as teeth or the gut.

## WHY ARE BIOFILMS SO UBIQUITOUS?

The Freter model also explains another more general observation, that biofilms are ubiquitous, present on almost any surface that is at least occasionally wet and provides abiotic conditions allowing bacterial growth. The common explanation that microbes form biofilms because they provide better growth conditions than the bulk liquid neglects the intense competition for resources diffusing into the biofilm, allowing only the top layer of cells to grow. Rather, the point is that microorganisms attached to surfaces only need to grow fast enough to compensate the rate of loss due to detachment, whereas planktonic cells usually must grow faster to compensate washout with the flow of bulk liquid (43).

In the small intestine, the rate of microbial removal is greater than the growth rates achieved under in situ conditions; thus, without wall growth, washout would occur (29). In contrast, the rate of removal in the large intestine is below the growth rates that the bacterial flora can sustain, so wall growth in the large intestine is not necessary for the existence of the flora, even though it is necessary for colonization resistance (29). As a corollary, it is much easier to eradicate a planktonic population by reducing the growth rate below the washout rate than it is to reduce the growth rate in the biofilm below the detachment rate (43).

## THE FRETER MODEL REVISITED: PFRs

Since many parts of the gastrointestinal tracts of animals are more like PFRs than chemostats or continuous-flow stirred-tank reactors (CSTRs) (see below for selective advantages of PFRs), Smith and coworkers have adapted the Freter model, originally couched in terms of the CSTR (Fig. 1), to the physical transport processes in a PFR (Fig. 9) (6). While the biology (growth and substrate consumption and attachment to and detachment from the wall) is independent of reactor type, the physics of mass transfer is different. Importantly, a gradient of decreasing substrate concentrations develops along the PFR (Fig. 9).

Given that the PFR differs from the CSTR in material flow and spatial gradients, it is surprising that the predictions of the PFR and CSTR models are in close agreement if the parameters and initial conditions of the PFR are chosen to reflect those of the CSTR. If the steady-state distribution of a population of residents only in the PFR is chosen as the initial condition for invasion, where wall-bound residents concentrate towards the upstream end, the PFR excludes the invader even more effectively than the CSTR (6).

## PLUG FLOW MODELS AND COEXISTENCE

Now that I have introduced the plug flow version of the Freter model developed by Smith and colleagues (6), I can briefly come back to the competitive exclusion principle. First note that the Freter CSTR model predicts that two or more bacterial strains that compete in the gut for the same limiting nutrient can coexist, if the metabolically less efficient strains have specific adhesion sites available (30). This mechanism allows more species in the gut than limiting nutrients. But strains competing for the same resource and the same wall attachment sites cannot coexist in a CSTR. Yet in a PFR, they can coexist in steady state because a gradient of decreasing substrate concentrations develops due to the lack of axial mixing. Spatial structure is key to this coexistence; the bacterial strains segregate into separate nonoverlapping segments along the wall of the reactor, corresponding to that range of nutrient concentrations for which each strain is the superior competitor (6). Thus, spatial separation into virtually pure cultures along the gut wall can be explained without having to invoke specific adhesion sites or local immune reactions (6).

## ARE IDEAL GUTS PFRs?

Penry and Jumars (78) compared different reactor designs and operating strategies known from chemical engineering to determine

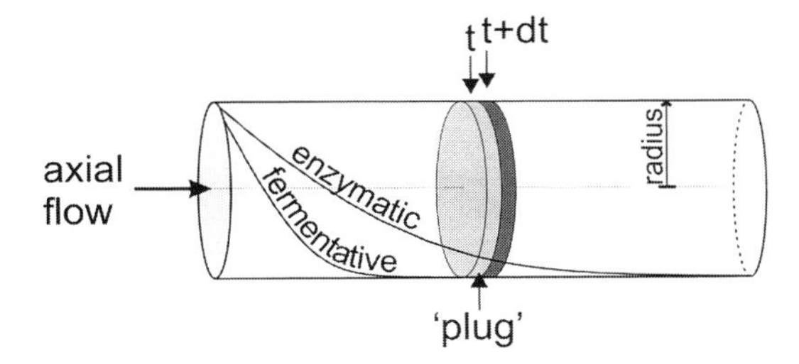


**FIGURE 9** Schematic drawing of a PFR. A "plug" or disk-shaped volume element of material enters the PFR tube on one end, being transported with constant volumetric flow rate along the length of the tube without axial mixing of the liquid (diffusive spread of the plug's components in the axial direction might be included). On the other hand, radial mixing is assumed to be perfect, so diffusion gradients cannot form in the radial direction. The plug's material will "age" during transport, e.g., microbes grow and nutrients deplete, so at a certain position along the length of the tube the plug's material will have a constant age and composition at steady state. Axial profiles of substrate concentrations resulting from enzymatic breakdown (assuming Michaelis-Menten kinetics and uniform enzyme concentration) or autocatalytic fermentation (assuming Monod kinetics for growth of the biomass) are shown. Note that a PFR can be approximated by linking CSTRs in series, since this reduces the dilution effect of a single CSTR and localizes the mixing. Indeed, as the number of CSTRs in the series increases, the behavior of the system rapidly approaches that of plug flow.

which combination of reactor type and operating strategy would optimize digestion. As the objective or goal function, which in optimization algorithms defines what is to be minimized or maximized, they chose to maximize conversion of ingested food in the shortest amount of time (or, equivalently, reactor volume).

For catalytic (e.g., enzymatic digestion) systems, they found that the PFR was the optimal design for the gut, outperforming a batch reactor because of its continuous operation. The PFR was also better than a CSTR because the lack of mixing in plug flow maintains a concentration gradient with maximal concentration at the inlet, whereas the CSTR immediately dilutes the inflowing material to some lower, constant level. Dilution of inflowing reactants reduces the efficiency of the CSTR because reaction rates decrease monotonically with the concentration of the reactants (unless they are toxic and inhibit the reaction at a high concentration). Thus, for enzymatic digestion, the PFR is the ideal gut.

For microbial fermentation, which is an autocatalytic reaction due to the growth of catalytically active biomass, the PFR takes longer (or requires more reactor volume) than the CSTR for a low degree of conversion. This is because the catalytic biomass has to build up in the PFR with increasing age and axial position of the plug volume element, while it is already present in the CSTR from the beginning. Since this biomass buildup is exponential, it is slow initially but increasingly fast, so for higher conversion, the PFR becomes more efficient than the CSTR. To achieve high conversions of reactants to products in minimum throughput time (or reactor volume), a combination of a CSTR followed by a PFR is the optimal configuration.

For a mixed microbial fermentation and enzymatic digestion process, a combination of CSTR and PFR in series is optimal again. In foregut fermenters, acid treatment ends fermentation when the material leaves the foregut, which operates as a CSTR chamber, so the remaining digestion is enzymatic, which is better performed in a PFR. In hindgut fermenters, enzymatic digestion precedes fermentation. In this case, the reversed sequence of PFR first and then CSTR is optimal, and this is indeed the design used by hindgut fermenters.

## COMPETITION OF COLONIES ON AGAR PLATES

I use competition between colonies on agar plates to introduce the concept of cellular automata (CA), which are often used as a simple tool to qualitatively model local interactions in spatially structured systems (24, 39, 58). Although CA can also be used as quantitative models, the fact is that they can be very simple yet generate complex spatial patterns, giving insight into the qualitative behavior of a system and demonstrating that complexity can emerge from simple rules (58).

A CA model typically divides the spatial domain into a grid of squares, and all the interactions take place between neighboring grid elements. The number of neighbors considered can be the four nearest neighbors (north, south, east, and west), known as the von Neumann neighborhood. With inclusion of the diagonal neighbors, which are slightly further away ($\sqrt{2}$), the Moore neighborhood results. Which interactions take place between the neighbors is specified by a set of rules for updating the status of the focal grid element, depending on the status of the neighbors and the focal grid element itself. The states of grid elements in a CA are discrete, like space and time, so the grid element can only be in one of a finite number of states at any one time.

CA rules can be used to model diffusion and (traffic) flow, pattern formation, and much more, but I illustrate CA rules (Fig. 10) by going back to the example of C, R, and S strains of *E. coli*. Kerr et al. (47) used CA to show that the competitors can coexist only if they can chase each other in space. The CA was set up to correspond to their experimental studies of the dynamics of colonies on agar plates. For this spatially structured case, interactions were considered to be local, taking place in a Moore neighborhood (eight neighbors). To simulate the global interactions in a mixed liquid culture, the neighborhood was simply all but the focal patch. The grid cells or patches of their CA can be in four different states: empty or colonized by either the C, R, or S strain. Starting with a random initial distribution of these states, focal patches are chosen randomly and their state updated using a series of simple rules. Specifically, the probability of filling an empty patch with either the C, R, or S strain is given by the fraction of the neighborhood occupied by the C, R, or S strain, respectively. If the focal patch is occupied with a colony of the C, R, or S strain, the colony is killed with probability $\Delta_C$, $\Delta_R$, or $\Delta_S$, where $\Delta_S$ increases with the fraction of the C strain in the neighborhood and with the toxicity of the colicin, satisfying the order of decreasing mortality $\Delta_{S,C} > \Delta_C > \Delta_R > \Delta_{S,0}$.

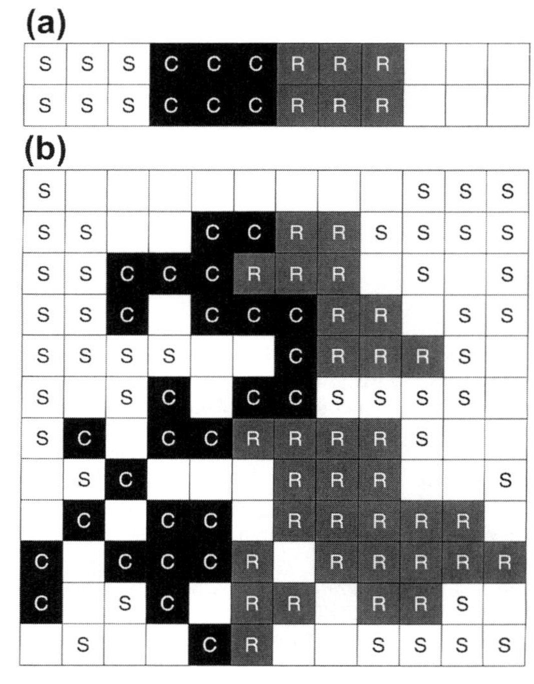


**FIGURE 10** An example of the spatial dynamics of colicin-producing (C), -resistant (R), and -sensitive (S) "colonies on agar plates" as simulated with the cellular automaton model described in reference 47, using the parameters of Fig. 1a to d of that reference. White patches are not colonized. (a) Initial distribution of colonies (only the bottom strip is shown since the distribution is uniform in the vertical direction). (b) After 10 steps (epochs) of the simulation. The small size of the CA lattice and uniformity of the initial lineup was chosen to show the workings of the CA rules described in the text. Since these rules are stochastic, runs will differ; however, the general trend is as shown: the C strain chases the S strain, the R strain chases the C strain, and the S strain chases the R strain. The empty patches can be colonized by the neighbors, here the S and R strains.

Thereby, it is guaranteed that the S strain without the C strain in the neighborhood is least likely to die ($\Delta_{S,0}$) but that the S strain surrounded by the C strain ($\Delta_{S,C}$) is most likely to die, leading to the cyclic order of competitiveness where the C strain displaces the S strain, the R strain displaces the C strain, and the S strain displaces the R strain in the absence of the C strain. As a result, when interactions are spatially localized (Moore neighborhood), the absence of the C strain in some locations allows the S strain to survive and replace the R strain (Fig. 10), but when the interactions are global due to mixing, no patches are free from colicin and the S strain dies out first, removing the advantage of the C strain so only the R strain survives in the end (47).

## DIFFUSION AND BIOFILMS

In contrast to the Freter model, which assumes that the biofilm is indeed just a (infinitesimally) thin film on the gut wall, smaller-scale models of biofilms aim to predict the development of the spatial structure of the biofilm and the gradients of substrate and metabolite concentrations, together with the ensuing physiological gradients. For all mechanistic small-scale biofilm models, reaction-diffusion equations are a key ingredient. I explain diffusion first; for a broader and deeper yet readable introduction, see reference 8.

On the microscopic scale, we find that molecules or individual particles execute a random walk known as Brownian motion, where the probability of motion in any direction is equally likely. For simplicity, we consider only one dimension (movement in the other dimensions is independent) in which the particle will step forwards or backwards with equal probability, so on average, it will remain where it started. Further, the distance traveled (the root-mean-square displacement) increases not linearly with time but with the square root of time, so it takes four times as long to get twice as far! This makes random walk counterintuitive because in our own experience, walking twice as long will get us twice as far. An important consequence is that a small molecule in water at room temperature (diffusion coefficient, $10^{-5}$ cm$^2$/s) takes only 0.5 ms to travel across a bacterial cell with a 1-μm radius, but to travel 1 cm from the gut lumen to the gut wall by random walk takes about 14 h.

To see how stochastic motion on the microscopic scale leads to deterministic behavior described by Fick's laws of diffusion on the macroscopic scale (equation 6), consider a large number of particles moving randomly and independently. A volume element with twice as many particles as a neighboring volume element will lose twice as many particles to the neighboring volume element per unit of time; hence, net movement (flux) of these undirected particles becomes directed down the concentration gradient and is proportional to the concentration gradient:

$$J_x = -d \frac{\partial S}{\partial x} \text{ (Fick's first law)} \qquad \textbf{(6a)}$$

where $J_x$ is specific flux (flux per area) in direction $x$ and $d$ is the diffusion coefficient.

The concentration increases with time when the flux is decreasing in $x$ direction (say, from left to right), so more flows in from the left side than leaves to the right side:

$$\frac{\partial S}{\partial t} = -\frac{\partial J_x}{\partial x} = d \frac{\partial^2 S}{\partial x^2} \text{ (Fick's second law)} \qquad \textbf{(6b)}$$

Diffusion as the macroscopic result of random motion of particles has also served as a simple way to describe animal movement or seed dispersal in ecological models (76).

Now we turn our attention to the consequences of diffusion for life in biofilms. Since substrate consumption in the biofilm generates a concentration gradient from biofilm sink to bulk liquid source, a cluster with a higher rate of substrate consumption, i.e., stronger sink, will create a steeper gradient and, according to Fick's first law (equation 6a), receive a higher influx of substrate (Fig. 11). Those who consume more resources will get more. Another consequence of substrate diffusion down the gradient is that the surface of the biofilm, particularly the surface facing the bulk liquid source, experiences the highest concentration of substrate and hence the highest growth rate. A small protrusion from the biofilm surface will experience an even higher substrate concentration, enabling the protruding part to grow faster, which will enlarge the protrusion, and so on. This positive feedback is the reason

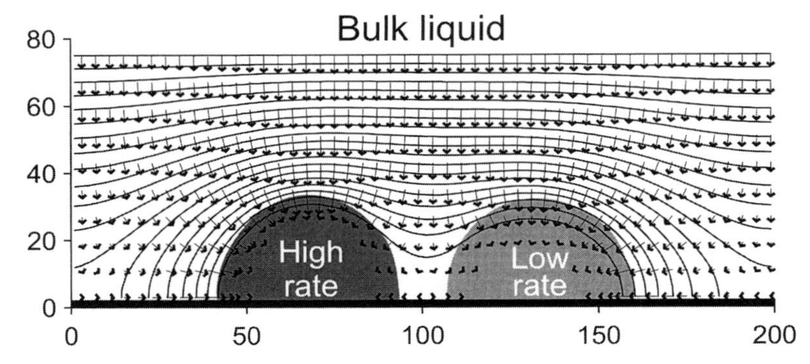


**FIGURE 11** Diffusion of substrate is driven by the substrate consumption in the biomass, which creates a concentration gradient from the bulk liquid (source) through the concentration boundary layer to the biomass (sink). Two equally sized biofilm clusters are shown as gray areas. The 20 contour lines indicate substrate concentration from 100% in the bulk liquid down to 5% near the cluster surfaces, in steps of 5%, and the arrows indicate substrate flux. Substrate concentration in the bulk liquid is assumed to be in steady state. The length scale is in micrometers. Note the steeper gradient towards the cluster with the higher rate of substrate consumption and that most of the substrate is consumed in a thin layer along the biofilm surface. Adapted from reference 54.

why a flat biofilm surface is unstable and any small random deviation will be amplified. This phenomenon, known as fingering instability because it creates finger-like structures (19, 82), is similar to what happens when plants compete for sunlight, where the higher plants absorb light that can no longer shine on their neighbors, resulting in lateral inhibition.

## DIVERSITY OF APPROACHES TO BIOFILM MODELING

The range of approaches that have been applied to biofilm modeling over the years can be perplexing. There are analytical versus numerical, deterministic versus stochastic, continuum versus discrete, and hybrid continuum/discrete models to choose from (93, 95). Those models with discrete biomass can be further subdivided into grid-based CA and particle-based models, where the particles are not constrained to move on a grid. The particle-based models have provided the basis for individual-based models (IbMs), but CA could in principle also be used as the basis for IbM if the discrete biomass blocks are allowed to have individuality.

The qualitative results obtained from various models should not depend on the actual mathematical approach chosen but only on the basic assumptions characterizing the system and expressed in the equations. Such model stability is indeed the case for single-species biofilms, where (for example) hybrid continuum/CA models (82), continuum models (19, 21), or IbMs (56) produce similar finger-like structures for reasons of diffusion limitation explained above. The correspondence of the overall biofilm structure (as a stand-in for a single-species biofilm) of the CA (82) and IbM (56) is illustrated in Fig. 12.

The finger-like biofilm structures in Fig. 12 are what we would expect a biofilm to look like if the only processes dictating biofilm structure are physics of mass transfer and a generic description of biomass growth, not including any biological factors such as cell-cell communication or motility.

While models agree on the structure of single-species biofilms, for two or more species, different models can generate divergent results because of the differences in the way biomass spreading is modeled, which affects

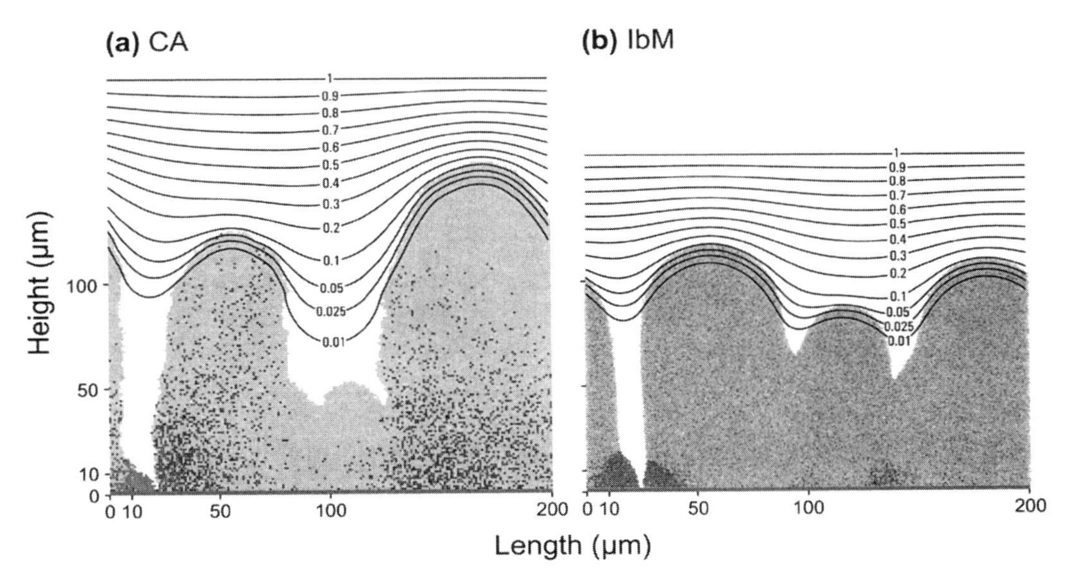


**FIGURE 12** Effect of different spreading mechanisms on spatial mixing in multispecies biofilms. Shown are nitrifying biofilms with ammonia-oxidizing bacteria (light gray) and nitrite oxidizing bacteria (darker gray) competing for oxygen. The contour lines show the oxygen concentration in milligrams per liter. (a) CA use stochastic rules for placing newly formed biomass blocks in the neighborhood. These sudden division events inside the biofilm trigger recursive random displacements of neighbors until the biofilm surface has been reached, resulting in stronger mixing than in IbM. (b) IbM uses a shoving algorithm where cells push each other away if they happen to come too close, and only as far as necessary. The stronger mixing evident in the CA model allows nitrite-oxidizing bacteria deep in the biofilm, where the oxygen concentration is so low that they hardly grow, to mix into the still-growing biomass of the ammonia-oxidizing competitors, so that they are carried along with the flow of the growing biomass upwards, where the oxygen comes from. Thus, spatial mixing can profoundly change competitive interactions. Adapted from Fig. 7c and d of reference 56. A movie of the time course can be found on my website (http://www.theobio.uni-bonn.de/people/jan_kreft/biofilms.html).

the extent of spatial mixing of the growing biomass (Fig. 12). Mixing changes the neighbors with which a cell competes, cooperates, or communicates, and it also changes the position of the cell in the gradients of substrate, product, signal, or antimicrobial concentration. Also, mixing lowers the relatedness of the cells in a neighborhood, and that is a key factor in the evolution of cooperative behavior (98).

## WHY OR WHEN TO USE IbMs?

Like CA, IbMs are bottom-up models and spatially explicit; i.e., space and spatial interactions are directly represented, so distinct locations and distances for interacting neighbors are defined. This is in contrast to the Freter model, where space is not modeled directly but the effects of space are implicit (e.g., in the wall occupancy fraction). The key difference from CAs is that in IbMs, individual members of a population are allowed to have their own distinct state, while all individuals of one species have the same potential states and activities (same "genotype"). The individual cells and their activities can be described by rules (like in a CA), by differential equations such as those used in structured models, or using a mixture of both (55). While a few simple rules and kinetic equations are all that is needed to describe cell division and growth, more sophisticated descriptions of internal dynamics such as metabolic pathways and networks of gene regulation or signal transduction networks can be embedded using the full range of approaches for modeling intracellular dynamics. Such networks can even be allowed to evolve.

Whereas the importance of individuality is beyond doubt in higher organisms, the activities of microorganisms have traditionally been measured as averages of large populations because their small size did not allow single-cell measurements. Recent technical advances have allowed us to study single cells, leading to the discovery that even cells in a population of simple unicellular prokaryotes differ from one another (10). The reasons for these differences are varied. For instance, phenotypic differences can result from environmental conditions, phase of the cell cycle, age, and changes in gene regulation or signaling (25, 51, 88). Also, the individual's state can depend on previous environments and intracellular changes. Last but not least, mutation provides the genotypic background for these phenotypic differences.

Stochastic gene regulation can result in some individuals of a population switching into an up-regulated steady state, while other isogenic individuals can be in a down-regulated steady state under the same external conditions (88). Mathematical analysis has shown that either positive-feedback or double-negative-feedback loops are required for such bistable behavior in gene regulation (25). Well-understood examples are sporulation in *Bacillus subtilis* and competence in *Streptococcus pneumoniae* (88). Also, signaling networks can be noisy, as is the case for chemotaxis. This noise is caused by the signaling network itself, and a change in the concentration of a key network component can suppress noise, suggesting that noisiness is a selected property of the network (51). Stochastic models of bistable networks can be embedded in IbMs.

## INDIVIDUAL VARIABILITY AS AN INSURANCE MECHANISM

The diversity of physiological states of individual members of a population has been proposed to provide an "insurance" mechanism for population survival in unpredictably changing environments (9, 102). An important example are persisters, which form a small fraction of the population that can survive treatment with a diverse group of antimicrobials, which rules out specific resistance mechanisms (33). In fact, these persisters are simply not killed because they are not growing, and most antibiotics target growing cells (33).

On the other hand, nongrowing cells will be outcompeted when the population is not challenged by an antimicrobial agent. Generally, maintaining a small percentage of the population in a physiological state that is better adapted to another environment comes at a fitness cost in the current environment. This is because this fraction of the population is ill adapted and will be outcompeted by the better-adapted fraction. Crucially, natural selection cannot act on these fitness differences if the differences between the subpopulations are phenotypic and therefore not inherited, and switching from growing to persister state or vice versa is a phenotypic change under genetic control (5). However, costly maintenance of cells preadapted to some potential future environment could be avoided if the cells could detect the new environment and react sufficiently quickly (100).

## SPATIAL STRUCTURE AND THE EVOLUTION OF SOCIAL BEHAVIOR IN BIOFILMS

Behavior in the sense of evolutionary biology is called social if it has consequences for the fitness of the actor and recipient, i.e., it does not require brains (Table 1). The realization that microbes show a variety of interesting social behaviors has led many evolutionary biologists to use microbes as model systems to study fundamental questions of evolutionary theory (15, 92, 98). It is particularly difficult to see how behavior can evolve that has a negative net effect on the direct fitness of the individual, i.e., altruism or spite (Table 1), because mutants without this behavior, called cheaters, will have a higher direct fitness, so their frequency should increase over time.

**TABLE 1** Hamilton's classification of social behavior according to the fitness effects of the behavior for actor and recipient(s)[a]

| Hamilton's classification of social behavior | Effect on recipient | | | |
|---|---|---|---|---|
| | Positive | | Negative | |
| Effect on actor | | | | |
| Positive | Cooperation | Mutual benefit | Harming | Selfishness |
| Negative | | Altruism | | Spite |

[a] A more thorough analysis of the variety of meanings of these and alternative terms is given in reference 97.

There are two different but mathematically equivalent ways to explain how altruism can evolve (89, 98). Kin selection theory is based on the realization that fitness benefits for recipients can increase the inclusive fitness (direct and indirect fitness) of the actor if their genetic relatedness is positive (relatedness can be negative because it is relative to the population average). This is captured in Hamilton's rule stating that kin selection favors the evolution of a trait if the benefit $b$ to the recipient, weighted by relatedness $r$, is higher than the cost $c$ to the actor: $rb > c$ (98).

Another way of looking at the same problem is group selection theory (89), which I emphasize here because it is often misunderstood in the microbiological literature. Specifically, old group selection theory, which dates from the 1960s, was based on the idea that individuals behave for the good of the population or species. But old group selection has been shown theoretically and experimentally not to work (96). In contrast, new group selection theory considers selection between groups within the same population rather than between different populations. In other words, selection in a group-structured population is considered, not the selection of group characters. This within-population group selection does work; in fact, it leads to the same results as kin selection theory (53, 89, 97, 98).

Group structure in the population can simply arise from a lack of spatial mixing, as is typical for the growth of microbes attached to a surface, or embedded in aggregates (such as clumps, sludge flocs, marine snow, etc.), where the cells can become immobilized in an extracellular matrix or motility can be switched off. Under such conditions, the multiplication of a single immotile or immobilized cell will give rise to a microcolony (a cluster of clonal cells), and thus the group is spatially coherent, a neighborhood consisting of relatives, so we come back to the kin selection perspective above.

How fitness advantages on the group level can lead to the success of group-beneficial traits despite the fitness costs for the individual can be explained by referring to a counterintuitive statistical result known as Simpson's paradox (Table 2) (89). Consider a group consisting of altruists $(A)$ and selfish individuals $(S)$, in which the total number of individuals is $N = A + S$, and the frequency of altruists is $p = A/N$. Assume for simplicity that the benefit derived from the behavior of an altruistic individual is shared equally among all members of the group rather than directed towards a particular individual. For example, investments into what is known as a public good, such as exoenzymes for the degradation of particulate organic matter, would provide undirected benefits because the exoenzymes and their breakdown products diffuse and are available to all. Within such a group, the fitness of the selfish fraction of the group $(F_S)$ is the sum of a basal fitness (the fitness in the absence of altruistic behavior, $F_0$) and the benefit $b$ from the behavior of the altruistic group members with frequency $p$: $F_S = F_0 + pb$. The fitness of the altruistic fraction $(F_A)$ is reduced by the cost of altruistic behavior $c$ and

**TABLE 2** An example of Simpson's paradox showing that the frequency of altruists can increase globally despite a decrease of the frequency of altruists within each group[a]

| Generation | Group 1 | | | Group 2 | | | Sum | | |
|---|---|---|---|---|---|---|---|---|---|
| | $N$ | $A$ | $p$ | $N$ | $A$ | $p$ | $N$ | $A$ | $p$ |
| $t$ | 100 | 10 | 0.10 | 100 | 90 | 0.90 | 200 | 100 | 0.50 |
| $t + 1$ | 1,040 | 95 | 0.09 | 1,360 | 1,215 | 0.89 | 2,400 | 1,310 | 0.55 |

[a] $N$, number of individuals; $A$, number of altruists; $p$, frequency of altruists. For parameters and equations, see the text.

hence always lower than $F_S$: $F_A = F_S - c = F_0 + pb - c$. The benefit is assumed to be larger than this cost; otherwise, the altruistic strategy would not be viable. The fitness of the group as a whole is the weighted sum of the fitnesses of the altruistic and selfish fractions: $F_G = pF_A + (1 - p)F_S$. From these fitness expressions, the number of individuals in the next generation can be found, e.g., $N_{t+1} = N_tF_G$.

Now consider a second group with a different frequency of altruists and therefore different fitnesses for altruists, nonaltruists, and the group as a whole, since all these fitnesses are frequency dependent. While the relative fitness of the altruists is lower than the fitness of the nonaltruists within each group, the overall number of altruists may nevertheless increase because the predominantly altruistic group has far more offspring, offsetting the reduction in the altruists' frequency. Using parameter values of $F_0 = 10$, $c = 1$, and $b = 5$, Table 2 summarizes this example of Simpson's paradox. Note that summing over all groups as done in Table 2 is only meaningful when groups compete, that is, if they are groups within the same population. An important consequence of the steadily declining frequency of altruists within each group is that altruists need some sort of "purification step" from time to time in order to survive. The easiest way to achieve this would be by dissolution of the group into single cells or by shedding of single cells followed by refounding of new groups (colonies) from single cells (52). If groups propagate as a unit, e.g., if the biofilm community as a whole would colonize a new patch, this purification step would be forgone, and the frequency of cheaters would rise continuously.

The example we now consider, a microbial "tragedy of the commons," illuminates the effect of spatial structure on competition between different groups of the population with different growth strategies (52, 79, 81). The rate strategy has a high $\mu_{max}$ but a low growth yield because of the trade-off between the yield and rate of ATP production (as explained in the section on kinetic theory). Conversely, the yield strategy has a high yield but low $\mu_{max}$. The specific affinity is assumed to be the same for both strategies, because otherwise further trade-offs between substrate affinity and yield or rate would muddy the waters, and we want to look at one trade-off at a time (52). Given the Monod functions of the two competing growth strategies (same initial slope and higher $\mu_{max}$ of the rate strategists), it is clear that the specific growth rate of the rate strategists is always higher. To see what happens in the chemostat, we only need to apply the $R^\star$ rule (Fig. 2b), which tells us that the rate strategists are the better competitors for all substrate concentrations (52). In biofilms, the higher substrate consumption rate of a group or cluster of rate strategists will lead to slightly steeper concentration gradients and thus a slightly higher substrate flux into the biofilm (Fig. 11). However, the group of yield strategists, while receiving slightly less substrate flux, converts the incoming amount of substrate with a higher yield or efficiency into biomass, so a group of slower-growing individuals can grow faster than a group of fast-growing individuals, provided that growth is limited by diffusion into the biofilm (Fig.

13). In the early phase of colonization of a surface, however, growth is not limited by diffusion and the cells grow exponentially, so the rate strategists do grow faster. If groups of rate strategists can overgrow the groups of yield strategists in this early phase of exponential growth, they can win the competition (Fig. 13). Rate strategists will be unable to overgrow all yield strategists if at least some are far enough away, e.g., because a low abundance of cells colonizing the surface means they are few and far between. Later, when more and more cells have piled up, the increased substrate consumption leads to the diffusion-limited growth phase, in which all substrate diffusing into the biofilm is consumed, so a constant flux and concentration gradient develops, leading to linear growth of the biofilm (52).

So, the key condition for yield strategists to win is a spatially structured population with limited mixing of the organisms, e.g., all cells immobilized in an extracellular matrix, so the different groups can remain more or less intact and spatially separated. As discussed above, the various approaches to biofilm modeling mainly differ in the way the biomass is spread; e.g., the individual or particle-based models lead to minimal mixing of cells in the biofilm (Fig. 12). It turns out that this is particularly relevant for the competition of rate and yield strategists.

As demonstrated by Simpson's paradox, a mutant or immigrant rate strategist within a yield group has a higher growth rate than the neighboring yield strategists, which experience the same substrate concentration. Hence, cheaters can overtake groups of yield strate-

**FIGURE 13** Competition between groups of rate (light) and yield (dark) strategists in a biofilm population. The surface was initially colonized by single cells at equal distances, alternating between rate and yield strategists. (a) 5 cells each; (b) 10 cells each; (c) 20 cells each. If cells were far enough apart initially (a), the rate strategists could not overgrow the yield strategists during the initial phase of unlimited growth, when they grow faster. When growth becomes limited by diffusion of substrate into the biofilm, the yield strategists convert the influx of substrate more efficiently into biomass and grow faster as a group. At high initial cell density (c), the rate strategists overgrow the yield strategists, but by chance some yield strategists are carried upwards, with the flow of the spreading biomass, and form a cluster of yield strategists at the biofilm surface. If these yield clusters at the surface are large enough, they can overgrow the neighboring rate strategists. The breadth of the domain was 200 μm. Movies of these simulations that show the spreading of the biomass can be seen on my website (http://www.theobio.uni-bonn.de/people/jan_kreft/altruism.html). Adapted from reference 52.

gists, so they need to be refounded from time to time, e.g., from a single yield individual.

Another necessary condition is that diffusion of substrate into the biofilm has to limit growth rate so the higher yield pays off. This condition is easily met, since a growing biofilm will sooner or later consume enough substrate to limit growth by diffusional influx of substrate.

Experiments with isogenic strains of yeast, which show the rate versus yield trade-off due to alternative pathways of glucose metabolism, are in excellent agreement with model predictions (63). (When fermentation of glucose is used in addition to glucose respiration, this respirofermentative metabolism supports faster growth but at a lower yield. Respiration alone leads to a higher yield but slows down growth, which is in accordance with the prediction of kinetic theory, that the longer the pathway, the lower the rate but the higher the yield of ATP production.) Another recent study (46) has shown that spatially restricted dispersal is essential for the survival of T4 phages with a strategy of efficient resource use (*E. coli* as a host or resource) at the cost of reduced growth rate, again in excellent agreement with model predictions.

## EPS, BIOFILM STRUCTURE, AND COOPERATION

The production of an extracellular matrix or extracellular polymeric substances (EPS) is typical for biofilms and has a large variety of effects; these effects are often assumed to be the basis upon which EPS functions evolved (99), which implies fitness benefits. Arguably the earliest and most pervasive such function of EPS production is the production of "volume" or distancing from neighbors, which could be competitors, cooperators, phages, or macrophages, since it is an unavoidable effect of EPS production. Production of volume per se already has various consequences for the structure of the biofilm, including spatial segregation of producing cells from one another or from nonproducing cells, spatial spreading of biomass and activity, and changes in the vertical structure (2, 57).

Add to this the effect of the structural changes on the fitness of the EPS producer and others, and volume production becomes exciting. Using an IbM of biofilms, Xavier and Foster (101) studied the competition of strains that differ in their levels of EPS production. EPS-producing cells have a lower fitness than nonproducers in pure biofilms because they have to divert part of their resources from growth into EPS production. However, an EPS-producing lineage has a strong fitness advantage in a biofilm of nonproducers, provided the costs of volume production are not too high (101).

A single polymer-producing cell in a neighborhood of nonproducers suffers a reduced rate of growth but will benefit cells that lie above by pushing them further up towards the substrate source, including some of its descendants. On the other hand, those cells on the side and below the focal EPS producer suffer reduced fitness because of increased substrate consumption above them, reducing their substrate supply. Intriguingly, EPS production is social behavior, in which altruism for the cells above and spite for the cells on the side and below are combined (101). EPS production is analogous to plant competition in which fast vertical growth towards the sunlight leads to shading of the lateral neighbors and those plants below (101).

## MUTUALISM: COOPERATION BETWEEN SPECIES

As occurs in the case of cooperation within species, cheaters are also a problem for the evolution of mutualism, i.e., cooperation between species. Mutualism resembles the prisoner's dilemma (27), where two players can benefit from cooperation but gain most from cheating their partners. In the prisoner's dilemma, cooperative behavior is favored when the interactions are repeated (iterated prisoner's dilemma [IPD]) and when the behavior is conditional on the partner's past behavior, e.g., tit for tat (20). However, when the investments in mutualistic interactions and the corresponding payoffs are allowed to evolve, they evolve towards zero in a nonspatial IPD

(20). For mutualism to evolve, increased investments in a partner must yield increased returns (20). Note that spatial structure in the IPD is required for the evolution of mutualism but not for the evolution of intraspecific cooperation (85). Interestingly, the partner with the higher evolutionary rate, which is typically the smaller symbiont, has a higher investment and a lower payoff (20).

There are three feedback mechanisms that allow an increased investment in a partner to yield an increased return (27). In the first, called cooperator association, an individual tends to associate with the more cooperative genotypes of the partner species. Partner fidelity feedback is when investment improves the fitness of the partner species and thereby improves its ability to return aid; this corresponds to repeated interactions in the IPD. Finally, partner choice and sanctions are behavioral adaptations in the partner species that result in aid being preferentially directed to the more mutualistic individuals. This corresponds to conditional behavior in the IPD. The last two are phenotypic feedbacks, which respond more rapidly and are therefore probably the more important of the three mechanisms (27). These feedbacks will only favor mutualism when most of the benefit returns to the individual that initiated the feedback—or its relatives. Therefore, within-species relatedness and between-species fidelity must be high (26). Ironically, the presence of cheaters is a necessary incentive for the evolutionary maintenance of partner choice (26).

Plenty of opportunities for mutualistic interaction can be found in the stable communities of the oral cavity and the gut. It has been suggested that the human host and its gut microbiota are a unit of selection in which the fitness of the host and its gut community are coupled (62). It has further been suggested that group selection theory, which has been successfully applied to selection between groups within a population of individuals of the same species, can also be applied to this unit of selection, which is a community of populations (62). It should be noted that one essential condition for group selection to work is that the group phenotype is heritable in some way, analogous to the individual level of selection, which requires heritability of the phenotype of the individual (53), and it has indeed been reported that the gut community of mice is inherited from the mother (62). However, a study of the temporal development of the gut community of 14 human babies over their first year of life has shown that the initial communities, which were determined by the specific bacteria the babies happened to be exposed to and therefore related to the mother's communities, do not persist. Rather, the communities of all babies converged towards a common community typical for adults (77). A further problem of this idea of host-microbiota group selection is that any mutant gut microbes with increased fitness could easily arise over the life of the host and among the huge number of resident microbes. These mutants would steadily increase in frequency, and assuming that the gut community is actually passed from mother to baby, this trend would likely continue in any offspring until the more fit mutant has replaced the wild-type strain. While this mutant could render the host more or less fit, this would not affect the mutant's fitness within the gut community or make it less likely to transfer to the gut of any descendants, as has been argued previously (61). As explained by Simpson's paradox, a refounding of the group from single cells would be necessary to stop the within-group rise of more fit mutants that reduce the fitness of the group. But it is hard to imagine how the whole gut community could be reassembled from single cells excluding any offspring of the mutant.

This suggests that the host, who cannot rely on group selection to favor a beneficial gut community, has evolved a mechanism of rewards or punishments to control the gut community in its favor. The immune system may be able to selectively punish those commensals that evolve aggressive traits. At least it is known that the immune system is nonresponsive to the commensal flora, so well-behaving microbes do not appear to be punished (42). Host sanctions have been demonstrated in mutualistic associations. Legumes reduce oxygen supply to root nodules that fix less nitrogen, so mutants of rhizobia that cheat will rise within the nodule group but the group's fitness is reduced due to the plant's sanctions (48). Dark mutants of *Vibrio fischeri,* which do not invest energy in bioluminescence, are unable to colonize the bobtail squid host (94). Such host sanctions must be preferentially directed towards the less mutualistic symbionts, and this requires maintenance of spatial separation of cheater and wild type.

## SPATIAL STRUCTURE AND QUORUM SENSING

One of the particularly fascinating social behaviors of microorganisms is communication by signal molecules, called quorum sensing (31). However, there are instances in nature that look like communication but are in fact chemical manipulation or just cues (45), especially for so-called communication between different species (17). Further, it has been pointed out that the information that cells obtain by measuring the concentration of a signal molecule is not simply related to cell density, since barriers to the diffusion of the signal (83) or spatial clustering of the cells (37) can drastically alter the signal concentration, irrespective of cell density (Fig. 14). Signal concentration in fact correlates with cell density only in well-mixed liquid cultures (37). This means that in most natural habitats, the cells can only measure the combination of factors influencing the concentration of the signal molecules. Comparing randomly distributed cells with clustered distributions, it can be shown that a cell in a microcolony communicates largely with other cells of the same microcolony rather than cells in other microcolonies or lone cells. In contrast, the signal concentration experienced by lone cells comes from a large catchment area which includes various clusters and lone cells (A. Dötsch, B. A. Hense, and J. U. Kreft, unpublished data). Interestingly, the positive feedback typically observed in signal production has two effects: it facilitates keeping communication in a microcolony private, and it synchronizes the response of all cells in a microcolony. Diverse species may actually use similar signal molecules. Consequently, in habitats having a high biodiversity

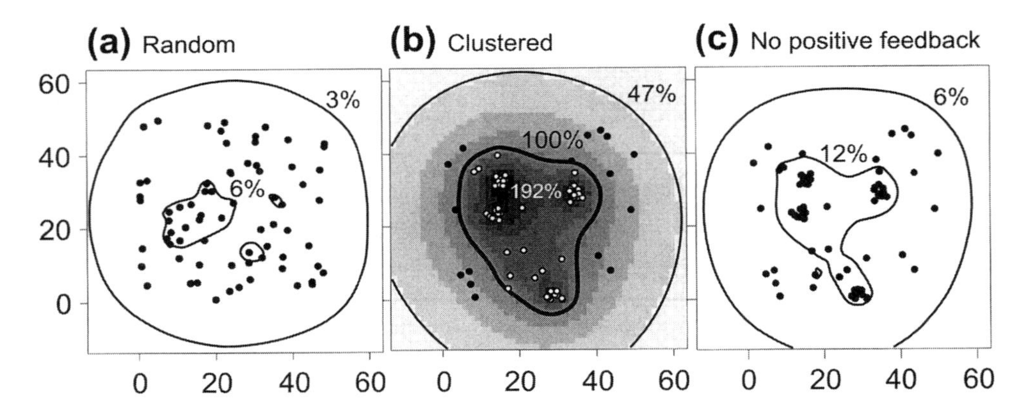


**FIGURE 14** A simple mathematical model of autoinducer production and diffusion shows the effect of spatial distribution of the same number of cells (same cell density) on the steady-state concentration of autoinducer obtained. (a) Random cell distribution; (b) clustered cell distribution; (c) same clustered distribution but without positive feedback in the production of autoinducer. The autoinducer concentration is indicated by contour lines in percentage of the threshold for up-regulation and also by increased darkness at increasing concentration. It is much higher in the clusters and their vicinity because cells in clusters become up-regulated (white inside), which due to positive feedback leads to higher rates of signal production in the clusters. The scale is in micrometers. Adapted from reference 37.

of signaling species, confusion created by the largely random mix of species in the vicinity is effectively avoided if cells grow in microcolonies (37).

This signaling system also faces evolutionary problems of exploitation by cheating and dishonest mutants (11, 18, 84). The production of the signal is costly; hence, cheater mutants not producing the signal but reaping the benefits can exploit the wild type (18). Further, responding to the signal, e.g., by switching on the production of exoenzymes at a high signal concentration, is also costly, so cheater mutants not listening to the signal can also exploit the wild type (18, 84). Moreover, the sender has an incentive to be dishonest by overproducing signal (pretending that more cells are nearby) if that would manipulate the recipient to overrespond. After all, the recipient has no means of checking the conveyed information and therefore cannot punish dishonest senders (11). Because of their clonal structure, microcolonies not only help cells ignore the noise from unrelated cells chattering in the vicinity but also "solve" these evolutionary problems (18, 37).

## CONCLUSIONS

In this chapter, I have made an effort to explain some basic and complex mathematical models, illustrating their components and applications. In so doing, I have attempted to explain why spatial structure is important to microbial ecology. In short, limited dispersal of organisms leads to clustered spatial distributions of siblings, and limited transport of chemical substances leads to diffusion gradients that critically affect competition, cooperation, and communication in natural systems, to name just a few common behaviors of microbes.

There are a variety of ways in which we can learn from models. Models organize our thinking and experiments in a logical way. They force us to describe processes in their entirety, thereby pointing a finger at any gaps in our knowledge. The language of mathematics allows us—or forces us—to state assumptions unequivocally. This enables us to identify those implicit assumptions that we have made unknowingly. Models simplify reality and prioritize processes. They can be used to test hypothetical scenarios that are difficult to realize experimentally. For example, what would a biofilm look like if only the physics of mass transfer and a generic description of growth shaped its structure? What effect would a specific biological process, such as twitching motility, have on biofilm structure and fitness? Models based on obvious assumptions may well produce unexpected or counterintuitive results. For example, the distance a random walker has progressed is proportional to the square root of time. Microcolonies of slowly but efficiently growing cells grow faster than microcolonies of fast-growing cells. Some of these surprising results have been given a name, such as Simpson's paradox. Sometimes, a model may allow us to explore areas where our intuition fails, such as general relativity. Models of entirely different processes, such as enzyme and population dynamics, may be mathematically equivalent, highlighting analogies between processes that might otherwise be missed, especially if they are subjects arising from different disciplines. Whatever flavor of model is used, consideration of spatial structure is critical if we are to predict the dynamics of microbial interactions in natural systems.

## ACKNOWLEDGMENTS

I thank my colleagues Ginger Booth, Julian Wimpenny, Cristian Picioreanu, Martin Rost, and Wolfgang Alt for helping me in various ways to learn mathematical modeling over the years.

## REFERENCES

1. **Adams, J.** 2004. Microbial evolution in laboratory environments. *Res. Microbiol.* **155:**311–318.
2. **Alpkvist, E., C. Picioreanu, M. C. van Loosdrecht, and A. Heyden.** 2006. Three-dimensional biofilm model with individual cells and continuum EPS matrix. *Biotechnol. Bioeng.* **94:**961–979.
3. **Armstrong, R. A., and R. McGehee.** 1980. Competitive exclusion. *Am. Nat.* **115:**151–170.

4. **Baer, S. M., B. T. Li, and H. L. Smith.** 2006. Multiple limit cycles in the standard model of three species competition for three essential resources. *J. Math. Biol.* **52:**745–760.
5. **Balaban, N. Q., J. Merrin, R. Chait, L. Kowalik, and S. Leibler.** 2004. Bacterial persistence as a phenotypic switch. *Science* **305:**1622–1625.
6. **Ballyk, M. M., D. A. Jones, and H. L. Smith.** 2001. Microbial competition in reactors with wall attachment: a mathematical comparison of chemostat and plug flow models. *Microb. Ecol.* **41:**210–221.
7. **Becks, L., F. M. Hilker, H. Malchow, K. Jürgens, and H. Arndt.** 2005. Experimental demonstration of chaos in a microbial food web. *Nature* **435:**1226–1229.
8. **Berg, H. C.** 1993. *Random Walks in Biology.* Princeton University Press, Princeton, NJ.
9. **Boles, B. R., M. Thoendel, and P. K. Singh.** 2004. Self-generated diversity produces "insurance effects" in biofilm communities. *Proc. Natl. Acad. Sci. USA* **101:**16630–16635.
10. **Brehm-Stecher, B. F., and E. A. Johnson.** 2004. Single-cell microbiology: tools, technologies, and applications. *Microbiol. Mol. Biol. Rev.* **68:**538–559.
11. **Brown, S. P., and R. A. Johnstone.** 2001. Cooperation in the dark: signalling and collective action in quorum-sensing bacteria. *Proc. R. Soc. Lond. Ser. B* **268:**961–965.
12. **Button, D. K.** 1991. Biochemical basis for whole-cell uptake kinetics—specific affinity, oligotrophic capacity, and the meaning of the Michaelis constant. *Appl. Environ. Microbiol.* **57:**2033–2038.
13. **Button, D. K., B. Robertson, E. Gustafson, and X. Zhao.** 2004. Experimental and theoretical bases of specific affinity, a cytoarchitecture-based formulation of nutrient collection proposed to supercede the Michaelis-Menten paradigm of microbial kinetics. *Appl. Environ. Microbiol.* **70:**5511–5521.
14. **Costa, E., J. Pérez, and J. U. Kreft.** 2006. Why is metabolic labour divided in nitrification? *Trends Microbiol.* **14:**213–219.
15. **Crespi, B. J.** 2001. The evolution of social behavior in microorganisms. *Trends Ecol. Evol.* **16:**178–183.
16. **Czaran, T. L., R. F. Hoekstra, and L. Pagie.** 2002. Chemical warfare between microbes promotes biodiversity. *Proc. Natl. Acad. Sci. USA* **99:**786–790.
17. **Diggle, S. P., A. Gardner, S. A. West, and A. S. Griffin.** 2007. Evolutionary theory of bacterial quorum sensing: when is a signal not a signal? *Philos. Trans. R. Soc. Lond. B* **362:**1241–1249.
18. **Diggle, S. P., A. S. Griffin, G. S. Campbell, and S. A. West.** 2007. Cooperation and conflict in quorum-sensing bacterial populations. *Nature* **450:**411–414.
19. **Dockery, J., and I. Klapper.** 2002. Finger formation in biofilm layers. *SIAM J. Appl. Math.* **62:**853–869.
20. **Doebeli, M., and N. Knowlton.** 1998. The evolution of interspecific mutualisms. *Proc. Natl. Acad. Sci. USA* **95:**8676–8680.
21. **Eberl, H. J., D. F. Parker, and M. C. M. van Loosdrecht.** 2001. A new deterministic spatio-temporal continuum model for biofilm development. *J. Theor. Med.* **3:**161–175.
22. **Edelstein-Keshet, L.** 1988. *Mathematical Models in Biology.* McGraw-Hill, Boston, MA.
23. **Egli, T.** 1995. The ecological and physiological significance of the growth of heterotrophic microorganisms with mixtures of substrates. *Adv. Microb. Ecol.* **14:**305–386.
24. **Ermentrout, G. B., and L. Edelstein-Keshet.** 1993. Cellular automata approaches to biological modeling. *J. Theor. Biol.* **160:**97–133.
25. **Ferrell, J. E., Jr.** 2002. Self-perpetuating states in signal transduction: positive feedback, double-negative feedback and bistability. *Curr. Opin. Cell Biol.* **14:**140–148.
26. **Foster, K. R., and H. Kokko.** 2006. Cheating can stabilize cooperation in mutualisms. *Proc. Biol. Sci.* **273:**2233–2239.
27. **Foster, K. R., and T. Wenseleers.** 2006. A general model for the evolution of mutualisms. *J. Evol. Biol.* **19:**1283–1293.
28. **Franchini, A. G., and T. Egli.** 2006. Global gene expression in *Escherichia coli* K-12 during short-term and long-term adaptation to glucose-limited continuous culture conditions. *Microbiology* **152:**2111–2127.
29. **Freter, R.** 1983. Mechanisms that control the microflora in the large intestine, p. 33–54. *In* D. Hentges (ed.), *Human Intestinal Microflora in Health and Disease.* Academic Press, New York, NY.
30. **Freter, R., H. Brickner, J. Fekete, M. M. Vickerman, and K. E. Carey.** 1983. Survival and implantation of *Escherichia coli* in the intestinal tract. *Infect. Immun.* **39:**686–703.
31. **Fuqua, W. C., S. C. Winans, and E. P. Greenberg.** 1994. Quorum sensing in bacteria: the LuxR-LuxI family of cell density-responsive transcriptional regulators. *J. Bacteriol.* **176:**269–275.
32. **Fussmann, G. F., S. P. Ellner, K. W. Shertzer, and N. G. Hairston, Jr.** 2000. Crossing the Hopf bifurcation in a live predator-prey system. *Science* **290:**1358–1360.
33. **Gilbert, P., T. Maira-Litran, A. J. McBain, A. H. Rickard, and F. W. Whyte.** 2002. The

physiology and collective recalcitrance of microbial biofilm communities. *Adv. Microb. Physiol.* **46:**202–256.

34. **Heinrich, R., and S. Schuster.** 1996. *The Regulation of Cellular Systems.* Chapman & Hall, New York, NY.
35. **Heinrich, R., S. Schuster, and H. G. Holzhütter.** 1991. Mathematical analysis of enzymic reaction systems using optimization principles. *Eur. J. Biochem.* **201:**1–21.
36. **Helling, R. B., C. N. Vargas, and J. Adams.** 1987. Evolution of *Escherichia coli* during growth in a constant environment. *Genetics* **116:**349–358.
37. **Hense, B. A., C. Kuttler, J. Müller, M. Rothballer, A. Hartmann, and J. U. Kreft.** 2007. Does efficiency sensing unify diffusion and quorum sensing? *Nat. Rev. Microbiol.* **5:**230–239.
38. **Hofbauer, J., and K. Sigmund.** 1998. *Evolutionary Games and Population Dynamics.* Cambridge University Press, Cambridge, England.
39. **Hogeweg, P.** 1988. Cellular automata as a paradigm for ecological modeling. *Appl. Math. Comput.* **27:**81–100.
40. **Huisman, J., and F. J. Weissing.** 1999. Biodiversity of plankton by species oscillations and chaos. *Nature* **402:**407–410.
41. **Huisman, J., and F. J. Weissing.** 2001. Fundamental unpredictability in multispecies competition. *Am. Nat.* **157:**488–494.
42. **Iweala, O. I., and C. R. Nagler.** 2006. Immune privilege in the gut: the establishment and maintenance of non-responsiveness to dietary antigens and commensal flora. *Immunol. Rev.* **213:** 82–100.
43. **Jones, D., H. V. Kojouharov, D. Le, and H. Smith.** 2003. The Freter model: a simple model of biofilm formation. *J. Math. Biol.* **47:**137–152.
44. **Kappler, O., P. H. Janssen, J. U. Kreft, and B. Schink.** 1997. Effects of alternative methyl group acceptors on the growth energetics of the O-demethylating anaerobe *Holophaga foetida. Microbiology* **143:**1105–1114.
45. **Keller, L., and M. G. Surette.** 2006. Communication in bacteria: an ecological and evolutionary perspective. *Nat. Rev. Microbiol.* **4:**249–258.
46. **Kerr, B., C. Neuhauser, B. J. Bohannan, and A. M. Dean.** 2006. Local migration promotes competitive restraint in a host-pathogen 'tragedy of the commons.' *Nature* **442:**75–78.
47. **Kerr, B., M. A. Riley, M. W. Feldman, and B. J. Bohannan.** 2002. Local dispersal promotes biodiversity in a real-life game of rock-paper-scissors. *Nature* **418:**171–174.
48. **Kiers, E. T., R. A. Rousseau, S. A. West, and R. F. Denison.** 2003. Host sanctions and the legume-rhizobium mutualism. *Nature* **425:** 78–81.
49. **Koch, A. L.** 1974. Competitive coexistence of two predators utilizing the same prey under constant environmental conditions. *J. Theor. Biol.* **44:** 387–395.
50. **Kolenbrander, P. E., R. N. Andersen, D. S. Blehert, P. G. Egland, J. S. Foster, and R. J. Palmer, Jr.** 2002. Communication among oral bacteria. *Microbiol. Mol. Biol. Rev.* **66:**486–505.
51. **Korobkova, E., T. Emonet, J. M. Vilar, T. S. Shimizu, and P. Cluzel.** 2004. From molecular noise to behavioural variability in a single bacterium. *Nature* **428:**574–578.
52. **Kreft, J. U.** 2004. Biofilms promote altruism. *Microbiology* **150:**2751–2760.
53. **Kreft, J. U.** 2004. Conflicts of interest in biofilms. *Biofilms* **1:**265–276.
54. **Kreft, J. U., and S. Bonhoeffer.** 2005. The evolution of groups of cooperating bacteria and the growth rate versus yield trade-off. *Microbiology* **151:**637–641.
55. **Kreft, J. U., G. Booth, and J. W. T. Wimpenny.** 1998. BacSim, a simulator for individual-based modelling of bacterial colony growth. *Microbiology* **144:**3275–3287.
56. **Kreft, J. U., C. Picioreanu, J. W. T. Wimpenny, and M. C. M. van Loosdrecht.** 2001. Individual-based modelling of biofilms. *Microbiology* **147:**2897–2912.
57. **Kreft, J. U., and J. W. T. Wimpenny.** 2001. Effect of EPS on biofilm structure and function as revealed by an individual-based model of biofilm growth. *Water Sci. Technol.* **43:**135–141.
58. **Langton, C. G.** 1989. Artificial life, p. 1–47. *In* C. G. Langton (ed.), *Artificial Life.* Addison-Wesley, Reading, MA.
59. **Law, A. T., and D. K. Button.** 1977. Multiple-carbon-source-limited growth kinetics of a marine coryneform bacterium. *J. Bacteriol.* **129:**115–123.
60. **Lendenmann, U., M. Snozzi, and T. Egli.** 2000. Growth kinetics of *Escherichia coli* with galactose and several other sugars in carbon-limited chemostat culture. *Can. J. Microbiol.* **46:**72–80.
61. **Levins, R.** 1979. Coexistence in a variable environment. *Am. Nat.* **114:**765–783.
62. **Ley, R. E., D. A. Peterson, and J. I. Gordon.** 2006. Ecological and evolutionary forces shaping microbial diversity in the human intestine. *Cell* **124:**837–848.
63. **MacLean, R. C., and I. Gudelj.** 2006. Resource competition and social conflict in experimental populations of yeast. *Nature* **441:**498–501.
64. **Maynard Smith, J.** 1982. *Evolution and the Theory of Games.* Cambridge University Press, Cambridge, England.
65. **Monod, J.** 1942. Recherches sur la croissance des cultures bactériennes. Hermann & Co., Paris, France.

66. **Monod, J.** 1945. Sur la nature du phenomene de diauxie. *Ann. Inst. Pasteur* **71:**37–40.
67. **Monod, J.** 1950. La technique de culture continue, théorie et applications. *Ann. Inst. Pasteur* **79:**390–410.
68. **Mulchandani, A., and J. H. T. Luong.** 1989. Microbial inhibition kinetics revisited. *Enzyme Microb. Technol.* **11:**66–73.
69. **Narang, A.** 1998. The dynamical analogy between microbial growth on mixtures of substrates and population growth of competing species. *Biotechnol. Bioeng.* **59:**116–121.
70. **Narang, A.** 1998. The steady states of microbial growth on mixtures of substitutable substrates in a chemostat. *J. Theor. Biol.* **190:**241–261.
71. **Narang, A.** 2006. Comparative analysis of some models of gene regulation in mixed-substrate microbial growth. *J. Theor. Biol.* **242:**489–501.
72. **Narang, A., A. Konopka, and D. Ramkrishna.** 1997. The dynamics of microbial growth on mixtures of substrates in batch reactors. *J. Theor. Biol.* **184:**301–317.
73. **Narang, A., and S. S. Pilyugin.** 2007. Bacterial gene regulation in diauxic and non-diauxic growth. *J. Theor. Biol.* **244:**326–348.
74. **Novak, M., T. Pfeiffer, R. E. Lenski, U. Sauer, and S. Bonhoeffer.** 2006. Experimental tests for an evolutionary trade-off between growth rate and yield in *E. coli*. *Am. Nat.* **168:** 242–251.
75. **Novick, A., and L. Szilard.** 1950. Experiments with the chemostat on spontaneous mutations of bacteria. *Proc. Natl. Acad. Sci. USA* **36:**708–719.
76. **Okubo, A., and S. A. Levin.** 2001. *Diffusion and Ecological Problems: Modern Perspectives*. Springer, New York, NY.
77. **Palmer, C., E. M. Bik, D. B. Digiulio, D. A. Relman, and P. O. Brown.** 2007. Development of the human infant intestinal microbiota. *PLoS Biol.* **5:**e177.
78. **Penry, D. L., and P. A. Jumars.** 1987. Modeling animal guts as chemical reactors. *Am. Nat.* **129:**69–96.
79. **Pfeiffer, T., and S. Bonhoeffer.** 2003. An evolutionary scenario for the transition to undifferentiated multicellularity. *Proc. Natl. Acad. Sci. USA* **100:**1095–1098.
80. **Pfeiffer, T., and S. Bonhoeffer.** 2004. Evolution of cross-feeding in microbial populations. *Am. Nat.* **163:**E126–E135.
81. **Pfeiffer, T., S. Schuster, and S. Bonhoeffer.** 2001. Cooperation and competition in the evolution of ATP-producing pathways. *Science* **292:** 504–507.
82. **Picioreanu, C., M. C. M. van Loosdrecht, and J. J. Heijnen.** 1998. Mathematical modeling of biofilm structure with a hybrid differential-discrete cellular automaton approach. *Biotechnol. Bioeng.* **58:**101–116.
83. **Redfield, R. J.** 2002. Is quorum sensing a side effect of diffusion sensing? *Trends Microbiol.* **10:** 365–370.
84. **Sandoz, K. M., S. M. Mitzimberg, and M. Schuster.** 2007. Social cheating in *Pseudomonas aeruginosa* quorum sensing. *Proc. Natl. Acad. Sci. USA* **104:**15876–15881.
85. **Scheuring, I.** 2005. The iterated continuous prisoner's dilemma game cannot explain the evolution of interspecific mutualism in unstructured populations. *J. Theor. Biol.* **232:**99–104.
86. **Senn, H., U. Lendenmann, M. Snozzi, G. Hamer, and T. Egli.** 1994. The growth of *Escherichia coli* in glucose-limited chemostat cultures: a re-examination of the kinetics. *Biochim. Biophys. Acta* **1201:**424–436.
87. **Shoemaker, J., G. T. Reeves, S. Gupta, S. S. Pilyugin, T. Egli, and A. Narang.** 2003. The dynamics of single-substrate continuous cultures: the role of transport enzymes. *J. Theor. Biol.* **222:** 307–322.
88. **Smits, W. K., O. P. Kuipers, and J. W. Veening.** 2006. Phenotypic variation in bacteria: the role of feedback regulation. *Nat. Rev. Microbiol.* **4:**259–271.
89. **Sober, E., and S. D. Wilson.** 1998. *Unto Others: the Evolution and Psychology of Unselfish Behaviour.* Harvard University Press, Cambridge, MA.
90. **Stouthamer, A. H.** 1979. The search for correlation between theoretical and experimental growth yields. *Int. Rev. Biochem.* **21:**1–47.
91. **Tilman, D.** 1982. *Resource Competition and Community Structure.* Princeton University Press, Princeton, NJ.
92. **Travisano, M., and G. J. Velicer.** 2004. Strategies of microbial cheater control. *Trends Microbiol.* **12:**72–78.
93. **van Loosdrecht, M. C. M., J. J. Heijnen, H. J. Eberl, J. U. Kreft, and C. Picioreanu.** 2002. Mathematical modelling of biofilm structures. *Antonie van Leeuwenhoek* **81:**245–256.
94. **Visick, K. L., J. Foster, J. Doino, M. Fall-Ngai, and E. G. Ruby.** 2000. *Vibrio fischeri lux* genes play an important role in colonization and development of the host light organ. *J. Bacteriol.* **182:**4578–4586.
95. **Wanner, O., H. J. Eberl, E. Morgenroth, D. R. Noguera, C. Picioreanu, B. E. Rittmann, and M. C. M. van Loosdrecht.** 2006. *Mathematical Modeling of Biofilms.* IWA Publishing, London, England.
96. **Watnick, P., and R. Kolter.** 2000. Biofilm, city of microbes. *J. Bacteriol.* **182:**2675–2679.

**97. West, S. A., A. S. Griffin, and A. Gardner.** 2007. Social semantics: altruism, cooperation, mutualism, strong reciprocity and group selection. *J. Evol. Biol.* **20:**415–432.

**98. West, S. A., A. S. Griffin, A. Gardner, and S. P. Diggle.** 2006. Social evolution theory for microorganisms. *Nat. Rev. Microbiol.* **4:**597–607.

**99. Wingender, J., T. R. Neu, and H.-C. Flemming (ed.).** 1999. *Microbial Extracellular Polymeric Substances.* Springer-Verlag, Berlin, Germany.

**100. Wolf, D. M., V. V. Vazirani, and A. P. Arkin.** 2005. Diversity in times of adversity: probabilistic strategies in microbial survival games. *J. Theor. Biol.* **234:**227–253.

**101. Xavier, J. B., and K. R. Foster.** 2007. Cooperation and conflict in microbial biofilms. *Proc. Natl. Acad. Sci. USA* **104:**876–881.

**102. Yachi, S., and M. Loreau.** 1999. Biodiversity and ecosystem productivity in a fluctuating environment: the insurance hypothesis. *Proc. Natl. Acad. Sci. USA* **96:**1463–1468.

# INDEX